Springer Texts in Statistics

Series Editors:
G. Casella
S.E. Fienberg
I. Olkin

For further volumes:
http://www.springer.com/series/417

Andrzej Gałecki • Tomasz Burzykowski

Linear Mixed-Effects Models Using R

A Step-by-Step Approach

Andrzej Gałecki
University of Michigan
300 North Ingalls Building
Ann Arbor
Michigan
USA

Tomasz Burzykowski
Center for Statistics
Hasselt University
Agoralaan D
Diepenbeek
Belgium

ISSN 1431-875X
ISBN 978-1-4899-9667-1 ISBN 978-1-4614-3900-4 (eBook)
DOI 10.1007/978-1-4614-3900-4
Springer New York Heidelberg Dordrecht London

Printed on acid-free paper

Springer is part of Springer Science+Business Media (www.springer.com)

Bliskim mojemu sercu

Oli i Łukaszowi

Rodzinie i Nauczycielom

Dekadentom

– A.T.G.

Moim najbliższym i przyjaciołom

– T.B.

In memory of Tom Ten Have

Preface

Linear mixed-effects models (LMMs) are powerful modeling tools that allow for the analysis of datasets with complex, hierarchical structures. Intensive research during the past decade has led to a better understanding of their properties. The growing body of literature, including recent monographs, has considerably increased their popularity among applied researchers. There are several statistical software packages containing routines for LMMs. These include, for instance, SAS, SPSS, STATA, S+, and R. The major advantage of R is that it is a freely available, dynamically developing, open-source environment for statistical computing and graphics.

The goal of our book is to provide a description of tools available for fitting LMMs in R. The description is accompanied by a presentation of the most important theoretical concepts of LMMs. Additionally, examples of applications from various research areas illustrate the main features of both theory and software. The presented material should allow readers to obtain a basic understanding of LMMs and to apply them in practice. In particular, we elected to present several theoretical concepts and their practical implementation in R in the context of simpler, more familiar classes of models such as e.g., the classical linear regression model. Based on these concepts, more advanced classes of models, such as models with heterogenous variance and correlated residual errors, along with related concepts are introduced. In this way, we incrementally set the stage for LMMs, so that the exposition of the theory and R tools for these models becomes simpler and clearer. This structure naturally corresponds to the object-oriented programming concept, according to which R functions/methods for simpler models are also applicable to the more complex ones.

We assume that readers are familiar with intermediate linear algebra, calculus, and the basic theory of statistical inference and linear modeling. Thus, the intended audience for this book is graduate students of statistics and applied researchers in other fields.

Our exposition of the theory of various classes of models presented in the book focuses on concepts, which are implemented in the functions available in R. Readers interested in a more detailed description of the theory are referred to appropriate theoretical monograph books, which we indicate in the text.

There are a large number of R packages that can be used to fit LMMs. Rather than attempting to describe all of these packages, we focus mainly on two of them, namely, **nlme** and **lme4.0**. In this way, we can provide a more detailed account of the tools offered by the two packages, which include a wide variety of functions for model fitting, diagnostics, inference, etc.

The package **nlme** includes functions, which allow fitting of a wide range of linear models and LMMs. Moreover, it has been available for many years and its code has been stable for some time now. Thus, it is a well-established R tool.

In turn, **lme4.0** is a developmental branch version of the **lme4** package. The latter has been under development for several years. Both packages offer an efficient computational implementation and an enhanced syntax, though at the cost of a more restricted choice of LMMs, as compared to the **nlme** package. At the time of writing of our book, the implementation of LMMs in **lme4** has undergone major changes in terms of internal representation of the objects representing fitted models. Consequently, at the beginning of 2012, a snapshot version of **lme4** has been made available to the R users under the name of **lme4.0**. As we anticipate that **lme4.0** will not undergo any major changes, we decided to present it in more detail in our book. We would like to underscore, however, that the major part of the syntax, presented in the book, will be applicable both to **lme4** and **lme4.0**.

All classes of linear models presented in the book are illustrated using data from a particular dataset. In this way, the differences between the various classes of models, as well as differences in the R software, can be clearly delineated. LMMs, which are the main focus of the book, are also illustrated using three additional datasets, which extend the presentation of various aspects of the models and R functions. We have decided to include the direct output of R commands in the text. In this way, readers who would like to repeat the analyses conducted in the book can directly check their own output. However, in order to avoid the risk of incompatibility with updated versions of the software, the results of the analyses have also been summarized in the form of edited tables.

To further support those readers who are interested in actively using the material presented in the book, we have developed the package **nlmeU**. It contains all the datasets and R code used in the book. The package is downloadable at `http://www-personal.umich.edu/~agalecki/`.

We hope that our book, which aims to provide a state-of-the-art description of the details of implementing of LMMs in R, will support a widespread use of the models by applied researchers in a variety of fields including biostatistics, public health, psychometrics, educational measurement, and sociology.

When working on the text, we received considerable assistance and valuable comments from many people. We would like to acknowledge Geert Molenberghs (Hasselt University and the Catholic University of Leuven), Geert Verbeke (Catholic University of Leuven), José Pinheiro (Novartis AG), Paul Murrell (Auckland University), Przemysław Biecek (Warsaw University), Fabian Scheipl (Ludwig Maximilian University of Munich), Joshua Wiley (University of California, Los

Angeles), Tim Harrold (NSW Ministry of Health), Jeffrey Halter (University of Michigan), Shu Chen (University of Michigan), Marta Gałecka (Weill Cornell Medical College), anonymous reviewers and members of the R-sig-ME discussion group led by Douglas Bates (University of Wisconsin-Madison), and Ben Bolker (McMaster University) for their comments and discussions at various stages during the preparation of the book. We also acknowledge a formidable effort on the part of the developers of the **nlme** and **lme4** packages. Without them this book would not have been written. In particular, Ben Bolker's contribution was invaluable to ensure that the majority of the **lme4.0** syntax used in the text can also be used with the **lme4** package. We are grateful to John Kimmel for encouraging us to consider writing the book and to Marc Strauss, Hannah Bracken, and Brian Halm from Springer for their editorial assistance and patience. Finally, we gratefully acknowledge financial support from the Claude Pepper Center grants AG08808 and AG024824 from the National Institute of Aging and from the IAP Research Network P7/06 of the Belgian Government (Belgian Science Policy).

Ann Arbor, MI, USA — Andrzej Gałecki
Diepenbeek, Belgium, and Warszawa, Poland — Tomasz Burzykowski

Contents

Part I Introduction

List of Tables

List of Figures

List of R Panels

Part I
Introduction

Chapter 1
Introduction

1.1 The Aim of the Book

Linear mixed-effects models (LMMs) are an important class of statistical models that can be used to analyze correlated data. Such data include *clustered observations*, *repeated measurements*, *longitudinal measurements*, *multivariate observations*, etc.

The aim of our book is to help readers in fitting LMMs using R software. R (`www.r-project.org`) is a language and an environment aimed at facilitating implementation of statistical methodology and graphics. It is an open-source software, which can be freely downloaded and used under the GNU General Public License. In particular, users can define and share their own functions, which implement various methods and extend the functionality of R. This feature makes R a very useful platform for propagating the knowledge and use of statistical methods.

We believe that, by describing selected tools available in R for fitting LMMs, we can promote the broader application of the models. To help readers less familiar with this class of linear models (LMs), we include in our book a description of the most important theoretical concepts and features of LMMs. Moreover, we present examples of applications of the models to real-life datasets from various areas to illustrate the main features of both theory and software.

1.2 Implementation of Linear Mixed-Effects Models in R

There are many packages in R, which contain functions that allow fitting various forms of LMMs. The list includes, but is not limited to, packages **amer**, **arm**, **gamm**, **gamm4**, **GLMMarp**, **glmmAK**, **glmmBUGS**, **heavy**, **HGLMMM**, **lme4.0**, **lmec**, **lmm**, **longRPart**, **MASS**, **MCMCglmm**, **nlme**, **PSM**, and **pedigreemm**. On the one hand, it would seem that the list is rich enough to allow for

A. Gałecki and T. Burzykowski, *Linear Mixed-Effects Models Using R: A Step-by-Step Approach*, Springer Texts in Statistics, DOI 10.1007/978-1-4614-3900-4_1,

a widespread use of LMMs. On the other hand, the number of available packages leads to difficulty in evaluating their relative merits and making the most suitable choice.

It is virtually impossible to describe the contents of all of the packages mentioned above. To facilitate and promote the use of LMMs in practice, it might be more useful to provide details for a few of them, so that they could be used as a starting point. Therefore, we decided to focus on the packages **nlme** and **lme4.0**, for several reasons. First, they contain the functions `lme()` and `lmer()`, respectively, which are specifically designed for fitting a broad range of LMMs. Second, they include many tools useful for applications such as model diagnostics. Finally, many other packages, which add new LMM classes or functionalities, depend on and are built around **nlme** and/or **lme4.0**. Examples include, but are not limited to, packages **amer**, **gamm**, **gamm4**, or **RLRsim**.

The reader may note that we focus more on the package **nlme** than on **lme4.0**. The main reason is that the former has already been around for some time. Thus, its code is stable. On the other hand, the package **lme4.0** is a development version of **lme4** made available at the beginning of 2012. At that time **lme4**'s code underwent major changes in terms of internal representation of the objects representing fitted models. Hence, the developers of **lme4** decided to make available the snapshot version of **lme4**, under the name of **lme4.0**, containing the functionalities preceding the changes. It is these dynamics of the development of the code of **lme4** and **lme4.0** which prompted us to focus more on **nlme**. However, it is expected that **lme4.0** will not undergo any major modifications, either. Given that it offers interesting tools for fitting LMMs, we decided to include a presentation of it in our book. The presentation should also be of help for **lme4** users. In particular, the major part of the **lme4.0** syntax used in the book should also be applicable to **lme4**.

An important feature that distinguishes R from many other existing statistical software packages implementing LMMs is that it incorporates several concepts of an *object-oriented* (O-O) programming, such as *classes* of *objects* and *methods* operating on those classes. There are two O-O systems that have been implemented in R, namely, S3 and S4. They incorporate the O-O concepts to a different degree, with S3 being a less formal and S4 being a more stringent implementation. In both systems, the O-O concepts are implemented by defining special type of functions called *generic* functions. When such a function is applied to an object, it dispatches an appropriate *method* based on object's class. The system S3 has been used in the package **nlme**, while S4 has been used in the package **lme4.0**.

The O-O programming approach is very attractive in the context of statistical modeling because models can often be broken down into separable (autonomous) components such as data, mean structure, variance function, etc. Moreover, components defined for one type of model can also be used as building blocks for a different type of model.

1.3 The Structure of the Book

As it was mentioned in the previous section, an inherent feature of the O-O programming approach is that concepts and methods used for simpler objects or models are applicable to the more complex ones. For this reason, in our book we opted for an incremental build-up of the knowledge about the implementation of LMMs in the functions from packages **nlme** and **lme4.0**. In particular, in the first step, we decided to introduce theoretical concepts and their practical implementation in the R code in the context of simpler classes of LMs, like the classical linear regression model. The concepts are then carried over to more advanced classes of models, including LMMs. This step-by-step approach offers a couple of advantages. First, we believe that it makes the exposition of the theory and R tools for LMMs simpler and clearer. In particular, the presentation of the key concepts in the context of a simpler model makes them easier to explain and become familiar with. Second, the step-by-step approach is helpful in the use of other R packages, which rely on classes of objects defined in the **nlme** and/or **lme4.0** packages.

As a result of this conceptual approach, we divided our book into four parts. Part I contains the introduction to the datasets used in the book. Parts II, III, and IV focus on different classes of LMs of increasing complexity. The structure of the three parts is, to a large extent, similar. First, a review of the main concepts and theory of a particular class of models is presented. Special attention is paid to the presentation of the link between similar concepts used for different classes. Then, the details of how to implement the particular class of models in the packages **nlme** and/or **lme4.0** are described. The idea is to present the key concepts in the context of simpler models, in order to enhance the understanding of them and facilitate their use for the more complex models. Finally, in each part, the particular class of LMs and the corresponding R tools are illustrated by analyzing real-life datasets.

In a bit more detail, the contents of the four parts are as follows:

Chapter 2 of Part I contains a description of four case studies, which are used to illustrate various classes of LMs and of the corresponding R tools. Chapter 3 contains results of exploratory analyses of the datasets. The results are used in later chapters to support model-based analyses. Note that one of the case studies, the Age-Related Macular Degeneration (ARMD) clinical trial, is used repeatedly for the illustration of all classes of LMs. We believe that in this way the differences between the models concerning, e.g., the underlying assumptions, may become easier to appreciate.

Part II focuses on LMs for independent observations. In Chap. 4, we recall the main concepts of the theory of the classical LMs with homoscedastic residual errors. Then, in Chap. 5, we present the tools available in R to fit such models. This allows us to present the fundamental concepts used in R for statistical model building, like *model formula*, *model frame*, etc. The concepts are briefly illustrated in Chap. 6 using the data from the ARMD trial.

Subsequently, we turn our attention to models with heteroscedastic residual errors. In Chap. 7, we review the basic elements of the theory. Chapter 8 presents the function `gls()` from the package **nlme**, which can be used to fit the models. In particular, the important concept of the *variance function* is introduced in the chapter. The use of the function `gls()` is illustrated using data from the ARMD trial in Chap. 9.

In Part III, we consider general LMs, i.e., LMs for correlated observations. In Chap. 10, we recall the basic elements of the theory of the models. In particular, we explain how the concepts used in the theory of the LMs with heteroscedastic residual errors for independent observations, presented in Chap. 7, are extended to the case of models for correlated observations. In Chap. 11, we describe additional features of the function `gls()`, which allow its use for fitting general LMs. In particular, we introduce the key concept of the *correlation structure*. The use of the function `gls()` is illustrated in Chap. 12 using the data from the ARMD trial.

Finally, Part IV is devoted to LMMs. Chapter 13 reviews the fundamental elements of the theory of LMMs. In the presentation, we demonstrate the links between the concepts used in the theory of LMMs with those developed in the theory of general LMs (Chap. 10). We believe that, by pointing to the links, the exposition of the fundamentals of the LMM theory becomes more transparent and easier to follow.

In Chap. 14, we describe the features of the function `lme()` from the package **nlme**. This function is the primary tool in the package used to fit LMMs. In particular, we describe in detail the representation of positive-definite matrices, which are instrumental in the implementation of the routines that allow fitting LMMs. Note that the concepts of the variance function and correlation structure, introduced in Chaps. 8 and 11, respectively, are also important for the understanding of the use of the function `lme()`.

In Chap. 15, we present the capabilities of the function `lmer()` from the package **lme4.0**. In many aspects, the function is used similarly to `lme()`, but there are important differences, which we discuss. The basic capabilities of both of the functions are illustrated by application of LMMs to the analysis of the ARMD trial data in Chap. 16. More details on the use of the function `lme()` are provided in Chaps. 17, 18, and 19, in which we apply LMMs to analyze the data from the progressive resistance training (PRT) study, the study of instructional improvement (SII), and the Flemish Community Attainment-Targets (FCAT) study, respectively. Finally, in Chap. 20, we present somewhat more advanced material on the additional R tools for LMMs, including the methods for power calculations, influence diagnostics, and a new class of positive-definite matrices. The latter can be used to construct LMMs with random effects having a variance–covariance matrix defined as a Kronecker product of two or more matrices. Note that the newly defined class is used in the analysis presented in Chap. 17.

Table 1.1 summarizes the successive classes of LMs, described in our book, together with the concepts introduced in the context of the particular class. The classes are identified by the assumptions made about the random part of the model.

Our book contains 67 figures, 46 tables, and 187 panels with R code.

Table 1.1 Classes of linear models with the corresponding components (building blocks) presented in the book. The R classes refer to the package **nlme**

Linear model			Model component	
Class (residual errors)	Theory	Syntax	Name	R class
Homoscedastic, indep.	Ch. 4	Ch. 5	Data	*data.frame*
			Mean structure	*formula*
Heteroscedastic, indep.	Ch. 7	Ch. 8	Variance structure	*varFunc*
Correlated	Ch. 10	Ch. 11	Correlation structure	*corStruct*
Mixed effects (LMM)	Ch. 13	Ch. 14	Random-effects structure	*reStruct*

Finally, we would like to outline the scope of the contents of the book:

- The book is aimed primarily at providing explanations and help with respect to the tools available in R for fitting LMMs. Thus, we do not provide a comprehensive account of the methodology of LMMs. Instead, we limit ourselves to the main concepts and techniques, which have been implemented in the functions `lme()` and `lmer()` from the packages **nlme** and **lme4.0**, respectively, and which are important to the understanding of the use of the functions. A detailed exposition of the methodology of LMMs can be found in books by, e.g., Searle et al. (1992), Davidian and Giltinan (1995), Vonesh and Chinchilli (1997), Pinheiro and Bates (2000), Verbeke and Molenberghs (2000), Demidenko (2004), Fitzmaurice et al. (2004), or West et al. (2007).
- In our exposition of methodology, we focus on the likelihood-based estimation methods, as they are primarily used in `lme()` and `lmer()`. Thus, we do not discuss, e.g., Bayesian approaches to the estimation of LMMs.
- We describe the use of various functions, which are available in the packages **nlme** and **lme4.0**, in sufficient detail. In our presentation, we focus on the main, or most often used, arguments of the functions. For a detailed description of all of the arguments, we refer the readers to R's help system.
- It is worth keeping in mind that, in many instances, the same task can be performed in R in several different ways. To some extent, the choice between the different methods is a matter of individual preference. In our description of the R code, we present methods, which we find to be the most useful. If alternative solutions are possible, we may mention them, but we are not aiming to be exhaustive.
- The analyses of the case studies aim principally at illustrating various linear models and the possibility of fitting the models in R. While we try to conduct as meaningful analyses as possible, they are not necessarily performed in the most optimal way with respect to, e.g., the model-building strategy. Thus, their results should not be treated as our contribution to the subject-matter discussion related to the examples. However, whenever possible or useful, we make an attempt to provide quantitative and/or qualitative interpretation of the results. We also try to formulate practical recommendations or guidance regarding model-building strategies, model diagnostics, etc. As mentioned earlier, however, the book is not meant to serve as a complete monograph on statistical modeling. Thus, we limit ourselves to providing recommendations or guidance for the topics which appear to be of interest in the context of the analyzed case studies.

1.4 Technical Notes

The book is aimed at helping readers in fitting LMMs in R. We do assume that the reader has a basic knowledge of R. An introduction to R can be found in the book by Dalgaard (2008). A more advanced exposition is presented by Venables and Ripley (2010).

To allow readers to apply the R code presented in the book, we have created the R package **nlmeU**. The package contains all the datasets and the code that we used in the text. It also includes additional R functions, which we have developed.

We tried to use short lines of the R code to keep matters simple, transparent, and easy to generalize. To facilitate locating the code, we placed it in *panels*. The panels are numbered consecutively in each chapter and referred to, e.g., as R2.3, where "2" gives the number of the chapter and "3" is the consecutive number of the panel within the chapter. Each panel was given a caption explaining the contents. In some cases, the contents of a panel were logically split into different subpanels. The subpanels are then marked by consecutive letters and referred to by adding the appropriate letter to panel's number, e.g., R2.3a or R2.3b. Tables and figures are numbered in a similar fashion.

Only in rare instances were a few lines of R code introduced directly into the text. In all these cases (as in the examples given later in this section), the code was written using the `true type` font and placed in separate lines marked with ">", mimicking R's command-window style.

To limit the volume of the output presented in the panels, in some cases we skipped a part of it. These interventions are indicated by the "... [*snip*]" string. Also, long lines in the output were truncated and extra characters were replaced with the "..." string.

The R functions are referred to in the text as `function()`, e.g., `lme()`. Functions' arguments and objects are marked using the same font, e.g., `argument` and `object`. For the R classes, we use italic, e.g., the *lme* class.

For the proper execution of the R code used in the book, the following packages are required: **lattice**, **lme4.0**, **nlme**, **Matrix**, **plyr**, **reshape**, **RLRsim**, **splines**, and **WWGbook**. Additionally, **nlmeU** is needed. Packages **lattice**, **nlme**, **Matrix**, and **splines** come with basic distribution of R and do not need to be installed. The remaining packages can be installed using the following code:

```
> pckgs  <-
+   c("lme4.0", "nlmeU", "plyr", "reshape", "RLRsim", "WWGbook",
+     "ellipse")
> install.packages(pckgs)
```

There are additional utility functions, namely, `Sweave()` (Leisch, 2002) and `xtable()` in **utils** and **xtable** (Dahl, 2009) packages, respectively, which are not needed to execute the code presented in the book, but which were extensively used by us when preparing this manuscript.

It is worth noting that there are functions that bear the same name in the packages **nlme** and **lme4.0**, but which have different definitions. To avoid unintentional masking of the functions, the packages should not be attached simultaneously. Instead, it is recommended to switch between the packages. For example, when using **nlme** in a hypothetical R session, we attach the package by using the `library()` or `require()` functions and execute statements as needed. Then, before switching to **lme4.0**, it is mandatory to detach the **nlme** package by using the `detach()` function. We also note that the `conflicts()` function, included for illustration below, is very useful to identify names' conflicts:

```
> library(nlme)                   # Attach package
> conflicts(detail = TRUE)        # Identifies names' conflicts
 ... statements omitted
> detach(package:nlme)            # Detach package
```

A similar approach should be applied when using the package **lme4.0**:

```
> library(lme4.0)
 ... statements omitted
> detach(package:lme4.0)
> detach(package:Matrix)         # Recommended
```

Note that detaching **Matrix** is less critical, but recommended.

In the examples presented above, we refer to the packages **nlme** and **lme4.0**. However, to avoid unintentional masking of objects, the same strategy may also be necessary for other packages, which may cause function names' conflicts.

When creating figures, we used `"CMRoman"` and `"CMSans"` Computer Modern font families available in **cmrutils** package. These fonts are based on the CM-Super and CMSYASE fonts (Murrell and Ripley, 2006). The full syntax needed to create figures presented in the book is often extensive. In many cases, we decided to present a shortened version of the code. A full version is available in the **nlmeU** package.

Finally, the R scripts in our book were executed by using `R version 2.15.0 (2012-03-30)` under the Windows 7 operating system. We used the following global options:

```
> options(width = 65, digits = 5, show.signif.stars = FALSE)
```

Chapter 2
Case Studies

2.1 Introduction

In this chapter, we introduce the case studies that will be used to illustrate the models and R code described in the book.

The case studies come from different application domains; however, they share a few features. For instance, in all of them the study and/or sampling design generates the observations that are *grouped* according to the levels of one or more *grouping* factors. More specifically, the levels of grouping factors, i.e., subjects, schools, etc., are assumed to be randomly selected from a population being studied. This means that observations within a particular group are likely to be correlated. The correlation should be taken into account in the analysis. Also, in each case there is one (or more) continuous measurement, which is treated as the dependent variable in the models considered in this book.

In particular, we consider the following datasets:

- *Age-Related Macular Degeneration (ARMD) Trial*: A clinical trial comparing several doses of interferon-α and placebo in patients with ARMD. Visual acuity of patients participating in the trial was measured at baseline and at four post-randomization timepoints. The resulting data are an example of *longitudinal data* with observations grouped by subjects. We describe the related datasets in more detail in Sect. 2.2.
- *Progressive Resistance Training (PRT) Trial*: A clinical trial comparing low- and high-intensity training for improving the muscle power in elderly people. For each participant, characteristics of two types of muscle fibers were measured at two occasions, pre- and post-training. The resulting data are an example of *clustered* data, with observations grouped by subjects. We present more detailed information about the dataset in Sect. 2.3.
- *Study of Instructional Improvement (SII)*: An educational study aimed at assessing improvement in mathematics grades of first-grade pupils, as compared to their kindergarten achievements. It included pupils from randomly selected

A. Gałecki and T. Burzykowski, *Linear Mixed-Effects Models Using R: A Step-by-Step Approach*, Springer Texts in Statistics, DOI 10.1007/978-1-4614-3900-4_2,

classes in randomly selected elementary schools. The dataset is an example of *hierarchical* data, with observations (pupils' scores) grouped within classes, which are themselves grouped in schools. We refer to Sect. 2.4 for more details about the data.

- *Flemish Community Attainment-Targets (FCAT) Study*: An educational study, in which elementary school graduates were evaluated with respect to reading comprehension in Dutch. Pupils from randomly selected schools were assessed for a set of nine attainment targets. The dataset is an example of *grouped* data, for which the grouping factors are *crossed*. We describe the dataset in more detail in Sect. 2.5.

The data from the ARMD study will be used throughout the book to illustrate various classes of LMs and corresponding R tools. The remaining case studies will be used in Part IV only, to illustrate R functions for fitting LMMs.

For each of the aforementioned case studies there is one or more datasets included into the package **nlmeU**, which accompanies this book. In the next sections of this chapter, we use the R syntax to describe the contents of these datasets. Results of exploratory analyses of the case studies are presented in Chap. 3. Note that, unlike in the other parts of the book, we are not discussing the code in much detail, as the data-processing functionalities are not the main focus of our book. The readers interested in the functionalities are referred to the monograph by Dalgaard (2008).

The R language is not particularly suited for data entry. Typically, researchers use raw data created using other software. Data are then stored in external files, e.g., in the .csv format, read into R, and prepared for the analysis. To emulate this situation, we assume, for the purpose of this chapter, that the data are stored in a .csv-format file in the "C:\temp" directory.

2.2 Age-Related Macular Degeneration Trial

The ARMD data arise from a randomized multi-center clinical trial comparing an experimental treatment (interferon-α) *versus* placebo for patients diagnosed with ARMD. The full results of this trial have been reported by Pharmacological Therapy for Macular Degeneration Study Group (1997). We focus on the comparison between placebo and the highest dose (6 million units daily) of interferon-α.

Patients with macular degeneration progressively lose vision. In the trial, visual acuity of each of 240 patients was assessed at baseline and at four post-randomization timepoints, i.e., at 4, 12, 24, and 52 weeks. Visual acuity was evaluated based on patient's ability to read lines of letters on standardized vision charts. The charts display lines of five letters of decreasing size, which the patient must read from top (largest letters) to bottom (smallest letters). Each line with at least four letters correctly read is called one "line of vision." In our analyses, we will focus on the visual acuity defined as the total number of *letters* correctly read.

Another possible approach would be to consider visual acuity measured by the number of *lines* correctly read. Note that the two approaches are closely linked, as each line of vision contains five letters.

It follows that, for each of 240 patients, we have *longitudinal data* in the form of up to five visual acuity measurements collected at different, but common to all patients, timepoints. These data will be useful to illustrate the use of LMMs for continuous, longitudinal data. We will also use them to present other classes of LMs considered in our book.

2.2.1 Raw Data

We assume that the raw ARMD data are stored in the "C:\temp" directory in a .csv-format file named `armd240.data.csv`. In what follows, we also assume that our goal is to verify the contents of the data and prepare them for analysis in R.

In Panel R2.1, the data are loaded into R using the `read.csv()` function and are stored in the data frame object `armd240.data`. Note that this data frame is not included in the **nlmeU** package.

The number of rows (records) and columns (variables) in the object `armd240.data` is obtained using the function `dim()`. The data frame contains 240 observations and 9 variables. The names of the variables are displayed using the `names()` function. All the variables are of class *integer*. By applying the function `str()`, we get a summary description of variables in the `armd240.data` data. In particular, for each variable, we get its class and a listing of the first few values.

The variable `subject` contains patients' identifiers. Treatment identifiers are contained in the variable `treat`. Variables `visual0`, `visual4`, `visual12`, `visual24`, and `visual52` store visual acuity measurements obtained at baseline and week 4, 12, 24, and 52, respectively. Variables `lesion` and `line0` contain additional information, which will not be used for analysis in our book.

Finally, at the bottom of Panel R2.1, we list the first three rows of the data frame `armd240.data` with the help of the `head()` function. To avoid splitting lines of the output and to make the latter more transparent, we shorten variables' names using the `abbreviate()` function. After printing the contents of the first three rows and before proceeding further, we reinstate the original names. Note that we apply a similar sequence of R commands in many other R panels across the book to simplify the displayed output.

Based on the output, we note that the data frame contains one record for each patient. The record includes all information obtained for the patient. In particular, each record contains five variables with visual acuity measurements, which are, essentially, of the same format. This type of data storage, with one record per subject, is called the "wide" format. An alternative is the "long" format with multiple records per subject. We will discuss the formats in the next section.

R2.1 *ARMD Trial*: Loading raw data from a .csv-format file into the armd240.data object and checking their contents

```
> dataDir <- file.path("C:", "temp")          # Data directory
> fp <-                                       # File path
+   file.path(dataDir, "armd240.data.csv")
> armd240.data <-                             # Read data
+   read.csv(fp, header = TRUE)
> dim(armd240.data)                           # No. of rows and cols
  [1] 240   9
> (nms <- names(armd240.data))                # Variables' names
  [1] "subject"  "treat"    "lesion"   "line0"    "visual0"
  [6] "visual4"  "visual12" "visual24" "visual52"
> unique(sapply(armd240.data, class))         # Variables' classes
  [1] "integer"
> str(armd240.data)                           # Data structure
 'data.frame':   240 obs. of  9 variables:
  $ subject : int  1 2 3 4 5 6 7 8 9 10 ...
  $ treat   : int  2 2 1 1 2 2 1 1 2 1 ...
  $ lesion  : int  3 1 4 2 1 3 1 3 2 1 ...
  $ line0   : int  12 13 8 13 14 12 13 8 12 10 ...
  $ visual0 : int  59 65 40 67 70 59 64 39 59 49 ...
  $ visual4 : int  55 70 40 64 NA 53 68 37 58 51 ...
  $ visual12: int  45 65 37 64 NA 52 74 43 49 71 ...
  $ visual24: int  NA 65 17 64 NA 53 72 37 54 71 ...
  $ visual52: int  NA 55 NA 68 NA 42 65 37 58 NA ...
> names(armd240.data) <- abbreviate(nms) # Variables' names shortened
> head(armd240.data, 3)                       # First 3 records
   sbjc tret lesn lin0 vsl0 vsl4 vs12 vs24 vs52
  1   1    2    3   12   59   55   45   NA   NA
  2   2    2    1   13   65   70   65   65   55
  3   3    1    4    8   40   40   37   17   NA
> names(armd240.data) <- nms                  # Variables' names reinstated
```

2.2.2 *Data for Analysis*

In this section, we describe auxiliary data frames, namely, armd.wide, armd0, and armd, which were derived from armd240.data for the purpose of analyses of the ARMD data that will be presented later in the book. The data frames are included in the package **nlmeU**. In what follows, we present the structure, contents, and for illustration purposes, how the data were created.

2.2.2.1 *Data in the "Wide" Format: The Data Frame* armd.wide

Panel R2.2 presents the structure and the contents of the armd.wide data frame.

Note that the data are loaded into R using the data() function, without the need for attaching the package **nlmeU**. The data frame contains 10 variables. In particular, it includes variables visual0, visual4, visual12, visual24, visual52, lesion, and line0, which are exactly the same as those in the armd240.data. In contrast to the armd240.data data frame, it contains three factors: subject, treat.f, and miss.pat. The first two contain patient's identifier and treatment. They are constructed from the corresponding numeric variables available in armd240.data. The factor miss.pat is a new variable and contains a missing-pattern identifier, i.e., a character string that indicates which of the four post-randomization measurements of visual acuity are missing for a particular patient. The missing values are marked by X. Thus, for instance, for the patient with the subject identifier equal to 1, the pattern is equal to --XX, because there is no information about visual acuity at weeks 24 and 52. On the other hand, for the patient with the subject identifier equal to 6, there are no missing visual acuity

R2.2 *ARMD Trial*: The structure and contents of data frame armd.wide stored in the "wide" format

```
> data(armd.wide, package = "nlmeU")          # armd.wide loaded
> str(armd.wide)                              # Structure of data
 'data.frame':   240 obs. of  10 variables:
  $ subject : Factor w/ 240 levels "1","2","3","4",..: 1 2 3 4 5 6 ...
  ...      [snip]
  $ treat.f : Factor w/ 2 levels "Placebo","Active": 2 2 1 1 2 2 1 ...
  $ miss.pat: Factor w/ 9 levels "----","---X",..: 4 1 2 1 9 1 1 1 ...
> head(armd.wide)                             # First few records
   subject lesion line0 visual0 visual4 visual12 visual24
 1       1      3    12      59      55       45       NA
 ...      [snip]
 6       6      3    12      59      53       52       53
   visual52 treat.f miss.pat
 1       NA  Active     --XX
 ...      [snip]
 6       42  Active     ----
> (facs <- sapply(armd.wide, is.factor))      # Factors indicated
  subject   lesion    line0  visual0  visual4 visual12 visual24
     TRUE    FALSE    FALSE    FALSE    FALSE    FALSE    FALSE
 visual52  treat.f miss.pat
    FALSE     TRUE     TRUE
> names(facs[facs == TRUE])                   # Factor names displayed
 [1] "subject"  "treat.f"  "miss.pat"
```

measurements, and hence the value of the `miss.pat` factor is equal to `----`. At the bottom of Panel R2.2, we demonstrate how to extract the names of the factors from a data frame.

Panel R2.3 presents the syntax used to create factors `treat.f` and `miss.pat` in the `armd.wide` data frame. The former is constructed in Panel R2.3a from the variable `treat` from the data frame `armd240.data` using the function `factor()`. The factor `treat.f` has two levels, `Placebo` and `Active`, which correspond to the values of 1 and 2, respectively, of `treat`.

The factor `miss.pat` is constructed in Panel R2.3b with the help of the function `missPat()` included in the **nlmeU** package. The function returns a character vector of length equal to the number of rows of the matrix created by column-wise concatenation of the vectors given as arguments to the function. The elements of the resulting vector indicate the occurrence of missing values in the rows of the matrix. In particular, the elements are character strings of the length equal to the number of the columns (vectors). As shown in Panel R2.2, the strings contain characters "-" and "X", where the former indicates a nonmissing value in the corresponding column of the matrix, while the latter indicates a missing value. Thus, application

---

R2.3 *ARMD Trial*: Construction of factors `treat.f` and `miss.pat` in the data frame `armd.wide`. The data frame `armd240.data` was created in Panel R2.1

(a) *Factor* `treat.f`

```
> attach(armd240.data)                          # Attach data
> treat.f <-                                    # Factor created
+   factor(treat, labels = c("Placebo", "Active"))
> levels(treat.f)                               # (1) Placebo, (2) Active
  [1] "Placebo" "Active"
> str(treat.f)
   Factor w/ 2 levels "Placebo","Active": 2 2 1 1 2 2 1 1 2 1 ...
```

(b) *Factor* `misspat`

```
> miss.pat <-                                   # Missing patterns
+   nlmeU::missPat(visual4, visual12, visual24, visual52)
> length(miss.pat)                              # Vector length
  [1] 240
> mode(miss.pat)                                # Vector mode
  [1] "character"
> miss.pat                                      # Vector contents
    [1] "--XX" "----" "---X" "----" "XXXX" "----" "----" "----"
  ...     [snip]
  [233] "----" "----" "----" "----" "----" "----" "----" "----"
> detach(armd240.data)                          # Detach armd240.data
```

---

of the function to variables visual4, visual12, visual24, and visual52 from the data frame armd240.data results in a character vector of length 240 with strings containing four characters as the elements. The elements of the resulting miss.pat vector indicate that, for instance, for the first patient in the data frame armd240.data visual acuity measurements at week 24 and 52 were missing, while for the fifth patient, no visual acuity measurements were obtained at any post-randomization visit.

Note that we used the nlmeU:::missPat() syntax, which allowed us to invoke the missPat() function without attaching the **nlmeU** package.

2.2.2.2 *Data in the "Long" Format: The Data Frame* armd0

In addition to the armd.wide data stored in the "wide" format, we will need data in the "longitudinal" (or "long") format. In the latter format, for each patient, there are multiple records containing visual acuity measurements for separate visits. An example of data in "long" format is stored in the data frame armd0. It was obtained from the armd.wide data using functions melt() and cast() from the package **reshape** (Wickham, 2007).

Panel R2.4 presents the contents and structure of the data frame armd0. The data frame includes eight variables and 1,107 records. The contents of variables subject, treat.f, and miss.pat are the same as in armd.wide, while visual0 contains the value of the visual acuity measurement at baseline. Note that the values of these four variables are repeated across the multiple records corresponding to a particular patient. On the other hand, the records differ with respect to the values of variables time.f, time, tp, and visual. The first three of those four variables are different forms of an indicator of the visit time, while visual contains the value of the visual acuity measurement at the particular visit. We note that having three variables representing time visits is not mandatory, but we created them to simplify the syntax used for analyses in later chapters.

The numerical variable time provides the actual week, at which a particular visual acuity measurement was taken. The variable time.f is a corresponding ordered factor, with levels Baseline, 4wks, 12wks, 24wks, and 52wks. Finally, tp is a numerical variable, which indicates the position of the particular measurement visit in the sequence of the five possible measurements. Thus, for instance, tp=0 for the baseline measurement and tp=4 for the fourth post-randomization measurement at week 52.

Interestingly enough, visual acuity measures taken at baseline are stored both in visual0 and in selected rows of the visual variables. This structure will prove useful when creating the armd data frame containing rows with post-randomization visual acuity measures, while keeping baseline values.

The "long" format is preferable for storing longitudinal data over the "wide" format. We note that storing of the visual acuity measurements in the data frame armd.wide requires the use of six variables, i.e., subject and the five variables containing the values of the measurements. On the other hand, storing the same

R2.4 *ARMD Trial*: The structure and contents of the data frame armd0 stored in the "long" format

```
> data(armd0, package = "nlmeU")              # From nlmeU package
> dim(armd0)                                  # No. of rows and cols
  [1] 1107    8
> head(armd0)                                 # First six records
    subject treat.f visual0 miss.pat  time.f time visual tp
  1       1  Active      59     --XX Baseline    0     59  0
  2       1  Active      59     --XX     4wks    4     55  1
  3       1  Active      59     --XX    12wks   12     45  2
  4       2  Active      65     ---- Baseline    0     65  0
  5       2  Active      65     ----     4wks    4     70  1
  6       2  Active      65     ----    12wks   12     65  2
> names(armd0)                                # Variables' names
  [1] "subject"  "treat.f"  "visual0"  "miss.pat" "time.f"
  [6] "time"     "visual"   "tp"
> str(armd0)                                  # Data structure
  'data.frame': 1107 obs. of  8 variables:
   $ subject : Factor w/ 240 levels "1","2","3","4",..: 1 1 1 2 2 2 ...
   $ treat.f : Factor w/ 2 levels "Placebo","Active": 2 2 2 2 2 2 2 ...
   $ visual0 : int  59 59 59 65 65 65 65 65 40 40 ...
   $ miss.pat: Factor w/ 9 levels "----","---X",..: 4 4 4 1 1 1 1 1 ...
   $ time.f  : Ord.factor w/ 5 levels "Baseline"<"4wks"<..: 1 2 3 1 ...
   $ time    : num  0 4 12 0 4 12 24 52 0 4 ...
   $ visual  : int  59 55 45 65 70 65 65 55 40 40 ...
   $ tp      : num  0 1 2 0 1 2 3 4 0 1 ...
```

information in the data frame armd0 requires only three variables, i.e., subject, time, and visual. Of course, this is achieved at the cost of including more rows in the armd0 data frame, i.e., 1,107, as compared to 240 records in armd.wide.

We also note that variables, with values invariant within subjects, such as treat.f, visual0, are referred to as *time-fixed*. In contrast, time, tp, and visual are called *time-varying*. This distinction will have important implications for the specification of the models and interpretation of the results.

2.2.2.3 Subsetting Data in the "Long" Format: The Data Frame armd

Data frame armd was also stored in a "long" format and was created from the armd0 data frame by omitting records corresponding to the baseline visual acuity measurements.

Panel R2.5 presents the syntax used to create the data frame armd. In particular, the function subset() is used to remove the baseline measurements, by selecting

R2.5 *ARMD Trial*: Creation of the data frame armd from armd0

```
> auxDt <- subset(armd0, time > 0)          # Post-baseline measures
> dim(auxDt)                                # No. of rows & cols
  [1] 867   8
> levels(auxDt$time.f)                      # Levels of treat.f
  [1] "Baseline" "4wks"     "12wks"    "24wks"    "52wks"
> armd <- droplevels(auxDt)                 # Drop unused levels
> levels(armd$time.f)                       # Baseline level dropped
  [1] "4wks"  "12wks" "24wks" "52wks"
> armd <-                                   # Data modified
+   within(armd,
+          {
+           contrasts(time.f) <-            # Contrasts assigned
+             contr.poly(4, scores = c(4, 12, 24, 52))
+          })
> head(armd)                                # First six records
    subject treat.f visual0 miss.pat time.f time visual tp
  2       1  Active      59     --XX   4wks    4     55  1
  3       1  Active      59     --XX  12wks   12     45  2
  5       2  Active      65     ----   4wks    4     70  1
  6       2  Active      65     ----  12wks   12     65  2
  7       2  Active      65     ----  24wks   24     65  3
  8       2  Active      65     ----  52wks   52     55  4
```

only the records, for which `time>0`, from the object `armd0`. By removing the baseline measurements, we reduce the number of records from 1,107 (see Panel R2.4) to 867.

While subsetting the data, care needs to be taken regarding the levels of the `time.f` and, potentially, other factors. In the data frame `armd0`, the factor had five levels. In Panel R2.5, we extract the factor `time.f` from the auxiliary data frame `auxDt`. Note that, in the data frame, the level `Baseline` is not used in any of the rows. For many functions in R it would not be a problem, but sometimes the presence of an unused level in the definition of a factor may lead to unexpected results. Therefore, it is prudent to drop the unused level from the definition of the `time.f` factor, by applying the function `droplevels()`. It is worth noting that, using the `droplevels()` function, the number of levels of the factors `subject` and `miss.pat` is also affected (not shown).

After modifying the aforementioned factors, we store the resulting data in the data frame `armd`. We also assign orthogonal polynomial contrasts to the factor `time.f` using syntax of the form "`contrasts(`*factor*`)<-`*contr.function*". We will revisit the issue of assigning contrasts to a factor in Panel R5.9 (Sect. 5.3.2).

The display of the first six records of `armd` in Panel R2.5 confirms that the data do not include the records corresponding to the baseline measurements of visual acuity.

Of course, the information about the values of the measurements is still available in the variable `visual0`.

Both data frames `armd0` and `armd`, introduced in this section, are stored in "long" format. The `armd0` will be primarily used for exploratory data analyses (Sect. 3.2). On the other hand, `armd` will be the primary data frame used for the analyses throughout the entire book.

2.3 Progressive Resistance Training Study

The PRT data originate from a randomized trial aimed for devising evidence-based methods for improving and measuring the mobility and muscle power of elderly men and women in the 70+ age category (Claflin *et al.*, 2011). The working hypothesis was that a 12-week program of PRT would increase: (a) the power output of the overall musculature associated with movements of the ankles, knees, and hips; (b) the cross-sectional area and the force and power of permeabilized single fibers obtained from the *vastus lateralis* muscle; and (c) the ability of young and elderly men and women to safely arrest standardized falls. The training consisted of repeated leg extensions by shortening contractions of the leg extensor muscles against a resistance that was increased as the subject trained using a specially designed apparatus.

In the trial, healthy young (21–30 years) and older (65–80 years) male and female subjects were randomized between a "high" and "low" intensity of a 12-week PRT intervention. Randomization was stratified by age group (young or old) and sex. In total, the dataset used in our book includes 63 subjects.

For each subject, multiple measurements characterizing two types of muscle fibers were obtained before and after the 12-week PRT. The resulting data are thus an example of *clustered* data. In particular, the measurements for a given characteristic of muscle fibers for each subject correspond to a 2×2 factorial design, with fiber type $(1, 2)$ and occasion (pre-training, post-training) as the two design factors, which has important implications for the data analysis (Chap. 17).

2.3.1 Raw Data

We assume that subjects' characteristics and experimental measurements are contained in external files named `prt.subjects.data.csv` and `prt.fiber.data.csv`, respectively.

In Panel R2.6, we present the syntax for loading and inspecting the two datasets. As can be seen from the output presented in Panel R2.6a, the file `prt.subjects.data.csv` contains information about 63 subjects, with one record per subject. It includes one character variable and five numeric variables, three of which are integer-valued. The variable `id` contains subjects' identifiers, `gender`

R2.6 *PRT Trial*: Loading raw data from .csv files into objects `prt.subjects.data` and `prt.fiber.data`. The object `dataDir` was created in Panel R2.1

(a) *Loading and inspecting data from the* `prt.subjects.data.csv` *file*

```
> fp <- file.path(dataDir, "prt.subjects.data.csv")
> prt.subjects.data <- read.csv(fp, header = TRUE, as.is = TRUE)
> dim(prt.subjects.data)
  [1] 63  6
> names(prt.subjects.data)
  [1] "id"       "gender"   "ageGrp"   "trainGrp" "height"
  [6] "weight"
> str(prt.subjects.data)
  'data.frame':   63 obs. of  6 variables:
   $ id      : int  5 10 15 20 25 35 45 50 60 70 ...
   $ gender  : chr  "F" "F" "F" "F" ...
   $ ageGrp  : int  0 0 1 1 1 0 0 1 0 0 ...
   $ trainGrp: int  0 1 1 1 1 0 0 0 0 1 ...
   $ height  : num  1.56 1.71 1.67 1.55 1.69 1.69 1.72 1.61 1.71 ...
   $ weight  : num  61.9 66 70.9 62 79.1 74.5 89 68.9 62.9 68.1 ...
> head(prt.subjects.data, 4)
    id gender ageGrp trainGrp height weight
  1  5      F      0        0   1.56   61.9
  2 10      F      0        1   1.71   66.0
  3 15      F      1        1   1.67   70.9
  4 20      F      1        1   1.55   62.0
```

(b) *Loading and inspecting data from the* `prt.fiber.data.csv` *file*

```
> fp <- file.path(dataDir, "prt.fiber.data.csv")
> prt.fiber.data <- read.csv(fp, header = TRUE)
> str(prt.fiber.data)
  'data.frame':   2471 obs. of  5 variables:
   $ id           : int  5 5 5 5 5 5 5 5 5 5 ...
   $ fiber.type   : int  1 1 2 1 2 1 1 1 2 1 ...
   $ train.pre.pos: int  0 0 0 0 0 0 0 0 0 0 ...
   $ iso.fo       : num  0.265 0.518 0.491 0.718 0.16 0.41 0.371 ...
   $ spec.fo      : num  83.5 132.8 161.1 158.8 117.9 ...
> head(prt.fiber.data, 4)
    id fiber.type train.pre.pos iso.fo spec.fo
  1  5          1             0  0.265    83.5
  2  5          1             0  0.518   132.8
  3  5          2             0  0.491   161.1
  4  5          1             0  0.718   158.8
```

identifies sex, `ageGrp` indicates the age group, and `trainGrp` identifies the study group. Finally, `height` and `weight` contain the information of subjects' height and weight at baseline.

Note that the `as.is` argument used in the `read.csv()` function is set to `TRUE`. Consequently, it prevents the creation of a factor from a character variable. This applies to the `gender` variable, which is coded using the "F" and "M" characters.

The output in Panel R2.6b presents the contents of the file `prt.fiber.data.csv`. The file contains 2,471 records corresponding to individual muscle fibers. It includes five numeric variables, three of which are integer-valued. The variable `id` contains subjects' identifiers, `fiber.type` identifies the type of fiber, while `train.pre.pos` indicates whether the measurement was taken pre- or post-training. Finally, `iso.fo` and `spec.fo` contain the measured values of two characteristics of muscle fibers. These two variables will be treated as outcomes of interest in the analyses presented in Part IV of the book.

2.3.2 Data for Analysis

In Panels R2.7 and R2.8, we present the syntax used to create the `prt` dataset that will be used for analysis.

First, in Panel R2.7, we prepare data for merging. Specifically, in Panel R2.7a, we create the data frame `prt.subjects`, corresponding to `prt.subjects.data`, with several variables added and modified. Toward this end, we use the function `within()`, which applies all the modifications to the data frame `prt.subjects.data`. In particular, we replace the variable `id` by a corresponding factor. We also define the numeric variable `bmi`, which contains subject's body mass index (BMI), expressed in units of kg/m^2. Moreover, we create the factors `sex.f`, `age.f`, and `prt.f`, which correspond to the variables `gender`, `ageGrp`, and `trainGrp`, respectively. Finally, we remove the variables `weight`, `height`, `trainGrp`, `ageGrp`, and `gender`, and store the result as the data frame `prt.subjects`. The contents of the data frame is summarized using the `str()` function.

In Panel R2.7b, we create the data frame `prt.fiber`. It corresponds to `prt.fiber.data`, but instead of the variables `fiber.type` and `train.pre.pos`, it includes the factors `fiber.f` and `occ.f`. Also, a subject's identifier `id` is stored as a factor.

In Panel R2.8, we construct the data frame `prt` by merging the data frames `prt.subjects` and `prt.fiber` created in Panel R2.7. As a result, we obtain data stored in the "long" format with 2,471 records and nine variables. The contents of the first six rows of the data frame `prt` are displayed with the help of the `head()` function.

R2.7 *PRT Trial*: Construction of the data frame `prt`. Creating data frames `prt.subjects` and `prt.fiber` containing subjects' and fiber measurements. Data frames `prt.subjects.data` and `prt.fiber.data` were created in Panel R2.6

(a) *Subjects' characteristics*

```
> prt.subjects <-
+   within(prt.subjects.data,
+          {
+           id <- factor(id)
+           bmi <- weight/(height^2)
+           sex.f <- factor(gender, labels = c("Female", "Male"))
+           age.f <- factor(ageGrp, labels = c("Young", "Old"))
+           prt.f <-
+             factor(trainGrp, levels = c("1", "0"),
+                    labels = c("High", "Low"))
+           gender <- ageGrp <- trainGrp <- height <- weight <- NULL
+          })
> str(prt.subjects)
 'data.frame':    63 obs. of  5 variables:
  $ id    : Factor w/ 63 levels "5","10","15",..: 1 2 3 4 5 6 7 8 9 ...
  $ prt.f: Factor w/ 2 levels "High","Low": 2 1 1 1 1 2 2 2 2 1 ...
  $ age.f: Factor w/ 2 levels "Young","Old": 1 1 2 2 2 1 1 2 1 1 ...
  $ sex.f: Factor w/ 2 levels "Female","Male": 1 1 1 1 1 1 2 1 2 2 ...
  $ bmi   : num  25.4 22.6 25.4 25.8 27.7 ...
```

(b) *Fiber measurements*

```
> prt.fiber  <-
+   within(prt.fiber.data,
+          {
+           id <- factor(id)
+           fiber.f <-
+             factor(fiber.type, labels = c("Type 1", "Type 2"))
+           occ.f <-
+             factor(train.pre.pos, labels = c("Pre", "Pos"))
+           fiber.type <- train.pre.pos <- NULL
+          })
> str(prt.fiber)
 'data.frame':    2471 obs. of  5 variables:
  $ id      : Factor w/ 63 levels "5","10","15",..: 1 1 1 1 1 1 1 1 ...
  $ iso.fo  : num  0.265 0.518 0.491 0.718 0.16 0.41 0.371 0.792 ...
  $ spec.fo: num  83.5 132.8 161.1 158.8 117.9 ...
  $ occ.f   : Factor w/ 2 levels "Pre","Pos": 1 1 1 1 1 1 1 1 1 1 ...
  $ fiber.f: Factor w/ 2 levels "Type 1","Type 2": 1 1 2 1 2 1 1 1 ...
```

R2.8 *PRT Trial*: Construction of the data frame prt by merging prt.subjects with prt.fiber containing subjects' and fiber data. Data prt.subjects and prt.fiber were created in Panel R2.7

```
> prt <- merge(prt.subjects, prt.fiber, sort = FALSE)
> dim(prt)
  [1] 2471    9
> names(prt)
  [1] "id"       "prt.f"    "age.f"    "sex.f"    "bmi"      "iso.fo"
  [7] "spec.fo" "occ.f"    "fiber.f"
> head(prt)
    id prt.f age.f   sex.f    bmi iso.fo spec.fo occ.f fiber.f
  1  5   Low Young Female 25.436  0.265    83.5   Pre  Type 1
  2  5   Low Young Female 25.436  0.518   132.8   Pre  Type 1
  3  5   Low Young Female 25.436  0.491   161.1   Pre  Type 2
  4  5   Low Young Female 25.436  0.718   158.8   Pre  Type 1
  5  5   Low Young Female 25.436  0.160   117.9   Pre  Type 2
  6  5   Low Young Female 25.436  0.410    87.8   Pre  Type 1
```

2.4 The Study of Instructional Improvement Project

The SII was carried out to assess the math achievement scores of first- and third-grade pupils in randomly selected classrooms from a national US sample of elementary schools (Hill et al., 2005). The dataset includes results for 1,190 first-grade pupils sampled from 312 classrooms in 107 schools.

The SII data exhibit a *hierarchical* structure. That is, pupils are grouped in classes, which, in turn, are grouped within schools. This structure implies that, e.g., scores for pupils from the same class are likely correlated. The correlation should be taken into account in the analysis.

2.4.1 Raw Data

As a starting point, we use the data frame classroom, which can be found in the **WWGbook** package.

In Panel R2.9, we investigate the structure and contents of the data frame. As it can be seen from the results of application of the dim() function, the data frame contains 1,190 records and 12 variables.

The names of the variables are listed with the help of the names() function. The contents of the variables, described on p. 118 of the book by West et al. (2007), are as follows:

R2.9 *SII Project*: The structure and contents of the data frame `classroom` from the **WWGbook** package

```
> data(classroom, package = "WWGbook")
> dim(classroom)                          # Number of rows & variables
  [1] 1190   12
> names(classroom)                        # Variable names
   [1] "sex"       "minority" "mathkind" "mathgain" "ses"
   [6] "yearstea" "mathknow" "housepov" "mathprep" "classid"
  [11] "schoolid" "childid"
> classroom                               # Raw data
        sex minority mathkind mathgain   ses yearstea mathknow
  1       1        1      448       32  0.46        1       NA
  2       0        1      460      109 -0.27        1       NA
  3       1        1      511       56 -0.03        1       NA
  ...   [snip]
  1189    0        0      473       44 -0.03       25     0.50
  1190    1        0      453       69 -0.37       25     0.50
        housepov mathprep classid schoolid childid
  1        0.082     2.00     160        1       1
  2        0.082     2.00     160        1       2
  3        0.082     2.00     160        1       3
  ...   [snip]
  1189     0.177     2.00     239      107    1189
  1190     0.177     2.00     239      107    1190
> str(classroom)
  'data.frame':    1190 obs. of  12 variables:
   $ sex      : int  1 0 1 0 0 1 0 0 1 0 ...
   $ minority: int  1 1 1 1 1 1 1 1 1 1 ...
   $ mathkind: int  448 460 511 449 425 450 452 443 422 480 ...
   $ mathgain: int  32 109 56 83 53 65 51 66 88 -7 ...
   $ ses      : num  0.46 -0.27 -0.03 -0.38 -0.03 0.76 -0.03 0.2 0.64 ...
   $ yearstea: num  1 1 1 2 2 2 2 2 2 2 ...
   $ mathknow: num  NA NA NA -0.11 -0.11 -0.11 -0.11 -0.11 -0.11 ...
   $ housepov: num  0.082 0.082 0.082 0.082 0.082 0.082 0.082 0.082 ...
   $ mathprep: num  2 2 2 3.25 3.25 3.25 3.25 3.25 3.25 3.25 ...
   $ classid : int  160 160 160 217 217 217 217 217 217 217 ...
   $ schoolid: int  1 1 1 1 1 1 1 1 1 1 ...
   $ childid : int  1 2 3 4 5 6 7 8 9 10 ...
```

- School-level variables:
  - `schoolid`: school's ID number
  - `housepov`: % of households in the neighborhood of the school below the poverty level

- Classroom-level variables:
  - `classid`: classroom's ID number
  - `yearstea`: years of teacher's experience in teaching in the first grade
  - `mathprep`: the number of preparatory courses on the first-grade math contents and methods followed by the teacher
  - `mathknow`: teacher's knowledge of the first-grade math contents (higher values indicate a higher knowledge of the contents)
- Pupil-level variables:
  - `childid`: pupil's ID number
  - `mathgain`: pupil's gain in the math achievement score from the spring of kindergarten to the spring of first grade
  - `mathkind`: pupil's math score in the spring of the kindergarten year
  - `sex`: an indicator variable for sex
  - `minority`: an indicator variable for the minority status
  - `ses`: pupil's socioeconomic status

The outcome of interest is contained in the variable `mathgain`.

The abbreviated display of the contents of the `classroom` data frame shows that the data are stored with one record for each pupil. The output of the `str()` function indicates that the variables, contained in the data frame, are all either numeric or integer-valued. Note, however, that we do not have information about, e.g., the number of distinct levels of the integer-valued variables.

2.4.2 *Data for Analysis*

In the analyses presented later in the book, we will be using the data frame `SIIdata`, which is included in the **nlmeU** package. It was constructed from the data frame `classroom` using the syntax shown in Panel R2.10.

Essentially, the data frame `SIIdata` contains all the variables from `classroom`. However, variables `sex`, `minority`, `schoolid`, `classid`, and `childid` are replaced by corresponding factors. Note that, in Panel R2.10, we illustrate various forms of the syntax for the function `factor()`, which can be used to create a factor. In this way, we can explain the process of construction of a factor in more detail.

For the variable `sex`, we explicitly use both the `levels` and `labels` arguments of the function `factor()`. In this way, we fully control the mapping of the values of the original variable to the factor levels and to their labels. In the syntax shown in Panel R2.10, the value `0` of the variable `sex` from the `classroom` data is considered the first level and is assigned the label `M`. On the other hand, the value `1` is considered the second level and is labeled `F`.

R2.10 *SII Project*: Creation of the data frame `SIIdata` from the `classroom` data

```
> SIIdata <-
+   within(classroom,
+          {
+            sex <-                              # 0 -> 1(M), 1 -> 2(F)
+              factor(sex, levels = c(0, 1), labels = c("M", "F"))
+            minority <-                         # 0 -> 1(No), 1 -> 2(Yes)
+              factor(minority, labels = c("Mnrt:No", "Mnrt:Yes"))
+            schoolid <- factor(schoolid)
+            classid <- factor(classid)
+            childid <- factor(childid)
+            })
> str(SIIdata)
 'data.frame':    1190 obs. of  12 variables:
  $ sex      : Factor w/ 2 levels "M","F": 2 1 2 1 1 2 1 1 2 1 ...
  $ minority: Factor w/ 2 levels "Mnrt:No","Mnrt:Yes": 2 2 2 2 2 2 ...
 ...      [snip]
  $ classid  : Factor w/ 312 levels "1","2","3","4",..: 160 160 160 ...
  $ schoolid: Factor w/ 107 levels "1","2","3","4",..: 1 1 1 1 1 1 ...
  $ childid  : Factor w/ 1190 levels "1","2","3","4",..: 1 2 3 4 5 6 ...
```

It is worth noting that, in the printout of the structure of `SIIdata`, the variable `sex` is defined as a factor with two levels: `M` (first) and `F` (second). In the listing of the first values of the variable, obtained using the `str()` function, we only see the numerical representation (the ranks) of the levels, i.e., `1` or `2`. Thus, the information about the coding, `0` and `1`, of the original variable `sex` from the `classroom` data frame is lost. Of course, if needed, we could recover it based on the specified value of the `levels` argument.

For the variable `minority`, we only use the `labels` argument of the function `factor()`. Thus, by default, the `levels` argument is obtained by taking the unique values of the variable, i.e., `0` and `1`; representing them as characters "0" and "1", respectively; and then sorting them according to an increasing order of the numeric values of the variable. Thus, the assumed (ordered) levels are "0" (first) and "1" (second). Subsequently, the `labels` argument assigns the label `"Mnrt:No"` to the first level ("0") and `"Mnrt:Yes"` to the second level ("1"). In the printout of the structure of `SIIdata`, the listing of the first values of `minority` includes only the value 2, i.e., the second level. Hence, we could conclude that, in the `classroom` data frame, the numeric value of `minority` for the first observations was equal to `1`, which is in agreement with the printout shown in Panel R2.9.

When converting variables `schoolid`, `childid`, and `classid` into factors, we use neither the `levels` nor `labels` argument. Thus, by default, the levels of the constructed factors are defined by taking the unique numeric values of each of the variables, representing the values as character strings, and sorting the strings in an

R2.11 *SII Project*: Saving the `SIIdata` data in an external file

```
> rdaDir <- file.path("C:", "temp")                # Dir path
> fp <- file.path(rdaDir, "SIIdata.Rdata")  # External file path
> save(SIIdata, file = fp)                          # Save data
> file.exists(fp)
  [1] TRUE
> (load(fp))                                        # Load data
  [1] "SIIdata"
```

increasing order according to the numeric values. On the other hand, the labels are defined, by default, as equal to the (character) levels of the factor. Hence, for instance, for the variable `schoolid`, the ordered (character) levels are "`1`", "`2`", ..., "`107`", with the same sequence used to create the corresponding set of labels (see Panel R2.10).

For illustration purposes, in Panel R2.11, we present a syntax that allows saving data in an external file for later use and then loading them back from that file. It is recommended to perform these steps at the end of an R session. In our book, we do not have to do it, because the data are already saved in the **nlmeU** package.

2.4.3 Data Hierarchy

In practice, we often want to verify whether identifying variables, contained in a dataset, were properly coded, so that they correctly reflect the intended data hierarchy. In this section, we present the R tools that can be used for this purpose. As an example, we use the data stored in the data frame `SIIdata`. In this way, we provide additional information about the structure of the data frame.

Toward this end, we create, in Panel R2.12, an auxiliary data frame `dtId`, which contains the school, class, and pupil identifiers from `SIIdata`. We then apply the function `duplicated()` to the auxiliary data frame. The function looks for duplicated rows in the data frame and returns a logical vector that indicates which rows are duplicates. By applying the function `any()` to the resulting logical vector, we check if any of the elements of the vector contains the logical value of `TRUE`. It turns out that there are no such elements, i.e., that there are no duplicated combinations of the three identifiers in the `SIIdata` data frame. This indicates that individual pupils in the data are uniquely identified by these variables, as intended.

Next, we apply the function `gsummary()` from the package **nlme**. The function provides a summary of variables, contained in a data frame, by groups of rows. In particular, the function can be used to determine whether there are variables that are invariant within the groups. Note that the groups are defined by the factors specified on the right-hand side of the formula specified in the argument `form` (more information on the use of formulae in R will be provided in Chap. 5).

R2.12 *SII Project*: Investigation of the data hierarchy in the data frame SIIdata

```
> data(SIIdata, package = "nlmeU")           # Load data
> dtId <- subset(SIIdata, select = c(schoolid, classid, childid))
> names(dtId)                                # id names
  [1] "schoolid" "classid"  "childid"
> any(duplicated(dtId))                      # Any duplicate ids?
  [1] FALSE
> require(nlme)
> names(gsummary(dtId, form = ~childid, inv = TRUE))
  [1] "schoolid" "classid"  "childid"
> names(gsummary(dtId, form = ~classid, inv = TRUE))
  [1] "schoolid" "classid"
> names(gsummary(dtId, form = ~schoolid, inv = TRUE))
  [1] "schoolid"
```

We first apply the function gsummary() to the data frame dtId, with groups defined by childid. We also use the argument inv = TRUE. This means that only those variables, which are invariant within each group, are to be summarized. By applying the function names() to the data frame returned by the function gsummary(), we learn that, within the rows sharing the same value of childid, the values of variables schoolid and classid are also constant. In other words, variable childid is *inner* to both classid and schoolid. In particular, this implies that no pupil is present in more than one class or school. Hence, we can say that pupils are *nested* within both schools and classes. If some pupils were enrolled in, e.g., more than one class, then we could say that pupils were *crossed* with classes. In such case, the values of the classid identifier would not be constant within the groups defined by the levels of the childid variable.

Application of the function gsummary() to the data frame dtId with groups defined by classid allows us to conclude that, within the rows sharing the same value of classid, the values of schoolid are also constant. This confirms that, in the data, classes are coded as nested within schools. Equivalently, we can say that the variable classid is inner to schoolid.

Finally, there are no invariant identifiers within the groups of rows defined by the same value of schoolid, apart from schoolid itself.

In a similar fashion, in Panel R2.13, we use the function gsummary() to investigate, which covariates are defined at the school, class, or pupil level. In Panel R2.13a, we apply the function to the data frame SIIdata, with groups defined by schoolid. The displayed result of the function names() implies that the values of the variable housepov are constant (invariant) within the groups of rows with the same value of schoolid. Hence, housepov is the only school-level covariate, in accordance with the information given in Sect. 2.4.1.

In Panel R2.13b, we apply the function gsummary() with groups defined by classid. We store the names of invariant variables in the character vector nms2a.

R2.13 *SII Project*: Identification of school-, class-, and pupil-level variables in the data frame SIIdata

(a) *School-level variables*

```
> (nms1 <-
+   names(gsummary(SIIdata,
+                  form = ~schoolid, # schoolid-specific
+                  inv = TRUE)))
  [1] "housepov" "schoolid"
```

(b) *Class-level variables*

```
> nms2a <-
+   names(gsummary(SIIdata,
+                  form = ~classid,  # classid- and schoolid-specific
+                  inv = TRUE))
> idx1  <- match(nms1, nms2a)
> (nms2 <- nms2a[-idx1])                    # classid-specific
  [1] "yearstea" "mathknow" "mathprep" "classid"
```

(c) *Pupil-level variables*

```
> nms3a <-
+   names(gsummary(SIIdata,
+                  form = ~childid,  # All
+                  inv = TRUE))
> idx12 <- match(c(nms1, nms2), nms3a)
> nms3a[-idx12]                             # childid-specific
  [1] "sex"      "minority" "mathkind" "mathgain" "ses"
  [6] "childid"
```

We identify the names of variables, which are constant both at the school and class level, by matching the elements of vectors nms1 and nms2a. After removing the matching elements from the vector nms2a, we store the result in the vector nms2. The latter vector contains the names of variables, which are invariant at the class level, namely, yearstea, mathknow, and mathprep.

Finally, in Panel R2.13c, we look for pupil-level variables. The syntax is similar to the one used in R2.13b. As a result, we identify variables sex, minority, mathkind, mathgain, and ses, again consistent with variables listed in Sect. 2.4.1.

Considerations, presented in Panel R2.13, aimed at identifying grouping factor(s) for which a given covariate is invariant. The resulting conclusions have important implications for computations of the number of denominator degrees of freedom for the conditional F-tests applied to fixed effects in LMMs (see Sect. 14.7 and Panel R18.5 in Sect. 18.2.2).

2.4.3.1 Explicit and Implicit Nesting

The `SIIdata` data frame is an example of data having *nested* structure. This structure, with classes being nested within schools, can be represented in the data in two different ways, depending on how the two relevant factors, namely, `schoolid` and `classid`, are coded.

First, we consider the case when the levels of `classid` are explicitly coded as *nested* within the levels of the `schoolid` grouping factor. This way of coding is referred to as *explicit nesting* and is consistent with that used in `SIIdata`, as shown in Panel R2.12. More specifically, the nesting was accomplished by using *different* levels of the `classid` factor for different levels of the `schoolid` factor. Consequently, the intended nested structure of data is explicitly reflected by the levels of the factors. This is the preferred and natural approach.

The nested structure could also be represented by using *crossed* grouping factors. Taking the `SIIdata` data as an example, we might consider the case when, by mistake or for any other reason, two different classrooms from two different schools would have *the same* code. In such a situation, and without any additional information about the study design, the factors would be incorrectly interpreted as (partially) crossed. To specify the intended nested structure, we would need to cross `schoolid` and `classid` factors using, e.g., the command `factor(schoolid:classid)`. The so-obtained grouping factor, together with `schoolid`, would specify the desired nested structure. Such an approach to data coding is referred to as *implicit nesting*.

Although the first way of representing the nested structure is simpler and more natural, it requires caution when coding the levels of grouping factors. The second approach is more inclusive, in the sense that it can be used both for crossed *and* nested factors.

We raise the issue of the different representations of nested data because it has important implications for a specification of an LMM. We will revisit this issue in Chap. 15.

2.5 The Flemish Community Attainment-Targets Study

The FCAT data results from an educational study, in which elementary-school graduates were evaluated with respect to reading comprehension in Dutch. The evaluation was based on a set of attainment targets, which were issued by the Flemish Community in Belgium. These attainment targets can be characterized by the text type and by the level of processing. We use data which consist of the responses of a group of 539 pupils from 15 schools who answered 57 items assumed to measure nine attainment targets. In Table 2.1, the nine attainment targets are described by the type of text and by the level of processing. In addition, we indicate the number of items that were used to measure each one of the targets.

Table 2.1 *FCAT Study*: FCAT Study: Attainment targets for reading comprehension in Dutch. Based on Janssen et al. (2000). Reproduced with permission from the copyright owner

Target	Text type	Level of processing	No. of items
1	Instructions	Retrieving	4
2	Articles in magazine	Retrieving	6
3	Study material	Structuring	8
4	Tasks in textbook	Structuring	5
5	Comics	Structuring	9
6	Stories, novels	Structuring	6
7	Poems	Structuring	8
8	Newspapers for children, textbooks, encyclopedias	Evaluating	6
9	Advertising material	Evaluating	5

These data were analyzed previously by, e.g., Janssen et al. (2000) and Tibaldi et al. (2007). In our analyses we will use two types of outcomes. First, we will consider total target scores, i.e., the sum of all positive answers for a target. Second, we will consider average target scores, i.e., the sum of all positive answers for a category divided by the number of items within the target. In both cases, we will treat the outcome as a continuous variable.

2.5.1 Raw Data

We assume that the raw data for the FCAT study are stored in an external file named `crossreg.data.csv`.

In Panel R2.14, we present the syntax for loading and inspecting the data. As seen from the output presented in the panel, the file `crossreg.data.csv` contains 4,851 records and three variables. The variable `id` contains pupils' identifiers, `target` identifies the attainment targets (see Table 2.1), and `scorec` provides the total target score for a particular pupil. Note that the data are stored using the "long" format, with multiple records per pupil.

In Panel R2.15, we investigate the contents of the `crossreg.data` data frame in more detail. In particular, by applying the function `unique()` to each of the three variables contained in the data frame, we conclude that there are 539 unique values for `id`, nine unique values for `target`, and 10 unique values for `scorec`. Thus, the data frame includes scores for nine targets for each of 539 pupils. Note that 9×539 =4,851, i.e., the total number of records (rows). Because the maximum number of items for a target is nine (see Table 2.1), the variable `scorec` contains integer values between 0 and 9.

R2.14 *FCAT Study*: Loading raw data from the .csv file into the object `crossreg.data`. The object `dataDir` was created in Panel R2.1

```
> fp <- file.path(dataDir, "crossreg.data.csv")
> crossreg.data <- read.csv(fp, header = TRUE)
> dim(crossreg.data)                          # No. of rows and columns
  [1] 4851    3
> names(crossreg.data)                        # Variable names
  [1] "target" "id"     "scorec"
> head(crossreg.data)                         # First six records
    target id scorec
  1      1  1      4
  2      2  1      6
  3      3  1      4
  4      4  1      1
  5      5  1      7
  6      6  1      6
> str(crossreg.data)                          # Data structure
  'data.frame':    4851 obs. of  3 variables:
   $ target: int  1 2 3 4 5 6 7 8 9 1 ...
   $ id    : int  1 1 1 1 1 1 1 1 1 2 ...
   $ scorec: int  4 6 4 1 7 6 6 5 5 3 ...
```

R2.15 *FCAT Study*: Inspection of the contents of the raw data. The data frame `cressreg.data` was created in Panel R2.14

```
> unique(crossreg.data$target)            # Unique values for target
  [1] 1 2 3 4 5 6 7 8 9
> (unique(crossreg.data$id))              # Unique values for id
    [1]   1   2   3   4   5   6   7   8   9  10  11  12  13 14 15
  ...      [snip]
  [526] 526 527 528 529 530 531 532 533 534 535 536 537 538 539
> unique(crossreg.data$scorec)            # Unique values for scorec
   [1] 4 6 1 7 5 3 2 8 0 9
> summary(crossreg.data$scorec)           # Summary statistics for scorec
     Min. 1st Qu.  Median    Mean 3rd Qu.    Max.
      0.0     3.0     4.0     3.9     5.0     9.0
```

2.5.2 *Data for Analysis*

In the analyses presented later in the book, we will be using the data frame `fcat`, which is constructed based on the data frame `crossreg.data`. In Panel R2.16, we present the syntax used to create the `fcat` data and to investigate data grouping structure. First, in Panel R2.16a, we replace the variables `id` and `target` by corresponding factors. For the factor `target`, the labels given in parentheses indicate the number of items for a particular target.

In Panel R2.16b, we cross-tabulate the factors `id` and `target` and store the resulting table in the object `tab1`. Given the large number of levels of the factor `id`, it is difficult to verify the values of the counts for all cells of the table. By applying the function `all()` to the result of the evaluation of expression `tab1>0`, we check that all counts of the table are nonzero. On the other hand, with the help of the `range()` function, we verify that all the counts are equal to 1. This indicates that, in the data frame `fcat`, the levels of the factor `target` are crossed with the levels of the factor `id`. Moreover, the data are balanced, in the sense that there is the same number of observations, namely, one observation for each combination of the levels of the two factors. Because all counts in the table are greater than zero, we can say that the factors are *fully crossed*.

2.6 Chapter Summary

In this chapter, we introduced four case studies, which will be used for illustration of LMs described in our book.

We started the presentation of each case study by describing study design and considering that raw data are stored in a .csv file. We chose this approach in an attempt to emulate a common situation of using external data files when analyzing data using R. In the next step, we prepared the data for analysis by creating the necessary variables and, in particular, factors. Including factors as part of data is a feature fairly unique to R. It affects how a given variable is treated by graphical and modeling functions. This approach is recommended, but not obligatory. In particular, creating factors can be deferred to a later time, when, e.g., *model formula* is specified. We will revisit this issue in Chap. 5.

The data frames, corresponding to the four case studies, are included in the package **nlmeU**. As with other packages, the list of datasets available in the package can be obtained by using the `data(package = "nlmeU")` command. For the reader's convenience, the datasets are summarized in Table 2.2. The table includes the information about the R-session panels, which present the syntax used to create the data frames, grouping factors, and number of rows and variables.

The four case studies introduced in this chapter are conducted by employing different study designs. All of them lead to grouped data defined by one or more nested or crossed grouping factors. The preferable way of storing this type of data

R2.16 *FCAT Study*: Construction and inspection of the contents of the data frame `fcat`. The data frame `crossreg.data` was created in Panel R2.14

(a) *Construction of the data frame* `fcat`

```
> nItms <- c(4, 6, 8, 5, 9, 6, 8, 6, 5)          # See Table 2.1
> (lbls <- paste("T", 1:9, "(", nItms, ")", sep = ""))
  [1] "T1(4)" "T2(6)" "T3(8)" "T4(5)" "T5(9)" "T6(6)" "T7(8)"
  [8] "T8(6)" "T9(5)"
> fcat <-
+   within(crossreg.data,
+          {
+           id <- factor(id)
+           target <- factor(target, labels = lbls)
+          })
> str(fcat)
  'data.frame':    4851 obs. of  3 variables:
   $ target: Factor w/ 9 levels "T1(4)","T2(6)",..: 1 2 3 4 5 6 7 8 ...
   $ id    : Factor w/ 539 levels "1","2","3","4",..: 1 1 1 1 1 1 1 ...
   $ scorec: int  4 6 4 1 7 6 6 5 5 3 ...
```

(b) *Investigation of the data grouping structure*

```
> (tab1 <- xtabs(~ id + target, data = fcat))    # id by target table
       target
  id    T1(4) T2(6) T3(8) T4(5) T5(9) T6(6) T7(8) T8(6) T9(5)
    1       1     1     1     1     1     1     1     1     1
    2       1     1     1     1     1     1     1     1     1
  ...    [snip]
    539     1     1     1     1     1     1     1     1     1
> all(tab1 > 0)                                  # All counts > 0?
  [1] TRUE
> range(tab1)                                    # Range of counts
  [1] 1 1
```

is to use the "long" format with multiple records per subject. Although this term is borrowed from the literature pertaining to longitudinal data, it is also used in the context of other grouped data. Below, we describe the key features of the data in each study.

In the ARMD trial, the `armd.wide` data frame stores data in the "wide" format. Data frames `armd` and `armd0` store data in the "long" format and reflect the hierarchical data structure defined by a single grouping factor, namely, `subject`. For this reason, and following the naming convention used in the **nlme** package, we will refer to the data structure in our book as data with a *single level of grouping*. Note that, more traditionally, these data are referred to as *two-level data* (West et al., 2007).

Table 2.2 Data frames available in the **nlmeU** package

Study	Data frame	R-panel	Grouping factors	Rows × vars
ARMD Trial	`armd.wide`	R2.2	*None*	240 × 10
	`armd0`	R2.4	`subject`	1,107 ×8
	`armd`	R2.5	`subject`	867 × 8
PRT Trial	`prt.subjects`	R2.7a	*None*	63 × 5
	`prt.fiber`	R2.7b	`id`	2,471 ×5
	`prt`	R2.8	`id`	2,471 ×9
SII Project	`SIIdata`	R2.10	`classid` nested ...	1,190 ×12
			... in `schoolid`	
FCAT Study	`fcat`	R2.16	`id` crossed ...	4,851 ×3
			... with `target`	

The hierarchical structure of data contained in the data frame `SIIdata` is defined by two (nested) grouping factors, namely, `schoolid` and `classid`. Thus, in our book, this data structure will be referred to as data with *two levels of grouping*.

This naming convention works well for hierarchical data, i.e., for data with nested grouping factors. It is more problematic for structures with crossed factors. This is the case for the FCAT study, in which the data structure is defined by two crossed grouping factors, thus without a particular hierarchy.

As a result of data grouping, variables can be roughly divided into group- and measurement-specific categories. In the context of longitudinal data they are referred to as time-fixed and time-varying variables. The classification of the variables has important implications for the model specification.

To our knowledge, the *groupedData* class, defined in the **nlme** package, appears to be the only attempt to directly associate a hierarchical structure of the data with objects of the *data.frame* class. We do not describe this class in more detail, however, because it has some limitations. Also, its initial importance has diminished substantially over time. In fact, the data hierarchy is most often reflected indirectly by specifying the structure of the model fitted to the data. We will revisit this issue in Parts III and IV of our book.

When introducing the SII case study, we noted that the nested data structure can be specified by using two different approaches, namely, explicit and implicit nesting, depending on the coding of the levels of grouping factors. The choice of the approach is left to the researcher's discretion. The issue has important implications for the specification of LMMs, though, and it will be discussed in Chap. 15.

The different data structures of the cases studies presented in this chapter will allow us to present various aspects of LMMs in Part IV of the book. Additionally, the ARMD dataset will be used in the other parts to illustrate other classes of LMs and related R tools.

The main focus of this chapter was on the presentation of the data frames related to the case studies. In the presentation, we also introduced selected concepts related to grouped data and R functions, which are useful for data transformation and inspection of the contents of datasets. By necessity, our introduction was very brief and fragmentary; a more in-depth discussion of those and other functions is beyond the scope of our book. The interested readers are referred to, e.g., the book by Dalgaard (2008) for a more thorough explanation of the subject.

Chapter 3
Data Exploration

3.1 Introduction

In this chapter, we present the results of exploratory analyses of the case studies introduced in Chap. 2. The results will serve as a basis for building LMs for the data in the following parts of the book.

While exploring the case-study data, we also illustrate the use of selected functions and graphical tools which are commonly used to perform these tasks. Note, however, that, unlike in the other parts of the book, we are not discussing the functions and tools in much detail. The readers interested in the functionalities are referred to the monograph by Venables and Ripley (2010).

3.2 ARMD Trial: Visual Acuity

In the ARMD data, we are mainly interested in the effect of treatment on the visual acuity measurements. Thus, in Fig. 3.1, we first take a look at the measurements by plotting them against time for several selected patients from both treatment groups. More specifically, we selected every 10th patient from each group.

Based on the plots shown in Fig. 3.1, several observations can be made:

- In general, visual acuity tends to decrease in time. This is in agreement with the remark made in Sect. 2.2 that patients with ARMD progressively lose vision.
- For some patients, a linear decrease of visual acuity over time can be observed, but there are also patients for whom individual profiles strongly deviate from a linear trend.
- Visual acuity measurements adjacent in time are fairly well correlated, with the correlation decreasing with an increasing distance in time.
- Visual acuity at baseline seems to, at least partially, determine the overall level of the post-randomization measurements.
- There are patients for whom several measurements are missing.

A. Gałecki and T. Burzykowski, *Linear Mixed-Effects Models Using R: A Step-by-Step Approach*, Springer Texts in Statistics, DOI 10.1007/978-1-4614-3900-4_3,

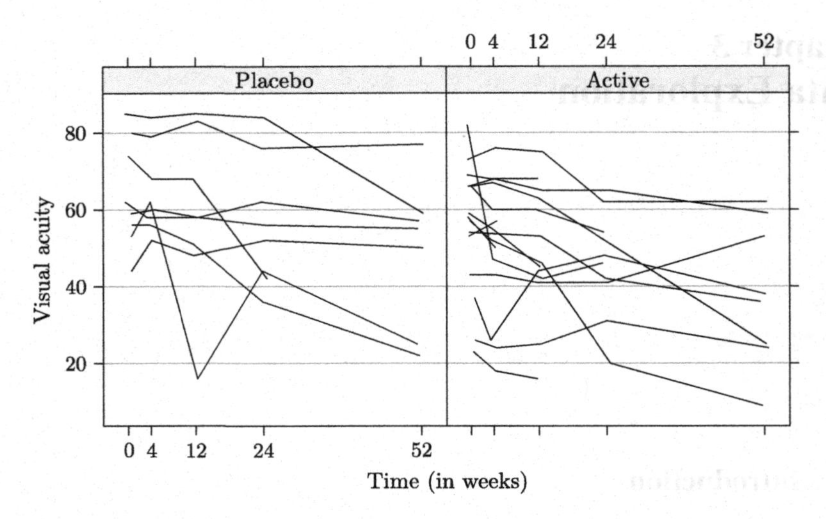

Fig. 3.1 *ARMD Trial*: Visual-acuity profiles for selected patients ("spaghetti plot")

These observations will be taken into account when constructing models for the data.

R3.1 *ARMD Trial*: Syntax for the plot of visual acuity profiles for selected patients in Fig. 3.1

```
> data(armd.wide, armd0, package = "nlmeU")          # Data loaded
> library(lattice)
> armd0.subset <-                                    # Subset
+   subset(armd0, as.numeric(subject) %in% seq(1, 240, 10))
> xy1 <-                                             # Draft plot
+   xyplot(visual ~ jitter(time) | treat.f,
+          groups = subject,
+          data = armd0.subset,
+          type = "l", lty = 1)
> update(xy1,                                        # Fig. 3.1
+        xlab = "Time (in weeks)",
+        ylab = "Visual acuity",
+        grid = "h")
> detach(package:lattice)
```

The syntax used to create Fig. 3.1 is shown in Panel R3.1. First, we load data to be used for exploration from the **nlmeU** package. Note that the code used to create figure employs the function `xyplot()` from the package **lattice** (Sarkar, 2008). The function is applied to the subset of the data frame `armd0` (Sect. 2.2.2). The formula used in the syntax indicates that the variables `visual` and `time` are to be used on the

y- and x-axis, respectively. These variables are plotted against each other in separate panels for different values of the `treat.f` factor. Within each panel, data points are grouped for each subject and connected using solid lines. The function `jitter()` is used to add a small amount of noise to the variable `time`, thereby reducing the number of overlapping points.

In the next sections, we explore particular features of the ARMD data in more detail.

3.2.1 Patterns of Missing Data

First, we check the number and patterns of missing visual acuity measurements. Toward this end, we use the data frame `armd.wide`. As mentioned in Sect. 2.2.2, the data frame contains the factor `miss.pat` that indicates which of the four post-randomization measurements are missing for a particular patient. For example, the pattern `--X-` indicates that the only missing measurement was at the third post-randomization timepoint, i.e., at 24 weeks.

In Panel R3.2, we use three different methods to tabulate the number of patients with different levels of the factor `miss.pat`. From the displayed results, we can conclude that, for instance, there were 188 patients for whom all four post-randomization visual acuity measurements were obtained. On the other hand, there were six patients for whom the four measurements were missing.

R3.2 *ARMD Trial*: Inspecting missing-data patterns in the `armd.wide` data for the post-randomization visual acuity measurements using three different methods

```
> table(armd.wide$miss.pat)
  ---- ---X --X- --XX -XX- -XXX X--- X-XX XXXX
   188   24    4    8    1    6    2    1    6
> with(armd.wide, table(miss.pat))
  miss.pat
  ---- ---X --X- --XX -XX- -XXX X--- X-XX XXXX
   188   24    4    8    1    6    2    1    6
> xtabs(~miss.pat, armd.wide)
  miss.pat
  ---- ---X --X- --XX -XX- -XXX X--- X-XX XXXX
   188   24    4    8    1    6    2    1    6
```

It is also worth noting that there are eight $(= 4 + 1 + 2 + 1)$ patients with four different nonmonotone missing-data patterns, i.e., with intermittent missing visual acuity measurements. When modeling data with such patterns, extra care is needed when specifying variance–covariance structures. We will come back to this issue in Sect. 11.4.2.

3.2.2 Mean-Value Profiles

In this section, we investigate the number of missing values and calculate the sample means of visual acuity measurements for different visits and treatment groups. Toward this end, in Panel R3.3, we use the "long"-format data frame `armd0`, which was described in Sect. 2.2.2.

R3.3 *ARMD Trial*: Sample means and medians for visual acuity by time and treatment

(a) *Counts of nonmissing visual acuity measurements*

```
> attach(armd0)
> flst <- list(time.f, treat.f)                          # "By" factors
> (tN <-                                                 # Counts
+   tapply(visual, flst,
+          FUN = function(x) length(x[!is.na(x)])))
         Placebo Active
 Baseline     119    121
 4wks         117    114
 12wks        117    110
 24wks        112    102
 52wks        105     90
```

(b) *Sample means and medians of visual acuity measurements*

```
> tMn <- tapply(visual, flst, FUN = mean)                # Sample means
> tMd <- tapply(visual, flst, FUN = median)              # Sample medians
> colnames(res  <- cbind(tN, tMn, tMd))                  # Column names
 [1] "Placebo" "Active"  "Placebo" "Active"  "Placebo" "Active"
> nms1 <- rep(c("P", "A"), 3)
> nms2 <- rep(c("n", "Mean", "Mdn"), rep(2, 3))
> colnames(res) <- paste(nms1, nms2, sep = ":")          # New column names
> res
         P:n A:n P:Mean A:Mean P:Mdn A:Mdn
 Baseline 119 121 55.336 54.579  56.0  57.0
 4wks     117 114 53.966 50.912  54.0  52.0
 12wks    117 110 52.872 48.673  53.0  49.5
 24wks    112 102 49.330 45.461  50.5  45.0
 52wks    105  90 44.438 39.100  44.0  37.0
> detach(armd0)
```

To calculate counts of missing values in Panel R3.3a, we use the function `tapply()`. In general, this function is used to apply a selected function to each (nonempty) group of values defined by a unique combination of the levels of one or more factors. In our case, the selected function, specified in the `FUN` argument, checks the length of the vector created by selecting nonmissing values from the vector passed as an argument to the function. Using the `tapply()` function, we apply it to the variable `visual` within the groups defined by combinations of the levels of factors `time.f` and `treat.f`. As a result, we obtain a matrix with the number of nonmissing visual acuity measurements for each visit and each treatment group. We store the matrix in the object `tN` for further use. The display of the matrix indicates that there were no missing measurements at baseline. On the other hand, at week 4, for instance, there were two and seven missing measurements in the placebo and active-treatment arms, respectively. In general, there are more missing measurements in the active-treatment group.

In Panel R3.3b, we use the function `tapply()` twice to compute the sample means and sample medians of visual acuity measurements for each combination of the levels of factors `time.f` and `treat.f`. We store the results in matrices `tMn` and `tMd`, respectively. We then create the matrix `res` by combining matrices `tN`, `tMn`, and `tMn` by columns. Finally, to improve the legibility of displays, we modify the names of the columns of `res`.

From the display of the matrix `res`, we conclude that, on average, there was very little difference in visual acuity between the two treatment groups at baseline. This is expected in a randomized study. During the course of the study, the mean visual acuity decreased with time in both arms, which confirms the observation made based on the individual profiles presented in Fig. 3.1. It is worth noting that the mean value is consistently higher in the placebo group, which suggests lack of effect of interferon-α.

Figure 3.2 presents box-and-whiskers plots of visual acuity for the five timepoints and the two treatment arms. The syntax to create the figure is shown in Panel R3.4. It uses the function `bwplot()` from the package **lattice**. Note that we first create a draft of the plot, which we subsequently enhance by providing labels for the horizontal axis. In contrast to Fig. 3.1, measurements for all subjects at all timepoints are plotted. A disadvantage of the plot is that it does not reflect the longitudinal structure of the data.

R3.4 *ARMD Trial*: Syntax for the box-and-whiskers plots in Fig. 3.2

```
> library(lattice)
> bw1 <-                                          # Draft plot
+   bwplot(visual ~ time.f | treat.f,
+          data = armd0)
> xlims <- c("Base", "4\nwks", "12\nwks", "24\nwks", "52\nwks")
> update(bw1, xlim = xlims, pch = "|")            # Final plot
> detach(package:lattice)
```

The box-and-whiskers plots illustrate the patterns implied by the sample means and medians, presented in Panel R3.3b. The decrease of the mean values in time is clearly seen for both treatment groups. It is more pronounced for the active-treatment arm. As there was a slightly higher dropout in that arm, a possible

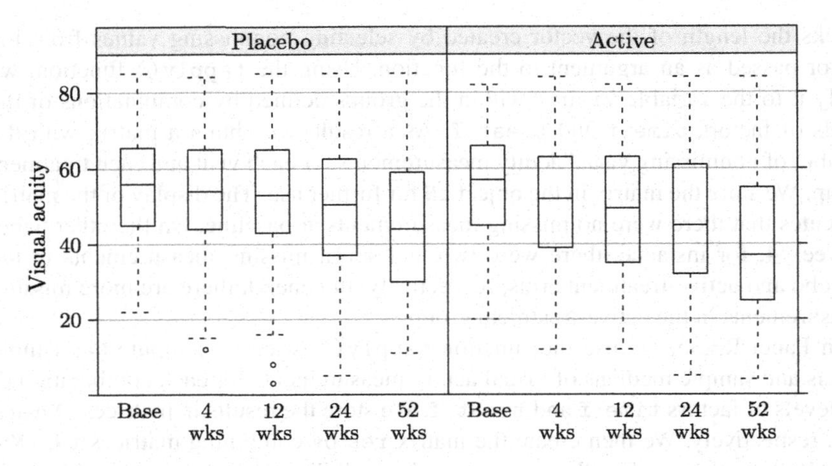

Fig. 3.2 *ARMD Trial*: Box-and-whiskers plots for visual acuity by treatment and time

explanation could be that patients whose visual acuity improved dropped out of the study. In such case, a faster progression of the disease in that treatment arm would be observed.

To check this possibility, we take a look at Fig. 3.3. It shows the mean values of visual acuity for patients with different monotone missing-data patterns. In addition, the number of subjects for each pattern is also given. We note that the number of subjects for the patterns with a larger number of missing values tends to be smaller. Note that, to save space, we do not present the syntax used to create the figure, as it is fairly complex.

The mean profiles, shown in Fig. 3.3, consistently decrease for the majority of the patterns. In general, they do not suggest an improvement in visual acuity before the drop off. Thus, they do not support the aforementioned explanation of a faster decrease of the mean visual acuity in the active-treatment arm.

In Panel R3.5, we present the syntax to investigate the number and form of monotone missing-data patterns for visual acuity. In particular, in Panel R3.5a, we create the data frame `armd.wide.mnt`, which contains data only for patients with monotone patterns. There are 232 such patients in total. Note that, despite the fact that some patterns are not present in the data frame `armd.wide.mnt`, they are still recognized as valid levels of the factor `miss.pat`. This might cause problems when using some R functions. Similarly to Panel R2.5, we could use the `droplevels()` function to remove the unused levels of the `miss.pat` variable. Instead, in Panel R3.5b, we modify the levels of the factor `miss.pat` in the `armd.wide.mnt` data with the help of the function `factor()`. Note that, instead of using the `levels` argument of the function, we could have used the argument `exclude` while indicating the levels to be excluded from the definition of the `miss.pat` factor.

Finally, in Panel R3.5c, we use the function `tapply()` to obtain a matrix containing the number of patients for each monotone missing-data pattern and for

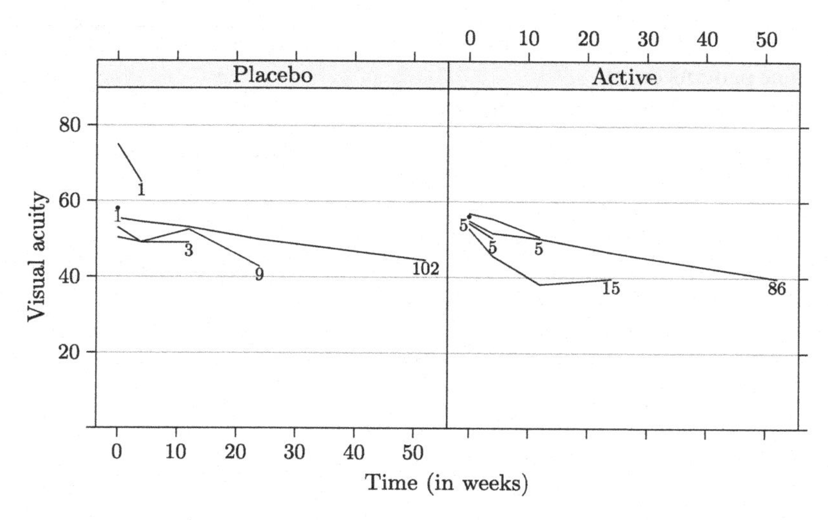

Fig. 3.3 *ARMD Trial*: Mean visual acuity profiles by missing pattern and treatment (monotone missing-data patterns only)

each treatment arm. The displayed results indicate that the mean-value profiles for missing-data patterns with a larger number of missing values, shown in Fig. 3.3, are based on measurements for a small number of patients. Thus, the variability of these profiles is larger than for the patterns with a smaller number of missing values. Therefore, Fig. 3.3 should be interpreted with caution.

3.2.3 Sample Variances and Correlations of Visual Acuity Measurements

Figure 3.4 shows a scatterplot matrix for the visual acuity measurements for those patients, for whom all post-randomization measurements are available. Scatterplots for corresponding pairs of variables are given below the diagonal. The size of the font for correlation coefficients reported above the diagonal is proportional to its value. We do not present the syntax for constructing the figure, as it is fairly complex. It can be observed that the measurements adjacent in time are strongly correlated. The correlation decreases with an increasing time gap. Worth noting is the fact that there is a substantial positive correlation between visual acuity at baseline and at the other post-randomization measurements. Thus, baseline values might be used to explain the overall variability of the post-randomization observations. This agrees with the observation made based on Fig. 3.1. It is worth noting that a scatterplot matrix of the type shown in Fig. 3.4 may not work well for longitudinal data with irregular time intervals.

R3.5 *ARMD Trial*: The number of patients by treatment and missing-data pattern (monotone patterns only)

(a) *Subset of the data with monotone missing-data patterns*

```
> mnt.pat<-                                        # Monotone patterns
+   c("----", "---X", "--XX", "-XXX", "XXXX")
> armd.wide.mnt <-                                 # Data subset
+   subset(armd.wide, miss.pat %in% mnt.pat)
> dim(armd.wide.mnt)                               # Number of rows and cols
  [1] 232  10
> levels(armd.wide.mnt$miss.pat)                   # Some levels not needed
  [1] "----" "---X" "--X-" "--XX" "-XX-" "-XXX" "X---" "X-XX"
  [9] "XXXX"
```

(b) *Removing unused levels from the* `miss.pat` *factor*

```
> armd.wide.mnt1 <-
+   within(armd.wide.mnt,
+          {
+            miss.pat <- factor(miss.pat, levels=mnt.pat)
+          })
> levels(armd.wide.mnt1$miss.pat)
  [1] "----" "---X" "--XX" "-XXX" "XXXX"
```

(c) *The number of patients with different monotone missing-data patterns*

```
> with(armd.wide.mnt1,
+      {
+        fl  <- list(treat.f, miss.pat)        # List of "by" factors
+        tapply(subject, fl, FUN=function(x) length(x[!is.na(x)]))
+      })
          ---- ---X --XX -XXX XXXX
  Placebo  102    9    3    1    1
  Active    86   15    5    5    5
```

In Panel R3.6, we provide the estimates of the variance–covariance and correlation matrices for visual acuity measurements. Toward this end, we create the data frame `visual.x` from `armd.wide` by selecting only the five variables containing the measurements. We then apply functions `var()` and `cor()` to estimate the variance–covariance matrix and the correlation matrix, respectively. Note that, for both functions, we specify the argument `use = "complete.obs"`, which selects only those rows of the data frame `visual.x` that do not contain any missing values. In this way, the estimated matrices are assured to be

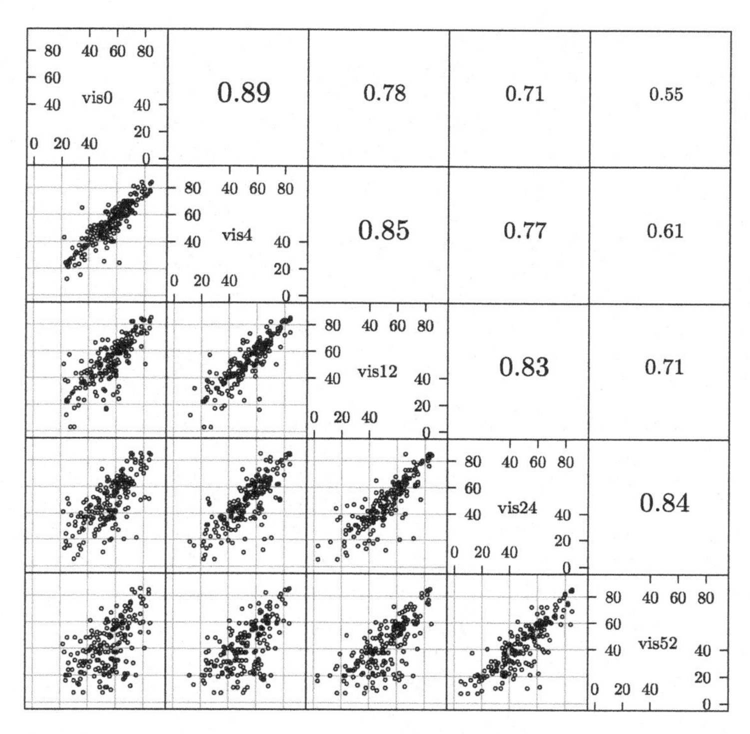

Fig. 3.4 *ARMD Trial*: Scatterplot matrix for visual acuity measurements. Scatterplots (*below diagonal*) and correlation coefficients (*above diagonal*) for complete cases only ($n = 188$)

positive semidefinite. An alternative (not shown) would be to specify `use = "pairwise.complete.obs"`. In that case, the elements of the matrices would be estimated using data for all patients with complete observations for the particular pair of visual acuity measurements. This could result in estimates of variance–covariance or correlation matrices, which might not be positive semidefinite.

The variance–covariance matrix for visual acuity measurements is stored in the `varx` matrix. It indicates an increase of the variance of visual acuity measurements obtained at later timepoints. The estimated correlation matrix suggests a moderate to strong correlation of the measurements. We also observe that the correlation clearly decreases with the time gap, as already concluded from Fig. 3.4.

At the bottom of Panel R3.6, we demonstrate how to extract the diagonal elements of the matrix `varx` using the `diag()` function. We also present the use of the function `cov2cor()` to compute a correlation matrix corresponding to the variance–covariance. Note that we do not display the result of the use of the function, as it is exactly the same as the one obtained for the function `cor()`, already shown in Panel R3.6.

R3.6 *ARMD Trial*: Variance–covariance and correlation matrices for visual acuity measurements for complete cases only ($n = 188$)

```
> visual.x <- subset(armd.wide, select = c(visual0:visual52))
> (varx <- var(visual.x, use = "complete.obs")) # Var-cov mtx
          visual0 visual4 visual12 visual24 visual52
 visual0   220.31  206.71   196.24   193.31   152.71
 visual4   206.71  246.22   224.79   221.27   179.23
 visual12  196.24  224.79   286.21   257.77   222.68
 visual24  193.31  221.27   257.77   334.45   285.23
 visual52  152.71  179.23   222.68   285.23   347.43
> print(cor(visual.x, use = "complete.obs"),     # Corr mtx
+       digits = 2)
          visual0 visual4 visual12 visual24 visual52
 visual0     1.00    0.89     0.78     0.71     0.55
 visual4     0.89    1.00     0.85     0.77     0.61
 visual12    0.78    0.85     1.00     0.83     0.71
 visual24    0.71    0.77     0.83     1.00     0.84
 visual52    0.55    0.61     0.71     0.84     1.00
> diag(varx)                        # Var-cov diagonal elements
  visual0  visual4 visual12 visual24 visual52
   220.31   246.22   286.21   334.45   347.43
> cov2cor(varx)                     # Corr mtx (alternative way)
 . . .      [snip]
```

3.3 PRT Study: Muscle Fiber Specific Force

In the PRT study, we are primarily interested in the effect of the intensity of the training on the muscle fiber specific force, measurements of which are contained in the variable `spec.fo` of the `prt` data frame (Sect. 2.3.2). In some analyses, we will also investigate the effect on the measurements of the isometric force, which are stored in the variable `iso.fo`.

First, however, we take a look at the information about subjects' characteristics, stored in the data frame `prt.subjects` (see Sect. 2.3.2). In Panel R3.7, we use the function `tapply()` to obtain summary statistics for the variable `bmi` for separate levels of the `prt.f` factor. The statistics are computed with the help of the `summary()` function. The displayed values of the statistics do not indicate any substantial differences in the distribution of BMI between subjects assigned to the low- or high-intensity training. Given that the assignment was randomized, this result is anticipated.

For illustration purposes, we also obtain summary statistics for all variables in the `prt.subjects` data frame, except for `id`, with the help of the function `by()`. The function splits the data frame according to the levels of the factor `prt.f` and applies the function `summary()` to the two data frames resulting from the split. As a result, we obtain summary statistics for variables `prt.f`, `age.f`, `sex.f`, and `bmi`

R3.7 *PRT Trial*: Summary statistics for subjects' characteristics

```
> data(prt.subjects, prt, package = "nlmeU") # Data loaded
> with(prt.subjects, tapply(bmi, prt.f, summary))
  $High
     Min. 1st Qu.  Median    Mean 3rd Qu.    Max.
     18.4    22.9    24.8    25.1    28.2    31.0

  $Low
     Min. 1st Qu.  Median    Mean 3rd Qu.    Max.
     19.0    23.1    24.8    24.7    26.3    32.3
> by(subset(prt.subjects, select = -id), prt.subjects$prt.f, summary)
  prt.subjects$prt.f: High
    prt.f        age.f        sex.f          bmi
   High:31   Young:15   Female:17   Min.   :18.4
   Low : 0   Old  :16   Male  :14   1st Qu.:22.9
                                    Median :24.8
                                    Mean   :25.1
                                    3rd Qu.:28.2
                                    Max.   :31.0
  ------------------------------------------------
  prt.subjects$prt.f: Low
    prt.f        age.f        sex.f          bmi
   High: 0   Young:15   Female:17   Min.   :19.0
   Low :32   Old  :17   Male  :15   1st Qu.:23.1
                                    Median :24.8
                                    Mean   :24.7
                                    3rd Qu.:26.3
                                    Max.   :32.3
```

for the two training-intensity groups. From the displayed values of the statistics, we conclude that there are no important differences in the distribution of sex and age groups between the two intervention groups. This is expected, given that the randomization was stratified by the two factors (see Sect. 2.3). Note that we should ignore the display for the factor `prt.f`, because it has been used for splitting the data.

In Panel R3.8, we take a look at fiber measurements stored in the data frame `prt`. In particular, in Panel R3.8a, we check the number of nonmissing measurements of the specific force per fiber type and occasion for selected subjects. Toward this aim, with the help of the function `tapply()`, we apply the function `length()` to the variable `spec.fo` for separate levels of the `id`, `fiber.f`, and `occ.f` factors. Note that, in the call to the function `tapply()`, we use a named list of the factors. The names of the components of the list are shortened versions of the factor names. In this way, we obtain a more legible display of the resulting array. In Panel R3.8a, we show the display for two subjects, `"5"` and `"335"`. For the latter, we see that no measurements of the specific force were taken for type-1 fibers before the training.

R3.8 *PRT Trial*: Extracting and summarizing the fiber-level information

(a) *Number of fibers per type and occasion for the subjects* "5" *and* "335"

```
> fibL <-
+    with(prt,
+         tapply(spec.fo,
+                list(id = id, fiberF = fiber.f, occF = occ.f),
+                length))
> dimnms <- dimnames(fibL)
> names(dimnms)            # Shortened names displayed
 [1] "id"     "fiberF" "occF"
> fibL["5", , ]            # Number of fiber measurements for subject 5
         occF
 fiberF   Pre Pos
   Type 1  12  18
   Type 2   7   4
> fibL["335", , ]          # Number of fiber measurements for subject 335
         occF
 fiberF   Pre Pos
   Type 1  NA   8
   Type 2  14  11
```

(b) *Mean value of* `spec.fo` *by fiber type and occasion for subject* "5"

```
> fibM <-
+    with(prt,
+         tapply(spec.fo,
+                list(id = id, fiberF = fiber.f, occF = occ.f),
+                mean))
> fibM["5", , ]
         occF
 fiberF      Pre    Pos
   Type 1 132.59 129.96
   Type 2 145.74 147.95
```

In Panel R3.8b, we take a look at the mean value of the specific force per fiber type and occasion for selected subjects. Toward this end, we use the function `tapply()` in a similar way as in Panel R3.8a, but in combination with the function `mean()`. In the panel, we display the mean values for the subject "5".

In Panel R3.9, we illustrate how to summarize the fiber-level information using functions from the package **reshape**. First, in Panel R3.9a, we use the generic function `melt()` to prepare the data for further processing. More specifically, we apply the function to the data frame `prt`, and we specify factors `id`, `prt.f`, `fiber.f`, and `occ.f` as "identifying variables." On the other hand, we indicate variables `spec.fo` and `iso.f` as "measured variables." In the resulting data frame, `prtM`, the values of the measured variables are "stacked" within the groups defined by the combinations of the levels of the identifying variables. The stacked values

R3.9 *PRT Trial*: Summarizing the fiber-level information with the help of functions `melt()` and `cast()` from the **reshape** package

(a) *Preprocessing of the data (melting)*

```
> library(reshape)
> idvar <- c("id", "prt.f", "fiber.f", "occ.f")
> meas.var <- c("spec.fo", "iso.fo")
> prtM <-                                              # Melting data
+    melt(prt, id.var = idvar, measure.var = meas.var)
> dim(prtM)
  [1] 4942    6
> head(prtM, n = 4)                                    # First four rows
    id prt.f fiber.f occ.f variable value
  1  5   Low  Type 1   Pre  spec.fo  83.5
  2  5   Low  Type 1   Pre  spec.fo 132.8
  3  5   Low  Type 2   Pre  spec.fo 161.1
  4  5   Low  Type 1   Pre  spec.fo 158.8
> tail(prtM, n = 4)                                    # Last four rows
        id prt.f fiber.f occ.f variable value
  4939 520  High  Type 2   Pos   iso.fo 0.527
  4940 520  High  Type 1   Pos   iso.fo 0.615
  4941 520  High  Type 2   Pos   iso.fo 0.896
  4942 520  High  Type 2   Pos   iso.fo 0.830
```

(b) *Aggregating data (casting)*

```
> prtC <- cast(prtM, fun.aggregate = mean)             # Casting data
> names(prtC)
  [1] "id"       "prt.f"    "fiber.f" "occ.f"    "spec.fo" "iso.fo"
> names(prtC)[5:6] <- c("spec.foMn", "iso.foMn") # Names modified
> head(prtC, n = 4)
    id prt.f fiber.f occ.f spec.foMn iso.foMn
  1  5   Low  Type 1   Pre    132.59  0.51500
  2  5   Low  Type 1   Pos    129.96  0.72289
  3  5   Low  Type 2   Pre    145.74  0.47057
  4  5   Low  Type 2   Pos    147.95  0.71175
```

are stored in a single variable named, by default, `value`. They are identified by the levels of factor named, by default, `variable`, which contain the names of the measured variables.

The display, shown in Panel R3.9a, indicates that the number of records in the data frame `prtM` increases to 4,942, as compared to 2,471 records in the data frame `prt` (see Panel R2.7). The increase results from the stacking of the values of `spec.fo` and `iso.fo` in the variable `value`. The outcome of the process is further illustrated by the display of the first and last four rows of the data frame `prtM`.

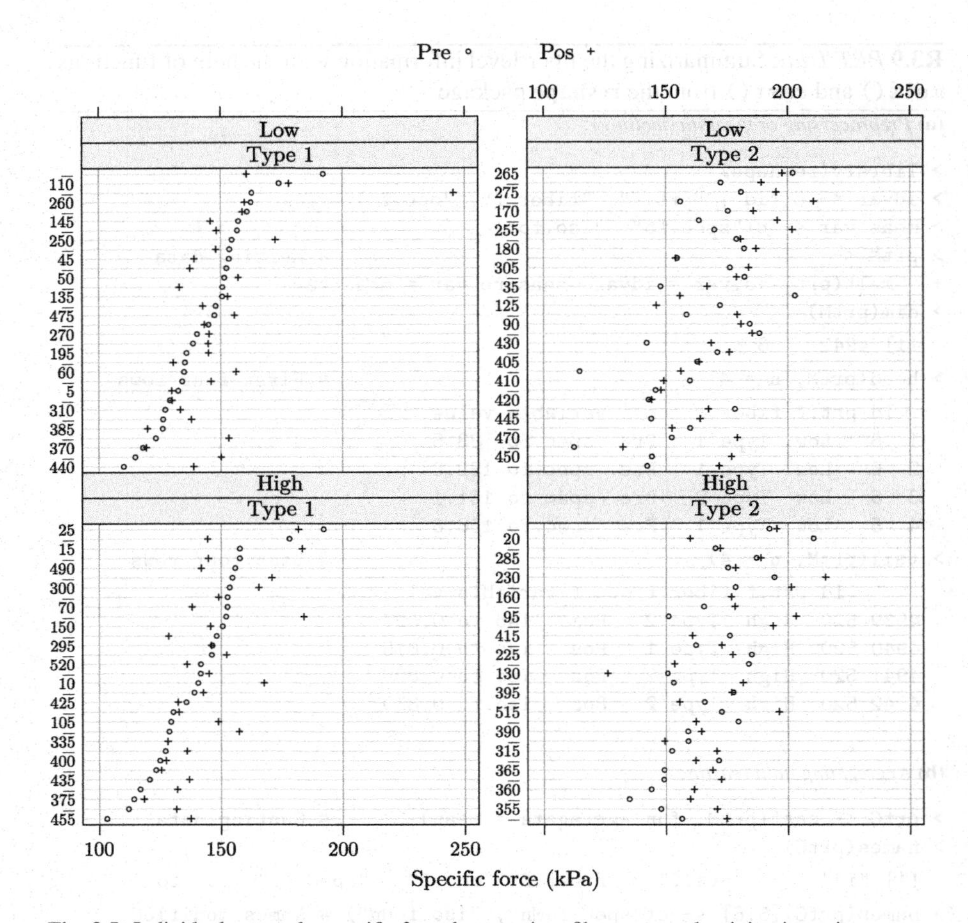

Fig. 3.5 Individual means for specific force by occasion, fiber type, and training intensity

In Panel R3.9b, we apply the function `cast()` to the data frame `prtM` to compute the mean values of the measured variables, i.e., `spec.fo` and `iso.fo`, within the groups defined by the combinations of the levels of the identifying variables. To indicate that we want to compute the mean values, we use the argument `fun.aggregate=mean`. The resulting data frame is stored in the object `prtC`. Before displaying the contents of the object, we modify the names of the two last variables, which contain the mean values of `spec.fo` and `iso.fo`. The display of the first four records of `prtC` shows the means per fiber type and occasion for the subject `"5"`. Note that, for `spec.fo`, the mean values correspond to the values reported at the end of Panel R3.8.

Figure 3.5 shows the pre- and post-training mean values of the specific force for all subjects separately for the two fiber types and training intensities. The figure was created using the function `dotplot()` from the package **lattice**. To increase interpretability of this figure, we ordered the subjects on the y-axis within each study group by mean values of the pre-training `spec.fo` for type-1 fibers. If for a given subject like, e.g., `"335"`, the pre-training measures were not available, the post-training measures were used instead. For brevity, we do not show the syntax used to create the figure.

Several observations can be made based on the figure:

- There is no clear effect of the training intensity
- In general, measurements of the specific force are higher for type-2 than for type-1 fibers
- On average, post-training values are larger than pre-training measurements
- For both types of fibers, there is a considerable variability between subjects with respect to the overall level of measurements and with respect to the magnitude of the post-pre differences
- There is a correlation between the mean measurements observed for the same individual, as seen, e.g., from the similar pattern of measurements for both types.

These observations will be taken into account when modeling the data in Part IV of the book.

Note that the plot in the lower-left panel of Fig. 3.5 confirms the missing pre-training measurements for type-1 fibers for the subject `"335"`.

Figure 3.6 presents information for the specific force for the type-1 fibers. More specifically, it shows box-and-whiskers plots for the individual measurements of the specific force for the two measurement occasions and training intensities. All 63 subjects on the y-axis are ordered in the same way as in Fig. 3.5. Figure 3.6 was created using the function `bw()` from the package **lattice**. Note, however, that we do not present the detailed code. The plots suggest that the subject-specific variances of the pre-training measurements are somewhat smaller than the post-training ones. There is also a considerable variability between the subjects with respect to the variance of the measurements.

Figure 3.7 presents the individual pre-post differences of the mean values for the specific force for the type-1 fibers for the two training-intensity groups. The differences were ordered according to increasing values within each training group. To conserve space, we do not show the syntax used to create the figure. The plots indicate an outlying value of the difference for the subject `"275"` in the low-intensity training group.

3.4 SII Project: Gain in the Math Achievement Score

In this section, we conduct an exploratory analysis of the SII data that were described in Sect. 2.4. We focus on the measurements of the gain in the math achievement score, stored in the variable `mathgain` (see Sect. 2.4.1). Given the hierarchical structure of the data, we divide the analysis into three parts, in which we look separately at the school-, class-, and child-level data.

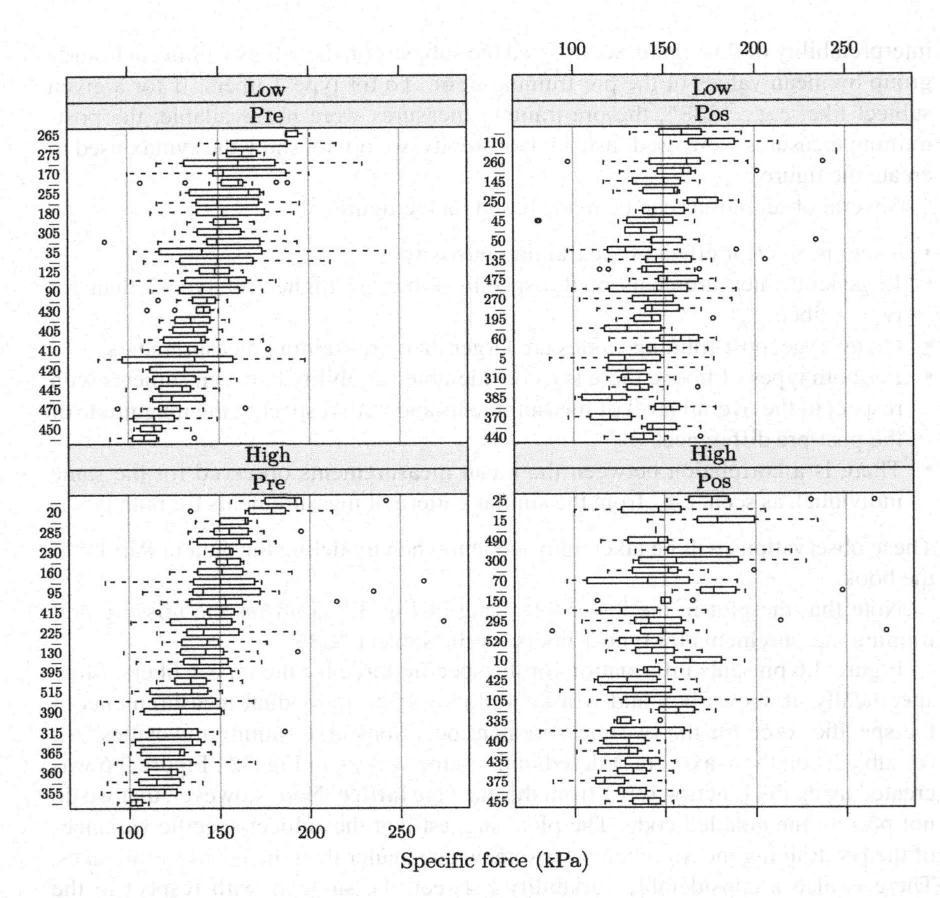

Fig. 3.6 *PRT Trial*: Subject-specific box-and-whiskers plots for the specific force by training intensity and measurement occasion (type-1 fibers only)

First, however, we check whether the data frame `SIIdata` contains complete information for all variables for all pupils. Toward this end, in Panel R3.10, we use the function `sapply()`. It applies the function, specified in the `FUN` argument, to each column (variable) of the data frame `SIIdata`. The latter function checks whether any value in a particular column is missing. The displayed results indicate that only the variable `mathknow` contains missing values. By applying the function `sum()` to the vector resulting from the transformation of a logical vector indicating the location of missing values in the variable `mathknow` to a numeric vector, we check that the variable contains 109 missing values. The nonmissing values range from −2.50 to 2.61.

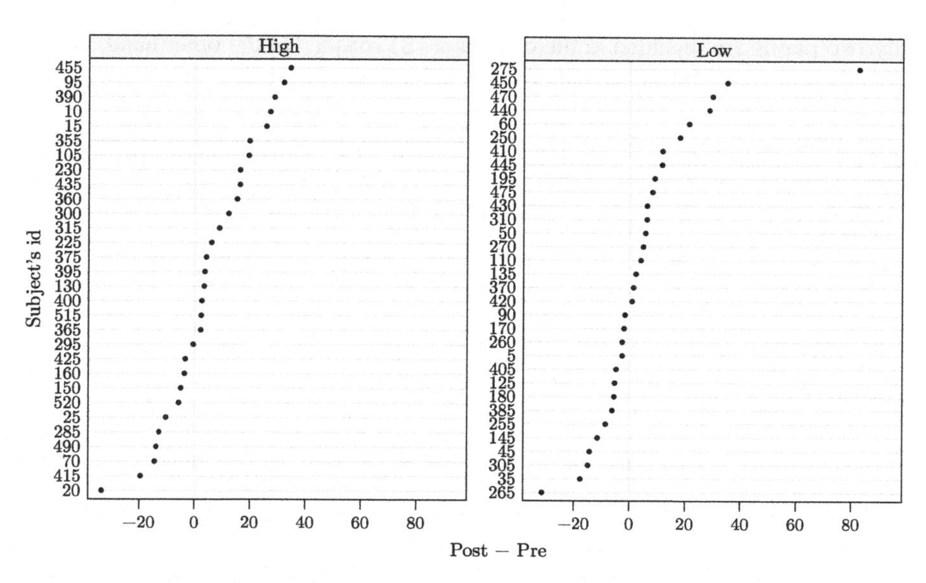

Fig. 3.7 *PRT Trial*: Individual post-pre differences of the mean values for the specific force, ordered by an increasing value, for the two training intensity groups (type-1 fibers only)

R3.10 *SII Project*: The number of missing values for variables included in the `SIIdata` data frame

```
> data(SIIdata, package = "nlmeU")
> sapply(SIIdata, FUN = function(x) any(is.na(x)))
      sex minority mathkind mathgain      ses yearstea mathknow
    FALSE    FALSE    FALSE    FALSE    FALSE    FALSE     TRUE
 housepov mathprep  classid schoolid  childid
    FALSE    FALSE    FALSE    FALSE    FALSE
> sum(as.numeric(is.na(SIIdata$mathknow)))
 [1] 109
> range(SIIdata$mathknow, na.rm = TRUE)
 [1] -2.50  2.61
```

3.4.1 School-Level Data

In this section, we investigate the school-level data.

First, in Panel R3.11, we use the function `xtabs()` to tabulate the number of pupils per school. The result is stored in the array `schlN`. The display of the array is difficult to interpret. By applying the function `range()`, we check that the number of pupils per school varied between 2 and 31. By applying the function `xtabs()` to the array `schlN`, we obtain the information about the number of schools with a particular number of pupils. For instance, there were two schools for which data for only two pupils are included in the data frame `SIIdata`. On the other hand, there was only one school for which data for 31 pupils were collected.

R3.11 *SII Project*: Extracting the information about the number of pupils per school

```
> (schlN <- xtabs(~schoolid, SIIdata))  # Number of pupils per school
  schoolid
   1   2   3   4   5   6   7   8   9  10  11  12  13  14  15  16
  11  10  14   6   6  12  14  16   6  18  31  27   9  15  13   6
  ...     [snip]
  97  98  99 100 101 102 103 104 105 106 107
   6   2  19  13  16  11   8   6  10   2  10
> range(schlN)
  [1]  2 31
> xtabs(~schlN) # Distribution of the number of pupils over schools
  schlN
   2  3  4  5  6  7  8  9 10 11 12 13 14 15 16 17 18 19 20 21 22 24 27 31
   2  4  6  5  8  5  9  9 10  7  7  6  3  5  4  2  2  3  1  2  2  1  3  1
```

In Panel R3.12, we obtain the information about the mean value of variables `mathkind` and `mathgain` for each school (see Sect. 2.4.1). Toward this end, with the help of the function `by()`, we apply the function `colMeans()` to the values of the two variables within the groups defined by the same level of the factor `schoolid`, i.e., within each school. Note that the resulting output has been abbreviated.

R3.12 *SII Project*: Computation of the mean value of pupils' math scores for each school

```
> attach(SIIdata)
> (mthgM <- by(cbind(mathgain, mathkind), schoolid, colMeans))
  INDICES: 1
  mathgain mathkind
    59.636  458.364
  ------------------------------------------------
  ...     [snip]
  ------------------------------------------------
  INDICES: 107
  mathgain mathkind
      48.2    464.2
> detach(SIIdata)
```

Panel R3.13 shows the syntax for constructing the data frame `schlDt`, which contains the school-specific means of variables `mathgain`, `mathkind`, and `housepov`. In particular, in Panel R3.13a, we use the functions `melt()` and `cast()` (for an explanation of the use of the functions, see the description of Panel R3.9) to create the data frame `cst1`, which contains the number of classes and children for each school. On the other hand, in Panel R3.13b, we use the functions to create the data frame `cst2` with the mean values of variables `mathgain`, `mathkind`, and `housepov` for each school. Finally, in Panel R3.13c, we merge the two data frames to create `schlDt`. Note that, after merging, we remove the two auxiliary data frames.

R3.13 *SII Project*: Constructing a data frame with summary data for schools

(a) *Creating a data frame with the number of classes and children for each school*

```
> library(reshape)
> idvars <- c("schoolid")
> mvars <- c("classid", "childid")
> dtm1 <- melt(SIIdata, id.vars = idvars, measure.vars = mvars)
> names(cst1  <-
+   cast(dtm1,
+        fun.aggregate = function(el) length(unique(el))))
  [1] "schoolid" "classid"  "childid"
> names(cst1) <- c("schoolid", "clssn", "schlN")
```

(b) *Creating a data frame with the school-specific means of selected variables*

```
> mvars <- c("mathgain", "mathkind", "housepov")
> dtm2 <- melt(SIIdata, id.vars = idvars, measure.vars = mvars)
> names(cst2 <- cast(dtm2, fun.aggregate = mean))
  [1] "schoolid" "mathgain" "mathkind" "housepov"
> names(cst2) <- c("schoolid", "mthgMn", "mthkMn", "housepov")
```

(c) *Merging the data frames created in parts (a) and (b) above*

```
> (schlDt  <- merge(cst1, cst2, sort = FALSE))
        schoolid clssn schlN mthgMn mthkMn housepov
  1            1     2    11 59.636 458.36    0.082
  2            2     3    10 65.000 487.90    0.082
  3            3     4    14 88.857 469.14    0.086
  4            4     2     6 35.167 462.67    0.365
  ...     [snip]
  107        107     2    10 48.200 464.20    0.177
> rm(cst1, cst2)
```

The data frame schlDt is used in Panel R3.14 to explore the school-specific mean values of variables housepov and mathgain. In particular, in Panel R3.14a, we use the function summary() to display the summary statistics for the mean values. On the other hand, in Panel R3.14b, we use the function xyplot() from the package **lattice** to construct scatterplots of the mean values of the variable mathgain *versus* variables housepov and mthkMn.

R3.14 *SII Project*: Exploring the school-level data. The data frame schlDt was created in Panel R3.13

(a) *Summary statistics for the school-specific mean values of* housepov

```
> summary(schlDt$housepov)
   Min. 1st Qu.  Median    Mean 3rd Qu.    Max.
 0.0120  0.0855  0.1480  0.1940  0.2640  0.5640
```

(b) *Scatterplots of the school-specific mean values for* housepov *and* mathkind

```
> library(lattice)
> xyplot(mthgMn ~ housepov,                          # Fig. 3.8a
+        schlDt, type = c("p", "smooth"), grid = TRUE)
> xyplot(mthgMn ~ mthkMn,                            # Fig. 3.8b
+        schlDt, type = c("p", "smooth"), grid = TRUE)
```

The scatterplots are shown in Fig. 3.8. The plot in Fig. 3.8a does not suggest a strong relationship between the school-specific mean values of mathgain and housepov. On the other hand, in Fig. 3.8b there is a strong negative relationship between the mean values of mathgain and mathkind: the larger the mean for the latter, the lower the mean for the former. The relationship suggests that the higher the mean grade of pupils in the kindergarden, the lower the mean gain in the math achievement score of pupils. Note that the plots in Fig. 3.8 should be interpreted with caution, as they show school-specific means, which were estimated based on different numbers of observations.

3.4.2 Class-Level Data

In this section, we investigate the class-level data.

First, in Panel R3.15, we use the function xtabs() to tabulate the number of pupils per class. The result is stored in the array clssN. By applying the function sum() to the array, we check that the total number of pupils is 1,190, in agreement with the information obtained, e.g., in Panel R2.10. With the help of the function range(), we find that the number of pupils per class varies between 1 and 10. By applying the function xtabs() to the array clssN, we obtain information about the

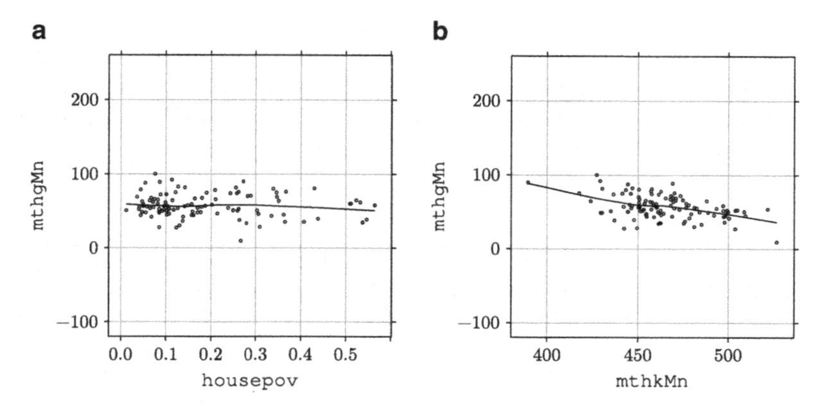

Fig. 3.8 *SII Project*: Scatterplots of the school-specific mean values of the variable `mathgain` *versus* variables (**a**) `housepov` and (**b**) `mthkMn`

number of classes with a particular number of pupils. The information is stored in the array `clssCnt`. The display of the array indicates that, for instance, there were 42 classes with only one pupil included in the data frame `SIIdata`. On the other hand, there were two classes for which data for 10 pupils were collected. Finally, by applying the function `sum()` to the array `clssCnt`, we verify that the data frame `SIIdata` contains information about 312 classes.

R3.15 *SII Project*: Extracting the information about the number of pupils per class

```
> (clssN <- xtabs(~ classid, SIIdata))
  classid
    1  2  3  4  5  6  7  8  9 10 11 12 13 14 15 16
    5  3  3  6  1  5  1  4  3  2  4  5  9  4  1  6
  ...     [snip]
  305 306 307 308 309 310 311 312
    4   4   4   3   3   3   2   4
> sum(clssN)                          # Total number of pupils
  [1] 1190
> range(clssN)
  [1]  1 10
> (clssCnt <- xtabs(~clssN)) # Distribution of no. of pupils/classes
  clssN
   1  2  3  4  5  6  7  8  9 10
  42 53 53 61 39 31 14 13  4  2
> sum(clssCnt)                        # Total number of classes
  [1] 312
```

In Panel R3.16, we present an abbreviated printout of the contents of the data frame `clssDt`. The data frame contains the mean values of variables `mathgain` and `mathkind` for each class, together with the count of pupils, `clssN`. It also includes the values of the class-level variables `mathknow` and `mathprep` and the school-level variable `housepov`. The data frame was created using a syntax (not shown) similar to the one presented in Panel R3.13.

R3.16 *SII Project*: Contents of the class-level data. The auxiliary data frame `clssDt` was created using a syntax similar to the one shown in Panel R3.13

```
> clssDt
    classid housepov mathknow mathprep clssN   mthgMn mthkMn
1         1    0.335    -0.72     2.50     5  47.8000 459.00
2         2    0.303     0.58     3.00     3  65.6667 454.00
3         3    0.040     0.85     2.75     3  15.6667 492.67
4         4    0.339     1.08     5.33     6  91.5000 437.00
...       [snip]
312     312    0.546    -1.37     2.00     4  47.5000 418.50
```

Figure 3.9 presents scatterplots of the class-specific means of the variable `mathgain` *versus* the values of the variable `housepov` and *versus* the class-specific means of the variable `mathkind`. The figure was created using a syntax similar to the one presented in Panel R3.14b based on the data from the data frame `clssDt`. Figure 3.9a does not suggest a strong relationship between the mean values of `mathgain` and `housepov`. On the other hand, as seen in Fig. 3.9b, there is a strong negative relationship between the mean values of `mathgain` and `mathkind`. These conclusions are similar to the ones drawn based on Fig. 3.8. As was the case for the latter figure, the plots in Fig. 3.9 should be interpreted with caution, as they show class-specific mean values estimated based on different numbers of observations.

3.4.3 Pupil-Level Data

In this section, specifically in Panel R3.17, we explore the pupil-level data.

First, in Panel R3.17a, we construct an auxiliary data frame `auxDt` by merging data frames `SIIdata` and `clssDt`. Note that the latter contains the class-level data, including the means of variables `mathgain` and `mathkind` and the number of pupils (see Panel R3.16). Next, with the help of the function `within()`, we add a new factor, `clssF`, to `auxDt` and store the resulting data frame in the object `auxDt2`. The factor `clssF` combines the information about the class and the school for each pupil. The information is stored in a character string of the form: `classid\n:schoolid\n(clssN)`. The particular format of the string will prove

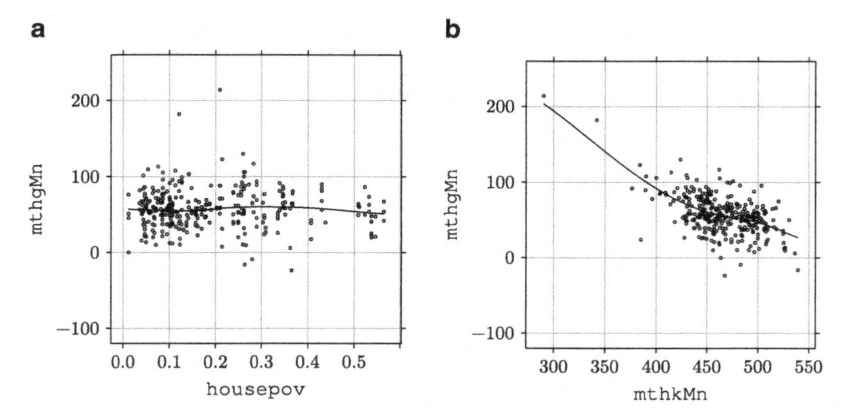

Fig. 3.9 *SII Project*: Scatterplots of the class-specific mean values of the variable `mathgain` *versus* variables (**a**) `housepov` and (**b**) `mthkMn`

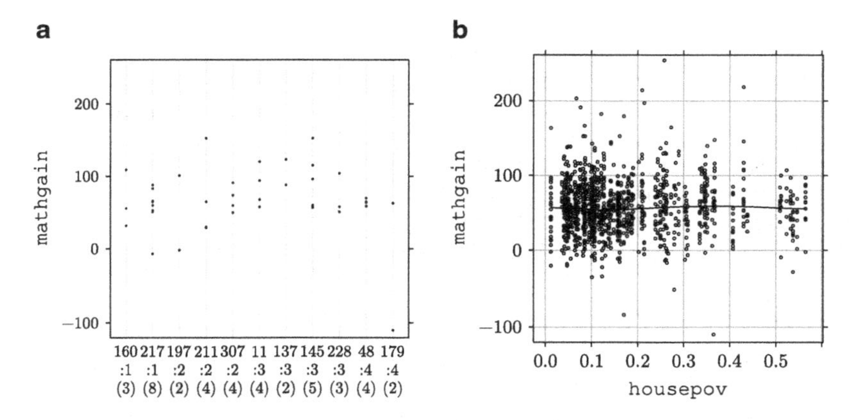

Fig. 3.10 *SII Project*: Scatterplots of the observed values of `mathgain` for individual pupils *versus* the (**a**) school/class indicator and (**b**) the variable `housepov`

useful in the construction of plots of the pupil-specific data. The format is illustrated in the display of the first and last four records of the data frame `auxDt2`. Note that we limit the display to variables `classid`, `schoolid`, `clssN`, and `clssF`.

In Panel R3.17b, we construct two plots of the pupil-level data. First, by applying the function `dotplot()` from the package **lattice** to the data frame `auxDt2`, we plot the values of the variable `mathgain` *versus* the levels of the factor `clssF` for the schools with `schoolid` between 1 and 4. Then, using the function `xyplot()`, we plot the values of the variable `mathgain` *versus* the values of the variable `houspov` for all pupils from the data frame `SIIdata`. The resulting plots are shown in Fig. 3.10.

The plot shown in Fig. 3.10a indicates considerable variability of the observed values of the gain in the math achievement score even between the classes belonging

R3.17 *SII Project*: Exploring the pupil-level data. The data frame `clssDt` was created in Panel R3.16

(a) *Adding the class-level data to the data frame* `SIIdata`

```
> auxDt <- merge(SIIdata, clssDt, sort = FALSE)
> auxDt2  <-
+   within(auxDt,
+          {
+           auxL  <- paste(classid, schoolid, sep = "\n:")
+           auxL1 <- paste(auxL, clssN, sep = "\n(")
+           auxL2 <- paste(auxL1, ")", sep = "")
+           clssF <-                               # Factor clssF created
+             factor(schoolid:classid, labels = unique(auxL2))
+          })
> tmpDt <- subset(auxDt2, select = c(classid, schoolid, clssN, clssF))
> head(tmpDt, 4)                               # First four records
  classid schoolid clssN          clssF
1     160        1     3 160\n:1\n(3)
2     160        1     3 160\n:1\n(3)
3     160        1     3 160\n:1\n(3)
4     217        1     8 217\n:1\n(8)
> tail(tmpDt, 4)                               # Last four records
     classid schoolid clssN          clssF
1187      96      107     8  96\n:107\n(8)
1188      96      107     8  96\n:107\n(8)
1189     239      107     2 239\n:107\n(2)
1190     239      107     2 239\n:107\n(2)
```

(b) *Scatterplots of the pupil-level data*

```
> library(lattice)
> dotplot(mathgain ~ clssF,                    # Fig. 3.10a
+         subset(auxDt2, schoolid %in% 1:4))
> xyplot(mathgain ~ housepov, SIIdata,         # Fig. 3.10b
+        type = c("p", "smooth"))
> detach(package:lattice)
```

to the same school. Note that the interpretation of the plot is much enhanced by the labels provided on the horizontal axis. The construction of the labels is facilitated by the chosen format of the levels of the factor `clssF`.

The plot shown in Fig. 3.10b indicates the lack of a relationship between the observed values of the gain in the math achievement score for individual pupils and the values of the variable `housepov`. Note that a similar conclusion was drawn for the school- and class-specific mean values of `mathgain` based on Figs. 3.8a and 3.9a, respectively.

3.5 FCAT Study: Target Score

The FCAT dataset have a rather simple structure and contents (see Sect. 2.5). The main interest pertains to the distribution of the scores for the nine attainment targets, which are stored in the variable `scorec` of the data frame `fcat` (Sect. 2.5.2). In Panel R3.18, we present syntax addressing this issue.

R3.18 *FCAT Study*: Summarizing the information about the total scores for attainment targets

(a) *Summarizing scores for each child and attainment target*

```
> data(fcat, package = "nlmeU")
> (scM <- with(fcat, tapply(scorec, list(id, target), mean)))
     T1(4) T2(6) T3(8) T4(5) T5(9) T6(6) T7(8) T8(6) T9(5)
 1       4     6     4     1     7     6     6     5     5
 2       3     4     6     2     7     4     6     3     3
 ...   [snip]
 539     0     3     5     1     6     3     5     2     4
```

(b) *Histograms of scores for different attainment targets*

```
> library(lattice)
> histogram(~scorec | target, data = fcat, # Fig. 3.11
+           breaks = NULL)
> detach(package:lattice)
```

In Panel R3.18a, we show how to obtain the mean value of the dependent variable for each combination of levels of the crossed factors, i.e., `id` and `target` in the `fcat` data frame. In particular, we use the function `tapply()` to apply the function `mean()` to the variable `scorec` for each combination of levels of the crossed factors. As a result, we obtain the matrix `scM`, which contains the mean value of the total score for each child and each attainment target. Obviously, in our case, there is only one observation for each child and target. Thus, by displaying a (abbreviated) summary of the matrix `scM`, we obtain, in fact, a tabulation of individual scores for all children.

In Panel R3.18b, we use the function `histogram()` from the package **lattice** to construct a histogram of the observed values of total scores for each attainment target. The resulting histograms are shown in Fig. 3.11. They clearly illustrate the differences in the measurement scale for different targets, which result from the varying number of items per target (Sect. 2.5). Some asymmetry of the distribution of the scores can also be observed.

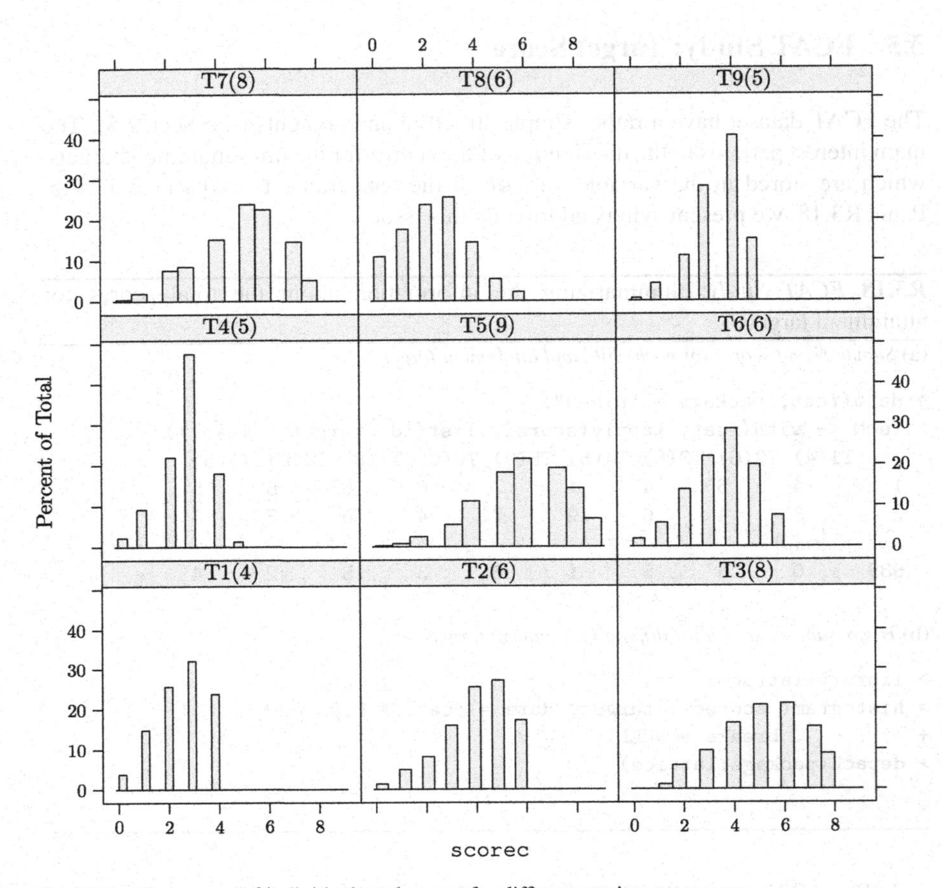

Fig. 3.11 Histograms of individual total scores for different attainment targets

3.6 Chapter Summary

In this chapter, we presented exploratory analyses of the four case studies introduced in Chap. 2. The results of the analyses will be used in the next parts of our book to build models for the case studies.

In parallel to the presentation of the results of the exploratory analyses, we introduced a range of R tools, which are useful for such analyses. For instance, functions `cast()` and `melt()` from the package **reshape** are very useful in transforming data involving aggregated summaries. The importance of using graphical displays is also worth highlighting. Toward this aim, the tools available in packages **graphics** (R Development Core Team, 2010) and **lattice** (Sarkar, 2008) are very helpful. The former package implements traditional graphical displays, whereas the latter offers displays based on a grid-graphics system (Murrell, 2005).

Due to space limitations, our presentation of the tools was neither exhaustive nor detailed. However, we hope that the syntax and its short description, which were provided in the chapter, can help the reader in finding appropriate methods applicable to the particular problem at hand.

Part II
Linear Models for Independent Observations

Chapter 4
Linear Models with Homogeneous Variance

4.1 Introduction

In Part II of this book, we consider the application of LMs to data originating from research studies, in which observations are *independent*.

In a broad sense, LMs are used to quantify the relationship between a dependent variable and a set of covariates with the use of a linear function depending on a small number of regression parameters.

In this chapter, we focus on the classical LM, suitable for analyzing data involving independent observations with a homogenous variance. The class of linear models outlined in this chapter includes standard linear regression, analysis of variance (ANOVA), and analysis of covariance (ANCOVA) models. In Chap. 7, we will relax the assumption of variance homogeneity and consider LMs that are appropriate for analyzing independent observations with nonconstant variance. Besides linear models, there are other parametric models that can be used for the analysis of data with independent observations. They include, e.g., generalized linear models (GLIMs) and nonlinear regression models, but they are beyond the scope of this book.

By outlining the basic concepts of LMs in Chaps. 4 and 7, we set the stage for fixed-effects LMs for correlated data (Part III) and LMMs (Part IV). In particular, we introduce several key concepts needed later in the context of LMMs, such as design matrix or likelihood estimation. We hope that, by introducing the concepts in a simpler and more familiar framework, their use in the context of more complex models may become easier to encompass.

In the current chapter, we provide theoretical concepts underlying the classical LM. Note that, in our presentation, we pay special attention to those concepts that are implemented in R. A more detailed treatment of the concepts can be found in, for instance, the monograph by Neter et al. (1990).

The chapter is structured as follows. In Sect. 4.2, we specify the classical LM in various ways. Section 4.3 introduces the concept of an offset. Section 4.4 contains a review of the estimation methods. In Sect. 4.5, we briefly discuss the diagnostic

A. Gałecki and T. Burzykowski, *Linear Mixed-Effects Models Using R: A Step-by-Step Approach*, Springer Texts in Statistics, DOI 10.1007/978-1-4614-3900-4_4,

tools, while in Sect. 4.6, inferential methods are presented. Model selection and reduction methods are outlined in Sect. 4.7. A summary of the chapter is provided in Sect. 4.8.

The implementation of the theoretical concepts and methods for the classical LMs in R will be discussed in Chap. 5.

4.2 Model Specification

In this section, we present the formulation of the classical LM. In particular, we look at the specification of the model both at the level of the observation and for all observations.

4.2.1 Model Equation at the Level of the Observation

The classical LM for independent, normally distributed observations y_i ($i = 1, \ldots, n$) with a constant variance can be specified in a variety of ways.

A commonly used specification assumes the following representation for the observation y_i:

$$y_i = x_i^{(1)}\beta_1 + \ldots + x_i^{(p)}\beta_p + \varepsilon_i, \tag{4.1}$$

where

$$\varepsilon_i \sim \mathcal{N}(0, \sigma^2), \tag{4.2}$$

$x_i^{(1)}, \ldots, x_i^{(p)}$ ($p < n$) are values of *known* covariates for the i-th observation, $\beta_1, \ldots, \beta_p$ are the corresponding (unknown) regression parameters, and where we assume that the residual random errors $\varepsilon_1, \ldots, \varepsilon_n$ are independent.

Upon defining the column vectors $\mathbf{x}_i \equiv (x_i^{(1)}, \ldots, x_i^{(p)})'$, and $\boldsymbol{\beta} \equiv (\beta_1, \ldots, \beta_p)'$, which contain the covariates' values for the i-th subject and fixed effects, respectively, (4.1) can be written as:

$$y_i = \mathbf{x}_i'\boldsymbol{\beta} + \varepsilon_i. \tag{4.3}$$

From (4.1)–(4.3), it follows that the expected value and variance of y_i are, respectively,

$$\mathrm{E}(y_i) \equiv \mu_i = \mathbf{x}_i'\boldsymbol{\beta}, \tag{4.4}$$

$$\mathrm{Var}(y_i) = \sigma^2. \tag{4.5}$$

4.2.2 *Model Equation for All Data*

The model specified in (4.1) and (4.2) can be expressed in a more compact way upon defining

$$\mathbf{y} \equiv \begin{pmatrix} y_1 \\ \vdots \\ y_n \end{pmatrix}, \boldsymbol{\varepsilon} \equiv \begin{pmatrix} \varepsilon_1 \\ \vdots \\ \varepsilon_n \end{pmatrix},$$

and

$$\boldsymbol{X} \equiv \begin{pmatrix} \mathbf{x}_1' \\ \vdots \\ \mathbf{x}_n' \end{pmatrix} = \begin{pmatrix} x_1^{(1)} & x_1^{(2)} & \dots & x_1^{(p)} \\ \vdots & \vdots & \ddots & \vdots \\ x_n^{(1)} & x_n^{(2)} & \dots & x_n^{(p)} \end{pmatrix} \equiv \left(\mathbf{x}^{(1)} \; \mathbf{x}^{(2)} \; \dots \; \mathbf{x}^{(p)} \right). \tag{4.6}$$

Then, (4.1) and (4.2) can be written as

$$\mathbf{y} = \boldsymbol{X\beta} + \boldsymbol{\varepsilon}, \tag{4.7}$$

where

$$\boldsymbol{\varepsilon} \sim \mathcal{N}_n(\mathbf{0}, \mathcal{R}), \tag{4.8}$$

with the variance–covariance matrix $\mathcal{R} = \sigma^2 \boldsymbol{I}_n$, where $\boldsymbol{I}_n$ is the $n \times n$ identity matrix.

For the sake of simplicity, we assume that the design matrix $\boldsymbol{X}$, defined in (4.6), is of full rank p or, equivalently, we assume that its columns $\mathbf{x}^{(1)}, \dots, \mathbf{x}^{(p)}$ are linearly independent. Note that the i-th row in the matrix $\boldsymbol{X}$ corresponds to the vector $\mathbf{x}_i$, used in (4.3) and (4.4).

4.3 Offset

Models (4.1) and (4.2) can be modified by introducing into (4.1) a *known* additional term x_i^0 for all i. This leads to

$$y_i = x_i^{(0)} + x_i^{(1)}\beta_1 + \dots + x_i^{(p)}\beta_p + \varepsilon_i, \tag{4.9}$$

where the distribution of the residual error ε_i is the same as that in (4.2).

Model equation (4.9) can be represented for all data as

$$\mathbf{y} = \mathbf{x}^{(0)} + \boldsymbol{X\beta} + \boldsymbol{\varepsilon}, \tag{4.10}$$

where $\mathbf{x}^{(0)} \equiv (x_1^{(0)}, \dots, x_n^{(0)})'$ and the distribution of the residual error vector $\boldsymbol{\varepsilon}$ is given by (4.8).

The term $x_i^{(0)}$ in (4.9) or, equivalently, the column vector $\mathbf{x}^{(0)}$ in (4.10), is referred to as an *offset*.

An LM with an offset can be represented as a classical LM specified by (4.7). Toward this end, the offset can be absorbed into the design matrix $\boldsymbol{X}$. That is, the offset is included as an additional (the first) column of the design matrix, and the corresponding parameter β_0 is assumed to be known and equal to 1.

It should be noted that an offset can easily be accommodated in the classical LM without the need for an explicit modification of the design matrix. By removing the term $x_i^{(0)}$ from the mean structure on the right-hand side of (4.9) and defining $\widetilde{y}_i \equiv y_i - x_i^{(0)}$, we obtain the classical LM, given by (4.1) and (4.2), with $\widetilde{y}_i$ as the dependent variable. For this reason, LMs with offsets are rarely used in practice. The concept is more commonly used in GLIMs, which will not be addressed in this book. However, the concept is important to illustrate various computational approaches used, for example, in the context of the generalized least squares algorithm (Sect. 7.8.1.1). Moreover, offsets are also used in R in the context of LMs as an argument of functions like, e.g., `lm()`, which are used to fit the models. For these reasons, it is convenient to introduce the concept now.

4.4 Estimation

Researchers are often interested in finding estimates of a set of parameters $\boldsymbol{\beta}$ and σ^2. In the context of the classical LM, specified in Sect. 4.2, the most common estimation technique is the method of ordinary least squares (OLS). We describe it in Sect. 4.4.1. However, OLS is less suitable for more complex LMs, including LMMs. Therefore, although it is not typically done in the context of the classical LM, in Sects. 4.4.2 and 4.4.3, we also introduce the maximum likelihood (ML) and the restricted maximum-likelihood (REML) estimation. These methods, in contrast to OLS, are more broadly applicable. Another reason for introducing the likelihood-based approaches this early is that they are implemented in the **nlme** package, namely, by the `gls()` function, which can be used to fit the classical LM as well.

4.4.1 Ordinary Least Squares

In OLS, the estimates of $\boldsymbol{\beta}$ are obtained by minimization, with respect to $\boldsymbol{\beta}$, of the residual sum of squares:

$$\sum_{i=1}^{n}(y_i - \mathbf{x}_i'\boldsymbol{\beta})^2. \tag{4.11}$$

The resulting estimator of $\boldsymbol{\beta}$ is expressed in a closed form as follows:

$$\widehat{\boldsymbol{\beta}}_{\text{OLS}} \equiv \left(\sum_{i=1}^{n} \mathbf{x}_i \mathbf{x}_i'\right)^{-1} \sum_{i=1}^{n} \mathbf{x}_i y_i = (\boldsymbol{X}'\boldsymbol{X})^{-1}\boldsymbol{X}'\mathbf{y}. \tag{4.12}$$

It is worth noting that the derivation of the OLS estimate does not require the normality assumption, specified by (4.2). Moreover, it is valid under the assumption of uncorrelated residual errors, which is a weaker assumption than the assumption of independence. This is in contrast to the ML and REML estimation, which are presented in the next two sections.

Although estimation of σ^2 is not part of OLS calculations, the following unbiased estimator of σ^2

$$\begin{aligned}\widehat{\sigma}^2_{\text{OLS}} &\equiv \frac{1}{n-p}\sum_{i=1}^{n}\left(y_i - \mathbf{x}_i'\widehat{\boldsymbol{\beta}}_{\text{OLS}}\right)^2 \\ &= \frac{1}{n-p}\left(\mathbf{y} - \boldsymbol{X}\widehat{\boldsymbol{\beta}}_{\text{OLS}}\right)'\left(\mathbf{y} - \boldsymbol{X}\widehat{\boldsymbol{\beta}}_{\text{OLS}}\right),\end{aligned} \tag{4.13}$$

is typically used.

4.4.2 Maximum-Likelihood Estimation

The classical LM defined in (4.1) and (4.2) implies that observations, y_i, are independent and normally distributed. Consequently, the likelihood function for this model given observed data is defined as follows:

$$L_{\text{Full}}(\boldsymbol{\beta},\sigma^2;\mathbf{y}) \equiv (2\pi\sigma^2)^{-n/2}\prod_{i=1}^{n}\exp\left[-\frac{(y_i - \mathbf{x}_i'\boldsymbol{\beta})^2}{2\sigma^2}\right]. \tag{4.14}$$

Its maximization over $\boldsymbol{\beta}$ and σ^2 is equivalent to maximization of the corresponding log-likelihood function:

$$\ell_{\text{Full}}(\boldsymbol{\beta},\sigma^2;\mathbf{y}) \equiv -\frac{n}{2}\log(\sigma^2) - \frac{1}{2\sigma^2}\sum_{i=1}^{n}(y_i - \mathbf{x}_i'\boldsymbol{\beta})^2. \tag{4.15}$$

Note that the contribution of data for observation i to the log-likelihood function, specified in (4.15), is equal to

$$\ell_{\text{Full}(i)}(\boldsymbol{\beta},\sigma^2;y_i) \equiv -\frac{1}{2}\log(\sigma^2) - \frac{(y_i - \mathbf{x}_i'\boldsymbol{\beta})^2}{2\sigma^2}. \tag{4.16}$$

Maximization of (4.14) provides an ML estimator of $\boldsymbol{\beta}$,

$$\widehat{\boldsymbol{\beta}}_{\text{ML}} \equiv \left(\sum_{i=1}^{n} \mathbf{x}_i\mathbf{x}_i'\right)^{-1} \sum_{i=1}^{n} \mathbf{x}_i y_i, \tag{4.17}$$

exactly the same as $\widehat{\boldsymbol{\beta}}_{\text{OLS}}$, displayed in (4.12). The ML estimator of σ^2 can be written explicitly as follows:

$$\widehat{\sigma}^2_{\text{ML}} \equiv \frac{1}{n}\sum_{i=1}^{n}\left(y_i - \mathbf{x}_i'\widehat{\boldsymbol{\beta}}_{\text{ML}}\right)^2. \tag{4.18}$$

Note that (4.18) differs from (4.13). Indeed, $\widehat{\sigma}^2_{\text{ML}}$ is biased downwards by a factor $(n-p)/n$. This is because the uncertainty in the estimation of $\boldsymbol{\beta}$ is not accounted for in (4.18). The bias is removed from (4.18) if the restricted maximum-likelihood (REML) estimation is used, as discussed in the next section.

4.4.3 Restricted Maximum-Likelihood Estimation

To obtain an unbiased estimate for σ^2, we will use an estimation approach that is orthogonal to the estimation of $\boldsymbol{\beta}$. This can be done by considering the likelihood function based on a set of $n-p$ independent contrasts of $\mathbf{y}$ (Verbeke and Molenberghs 2000, p. 43–46). The resulting log-restricted-likelihood function is given by

$$\ell_{\text{REML}}(\sigma^2;\mathbf{y}) \equiv -\frac{n-p}{2}\log(\sigma^2) - \frac{1}{2\sigma^2}\sum_{i=1}^{n} r_i^2, \tag{4.19}$$

where

$$r_i \equiv y_i - \mathbf{x}_i'\left(\sum_{i=1}^{n} \mathbf{x}_i\mathbf{x}_i'\right)^{-1} \sum_{i=1}^{n} \mathbf{x}_i y_i.$$

Maximization of (4.19) with respect to σ^2 leads to the following REML estimator:

$$\widehat{\sigma}^2_{\text{REML}} \equiv \frac{1}{n-p}\sum_{i=1}^{n} r_i^2. \tag{4.20}$$

Note that $\sum_{i=1}^{n} r_i^2$, used in $\widehat{\sigma}^2_{\text{REML}}$, is the same as in $\widehat{\sigma}^2_{\text{ML}}$, defined in (4.18). However, $n-p$ is used in the denominator in lieu of n. As a result, $\widehat{\sigma}^2_{\text{REML}}$ is an unbiased estimator of σ^2.

The REML objective function does not allow one to directly estimate the fixed-effects coefficients in $\boldsymbol{\beta}$. For this reason, we employ the formula for $\widehat{\boldsymbol{\beta}}_{\text{ML}}$, given in (4.17). The estimate $\widehat{\boldsymbol{\beta}}_{\text{REML}}$ of $\boldsymbol{\beta}$ obtained using this method is equal to $\widehat{\boldsymbol{\beta}}_{\text{ML}}$. This equality is true for the classical LM, given by (4.1) and (4.2), which assumes independent observations with homogeneous variance. However, it does not hold for models with less restrictive assumptions about the residual variance, discussed in subsequent chapters.

Finally, it is worth noting that the OLS estimators of $\boldsymbol{\beta}$ and σ^2, given in (4.12) and (4.13), respectively, are equivalent to the REML estimates. This OLS–REML equivalence for the classical LM with independent, homoscedastic (constant variance) residuals will not hold in general for more complex models considered later in the book.

4.4.4 Uncertainty in Parameter Estimates

The variance–covariance matrix of $\widehat{\boldsymbol{\beta}}$ equals

$$\text{Var}(\widehat{\boldsymbol{\beta}}) = \sigma^2 \left(\sum_{i=1}^{n} \mathbf{x}_i \mathbf{x}_i' \right)^{-1} = \sigma^2 \left(\boldsymbol{X}'\boldsymbol{X} \right)^{-1} \tag{4.21}$$

and is estimated by

$$\widehat{\text{Var}}(\widehat{\boldsymbol{\beta}}) \equiv \widehat{\sigma}^2 \left(\sum_{i=1}^{n} \mathbf{x}_i \mathbf{x}_i' \right)^{-1} = \widehat{\sigma}^2 \left(\boldsymbol{X}'\boldsymbol{X} \right)^{-1}, \tag{4.22}$$

where $\widehat{\sigma}^2$ is equal to $\widehat{\sigma}^2_{\text{OLS}}$, $\widehat{\sigma}^2_{\text{ML}}$, or $\widehat{\sigma}^2_{\text{REML}}$, depending on the estimation method used.

It is worth noting that OLS- and REML-based estimates together with their estimated variance–covariance matrices, computed by using (4.22), are identical.

On the other hand, even though the ML- and REML-based estimates of $\boldsymbol{\beta}$ are equal to each other, their estimated variance–covariance matrices are different. This is because the ML- and REML-based estimators of σ^2, defined in (4.18) and (4.20), respectively, differ. In fact, given the bias of $\widehat{\sigma}^2_{\text{ML}}$, one should consider the variance–covariance matrix of $\widehat{\boldsymbol{\beta}}$ based on $\widehat{\sigma}^2_{\text{REML}}$, especially in small sample size studies.

4.5 Model Diagnostics

After fitting an LM, and before making any inferences based upon it, it is important to check whether the model assumptions are met. The key assumptions for model (4.1) and (4.2) are that the residual errors, ε_i, are independent, homoscedastic, and that the effect of covariates can be written as a linear function of their values and the corresponding parameters. Note that the normality assumption is important in the context of the ML estimation, but not for OLS, as the latter does not require it. The main tools for checking the assumptions are based on the *estimated* residual errors or, simply, residuals.

Additionally, it might be of interest to check whether the fit of the model is sensitive to the inclusion or exclusion of certain observations. This process is called *influence diagnostics* and is described in Sect. 4.5.3.

4.5.1 Residuals

Commonly used devices in residual diagnostics are plots, which are informally evaluated with respect to the presence or absence of specific patterns and/or outlying data points. Such plots can be based on several types of residuals, which are presented in this section.

4.5.1.1 Raw Residuals

Various types of residuals can be considered. Here, we consider simple versions. These will later be generalized in Chaps. 7, 10, and 13, in the context of more advanced models, including LMMs.

The most basic residuals are the *raw* residuals, defined for the i-th observation as $\widehat{\varepsilon}_i \equiv y_i - \widehat{\mu}_i$, where $\widehat{\mu}_i \equiv \mathbf{x}_i'\widehat{\boldsymbol{\beta}}$ is referred to as the *fitted value*.

4.5.1.2 Scaled Residuals

The raw residuals are often *scaled*, i.e., divided by their true or estimated standard deviations, so that their interpretation does not depend on the measurement units of the dependent variable. It would be preferable to scale the residuals by their true standard deviations, i.e., by σ, to obtain *standardized residuals*. In practice, however, the true standard deviation is rarely known. Thus, scaling is done by using the estimated standard deviation, $\widehat{\sigma}$, instead. Residuals obtained in this manner are called *studentized residuals*. This category can be further subdivided into *internally studentized residuals* and *externally studentized residuals*. The former are obtained when the observation corresponding to the residual in question is included in the estimation of the standard deviation, while the latter are obtained when the observation is excluded from the estimation. Table 4.1 summarizes the basic forms of scaled residuals, along with the naming conventions used in the R syntax. Note that $\widehat{\sigma}$ denotes an estimate of σ based on all observations, while $\widehat{\sigma}_{(-i)}$ is an estimate obtained after excluding the i-th observation from the calculations.

Table 4.1 The basic forms of scaled residuals for linear models

Residual type	R naming convention	Mathematical formula
Standardized by σ		$\widehat{\varepsilon}_i/\sigma$
Internally studentized[a]	Standardized	$\widehat{\varepsilon}_i/\widehat{\sigma}$
Externally studentized[b]	Studentized	$\widehat{\varepsilon}_i/\widehat{\sigma}_{(-i)}$

[a] $\widehat{\sigma}$ is an estimate of σ based on all observations,
[b] $\widehat{\sigma}_{(-i)}$ is an estimate of σ obtained after excluding the i-th observation.

We note that by replacing $\widehat{\sigma}$ with $\widehat{\sigma}_{(-i)}$ the external studentization technique allows for outliers to stand out in a more prominent fashion compared to the internal one.

The scaling of raw residuals presented in Table 4.1 does not address an important issue, however, which is the fact that the variances of the residuals, $\widehat{\varepsilon}_i$, differ, even though the variances of the true errors, ε_i, are all equal.

To address the issue, a more advanced way of scaling of residuals is necessary. It is based on the $n \times n$ matrix $\boldsymbol{H}$, defined as

$$\boldsymbol{H} \equiv \boldsymbol{X}(\boldsymbol{X}'\boldsymbol{X})^{-1}\boldsymbol{X}'. \tag{4.23}$$

The matrix $\boldsymbol{H}$ represents a projection that maps the vector $\mathbf{y}$ on the subspace spanned by the columns of the design matrix $\boldsymbol{X}$. The matrix is referred to as *the leverage matrix* or *the hat matrix*.

Note that the vector of the predicted values of $\mathbf{y}$, $\widehat{\boldsymbol{\mu}} = \boldsymbol{X}\widehat{\boldsymbol{\beta}}$, can be expressed as $\widehat{\boldsymbol{\mu}} = \boldsymbol{H}\mathbf{y}$.

In what follows, we present a rationale for using the hat matrix to scale residuals. The vector of raw residuals $\widehat{\boldsymbol{\varepsilon}} = \mathbf{y} - \boldsymbol{X}\widehat{\boldsymbol{\beta}}$ for all data can be expressed with the use of the matrix $\boldsymbol{H}$ as follows:

$$\widehat{\boldsymbol{\varepsilon}} = \mathbf{y} - \boldsymbol{X}\widehat{\boldsymbol{\beta}} = (\boldsymbol{I}_n - \boldsymbol{H})\mathbf{y}, \tag{4.24}$$

where $\boldsymbol{I}_n$ denotes the $n \times n$ identity matrix. By simple algebra we obtain the following formula for the variance–covariance matrix of $\widehat{\boldsymbol{\varepsilon}}$:

$$\text{Var}(\widehat{\boldsymbol{\varepsilon}}) = \sigma^2(\boldsymbol{I}_n - \boldsymbol{H}). \tag{4.25}$$

In case the matrix $\boldsymbol{H}$ in (4.25) is not proportional to $\mathbf{I}_n$, the raw residuals are potentially *heteroscedastic* and/or correlated. Thus, direct interpretation of the raw residuals may not be straightforward. Moreover, as already mentioned, the scaled residuals, presented in Table 4.1, do not address the issue of heteroscedasticity and/or correlation.

To tackle the problem of unequal variances of the residuals from Table 4.1, a scaling that involves the $\boldsymbol{H}$ matrix can be used. Table 4.2 presents the residuals, corresponding to those shown in Table 4.1, which are scaled by standard error estimates involving diagonal elements $h_{i,i}$ of the $\boldsymbol{H}$ matrix. Note that the scaling addresses the problem of heteroscedasticity of the raw residuals, but does not remove the correlation between the scaled residuals. To address this, *error recovery* methods are used. They are briefly discussed next.

Error Recovery

Methods which aim at removing both the heteroscedasticity *and* correlation of the raw residuals $\widehat{\varepsilon}_i$ are referred to as *error recovery* methods (Schabenberger and Gotway 2005).

The general idea in these approaches is to transform the residuals in such a way that the transformed residuals have a zero mean, a constant variance, and become uncorrelated. The $n \times n$ matrix $\boldsymbol{P} \equiv \boldsymbol{I}_n - \boldsymbol{H}$, used in (4.24) and (4.25), plays a key role in this endeavor. Note that the matrix $\boldsymbol{P}$ is not of full rank. More specifically, assuming that $n > p$, the rank of $\boldsymbol{P}$ is equal to or less than $n - p$. Consequently, we may have at most $n - p$ transformed, uncorrelated residuals. In contrast to the raw and scaled residuals, residuals obtained by using error recovery methods may represent more than one observation, which makes their interpretation difficult. These types of residuals have been developed for the classical LM, described in Sect. 4.2, but do not generalize easily to more complex LMs. Therefore, we do not describe these residuals in more detail.

4.5.2 Residual Diagnostics

In the context of the LM, defined by (4.1) and (4.2), the most frequently used example of a diagnostic plot is the plot of raw residuals $\widehat{\varepsilon}_i$ against fitted values $\widehat{\mu}_i$ (see, e.g., Fig. 6.1a). The plot is assessed with respect to whether it displays a random pattern and constant variability along the x-axis. It is also used to detect outliers, i.e., observations with atypical values for the dependent variable and/or for covariates. For continuous covariates, a scatterplot of the residuals against the values of the covariate can also be used. A nonrandom pattern in the plot is interpreted as an indication of a misspecification of the functional form of the covariate.

Another useful plot is the normal quantile–quantile (Q–Q) plot of the residuals (see, e.g., Fig. 6.1b). In this plot, the quantiles of ordered residuals are plotted against the corresponding values for the standard normal distribution. If the residuals are (at least approximately) normally distributed, the shape of the plot should not deviate from a straight line. On the other hand, if the distribution of the residuals is, e.g., symmetric, but with "thicker" tails than the normal, the plot will look like a stretched S. A skewed distribution will result in a plot in the form of an arch.

However, from the discussion on different types of residuals presented earlier in this section, it follows that the raw residuals are intrinsically heteroscedastic and correlated. For this reason, the scatterplots and the Q–Q plot are preferably based

Table 4.2 Scaled residuals that involve $h_{i,i}$, the diagonal elements of the hat matrix

Residual type	Adjusted by $h_{i,i}$
Standardized	$(\widehat{\varepsilon}_i/\sigma)/\sqrt{1-h_{i,i}}$
Internally studentized[a]	$(\widehat{\varepsilon}_i/\widehat{\sigma})/\sqrt{1-h_{i,i}}$
Externally studentized[b]	$(\widehat{\varepsilon}_i/\widehat{\sigma}_{(-i)})/\sqrt{1-h_{i,i}}$

[a] $\widehat{\sigma}$ is an estimate of σ based on all observations
[b] $\widehat{\sigma}_{(-i)}$ is an estimate of σ obtained after excluding the i-th observation

on the scaled residuals, shown in Table 4.2, as they tend to remove not-desired heteroscedasticity carried by raw residuals.

If the plot of raw residuals reveals a nonlinear relationship between the dependent variable and a covariate, a suitable transformation of the dependent variable or the covariate may be considered to obtain a linear relationship (Neter et al. 1990, Sect. 4.6). Examples of transformations include a logarithmic transformation, square root, inverse, etc.

Instead of using a simple function, e.g., a logarithm, to transform a covariate, a more flexible transformation can be used. For instance, the use of a *spline* can be considered. In general, a spline is a sufficiently smooth piecewise-polynomial function. It allows for modeling a complex nonlinear relationship between the dependent variable and a covariate. More details on splines can be found in the monograph by, e.g., Hastie et al. (2009).

If a transformation is applied to the covariate, it should be noted that the interpretation of the estimated parameter β may become more difficult. This is due to the fact that it may be performed on a nontypical measurement scale, e.g., the square-root scale. In this context the advantage of using a logarithmic transformation is that the parameter estimates obtained on a logarithmic scale can be exponentiated and directly interpreted as multiplicative effects on the original scale.

A special class of transformations of the dependent variable are *variance-stabilizing* transformations. They can be used when the assumption of homogeneous variance of the observations seem to be violated. In particular, suppose that the variance can be expressed, at least approximately, as a function of the expected value $g(\mu)$. In that case, applying the transformation $h(y) = \int [g(y)]^{-1/2} dy$ to the observed values of the dependent variable should result in values with approximately homogeneous variance. For example, for $g(\mu) = a\mu$ we get $h(y) = 2\sqrt{y}/a$, i.e., a square-root transformation.

Note that it may be difficult to find a variance-stabilizing transformation that would alleviate the problem of the non-homogeneous-variance assumption. In this case, the use of an LM allowing for heterogeneous variance can be considered. Such models are presented in Chap. 7.

It should be kept in mind that if a transformation is applied to the dependent variable, the distribution of the transformed variable may change. Thus, after applying the transformation, the normal Q–Q plot of the scaled residuals should be checked for symptoms of the possible violation of the assumption of normality of the residual errors.

4.5.3 Influence Diagnostics

Influence diagnostics are formal techniques allowing for the identification of observations that influence estimates of $\boldsymbol{\beta}$ or σ^2. The idea of influence diagnostics for a given observation is to quantify the effect of omission of this observation from the data on the results of the analysis of the entire dataset. Although influence diagnostic methods are presented here for individual observations, they can be extended easily to a more general case in which the influence of multiple observations, e.g., pairs, triplets, etc., is investigated.

Influence diagnostics uses a variety of tools (Schabenberger 2004). In the context of the classical LM, a rather popular measure is Cook's distance, D_i. The measure is the scaled change, induced by the exclusion of a particular observation, in the estimated parameter vector. For fixed effects, the general formula for D_i for observation i is

$$\mathrm{D}_i \equiv \frac{\left(\widehat{\boldsymbol{\beta}} - \widehat{\boldsymbol{\beta}}_{(-i)}\right)' \left[\widehat{\mathrm{Var}}(\widehat{\boldsymbol{\beta}})\right]^{-1} \left(\widehat{\boldsymbol{\beta}} - \widehat{\boldsymbol{\beta}}_{(-i)}\right)}{\mathrm{rank}(\boldsymbol{X})}, \tag{4.26}$$

where $\widehat{\boldsymbol{\beta}}_{(-i)}$ is the estimate of the parameter vector $\boldsymbol{\beta}$ obtained by fitting an LM to the data with the i-th observation excluded. For the classical LM, defined in (4.1) and (4.2), D_i can be expressed as

$$\mathrm{D}_i = \frac{\widehat{\varepsilon}_i^2 h_{i,i}}{\widehat{\sigma}^2 (1 - h_{i,i})^2},$$

where $h_{i,i}$ is the i-th diagonal element of the matrix $\boldsymbol{H}$, defined in (4.23). The larger the value of D_i, the larger the influence of the i-th observation on the estimate of $\boldsymbol{\beta}$.

Note that Cook's distance is used to assess the influence of a given observation on $\widehat{\boldsymbol{\beta}}$ and does not take into account changes of $\widehat{\sigma}$. A basic tool to investigate the influence of a given observation on estimates of both $\boldsymbol{\beta}$ *and* σ^2 is the *likelihood displacement*. The likelihood displacement, LD_i, is defined as twice the difference between the log-likelihood computed at a maximum and displaced values of estimated parameters:

$$\mathrm{LD}_i \equiv 2 \times \left[\ell_{\mathrm{Full}}(\widehat{\Theta}; \mathbf{y}) - \ell_{\mathrm{Full}}(\widehat{\Theta}_{(-i)}; \mathbf{y})\right], \tag{4.27}$$

where $\widehat{\Theta} \equiv (\widehat{\boldsymbol{\beta}}', \widehat{\sigma}^2)'$ is the ML estimate of Θ obtained by fitting the classical LM, defined in Sect. 4.2, to all data, while $\widehat{\Theta}_{(-i)} \equiv (\widehat{\boldsymbol{\beta}}'_{(-i)}, \widehat{\sigma}^2_{(-i)})'$ is the ML estimate obtained by fitting the model to the data with the i-th observation *excluded*. Note that the value of the function $\ell_{\mathrm{Full}}(\widehat{\Theta}_{(-i)}; \mathbf{y})$, used in (4.27), is computed as in (4.15), i.e., with respect to all data, including the i-th observation. Verbeke and Molenberghs (2000, Sect. 11.2), following the work of Cook (1986), present more formal and general definitions of the likelihood displacement.

Formulae (4.26) and (4.27) for Cook's distance and the likelihood displacement, respectively, can be adapted for use in more advanced LMs, which will be considered in Parts II–IV of this book. More details about the measures can be found in, e.g., Chatterjee et al. (2000).

4.6 Inference

The main focus of inference in the classical LM, defined in (4.1) and (4.2), is the fixed parameters $\boldsymbol{\beta}$. To test hypotheses about the values of the parameters, three general testing paradigms are commonly used: the Wald, likelihood ratio, and score tests. In Sect. 4.6.1, we briefly outline the general principles of the construction of the tests, followed by considerations related to a linear case. We will refer to them in subsequent chapters of the book. In Sect. 4.6.2, we focus on the construction of confidence intervals for linear models.

4.6.1 *The Wald, Likelihood Ratio, and Score Tests*

4.6.1.1 *Nonlinear Case*

Assume that we have a sample of n independent, identically distributed observations from a distribution with density $f(y;\boldsymbol{\vartheta})$, where $\boldsymbol{\vartheta}$ is a p-dimensional vector of parameters. Let $\mathbf{y} \equiv (y_1,\ldots,y_n)$. Denote the log-likelihood function of the sample as $\ell(\boldsymbol{\vartheta};\mathbf{y})$. The score function $S(\boldsymbol{\vartheta};\mathbf{y})$ is defined as the vector of partial derivatives of $\ell(\boldsymbol{\vartheta};\mathbf{y})$ with respect to $\boldsymbol{\vartheta}$:

$$S(\boldsymbol{\vartheta};\mathbf{y}) \equiv \frac{\partial \ell(\boldsymbol{\vartheta};\mathbf{y})}{\partial \boldsymbol{\vartheta}}.$$

At the ML estimate (MLE) $\widehat{\boldsymbol{\vartheta}}$ of $\boldsymbol{\vartheta}$, we have $S(\widehat{\boldsymbol{\vartheta}};\mathbf{y}) = 0$.

The *observed* Fisher information matrix, $\mathbf{I}(\boldsymbol{\vartheta};\mathbf{y})$, is defined as the negative second derivative, i.e., the negative Hessian matrix of $\ell(\boldsymbol{\vartheta};\mathbf{y})$ with respect to $\boldsymbol{\vartheta}$:

$$\mathbf{I}(\boldsymbol{\vartheta};\mathbf{y}) \equiv -\frac{\partial^2 \ell(\boldsymbol{\vartheta};\mathbf{y})}{\partial \boldsymbol{\vartheta} \partial \boldsymbol{\vartheta}'}.$$

The observed Fisher information matrix evaluated at the MLE is $\mathbf{I}(\widehat{\boldsymbol{\vartheta}}) \equiv \mathbf{I}(\widehat{\boldsymbol{\vartheta}};\mathbf{y})$. Note that the variance of the ML estimator $\widehat{\boldsymbol{\vartheta}}$ of $\boldsymbol{\vartheta}$ can be estimated by the inverse of $\mathbf{I}(\widehat{\boldsymbol{\vartheta}})$.

The *expected* Fisher information matrix, $\mathcal{I}(\boldsymbol{\vartheta})$, is defined as

$$\mathcal{I}(\boldsymbol{\vartheta}) \equiv E[\mathbf{I}(\boldsymbol{\vartheta};\mathbf{y})],$$

where the expectation is taken over the distribution of $\mathbf{y}$.

Assume that we want to test the hypothesis

$$H_0 : G(\boldsymbol{\vartheta}) = 0 \quad \textit{versus} \quad H_A : G(\boldsymbol{\vartheta}) \neq 0, \tag{4.28}$$

where $G(\boldsymbol{\vartheta}) \equiv [g_1(\boldsymbol{\vartheta}), \ldots, g_q(\boldsymbol{\vartheta})]$ is a function with continuous first-order derivatives for all of its components. Let us denote by $\widehat{\boldsymbol{\vartheta}}_0$ and $\widehat{\boldsymbol{\vartheta}}_A$ the ML estimators of $\boldsymbol{\vartheta}$ under H_0 and H_A, respectively.

The statistic of the likelihood-ratio (LR) test is defined as

$$T_L \equiv -2[\ell(\widehat{\boldsymbol{\vartheta}}_0;\mathbf{y}) - \ell(\widehat{\boldsymbol{\vartheta}}_A;\mathbf{y})] \tag{4.29}$$

and is calculated based on the maximum value of the log-likelihood function obtained under the null and alternative hypotheses.

The Wald-test statistic is defined as

$$T_W = \left[G(\widehat{\boldsymbol{\vartheta}}_A)\right]' \left\{ \left[\frac{\partial G(\widehat{\boldsymbol{\vartheta}}_A)}{\partial \boldsymbol{\vartheta}}\right] \mathcal{I}(\widehat{\boldsymbol{\vartheta}}_A)^{-1} \left[\frac{\partial G(\widehat{\boldsymbol{\vartheta}}_A)}{\partial \boldsymbol{\vartheta}}\right]' \right\}^{-1} G(\widehat{\boldsymbol{\vartheta}}_A),$$

where $\frac{\partial G(\widehat{\boldsymbol{\vartheta}}_A)}{\partial \boldsymbol{\vartheta}}$ is a $q \times p$ Jacobian matrix for the function G evaluated at $\boldsymbol{\vartheta} = \widehat{\boldsymbol{\vartheta}}_A$. The statistic value is calculated based on the magnitude of the difference between the MLE of $G(\boldsymbol{\vartheta})$ and the value corresponding to H_0, i.e., 0, relative to the variability of the MLE.

Finally, the score-test statistic is defined as

$$T_S \equiv \left[S(\widehat{\boldsymbol{\vartheta}}_0)\right]' \mathcal{I}(\widehat{\boldsymbol{\vartheta}}_0)^{-1} S(\widehat{\boldsymbol{\vartheta}}_0).$$

The test statistic assesses the magnitude of the slope of the log-likelihood function relative to the curvature of the function at the restricted MLE.

Asymptotically, all three test statistics are distributed according to the χ^2 distribution with $p-q$ degrees of freedom. The asymptotic result also holds if in the definition of the score- and Wald-test statistics, the expected Fisher information matrices $\mathcal{I}(\widehat{\boldsymbol{\vartheta}}_0)$ and $\mathcal{I}(\widehat{\boldsymbol{\vartheta}})$ are replaced by the observed information matrices, $\mathbf{I}(\widehat{\boldsymbol{\vartheta}}_0)$ and $\mathbf{I}(\widehat{\boldsymbol{\vartheta}})$, respectively.

4.6.1.2 Linear Case

In the classical LM, defined by (4.1) and (4.2), linear hypotheses about fixed parameters $\boldsymbol{\beta}$ are often of interest. The hypotheses are of the form

$$H_0 : \boldsymbol{L\beta} = \mathbf{c}_0 \quad \textit{versus} \quad H_A : \boldsymbol{L\beta} \neq \mathbf{c}_0, \tag{4.30}$$

where $\boldsymbol{L}$ is a known matrix of rank q ($q \leq p$) and $\mathbf{c}_0$ is a known vector. Note that the hypotheses can be expressed as in (4.28) upon defining $G(\boldsymbol{\beta}) \equiv \boldsymbol{L}\boldsymbol{\beta} - \mathbf{c}_0$.

It follows that, when σ^2 is known, the statistics for the LR, Wald, and score test are exactly the same and are equal to

$$T \equiv \frac{(\boldsymbol{L}\widehat{\boldsymbol{\beta}} - \mathbf{c}_0)'[\boldsymbol{L}(\boldsymbol{X}'\boldsymbol{X})^{-1}\boldsymbol{L}']^{-1}(\boldsymbol{L}\widehat{\boldsymbol{\beta}} - \mathbf{c}_0)}{\sigma^2}. \tag{4.31}$$

In practice, we do not know σ^2. We can estimate it by using, e.g., the ML estimator, given in (4.18). However, in that case, the exact equivalence of the LR-, Wald-, and score-test statistic no longer holds. In particular, the LR-test statistic becomes equal to

$$T_L = n \log\left[1 + \frac{\text{rank}(\boldsymbol{L})}{n-p}F\right], \tag{4.32}$$

the Wald-test statistic is equal to

$$T_W = F\frac{n}{n-p}\text{rank}(\boldsymbol{L}), \tag{4.33}$$

and the score-test statistic takes the form

$$T_S = \frac{nF}{\frac{n-p}{\text{rank}(\boldsymbol{L})} + F}, \tag{4.34}$$

where

$$F \equiv \frac{(\boldsymbol{L}\widehat{\boldsymbol{\beta}} - \mathbf{c}_0)'[\boldsymbol{L}(\boldsymbol{X}'\boldsymbol{X})^{-1}\boldsymbol{L}']^{-1}(\boldsymbol{L}\widehat{\boldsymbol{\beta}} - \mathbf{c}_0)}{\widehat{\sigma}^2_{\text{REML}}\text{rank}(\boldsymbol{L})}, \tag{4.35}$$

with $\widehat{\sigma}^2_{\text{REML}}$ given by (4.20).

Formulae (4.32)–(4.34) show that the three test statistics, although different numerically, are monotonic functions of F, defined in (4.35), which is the well-known F-test statistic. Thus, exact tests for the test statistics would produce the same p-values. However, if the asymptotic χ^2 distribution is used, the p-values may differ. In this respect, it is worth noting that, under the null hypothesis $\text{H}_0 : \mathbf{L}\boldsymbol{\beta} = \mathbf{c}_0$, the statistic F, defined in (4.35), is distributed according to the central F-distribution with the numerator and denominator degrees of freedom equal to $\text{rank}(\mathbf{L})$ and $n-p$, respectively. Note that the distribution holds exactly for all sample sizes n. Thus, the use of the F-test statistic is preferred over the use of the asymptotic χ^2 distribution for the statistics, defined in (4.32)–(4.34), in the LM setting.

For future reference, it is worth noting that the statistic F, defined in (4.35), can be expressed as

$$F = \frac{(\boldsymbol{L}\widehat{\boldsymbol{\beta}} - \mathbf{c}_0)'[\boldsymbol{L}\widehat{\text{Var}}(\widehat{\boldsymbol{\beta}})\boldsymbol{L}']^{-1}(\boldsymbol{L}\widehat{\boldsymbol{\beta}} - \mathbf{c}_0)}{\text{rank}(\boldsymbol{L})}. \tag{4.36}$$

For the particular case of testing the null hypothesis about a *single* fixed effect parameter β, e.g., $\mathrm{H}_0 : \beta = c_0$ *versus* $\mathrm{H}_A : \beta \neq c_0$, the test based on the F-test statistic, given in (4.35), is equivalent to the test based on the following t-test statistic:

$$t \equiv \frac{\widehat{\beta} - c_0}{\sqrt{\widehat{\mathrm{Var}}(\widehat{\beta})}}, \tag{4.37}$$

where $\widehat{\mathrm{Var}}(\widehat{\beta})$ is the estimated variance of $\widehat{\beta}$, which can be obtained from (4.22). This is because, for the single parameter case, $\sqrt{F} = \mid t \mid$, and the two-sided p-values for the tests are identical. The null distribution of the t-test statistic is the t-distribution with $n - p$ degrees of freedom.

4.6.2 *Confidence Intervals for Parameters*

Confidence intervals for individual components of the parameter vector $\boldsymbol{\beta}$ can be constructed based on the fact that the test statistic, given in (4.37), has the t-distribution with $n - p$ degrees of freedom. It follows that the $(1 - \alpha)100\%$ confidence interval for a single parameter β is given by

$$\left[\widehat{\beta} - t_{1-\alpha/2,n-p}\sqrt{\widehat{\mathrm{Var}}(\widehat{\beta})},\ \widehat{\beta} + t_{1-\alpha/2,n-p}\sqrt{\widehat{\mathrm{Var}}(\widehat{\beta})}\right], \tag{4.38}$$

where $t_{1-\alpha/2,n-p}$ is the $(1 - \alpha/2)100$-th percentile of the t-distribution with $n - p$ degrees of freedom.

In some circumstances, a confidence interval for σ might be of interest. It can be constructed based on a χ^2-distribution. More specifically, a $(1 - \alpha)100\%$ confidence interval for σ, estimated by using the REML estimator (4.20), is

$$\left[\widehat{\sigma}_{\mathrm{REML}}\sqrt{\frac{n-p}{\chi^2_{1-\alpha/2,n-p}}},\ \widehat{\sigma}_{\mathrm{REML}}\sqrt{\frac{n-p}{\chi^2_{\alpha/2,n-p}}}\right], \tag{4.39}$$

where $\chi^2_{\alpha/2,n-p}$ is the $(\alpha/2)100$-th percentile of the χ^2-distribution with $n - p$ degrees of freedom. If the confidence interval is based on the ML estimator (4.18), $n - p$ in formula (4.39) should be replaced with n.

4.7 Model Reduction and Selection

In this section, we briefly discuss issues related to the choice of the most parsimonious form of a model, i.e., the form which contains the smallest possible number of parameters while enjoying an acceptable fit. In particular, in Sect. 4.7.1, we consider strategies to reduce the form of a particular model. Section 4.7.2 briefly summarizes

Table 4.3 The null and alternative models underlying the sequential (Type I) and marginal (Type III) approaches for tests of fixed-effects for a hypothetical model $Y = 1 + X_1 + X_2 + X_3$

	Sequential (Type I) tests		Marginal (Type III) tests	
Tested term	Null	Alternative	Null	Alternative
X_1	1	$1+X_1$	$1+X_2+X_3$	$1+X_1+X_2+X_3$
X_2	$1+X_1$	$1+X_1+X_2$	$1+X_1+X_3$	$1+X_1+X_2+X_3$
X_3	$1+X_1+X_2$	$1+X_1+X_2+X_3$	$1+X_1+X_2$	$1+X_1+X_2+X_3$

approaches to the choice of a "best" model from a set of models. The methods described are fairly general and are of interest also in the context of more complex models, which will be described in subsequent chapters.

4.7.1 Model Reduction

In practice, researchers are often interested in reducing the number of parameters in a fitted model without substantially affecting the goodness of fit of the model. This can be done by testing hypotheses about the parameters and by modifying the structure of the particular model depending on the outcome of the tests.

When testing a hypothesis that a given (set of) fixed-effect(s) coefficient(s) in an LM is equal to zero, we often consider two models: one without and with the coefficient(s) of interest. We refer to these models as *the null model* and *the alternative model*, respectively. The models are *nested*, in the sense that the model under the null hypothesis (the null model) could be viewed as a special case of the model under the alternative hypothesis (the alternative model).

The process of testing hypotheses about several terms in the model by comparing just two nested models at a time may be tedious and time consuming, especially for a large number of covariates. Therefore, to simplify this process, when fitting a given model, many software programs, including R, provide results of a series of tests for every coefficient/term separately. These tests are helpful in making a decision about whether a given coefficient/term should be kept in the model or not.

For the sake of simplicity, let us consider models (4.1) and (4.2) with, e.g., three terms/covariates: X_1, X_2, and X_3. One could consider testing of the null hypothesis for each of the terms that the effects of a given term are equal to 0. In Table 4.3, we demonstrate that the series of tests can be performed in at least two different ways. For both approaches, the null and alternative models involved in testing a corresponding term are included for reference.

In the first approach, we test the effects by "sequentially" adding tested terms to the null and alternative models involved. In particular, we test the effect of X_1 by comparing the alternative model, containing the intercept and X_1, with the null model that contains only the intercept. On the other hand, the effect of X_2 is tested by comparing a different alternative model, which contains the intercept, X_1, and X_2, with a null model that contains only the intercept and X_1. Finally, the effect of X_3 is tested by comparing the alternative model with the intercept and all three terms with the model that contains the intercept, X_1, and X_2. This strategy is called a "sequential" approach. In the literature, the resulting tests are often referred to as Type I tests.

In the second approach, the alternative model involved in testing any of the terms in a fitted model is the same and contains all the terms. The null hypothesis about the effect of X_1 can be tested by comparing the maximum log-likelihood of the (alternative) model containing all three terms and an intercept with that of a (null) model with X_1 omitted. The same strategy can evidently be followed for other terms as well. This strategy is called a "marginal" approach. In the literature, the resulting tests are often referred to as Type III tests.

Note that, in contrast to the marginal tests, the results of the tests in the sequential approach depend on the order of terms in the model. This is clearly seen from Table 4.3. In statistical software, in the case of tests about the mean structure parameters, the order is most often determined by the order of the terms that appear in the syntax defining the mean structure of the model. In R, the functions available for fitting linear (mixed-effects) models provide, by default, the sequential-approach tests.

From Table 4.3 it can be noted that, in contrast to the sequential approach, the results of the tests in the marginal one are not affected by the order of terms in the full model specification. It is also worth noting that, in both approaches, the results are equivalent for the last term listed, i.e., X_3, in the model. An important disadvantage of the marginal approach is that it includes tests that are not valid in some cases, e.g., when testing the main effect of a factor in the presence of interaction terms involving this factor.

4.7.2 Model Selection Criteria

Model reduction approaches, discussed in the previous section, considered the comparison of nested models. In the classical LM case, this is the most common situation. However, in the context of more complex models that will be discussed later in the book, a need may arise to discriminate between nonnested models. In such a situation, the use of *information criteria* is a possible solution.

The use of the criteria can be motivated by considering the procedure of the LR test (4.29). Denoted by ℓ_A and ℓ_0, the values of a log-likelihood function are computed by using the estimates obtained under the alternative and the null hypothesis, respectively. In the LR test, the null hypothesis is rejected if

$$\ell_A - \ell_0 > f(p_A) - f(p_0), \tag{4.40}$$

where p_A and p_0 are the number of unrestricted parameters in the models defined by the alternative and null hypotheses, respectively, and $f(\cdot)$ is a suitable function. For instance, for a test at the 5% significance level, the function may be chosen so that $f(p_A) - f(p_0) = 0.5\chi^2_{0.95,(p_A-p_0)}$, where $\chi^2_{0.95,(p_A-p_0)}$ is the 95th percentile of the χ^2 distribution with $p_A - p_0$ degrees of freedom.

Note that (4.40) can be expressed as

$$\ell_A - f(p_A) > \ell_0 - f(p_0). \tag{4.41}$$

Thus, the LR test can be viewed as a comparison of a suitably "corrected" log-likelihood function for two nested models.

The idea, expressed in (4.41), can be extended to the comparison of nonnested models. The question is, what "correction", in the form of function $f(\cdot)$, should be applied in such a case? Several choices have been proposed, leading to different information criteria. The main idea behind the criteria is to compare models based on their maximized log-likelihood value, while penalizing for the number of parameters.

The two most popular proposals are defined by using

$$f(p) = p$$

or

$$f(p) = 0.5\, p \log N^*,$$

where N^* is the effective sample size, defined as $N^* \equiv N$ for ML and $N^* \equiv N - p$ for REML. The first form of $f(\cdot)$ leads to the so-called *Akaike's information criterion* (AIC), while the second form defines the so-called *Schwartz* or *Bayesian information criterion* (BIC). The model with the largest AIC or BIC is deemed best. Note that sometimes the criteria are defined by using the negative of the differences, presented in (4.41). In this case, the model with the smallest criterion value is deemed best, and this convention is adopted in R.

Though the two criteria are developed based on the same underlying principle, they are based on different model-selection approaches. AIC aims to find the best approximating model to the true one. On the other hand, BIC aims to identify the true model. For $\log N^* > 2$, the penalty for the number of parameters used in BIC is larger than for AIC. Thus, the former criterion tends to select simpler models than the latter.

In view of the effective sample size, according to these criteria, differences in the likelihood need to be considered not only relative to the differences in numbers of parameters, but also relative to the number of observations included in the analysis. This feature is shared by several other information criteria that have been proposed in the literature (Verbeke and Molenberghs 2000, Sect. 6.4).

Finally, it should also be stressed that, in general, log-restricted-likelihoods are only fully comparable for LMs with the same mean structure. Hence, for comparing model fits with different mean structures, one should consider information criteria based on the ML estimation.

4.8 Chapter Summary

In this chapter, we briefly reviewed the theory of the classical LM, suitable for analyzing data involving independent observations with homogeneous variance. In Sects. 4.2 and 4.3, we introduced the specification of the model. Estimation methods were discussed in Sect. 4.4. Section 4.5 offered a review of the diagnostic methods, while in Sect. 4.6, we described the inferential tools available for the model. Finally, in Sect. 4.7, we summarized strategies that can be followed in order to reduce a model or to select one model from a set of several competing ones.

We did not aim to provide a detailed account of the theory. Such an account can be found, for instance, in the monograph by Neter et al. (1990). The purpose of our review was to introduce several key concepts, like model formulation, maximum-likelihood estimation, or model reduction/selection criteria, which will also be needed in the context of LMMs. We believe that introduction of the concepts in a simpler and more familiar framework should make their use in the context of the more complex models easier to present and to explain.

Chapter 5
Fitting Linear Models with Homogeneous Variance: The `lm()` and `gls()` Functions

5.1 Introduction

In Chap. 4, we outlined several concepts related to the classical LM. In the current chapter, we review the tools available in R for fitting the model.

More specifically, in Sects. 5.2–5.5, we present the details of the implementation of LMs in the function `lm()` from the base R distribution and in the function `gls()` from the **nlme** package. In particular, in Sect. 5.2, we describe the R syntax for the model structure. Section 5.3 explains the link between the syntax and the specification of the model. Section 5.4 describes the R functions available for fitting the LMs, while Sect. 5.5 explains how the details of the estimated form of the model can be accessed. Implementation of the tests of linear hypotheses about the mean-structure parameters is presented in Sect. 5.6. A summary of the chapter is provided in Sect. 5.7.

5.2 Specifying the Mean Structure Using a Model Formula

A model formula, or simply a formula, is an integral part of the R language. It is employed to symbolically and compactly represent various components of a wide range of models. In this section, we describe the use of a formula in the context of LMs, but the considerations are also useful in the context of other models, including the LMMs.

A linear structure, introduced in Sect. 4.2, is specified in R using a *two-sided* formula. The primary goal of the formula is to indicate the dependent variable $\mathbf{y}$ and to provide the information needed to create the design matrix $\boldsymbol{X}$, as specified in (4.6). Toward this end, we use an expression of the form

```
R expression ~ term.1 + term.2 + ··· + term.k.
```

A. Gałecki and T. Burzykowski, *Linear Mixed-Effects Models Using R: A Step-by-Step Approach*, Springer Texts in Statistics, DOI 10.1007/978-1-4614-3900-4_5,

Table 5.1 *R syntax*: Operators used when specifying an R formula

Operator	Is essential?	Role in the formula
+	Yes	Separates terms in the formula
:	Yes	Separates predictors in interaction terms
*, /, %in%, -, ^	No	Used to keep the formula short

The operator ~ (tilde) is an integral part of every formula, and separates its two sides. The R expression at the left-hand side of the formula defines the dependent variable. The right-hand side of the formula is used to specify the mean structure of the model. It contains terms, separated by the operator + (plus). Table 5.1 summarizes all operators used in R formulae. Their use will be explained in more detail later in this section.

It may be helpful to keep in mind that each term on the right-hand side of a model formula contributes one or more columns to the design matrix. The process of creating a design matrix from a formula is described in Sect. 5.3. The syntax for the formula follows the work presented in Wilkinson and Rogers (1973) and is explained in detail in Chap. 2 of Chambers and Hastie (1992).

5.2.1 The Formula Syntax

In this section, instead of formally presenting the syntax used to specify formulae, we present examples illustrating how a formula is constructed. In general, to construct terms in a formula, several operators from those listed in Table 5.1 can be used. Note that the operators + and : are *essential* for writing formulae; the remaining operators are primarily used to abbreviate the syntax.

To simplify presentation, we will focus on a hypothetical study with a dependent variable named y. Explanatory covariates include three continuous variables, named x1, x2, and x3, and three factors, named f1, f2, and f3. Note that formula considerations presented in this section refer to *symbolic* operations. Therefore, none of the objects f1, f2, f3, x1, x2, x3, nor any of the functions used in the formulae, need to be available for computations.

5.2.1.1 Operators Used in Formulae

As already mentioned, the operators used in formulae can be grouped into essential and nonessential operators. The two groups are described below.

Essential Operators

Panel R5.1 presents the syntax for several simple two-sided formulae, which involve just two operators, namely, the operator + to separate different terms and the operator : to separate predictors/factors within interaction terms. These two operators are *essential* operators.

R5.1 *R syntax*: Examples of basic formulae involving essential operators + and :

```
> y ~ x1                          # Univariate linear regression
> formula(y ~ x1)                 # ... equivalent specification
> y ~ 1 + x1                      # Explicit indication for intercept
> y ~ 0 + x1                      # No intercept using term 0
> y ~ f1 + x1                     # ANCOVA with main effects only
> y ~ f1 + x1 + f1:x1             # Main effects and ...
>                                 # ... factor by numeric interaction
> y ~ f1 + f2 + f1:f2             # Main effects and ...
>                                 # ... f1 by f2 two way interaction
> y ~ f1 + f1:f3                  # f3 nested within f1
> y ~ x1 + f1 + f2 +              # Main effects and ...
+       x1:f1+ x1:f2 + f1:f2      # ... two-way interactions
```

The reader may note that the function `formula()` does not have to be used explicitly in the formula specification. For example, the two following statements, `y ~ x1` and `formula(y ~ x1)`, are equivalent. Also, when a formula is created, an intercept is implicitly included by default. To explicitly specify the inclusion of an intercept in the model, we use `1` as a separate term, as in `y ~ 1 + x1`. On the other hand, to indicate that there is no intercept in the model, we can use `0` or `-1` as a separate term, as in `y ~ 0 + x1` or `y ~ -1 + x1` , respectively.

Nonessential Operators

The syntax of the formulae, displayed in Panel R5.1, can be extended by using additional operators, namely, `*`, `/`, `%in%`, `-`, and `^`. They are primarily used to abbreviate the syntax and hence are referred to as nonessential operators. Examples of formulae involving those additional operators are given in Panel R5.2.

The `*` operator, used in the first formula in Panel R5.2, denotes factor crossing, so that `f1*f2` is interpreted as `f1 + f2 + f1:f2`. The `%in%` operator denotes factor nesting. Thus, term `f3 %in% f1` implies that `f3` is nested in `f1` and it is interpreted as `f3:f1`. On the other hand, term `f1/f3` is interpreted as `f1 + f1:f3`. The `^` operator indicates crossing terms up to a specified degree. For example,

R5.2 *R syntax*: Examples of formulae employing nonessential operators, i.e., *, /, %in%,-, and ^

```
> y ~ f1*f2                 # ANOVA with two-way interaction
> y ~ f1 + f3 %in% f1       # f3 nested within f1
> y ~ f1/f3                 # ... equivalent specification
> y ~ (x1 + f1 + f2)^2      # Up to 2nd order interactions
> y ~ -1 + x1               # Intercept removed
```

Table 5.2 *R syntax*: Examples of expanding elementary formulae

Formula	Expanded formula
`y ~ f1*f2`	`y ~ f1 + f2 + f1:f2`
`y ~ f1 + f3 %in%f1`	`y ~ f1 + f1:f3`
`y ~ f1/f3`	`y ~ f1 + f1:f3`
`y ~ (f1 + f2 + f3)^2`	`y ~ f1 + f2 + f3+ f1:f2 + f1:f3 + f2:f3`

Table 5.3 *R syntax*: Interpretation of various nonessential formula-operators used in Panel R5.2

Operator	Exemplary term	Interpretation
*	`f1*f2`	`f1 + f2 + f1:f2`
%in%	`f3 %in% f1`	`f1:f3`
/	`f1/f3`	`f1 + f1:f3`
^	`(x1+f2+f3)^2`	`x1 + f2 + f3+ x1:f2 + x1:f3 + f2:f3`
-	`f1*f2-f1:f2`	`f1 + f2`

(x1 + f1 + f2)^2 is equivalent to (x1 + f1 + f2)*(x1 + f1 + f2), which, in turn, is equivalent to a formula containing the intercept, the main effects of x1, f1, and f2, together with their second-order interactions (but not the squares of the individual covariates/factors; see also the last formula in Table 5.2). The - operator removes the specified term, so that formula f1*f2-f1:f2 is equivalent to f1 + f2. In the last formula in Panel R5.2, the operator - is used to remove the intercept term. Thus, y~-1+x1 simply specifies a regression line through the origin.

Table 5.3 presents several terms containing nonessential operators, shown in Panel R5.2, along with their interpretation in terms of essential operators + and :.

5.2.1.2 Composite Terms

While the formulae, defined in Panels R5.1 and R5.2, employ just variable and factor names, they can also involve functions and arithmetic expressions, which offer another possibility of extending the formulae syntax. Several examples are shown in Panel R5.3.

R5.3 *R syntax*: Formulae with a more advanced syntax

(a) *Composite terms*

```
> y ~ sqrt(x1) + x2              # Square root transformation of x1
> y ~ ordered(x1, breaks)+       # Ordered factor created and ...
+             poly(x1, 2)        # ... second degree polynomial added
> y  ~ poly(x1, x2, 2)           # Bivariate quadratic surface ...
>                                # ... for x1 and x2
> log(y) ~ bs(x1, df = 3)        # log transform for y modeled ...
>                                # ... by using B-spline for x1
> y ~ f1*bs(x1, df = 3) - 1      # Factor by spline
                                 # interaction ...
>                                # ... with intercept omitted
```

(b) *Use of the* `I()` *and* `update()` *functions*

```
> form2   <- y ~ I(x1 + 100/x2) # I() function
> update(form2, . ~ . + x3)     # x3 predictor added to form2
> update(form2, . ~ . -1)       # Intercept omitted from form2
```

In the first set of formulae, presented in Panel R5.3a, we introduce composite terms created by applying various mathematical functions like, e.g., square-root `sqrt()` or logarithm `log()` to the dependent and/or explanatory variables. The second set of formulae, shown in Panel R5.3b, illustrates the use of functions `I()` and `update()`. The use of the function `I()` is described on page 94, where we explain potentially different meanings of operators used in a formula. The function `update()`, applied in the last two formulae of Panel R5.3b, is used to modify a formula, which was previously defined and stored in an R object. The use of the function makes the changes to formulae more explicit and allows constructing a more transparent and efficient R code.

5.2.1.3 Different Meanings of Operators and the Use of the `I()` *Function*

Note that all operators, i.e., `+`, `-`, `*`, `/`, `:`, `%in%`, and `^`, which were used in the formulae presented in Panels R5.1–R5.3, can have potentially two different meanings. We will refer to them as the *default* and the *arithmetic* meaning. The most common is the default meaning, related to the manipulation of terms in a formula. Thus, for instance, the default meaning of the operator `+` is the separation of terms, as in formula `y ~ x1 + x2`, while the default meaning of the operator `*` is the creation of an interaction, as in term `f1*f2`. On the other hand, the arithmetic meaning of these operators corresponds to their use as symbols of arithmetic operations. This was, e.g., the meaning of the operators `+` and `/` used in the formula `form2` in

Panel R5.3b. In particular, in the formula, the two operators were used as arithmetic operators to calculate the numerical value of the expression `x1 + 100/x2`.

Given the two possible meanings of the formula operators, it is prudent to indicate the intended meaning in the defined formula. Toward this end, the use of the function `I()` is recommended. The operators used within the scope of the function are given the arithmetic (nondefault) meaning.

5.2.1.4 Expansion of a Formula

In Panels R5.1–R5.3, we presented several formulae, in which we used various operators. In Table 5.2, we demonstrate how several of these formulae can be equivalently expanded using only the essential operators `+` and `:`. The table illustrates that, in general, these two operators, in combination with mathematical functions and the `I()` function, are sufficient to specify any formula. The other, nonessential operators are mainly used to abbreviate the formula syntax. The reader may note that, in the expanded formulae from Table 5.2, the terms representing lower-order interactions are listed before the terms for higher-order interactions. In the context of ordering terms in the expanded formula it is helpful to define the *interaction order*, or simply *order*, for each term as the number of variables/factors separated by the operator `:`. Following this definition, main effects like, e.g., `f1` or `f2` are of order 1, and they are listed in the expanded formula before the interaction term `f1:f2`, which is of order 2. Note that the interaction order of the intercept term is considered to be equal to zero.

5.2.2 Representation of R Formula: The terms Class

Creating an object of class *terms* is an important step in building the design matrix, i.e., matrix $\boldsymbol{X}$, of an LM. In this section, we introduce such objects, which constitute a different, more technical way of specifying a model formula. Objects of this class contain all information essential to create a *model frame* and a design matrix in the context of a given dataset, as will be described later in Sects. 5.3.1 and 5.3.2. Objects of class *terms* are typically created within other R functions like `model.frame()`, `model.matrix()`, or `lm()`, by applying the generic function `terms()` to a formula.

In Panel R5.4, the function `terms()` is applied to two formulae. As a result, two objects of class *terms*, `termsA` and `termsB`, are created.

First, in Panel R5.4a, we use the function `terms()` to create the object `termsA`. The object has several attributes, which contain the information about all terms used to build the formula `formA`. The names of the attributes are obtained by applying a superposition of two functions, namely, `names()` and `attributes()`, to the object `termsA`.

We will now describe the attributes of the object `termA`, which are most relevant in the context of creating the model frame and the design matrix.

R5.4 *R syntax*: Attributes of objects of class *terms* created from a formula using the `terms()` function

(a) *A formula with an intercept and interaction term*

```
> formA <- y ~ f1*f2                 # Formula A
> termsA <- terms(formA)             # Object of class terms
> names(attributes(termsA))          # Names of attributes
  [1] "variables"    "factors"       "term.labels"  "order"
  [5] "intercept"    "response"      "class"        ".Environment"
> labels(termsA)                     # Terms; interaction after main effects
  [1] "f1"    "f2"    "f1:f2"
> attr(termsA, "order")              # Interaction order for each term
  [1] 1 1 2
> attr(termsA, "intercept")          # Intercept present?
  [1] 1
> attr(termsA, "variables")          # Variable names
  list(y, f1, f2)
```

(b) *A formula without the intercept and interaction term*

```
> formB <- update(formA, . ~ . - f1:f2 -1)          # Formula B
> termsB <- terms(formB)
> labels(termsB)                     # Terms of formula B
  [1] "f1" "f2"
> attr(termsB, "intercept")          # Intercept omitted
  [1] 0
```

The attribute `term.labels` is a character vector representing a given formula in an expanded form. For example, in the expanded form, the formula `formA` includes three terms, namely, `f1`, `f2`, and `f1:f2`. They are given by the elements `"f1"`, `"f2"`, and `"f1:f2"`, respectively, of the character vector obtained by using the `labels()` function. The attribute `order` gives the interaction order for each term in the formula. For example, we easily find that the order of the term `f1:f2` in `form1` is equal to 2. The value of the `intercept` attribute provides the information whether an intercept is included into the model or not. In our case, its value is 1, indicating that an intercept is present. Another attribute, `variables`, indicates which variables are used in creating the model frame. More details on the model frame are provided in Sect. 5.3.1.

Description of the remaining attributes of objects of class *terms* can be obtained from R's help system by issuing the command `?terms.object`.

In Panel R5.4b, we use the `update()` function to create the formula `formB` from the formula `formA` by removing the intercept and interaction `f1:f2`. By checking the value of the `labels` attribute of the *terms*-class object `termsB`, corresponding to `formB`, we verify that the intercept and interaction were indeed removed from the formula. The removal of the intercept is also confirmed by the fact that the value of the attribute `intercept` of the formula changed to 0, as compared to the corresponding value of the attribute of the formula `formA` (see Panel R5.4a).

To conclude, we note that the specialized objects of class *terms* are rarely created by the data analyst. However, they may be useful to get additional insight into several features of a given formula, like, e.g., the names of variables involved in the formula specification, the expanded form of a formula, the interaction order of a particular term of a formula, etc.

5.3 From a Formula to the Design Matrix

In Sect. 5.2, we introduced the syntax of an R formula. We also described the concepts of an expanded formula and of an object of class *terms*, which represents a given formula in a more technical format.

In this section, we illustrate how a design matrix, X, based on a given formula *and* available data, is created. This, rather technical, process consists of two steps. First, a model frame is created based on available data. Then, the design matrix is itself constructed. Note that these steps are rarely performed in practice by the data analyst. Instead, they are carefully implemented inside many model-fitting functions, such as `lm()`, `gls()`, `lme()`, and others. We introduce the process of creating a design matrix for illustration purposes and to avoid the "black box" impression for the model fitting functions. Note that, in contrast to Sect. 5.2, where we dealt with *symbolic* operations, in the current and subsequent sections, all objects specified in the formula, including functions' definitions, need to be available for computations.

Figure 5.1 summarizes the steps necessary to obtain a design matrix from a model formula and data stored in a `data.frame`. By combining, with the use of the `model.frame()` function, the information stored in the object `terms.object` of class *terms* with the contents of the `data.frame`, a model frame is created. From it, with the help of the `model.matrix()` function, the design matrix is obtained. Note that creation of an object of class *terms* was already presented in Sect. 5.2.2. The construction of the model frame and of the design matrix is described in Sects. 5.3.1 and 5.3.2, respectively.

5.3.1 Creating a Model Frame

In the first step of the process aimed at the creation of the design matrix, a given formula is interpreted/evaluated in the context of specific data. As a result, a

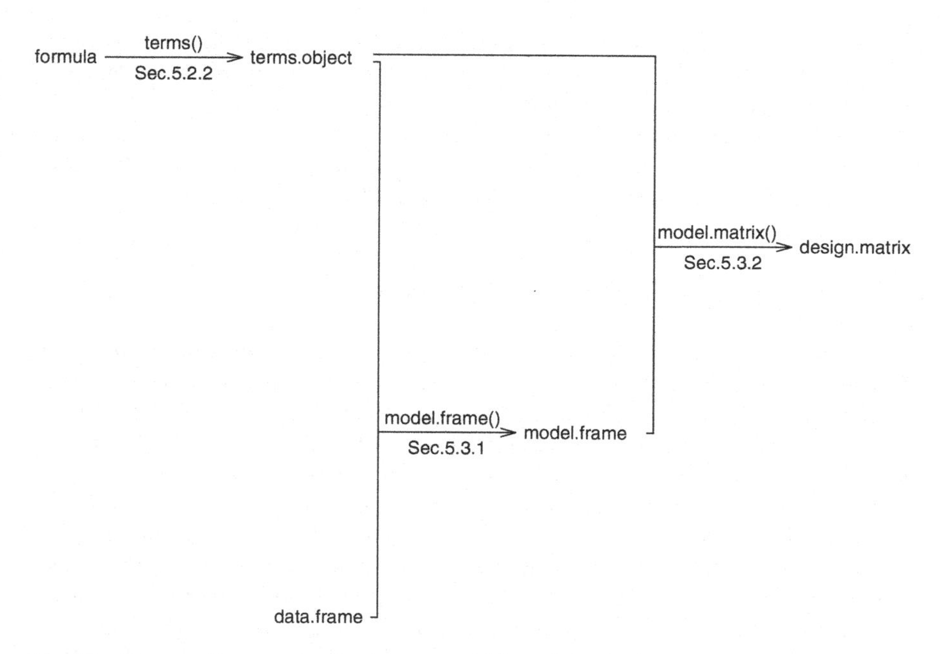

Fig. 5.1 *R syntax*: The steps leading from a model formula to the design matrix

specialized, working data frame, called a model frame, is created. The main function employed to perform this step is `model.frame()`.

5.3.1.1 Arguments of the `model.frame()` *Function*

The function `model.frame()` is an example of a function that interprets a formula in the context of arbitrary data. This type of functions typically uses at least four arguments: `formula`, `data`, `subset`, and `na.action`. The first argument specifies the model formula. The other three arguments, described briefly below, specify the data that are to be used for evaluation.

Arguments `data` and `subset` are used to tentatively define the subset of the data that are used to create the model frame. The function indicated in the argument `na.action` allows further modification of the data defined by the `data` and `subset` arguments. More specifically, the `na.action` argument points to a function, which indicates what should be done with a data record with missing values. The default value for the `na.action` argument is `na.omit`, which points to the generic function `na.omit()`. The function removes the records with missing values from the data. Another possible function is the `na.exclude()` function. Similarly to `na.omit()`, it removes the records with missing values, but its result differs when it comes to the computation of residuals or predicted values. A full list of available functions, and their properties, to be used as a value of the na.action argument can be obtained by issuing the ?na.action command.

The object generated by the model.frame() function is a model frame, an R object resembling a classical data frame. An example of how a model frame is created for a classical LM is presented in the next subsection.

5.3.1.2 *Creating a Model Frame: An Illustration*

In Panel R5.5, we illustrate how a model frame is created by evaluating a formula in the context of the armd.wide data (see Sect. 2.2.2) loaded from the **nlmeU** package.

First, in Panel R5.5a, we define the formula form1 with composite terms involving explanatory variables treat.f, visual0, visual24, line0, and lesion, stored in the armd.wide data frame. Note that the formula form1 is used only for illustration purposes and is not used anywhere else in the book.

The model frame armd.mf1 is created in Panel R5.5b by employing the model.frame() function. The argument data indicates that we evaluate the formula form1 with respect to the data contained in the armd.wide data frame. The argument subset indicates that we omit from the data two subjects with the subject identifiers equal to "1" and "2". Using the function dim() we check that the armd.wide data contains 240 records and 10 variables. However, owing to the use of the na.action and subset arguments, several rows are omitted from the data. As a consequence, the number of rows in the resulting model frame armd.mf1 is equal to 189, as shown in the result of the application of the function dim() to the model frame object. The model frame includes seven components; their names are listed using the names() function.

At the end of Panel R5.5b, we use the function head() to display the first four rows of the model frame. Note that the output of the head() function contains eight columns, not seven. This stems from the fact that the poly(visual0,2) term in form1 contributes two columns to the model frame, instead of one.

It is also worth noting that the output of the head() function indicates that the model frame contains a column labeled (SubjectId), which does not correspond to any of the variables involved in the specification of the model formula form1. This additional column was included in the model frame by setting the argument SubjectId = subject in the call of the model.frame() function. Thus, the column contains the values taken from the subject variable. In our example, it allows, for instance, direct verification of which subjects were omitted from the armd.wide data when creating the armd.mf1 model frame.

The specification of the argument SubjectId = subject in the call of the model.frame() function is an example of the argument assignment of the form *model.frame.var* = *data.var*. In the assignment, *data.var* is a variable that should be additionally included in the model frame. In the model frame, the additional variable is named *(model.frame.var)*, i.e., its name is enclosed in parentheses. This

R5.5 *R syntax*: Model frame created by evaluating a formula in the context of the `armd.wide` data

(a) *Formula*

```
> form1 <- formula(
+   visual52 ~                      # Dependent variable
+   sqrt(line0) +                   # Continuous explanatory variable
+   factor(lesion) +                # Factor with 4 levels
+   treat.f*log(visual24) +         # Crossing of two variables
+   poly(visual0, 2))               # Polynomial of 2nd degree
```

(b) *Model frame*

```
> data(armd.wide, package = "nlmeU")# Data loaded
> armd.mf1 <-
+    model.frame(
+       form1,                        # Formula
+       data = armd.wide,             # Data frame
+       subset =                      # Exclude two subjects
+          !(subject %in% c("1", "2")),
+       na.action = na.exclude,       # Dealing with missing data
+       SubjectId = subject)          # Identifier of data records
> class(armd.mf1)
  [1] "data.frame"
> dim(armd.wide)                      # Data frame dimensions
  [1] 240  10
> dim(armd.mf1)                       # Model frame dimensions
  [1] 189   7
> names(armd.mf1)                     # Components of the model frame
  [1] "visual52"         "sqrt(line0)"       "factor(lesion)"
  [4] "treat.f"          "log(visual24)"     "poly(visual0, 2)"
  [7] "(SubjectId)"
> head(armd.mf1, n = 4)               # First four records
    visual52 sqrt(line0) factor(lesion) treat.f log(visual24)
  4       68      3.6056              2 Placebo        4.1589
  6       42      3.4641              3  Active        3.9703
  7       65      3.6056              1 Placebo        4.2767
  8       37      2.8284              3 Placebo        3.6109
    poly(visual0, 2).1 poly(visual0, 2).2 (SubjectId)
  4          0.0523462         -0.0054435           4
  6          0.0175815         -0.0460843           6
  7          0.0393095         -0.0243944           7
  8         -0.0693302         -0.0091566           8
```

syntax plays an important role in the model-fitting functions, such as lm() or lme(). Specifically, it allows including into the model frame additional variables, which specify components of an LM other than design matrix, such as weights, offset, etc.

Although the model frame is an object of class *data.frame*, there are some differences between model frames and data frames. An important difference is that there are no restrictions for variable names included in a model frame. For example, poly(time,2) and log(visual0) are valid variable names in a model frame. However, such names are not valid in a data frame. Another difference, which we address next, is the presence of the terms attribute in the model frame.

5.3.1.3 Features of the terms *Attribute*

An important difference between the model frame and the data frame is the presence of the terms attribute. To demonstrate various features of the attribute, in Panel R5.6, we extract it from the model frame armd.mf1 and explore its contents.

R5.6 *R syntax*: The attribute terms of the armd.mf1 model frame. The model frame was created in Panel R5.5

```
> terms.mf1 <- attr(armd.mf1, "terms")         # terms attribute
> class(terms.mf1)
  [1] "terms"    "formula"
> names(attributes(terms.mf1))                 # Names of attributes
   [1] "variables"    "factors"      "term.labels"  "order"
   [5] "intercept"    "response"     "class"        ".Environment"
   [9] "predvars"     "dataClasses"
> attr(terms.mf1, "dataClasses")               # dataClasses attribute
          visual52       sqrt(line0)    factor(lesion)
         "numeric"         "numeric"          "factor"
           treat.f   log(visual24) poly(visual0, 2)
          "factor"         "numeric"       "nmatrix.2"
       (SubjectId)
          "factor"
> attr(terms.mf1, "predvars")                  # predvars attribute
  list(visual52, sqrt(line0), factor(lesion), treat.f, ...
      poly(visual0, 2, coefs = list(alpha = c(54.9541666666667,
      50.5097520799239), norm2 = c(1, 240, 52954.4958333333, ...
      ))))
> labels(terms.mf1)                            # Component names
  [1] "sqrt(line0)"           "factor(lesion)"
  [3] "treat.f"               "log(visual24)"
  [5] "poly(visual0, 2)"      "treat.f:log(visual24)"
```

First, using the `attr()` function, we obtain the `terms` attribute and store it in the object named `terms.mf1`. Note that `terms.mf1` is an object of class *terms* (Sect. 5.2.2), which inherits from the class *formula*.

We note that names of attributes, such as `variables`, `order`, and `term.labels`, are consistent with the attributes' names of objects of class *terms*, introduced in Panel R5.4 (Sect. 5.2.2). The difference lies in the presence of two additional attributes, namely, `dataClasses` and `predvars`. The attribute `dataClasses` contains the information about how each component of the model frame is interpreted in the design matrix, e.g., whether it is a numeric variable, a factor, or a matrix. The attribute `predvars` contains a list of expressions that are used to evaluate the components of the model frame when applied to a data frame.

5.3.1.4 Note on Data-Dependent Functions: `poly()`, `bs()`, `ns()`

In Panel R5.5, we defined the formula `form1` with the help of several *observation-by-observation* functions, such as `log()` and `sqrt()`. On the other hand, the function `poly()`, used in the same panel, is an example of a *data-dependent* function. Examples of other data-dependent functions include, but are not limited to, functions `bs()` and `ns()` from the **splines** package.

A working definition of a data-dependent function is that its returned value depends on *all* elements of the vector used as an argument. Thus, the function requires a pass over *all* rows of the data. To avoid ambiguity in what is meant by "all" values, it should be mentioned that it is neither advisable nor possible to apply these functions to vector(s) containing missing values. Moreover, the use of the argument `subset` in a call to the function `model.frame()` does *not* affect the argument used by the data-dependent functions. For instance, in Panel R5.5, the data-dependent function `poly()` is applied to the entire vector `visual0` stored in the `armd.wide` data frame. Note that the vector does not contain any missing values, and therefore, *all* 240 observations are used in deriving the coefficients of the polynomial specified by the `poly()` function, regardless of the values of the `subset` and `na.action` arguments used in the `model.frame()` function. As previously mentioned, the polynomial coefficients are stored in the `predvars` attribute of the `terms.mf1` object and can be *reused* by other functions. Examples of such functions include the `predict()` function, which may evaluate a model frame in the context of a dataset *different* from the one used to build the model frame. For this type of functions, it is essential to have the coefficients available through the `predvars` attribute. Otherwise, the functions would attempt to *reevaluate* the coefficients of the specified polynomial (or of another data-dependent function) using the new dataset. This could result in different numerical values of the coefficients. The issue of evaluating/reusing a given polynomial (or a spline) for a dataset different from the one used to define the model frame is discussed in more detail in the book by Venables and Ripley (2010) in the context of so-called "safe prediction."

5.3.2 *Creating a Design Matrix*

In Sect. 5.3.1, we presented the first step needed for the creation of the design matrix. The step involved the construction of a model frame for given data. In the current section, we explain how the design matrix is created based on a model formula *and* a model frame. As indicated in Fig. 5.1, the key function used in this step is `model.matrix()`.

5.3.2.1 *Illustration: Design Matrix*

In Panel R5.7, we illustrate how to create the design matrix based on the model frame `armd.mf1`, which corresponds to the formula `form1`, defined in Panel R5.5, and to the data frame `armd.wide`. The design matrix is stored in the object `Xmtx`.

Recall that (see Panels R5.5 and R5.6) the formula `form1`, in the expanded form, contains six terms (excluding intercept) and that the model frame `armd.mf1` contains 189 records. The resulting design matrix, named `Xmtx`, has the same number of rows, i.e., 189, and 10 columns. The column names are displayed in Panel R5.7 by applying the `colnames()` function. For reference, the first four rows of the `Xmtx` matrix are also displayed using the `head()` function.

The presence of the intercept in the first column, named `(Intercept)`, of the design matrix is worth noting. Factors `factor(lesion)` and `treat.f` are represented by three columns and one column, respectively. More details on how the columns representing factors are constructed will be provided later in Panel R5.8.

Note that the resulting matrix `Xmtx` has two additional attributes: `assign` and `contrasts`. The attribute `assign` provides a link between each column of the design matrix and a corresponding term in the expanded form of the model formula `form1`. For instance, based on the value of the attribute, we can confirm that columns 3, 4, and 5 of the design matrix correspond to the second term of the formula, i.e., `factor(lesion)`. Thus, the factor contributes three columns to the design matrix. Similarly, the fifth term, i.e., `poly(visual0,2)`, contributes two columns (the eighth and the ninth) to the design matrix.

The value of the attribute `contrasts` indicates that the function `contr.treatment()` was used to decode factors `factor(lesion)` and `treat.f` into corresponding columns of the design matrix `Xmtx`. We will discuss the issue of decoding factors next.

Note that, in the call of the `model.matrix()` function, we used the formula `form1` as the first argument. In general, objects of other classes can also be used in the argument as long as the function `terms()` returns for them an object of class *terms*. More information on the arguments of the `model.matrix()` function can be obtained by issuing the R help command `help(model.matrix)`.

R5.7 *R syntax*: Creating a design matrix based on a formula evaluated in a model frame. The model frame armd.mf1 was created in Panel R5.5

```
> Xmtx <-  model.matrix(form1, armd.mf1)          # Design matrix
> dim(Xmtx)                                         # No rows and cols
  [1] 189  10
> (nms <- colnames(Xmtx))                           # Col names ...
   [1] "(Intercept)"                   "sqrt(line0)"
   [3] "factor(lesion)2"               "factor(lesion)3"
   [5] "factor(lesion)4"               "treat.fActive"
   [7] "log(visual24)"                 "poly(visual0, 2)1"
   [9] "poly(visual0, 2)2"             "treat.fActive:log(visual24)"
> colnames(Xmtx) <- abbreviate(nms)                 # ... abbreviated
> print(head(Xmtx, n = 6), digits = 4)              # First 6 rows
     (In)  s(0) f()2 f()3 f()4 tr.A  l(24  p(0,2)1   p(0,2)2  t.A:
  4     1 3.606    1    0    0    0 4.159  0.05235 -0.005443 0.000
  6     1 3.464    0    1    0    1 3.970  0.01758 -0.046084 3.970
  7     1 3.606    0    0    0    0 4.277  0.03931 -0.024394 0.000
  8     1 2.828    0    1    0    0 3.611 -0.06933 -0.009157 0.000
  9     1 3.464    1    0    0    1 3.989  0.01758 -0.046084 3.989
  12    1 3.000    0    0    0    1 3.296 -0.03891 -0.044592 3.296
> names(attributes(Xmtx))                           # Attribute names
  [1] "dim"         "dimnames"  "assign"    "contrasts"
> attr(Xmtx, "assign")                              # Cols to terms map
   [1] 0 1 2 2 2 3 4 5 5 6
> attr(Xmtx, "contrasts")                           # Contrasts attribute
  $`factor(lesion)`
  [1] "contr.treatment"

  $treat.f
  [1] "contr.treatment"
```

5.3.2.2 Decoding Factors

In R, we typically use the `factor()` or `ordered()` functions to create unordered and ordered factors, respectively. To decode a given factor into the columns of a design matrix, it is necessary to associate the factor with an appropriate matrix of contrasts. In Panel R5.8, several examples of predefined contrast functions and of the corresponding contrast matrices are given for reference. The contrast functions include `contr.treatment()`, `contr.sum()`, `contr.helmert()`, `contr.poly()`, and `contr.SAS()`.

The first argument of all of the contrast functions indicates the number of the levels of the decoded factor. The contrast matrices, created during the session shown in Panel R5.8, are presented for the case of an artificial factor with three levels.

R5.8 *R syntax*: Predefined contrast functions and the corresponding contrast matrices for a hypothetical factor with three levels

```
> contr.treatment(3)               # Default base level = 1
  2 3
1 0 0
2 1 0
3 0 1
> contr.treatment(3, base = 3) # Base level = 3. Same as contr.SAS(3).
  1 2
1 1 0
2 0 1
3 0 0
> contr.sum(3)                     # Sum to zero
  [,1] [,2]
1    1    0
2    0    1
3   -1   -1
> contr.helmert(3)                 # Helmert contrasts
  [,1] [,2]
1   -1   -1
2    1   -1
3    0    2
> contr.poly(3, scores=c(1, 5, 7)) # Polynomial contrasts
            .L       .Q
[1,] -0.77152  0.26726
[2,]  0.15430 -0.80178
[3,]  0.61721  0.53452
```

The choice of the type of contrasts has implications for the interpretation of the parameters $\boldsymbol{\beta}$ of an LM. For instance, the contrasts defined by the function `contr.treatment()` imply that the elements of the vector $\boldsymbol{\beta}$ can be treated as differences of the expected values of the dependent variable between a reference level and every other level of the factor. On the other hand, the use of the `contr.sum()` contrasts implies that the elements can be interpreted as deviations between the expected values corresponding to the different levels of the factor and the overall mean of the dependent variable. Note that the statement `contr.SAS(3)` returns the same contrast matrix as the statement `contr.treatment(3, base=3)`. Thus, the `contr.SAS(3)` contrasts use the last level of the factor as the reference level, while `contr.treatment()`, by default, uses the first level as the reference.

R5.9 *R syntax*: Assigning and extracting a set of contrasts for a factor

(a) *Extracting default contrasts*

```
> options()$contrasts                          # Default contrasts
         unordered             ordered
  "contr.treatment"      "contr.poly"
> lesion.f <- factor(armd.wide$lesion)         # Factor created
> str(lesion.f)                                # Structure
  Factor w/ 4 levels "1","2","3","4": 3 1 4 2 1 3 1 3 2 1 ...
> names(attributes(lesion.f))                  # Names of factor attributes
  [1] "levels" "class"
> levels(lesion.f)                             # Levels extracted
  [1] "1" "2" "3" "4"
> contrasts(lesion.f)                          # Contrasts extracted
    2 3 4
  1 0 0 0
  2 1 0 0
  3 0 1 0
  4 0 0 1
```

(b) *Assigning a new set of contrasts using the* `C()` *function*

```
> lesion2.f <- C(lesion.f, contr.sum(4)) # New contrasts using C()
> names(attributes(lesion2.f))           # Names of factor attributes
  [1] "levels"    "class"     "contrasts"
> contrasts(lesion2.f)                   # Contrasts extracted
    [,1] [,2] [,3]
  1    1    0    0
  2    0    1    0
  3    0    0    1
  4   -1   -1   -1
```

(c) *Assigning contrasts using the "contrasts() <- contrast function" syntax*

```
> lesion2a.f <- factor(lesion.f)        # Factor copied
> contrasts(lesion2a.f) <- contr.sum(4)
```

In Panel R5.9, it is shown how to assign or extract a matrix of contrasts for a given factor.

In particular, in Panel R5.9a, we invoke the function `options()` to obtain a list of the current values of the global options. By displaying the component `contrasts` of the list, we verify the names of the default functions, which are used to create matrices of contrasts for ordered and unordered factors. Note that a similar result could be obtained using the command `getOption("contrasts")`.

The advantage of using the default choice of contrast matrices, i.e., contr.treatment() for unordered and contr.poly() for ordered factors, is that, in studies with a balanced design, columns of the design matrix $\boldsymbol{X}$ become orthogonal. Consequently, the matrix $\boldsymbol{X}'\boldsymbol{X}$, used in the estimation of the parameters $\boldsymbol{\beta}$ (see Sect. 4.4), becomes diagonal and the corresponding estimates of the elements of $\boldsymbol{\beta}$ are uncorrelated. Other choices of contrasts in LMs may introduce artificial correlations between the estimates, even for balanced designs. Thus, the choice of the contrasts involves a trade-off between the simplicity of the default choice and the interpretability of the estimates of the fixed effects, as discussed in the context of Panel R5.8.

To illustrate how to assign a new set of contrasts to a given factor, in Panel R5.9a, we create an unordered factor lesion.f with four levels based on the lesion variable extracted from the armd.wide data. Note that the factor has only two attributes: levels and class. As a result, when the function contrasts() is applied to the factor, it displays the contrast matrix defined by the default contrast function, i.e., contr.treatment().

A different matrix of contrasts can be assigned to a factor in at least two ways.

In Panel R5.9b, we illustrate the first method, which uses the C() function. The two main arguments of the function are object and contr. The first one gives the name of the factor for which the attribute is to be set. The second one indicates which contrasts to use. In the example shown in Panel R5.9b, the contrasts created by the function contr.sum() are used. Note that the newly created factor lesion2.f has an additional attribute, i.e., contrasts. As a result, when the function contrasts() is applied to the factor, it displays the contrast matrix created by the function contr.sum(). Thus, factors lesion.f and lesion2.f are essentially the same, except for the fact that they are associated with two different contrast matrices. Consequently, the factors are represented in two different ways in the design matrix $\boldsymbol{X}$.

In Panel R5.9c, we demonstrate the second method of assigning contrasts to a factor. Toward this aim, we create the factor lesion2a.f, which is fully equivalent to the factor lesion.f. Then, we use the function contrasts() to assign the contrasts, constructed by the function contr.sum(), to the newly created factor. As a result, we obtain a factor, which is fully equivalent to lesion2.f, with the same set of contrasts. Note that the method was already used in Panel R2.5 (Sect. 2.2.2) when creating the data frame armd.

It is worth noting that all the contrast matrices considered in Panels R5.8 and R5.9 have k rows and $k-1$ columns, where k is the number of levels of the corresponding factor. By choosing such a contrast matrix, we avoid collinearity in the design matrices containing an intercept. More generally, by assigning a contrast matrix with at most $k-1$ linearly independent columns, we avoid collinearity in a design matrix for any model containing all terms of lower order than a given factor or, more broadly, a term involving factor(s). However, in some cases like, e.g., of a model without an intercept, it is more appropriate to use a $k \times k$ identity matrix instead of a $k \times (k-1)$ contrast. Such a choice is possible using the contrasts=FALSE argument of the contrasts() function.

5.4 Using the lm() and gls() Functions to Fit a Linear Model

The primary function used to fit LMs in R is lm(). It comes with the basic distribution of R. It implements the OLS estimation method (Sect. 4.4.1). An alternative is to use the function gls() from the **nlme** package, which uses the ML and REML estimation (Sects. 4.4.2 and 4.4.3, respectively). For ANOVA and ANCOVA models, the aov() function may be preferable.

The use of the lm() function usually involves a call like lm(*formula*, *data*), where *formula* specifies the model to be fitted and *data* indicates the data frame containing the variables used to build the design matrix corresponding to the *formula*. A similar call is used for the gls() function, except that the formula is specified with the use of the argument model. Other often-used arguments of the lm() and gls() functions are displayed in Table 5.4.

The mean structure of the model is defined by the argument formula for lm() and model for gls(). The arguments specify a two-sided model formula that defines the dependent variable and the design matrix (see Sect. 5.2).

The definitions of the arguments data, subset, and na.action are consistent with the definitions of the similar arguments used by the model.frame() function, which were described in Sect. 5.3.1. These arguments, along with the formula or model arguments, are used to create the model frame necessary for fitting the model. An offset (Sect. 4.3) can be specified by the argument offset of the lm() function; there is no corresponding argument for the gls() function. In both functions, the estimation method is indicated using the method argument. By default, lm() uses the OLS estimation (Sect. 4.4.1), implemented with the help of the QR decomposition of the design matrix. The default estimation method of the gls() function is REML (Sect. 4.4.3).

It is a common practice to use update(), a generic function to modify and refit a given model, instead of specifying the model from scratch. The simplest use of the update() function, especially relevant in this chapter, is to modify a formula. Examples of updating formula and of modifying the mean structure of an

Table 5.4 *R syntax*: Selected arguments of the lm() and gls() functions used to fit a linear model with homogeneous variance. Default values are given for the na.action and method arguments

	Function arguments	
Component	lm()	gls()
Formula	formula	model
Offset	offset	–
Data	data	data
Subset	subset	subset
Missing values	na.action	na.action=na.fail
Estimation method	method="qr"[a]	method= "REML"

[a]OLS using the QR decomposition (Golub and Van Loan 1989)

LM are presented in Panel R5.3. Other functions, which can be considered in this context, are add1() and drop1(). They are used less often and we do not describe them here.

5.5 Extracting Information from a Model-Fit Object

Results of an LM, fitted using the lm() or gls() functions, are stored in an object of class *lm* or *gls*, respectively.

In general terms, there are at least two ways to extract the results from the objects representing model fits. An elegant and recommended way is using generic *extractor* functions such as print(), summary(), fitted(), coef(), vcov(), confint(), etc. However, if a method for extracting an interesting result of the model fit is not available, then it may need to be extracted directly from the appropriate component of the object (typically, a list), which represents the model fit.

In Table 5.5, we present the syntax that can be used to extract various components of an LM fitted using the lm() or gls() functions. Whenever possible, we show the use of an appropriate extractor function. Given that our focus is on LMMs, which cannot be fitted using the lm() nor gls() function, we do not describe the use of the extractor functions in detail; the necessary information can be found by calling R's help system. Note, however, that there are striking similarities of the syntax for extracting information from objects of class *lm* and *gls*. In fact, a syntax similar to that presented in Table 5.5 can be used in the case of other models, including LMMs. This is simply due to object-oriented approach, implemented using the appropriate generic functions mentioned earlier in this section.

It should be mentioned that the information provided in Table 5.5 implicitly assumes that all variables, used for fitting a model, were stored in a data frame. This approach follows a general recommendation to use data frames, not vectors nor matrices, for model fitting. If this is not the case, then the cl$data command will not extract the data name and the data frame will not be properly evaluated.

Sometimes we are interested in extracting a particular component of the fitted model, saving it in an intermediate object, and immediate printing of the object. Toward this end, we can enclose the syntax creating the intermediate object in parentheses, as it is shown, e.g., for the function summary() in Table 5.5. The use of the parentheses allows immediate printing of the contents of the newly created object.

Influence diagnostic measures for an LM with independent, homoscedastic residual errors (Sect. 4.5.3), fitted with the use of the lm() function, can be obtained by using the influence() generic function. Given that we are mainly interested in more complex LMs, we do not describe the use of the function. It should also be noted that the method is not developed for more complex models, which might be fitted using, e.g., the gls() or the lme() functions. In Chap. 20, however, we demonstrate how to overcome this shortcoming.

Table 5.5 *R syntax*: Extracting results from the `lm.fit` and `gls.fit` model-fit objects obtained using the `lm()` and `gls()` functions, respectively

Model fit component to be extracted	Function: `lm()` Package: **stats** Object: `lm.fit` Class: *lm*	Function: `gls()` Package: **nlme** Object: `gls.fit` Class: *gls*
Summary	`(summ <- summary(lm.fit))`	`(summ <- summary(gls.fit))`
Est. method		`gls.fit$method`
$\widehat{\boldsymbol{\beta}}$	`coef(lm.fit)`	`coef(gls.fit)`
$\widehat{\boldsymbol{\beta}}$, se($\widehat{\boldsymbol{\beta}}$), t-test	`coef(summ)`	`coef(summ)`
$\widehat{\text{Var}}(\widehat{\boldsymbol{\beta}})$	`vcov(lm.fit)`	`vcov(gls.fit)`

95% CI for $\boldsymbol{\beta}$	`confint(lm.fit)`	`confint(gls.fit)`
		`intervals(gls.fit,`
		`which="coef")`

$\widehat{\sigma}$	`summ$sigma`	`summ$sigma`
95% CI for σ		`intervals(gls.fit,` `which="var-cov")`
ML value	`logLik(lm.fit)`	`logLik(gls.fit, REML = FALSE)`
REML value	`logLik(lm.fit, REML=TRUE)`	`logLik(gls.fit, REML=TRUE)`
AIC	`AIC(lm.fit)`	`AIC(gls.fit)`
BIC	`BIC(lm.fit)`	`BIC(gls.fit)`
Fitted values	`fitted(lm.fit)`	`fitted(gls.fit)`
Raw residuals	`residuals(lm.fit,` `type="response")`	`residuals(gls.fit,` `type="response")`
Predicted	`predict(lm.fit, newdata)`	`predict(gls.fit, newdata)`
R-call	`(cl <- getCall(lm.fit))`	`(cl <- getCall(gls.fit))`
Formula for mean	`(form <- formula(lm.fit))`	`(form <- formula(gls.fit))`
Data name	`(df.name <- cl$data)`	`(df.name <- cl$data)`
Data frame	`eval(df.name)`	`eval(df.name)`
Model frame	`(mf <- model.frame(lm.fit))`	`mfDt <- getData(gls.fit)` `(mf <- model.frame(form, mfDt))`
Design matrix	`model.matrix(lm.fit)` `model.matrix(form, mf)`	`model.matrix(form, mf)`

5.6 Tests of Linear Hypotheses for Fixed Effects

Results of the hypothesis tests based on a fitted LM (Sect. 4.6) can be accessed in several ways. For each single estimated parameter of the model, the t-test, defined in (4.37), is provided by default if the result of the generic function `summary()` is displayed. Note that the provided results are for marginal tests (Sect. 4.7.1). Results of the t-tests can also be obtained using the `coef()` extractor function applied to object created by `summary()` function (Table 5.5).

Results of the F-tests (4.35) for continuous covariates and groups of contrasts corresponding to factors included in the model are obtained with the use of the generic `anova()` function. Note that, by default, results of the sequential testing approach (Sect. 4.7.1) are provided. The order of the tests is based on the order of terms in the expanded formula (Sect. 5.2.1). In the case of model-fit objects of class *gls*, the marginal F-tests can be obtained using the `type="marginal"` argument of the `anova()` function.

To conduct tests for user-defined linear hypotheses of the type defined in (4.30), the arguments `Terms` or `L` can be used when calling the `anova()` function. The former is applied to specify an integer or a character vector, which indicates the terms of the model, effects of which should be jointly tested to be equal to zero. Alternatively, the argument `L` can be used to specify a numeric vector or an array, which indicates the linear combinations of the coefficients of the model that should be tested to be equal to zero. Note that arguments `type`, `Terms`, and `L` are available only when the `anova()` function is applied to a model-fit object of class *gls*.

The `anova()` function can also be applied to more than one model-fit object. In that case, when applied to model-fit object of class *gls* it provides the LR tests for nested models (Sect. 4.6.1), as well as the values of AIC and BIC for each of the models (Sect. 4.7.2). The latter criteria can be used to choose the model with the best fit from a set of nonnested models.

5.7 Chapter Summary

The review of the theory of the classical LMs, presented in Chap. 4, allowed us introducing the basic ideas and tools available in R to fit the LMs in the current chapter. Many of these ideas and tools, like model formula, model frame, or model-fit extraction methods, are also utilized when fitting more complex models, including LMMs.

In Sect. 5.2, we introduced a two-sided R formula, typically used to specify the mean structure of the model. Numerous examples of two-sided R formulae were given in Sect. 5.2.1. Also, an important concept of the *expanded* formula was introduced (Table 5.2). We observed remarkable flexibility in defining different terms in a formula. In Sect. 5.2.2, we presented a more technical representation of the model formula using an object of the *terms* class.

It should be noted that, although all the formulae, defined in Panels R5.1–R5.3, are correct from a syntax point of view, not all of them may be useful in the context of a particular study. Ultimately, the responsibility to specify a meaningful, valid model, which correctly reflects the study design and allows answering specific research questions, lies with the researcher. Examples of study design considerations to be taken into account include, but are not limited to, proper crossing and/or nesting of factors.

It is also worth noting that the code used in Sect. 5.2 requires neither the variables nor functions used in formulae to be available for computations. Consequently, in the context of Sect. 5.2, it is not relevant, for instance, which of the `x1, x2, x3, f1, f2`, and `f3` covariates are continuous or factors. Generally speaking, we can say that the specification of formulae and creation of objects of the *terms* class are about *symbolic* operations, and *not* about numeric calculations.

The process of building the design matrix from a formula *and* data, common to many modeling functions in R, was described in Sect. 5.3 and summarized in Fig. 5.1. Broadly speaking, the process involves the following steps: expansion of a formula; creation of an object of class *model.frame*; and creation of the design matrix itself. In contrast to the syntax shown in Sect. 5.2, the code presented in Sect. 5.3 requires that the variables and functions are available for evaluation, so that the design matrix can be created.

In Sect. 5.4, we discussed the arguments of functions `lm()` and `gls()` that are available for fitting classical LMs in R. Finally, Sect. 5.5 introduced methods that are used for extracting results of a fitted model, while in Sect. 5.6 we described the tools that are available for inference.

We would like to stress that, ultimately, the detailed information about all of the aspects of the R syntax can be obtained from R's help system. In our description of the syntax, we focused on the most important and/or most often used features, functions, and arguments. Our goal was to provide extra insight that might increase the understanding of these tools and facilitate their use. This, in turn, should help in using the syntax necessary for fitting more complex models, including LMMs.

Chapter 6
ARMD Trial: Linear Model with Homogeneous Variance

6.1 Introduction

In this chapter, we illustrate the use of the R tools, described in Sects. 5.2–5.5. We apply them to fit an LM with independent, homoscedastic residual errors to the visual acuity measurements from the ARMD dataset. Note that the model is considered for software illustration purposes only. In view of the structure of the data and of the results of the exploratory analysis presented in Sect. 3.2, the assumptions of the independence and homoscedasticity of the visual acuity measurements are not correct. More advanced LMs, which properly take into account the structure of the data and do not require these assumptions, will be presented in Chaps. 12 and 16.

The chapter is structured as follows. In Sect. 6.2, we specify an LM with independent, homoscedastic residual errors. The model is fitted to the data using the function `lm()` in Sect. 6.3. Section 6.4 presents an alternative way of fitting the model with the use of the function `gls()`.

6.2 A Linear Model with Independent Residual Errors with Homogeneous Variance

We consider the following model for the visual acuity data:

$$\text{VISUAL}_{it} = \beta_{0t} + \beta_1 \times \text{VISUAL0}_i + \beta_{2t} \times \text{TREAT}_i + \varepsilon_{it}. \tag{6.1}$$

In the model specified in (6.1), VISUAL_{it} is the value of visual acuity measured for patient i ($i = 1, \ldots, 234$) at time t ($t = 1, 2, 3, 4$, corresponding to values of 4, 12, 24, and 52 weeks, respectively). In the explanatory (fixed) part of the model, VISUAL0_i is the baseline value of visual acuity, and TREAT_i is the treatment indicator (equal 1 for the active group and 0 otherwise), with β_{0t}, β_1, and β_{2t} denoting the timepoint-specific intercept, baseline visual acuity effect, and timepoint-specific treatment

A. Gałecki and T. Burzykowski, *Linear Mixed-Effects Models Using R: A Step-by-Step Approach*, Springer Texts in Statistics, DOI 10.1007/978-1-4614-3900-4_6,

effect, respectively. Thus, the model assumes a time-dependent treatment effect, with the time variable being treated as a factor. In what follows, we will be referring to the model, specified in (6.1), as model **M6.1**. The design matrix for the model is presented later in Panel R6.1.

Finally, the random part of the model includes a residual random error ε_{it}. Following the specification of the classical LM, defined in (4.1) and (4.2), we assume that ε_{it} is normally distributed with mean 0 and constant variance σ^2. Moreover, we assume that the errors at different timepoints are also independent. Obviously, these assumptions are not correct, as mentioned in Sect. 6.1 and implied by the matrix in Fig. 3.4. Thus, the analysis should be viewed only as an illustration of the use of the R functions.

6.3 Fitting a Linear Model Using the `lm()` Function

As mentioned in Sect. 5.4, the main function used to fit linear regression models with independent, homoscedastic errors in R is `lm()`. The use of the function typically involves a call like `lm(`*`formula`*`,`*`data`*`)`, where *`formula`* specifies the model we want to fit, and the *`data`* argument indicates the data frame containing the variables needed to fit the model.

In Panel R6.1, we demonstrate how to construct the design matrix for model **M6.1** based on a formula and the data used to fit the model.

The object `lm1.form` specifies the R formula (Sect. 5.2.1) corresponding to the mean structure, defined in (6.1). The formula defines `visual` as the dependent variable. The variables, used in the `formula()`-function call, were described in Sect. 2.2.2. In model **M6.1**, a separate treatment effect for each measurement occasion is specified. This is reflected in the formula `lm1.form` by including in its explanatory part, to the right of the ~ sign, time as a factor `time.f` and by adding an interaction of the time factor with treatment, `time.f:time.f`. To obtain timepoint-specific intercepts, the overall intercept is removed from the model by specifying the `-1` term (see Panel R5.2). Using this parameterization, the intercepts will be provided by the coefficients corresponding to the levels of `time.f`.

In the remainder of Panel R6.1, we present selected features of the design matrix corresponding to the formula `lm1.form` evaluated in the context of the `armd` data. The matrix contains nine columns: one for `visual0`, four (because the model does not include an intercept) for the four levels of `time.f`, and four for the four levels of the `time.f:treat.f` interaction. To simplify the display of the matrix, we abbreviate the names of the columns with the help of the `abbreviate()` function. Using the `attr()` function, we check the value of the `contrasts` attribute (Sect. 5.3.2) of the matrix. We find that, for the ordered factor `time.f`, the polynomial contrasts were used (see Panel R5.8). As mentioned in Sect. 5.3.2, these are the default contrasts for ordered factors. On the other hand, for the `treat.f` factor, the contrast matrix, defined by the (default) function `contr.treatment()`, was used (Sect. 5.3.2). The reference level for `treat.f` is

R6.1 *ARMD Trial*: The design matrix for the linear model **M6.1**

```
> lm1.form <-                          # Fixed effects formula:(6.1)
+     formula(visual ~ -1 + visual0 + time.f + treat.f:time.f )
> vis.lm1.mf <- model.frame(lm1.form, armd)         # Model frame
> vis.lm1.dm <- model.matrix(lm1.form, vis.lm1.mf) # Design matrix X
> dim(vis.lm1.dm)                      # Dimensions
  [1] 867   9
> (nms <- colnames(vis.lm1.dm)) # Long column names ...
  [1] "visual0"                    "time.f4wks"
  [3] "time.f12wks"                "time.f24wks"
  [5] "time.f52wks"                "time.f4wks:treat.fActive"
  [7] "time.f12wks:treat.fActive" "time.f24wks:treat.fActive"
  [9] "time.f52wks:treat.fActive"
> nms <- abbreviate(nms)               # ... abbreviated
> colnames(vis.lm1.dm) <- nms          # ... assigned.
> head(vis.lm1.dm, n = 6)              # X matrix. Six rows.
    vsl0 tm.4 tm.12 tm.24 tm.52 t.4: t.12: t.24: t.52:
  2   59    1     0     0     0    1     0     0     0
  3   59    0     1     0     0    0     1     0     0
  5   65    1     0     0     0    1     0     0     0
  6   65    0     1     0     0    0     1     0     0
  7   65    0     0     1     0    0     0     1     0
  8   65    0     0     0     1    0     0     0     1
> attr(vis.lm1.dm, "contrasts") # Contrasts attribute.
  $time.f
                .L       .Q        .C
  4wks   -0.522167  0.56505 -0.397573
  12wks  -0.302307 -0.16233  0.795147
  24wks   0.027482 -0.73674 -0.454369
  52wks   0.796992  0.33403  0.056796

  $treat.f
  [1] "contr.treatment"
> contrasts(armd$treat.f)              # Contrasts for treat.f
          Active
  Placebo      0
  Active       1
```

"Placebo" (Sect 2.2.2); thus, the interaction terms indicate the effect of the "Active" treatment, as compared to "Placebo", at different timepoints. Note that, because the intercept is removed from the formula, the values of the coefficients for polynomial contrasts are *not* used in the design matrix.

Panel R6.2 illustrates how to fit model **M6.1** using the lm() function. In Panel R6.2a, the formula lm1.form is used as an argument in the lm()-function call.

Table 6.1 *ARMD Trial*: The `lm()` and `gls()` estimates (with standard errors in parentheses) for model **M6.1**. For σ, the 95% confidence interval is provided

	Parameter	lm6.1	fm6.1
Panel with syntax		R6.2	R6.3
Object's class		*lm*	*gls*
Est. method		OLS[a]	REML
Log-REML value			−3400.81
Fixed effects:			
Visual acuity at $t = 0$	β_1	0.83(0.03)	0.83(0.03)
Time (4wks)	β_{01}	8.08(1.94)	8.08(1.94)
Time (12wks)	β_{02}	7.08(1.94)	7.08(1.94)
Time (24wks)	β_{03}	3.63(1.95)	3.63(1.95)
Time (52wks)	β_{04}	−1.75(1.99)	−1.75(1.99)
Tm(4wks):Trt(Actv)	β_{21}	−2.35(1.63)	−2.35(1.63)
Tm(12wks):Trt(Actv)	β_{22}	−3.71(1.64)	−3.71(1.64)
Tm(24wks):Trt(Actv)	β_{23}	−3.45(1.69)	−3.45(1.69)
Tm(52wks):Trt(Actv)	β_{24}	−4.47(1.78)	−4.47(1.78)
Scale	σ	12.38	12.38(11.82,12.99)

[a]OLS using the QR decomposition (Golub and Van Loan, 1989)

The resulting model fit is stored in the list-object `lm6.1`. The contents of the object are accessed with the help of the generic function `summary()`. The function returns a list with components containing the information about the model fit. In our case, the output produced by the function is stored in the object named `summ`. The default display of the contents of the object is too long, and therefore it is not printed to save space. Instead, we show selected components of the object `summ`. The names of all the components of the `summ` object (not shown) can be obtained by issuing the command `names(summ)`. Additional results can be found in Table 6.1.

In particular, by applying the `coef()` function (Sect. 5.5), we extract the estimated coefficients, their standard errors, values of the t-test statistics, and p values, and store them in the matrix-object `tT`. To obtain a more compact printout, we abbreviate the names of the rows of object `tT`, and then print the object using the `printCoefmat()` function. The use of the argument `P.values=TRUE` formats the values from the last column of the matrix as p values. From the printout we can conclude that the estimated coefficients for the `time.f:time.f` interaction indicate a negative effect of the "Active" treatment.

By issuing the command `summ$sigma`, we display the estimate of the residual standard deviation stored in the `sigma` component of the `summ` object.

In Panel R6.2b, we test the hypothesis of whether there is an overall treatment effect using the `anova()` function. As mentioned in Sect. 5.6, when the function is applied to a single model-fit object created by the `lm()` function, it provides results of the sequential-approach F-tests (Sect. 4.7.1) for the continuous covariates and groups of contrasts corresponding to the factors included in the model. Thus, the printout in Panel R6.2b shows the results of three F-tests: for the baseline visual

R6.2 *ARMD Trial*: The linear model **M6.1**, fitted using the lm() function. The formula-object lm1.form was defined in Panel R6.1

(a) *Model fit and parameter estimates*

```
> lm6.1 <- lm(lm1.form, data = armd)                # M6.1:(6.1)
> summ <- summary(lm6.1)                              # Summary
> tT <- coef(summ)                                    # β, se(β), t-test
> rownames(tT)                                        # Fixed effects (β) names
 [1] "visual0"                     "time.f4wks"
 [3] "time.f12wks"                 "time.f24wks"
 [5] "time.f52wks"                 "time.f4wks:treat.fActive"
 [7] "time.f12wks:treat.fActive"   "time.f24wks:treat.fActive"
 [9] "time.f52wks:treat.fActive"
> rownames(tT) <- abbreviate(rownames(tT))            # Abbreviated β names
> printCoefmat(tT, P.values = TRUE)
       Estimate Std. Error t value Pr(>|t|)
vsl0     0.8304     0.0284   29.21  < 2e-16
tm.4     8.0753     1.9434    4.16  3.6e-05
tm.12    7.0807     1.9407    3.65  0.00028
tm.24    3.6302     1.9532    1.86  0.06342
tm.52   -1.7464     1.9895   -0.88  0.38029
t.4:    -2.3528     1.6289   -1.44  0.14900
t.12:   -3.7085     1.6438   -2.26  0.02432
t.24:   -3.4492     1.6940   -2.04  0.04205
t.52:   -4.4735     1.7781   -2.52  0.01206
> summ$sigma                                          # σ
 [1] 12.376
```

(b) *Sequential-approach F-tests*

```
> anova(lm6.1)                                      # ANOVA table
 Analysis of Variance Table

 Response: visual
                 Df  Sum Sq Mean Sq  F value Pr(>F)
 visual0          1 2165776 2165776 14138.99 <2e-16
 time.f           4   14434    3608    23.56 <2e-16
 time.f:treat.f   4    2703     676     4.41 0.0016
 Residuals      858  131426     153
```

acuity visual0, for time factor time.f, and for the interaction time.f:treat.f. Note that the square of the value of the t-test statistic for visual0, $29.213^2 = 853.43$, does not equal the value of the F-test statistic, 14,138.99. This is because, as mentioned in Sect. 5.6, the results of the t-tests provided by the summary() function, pertain to the marginal tests. Meanwhile, the results of F-tests, produced

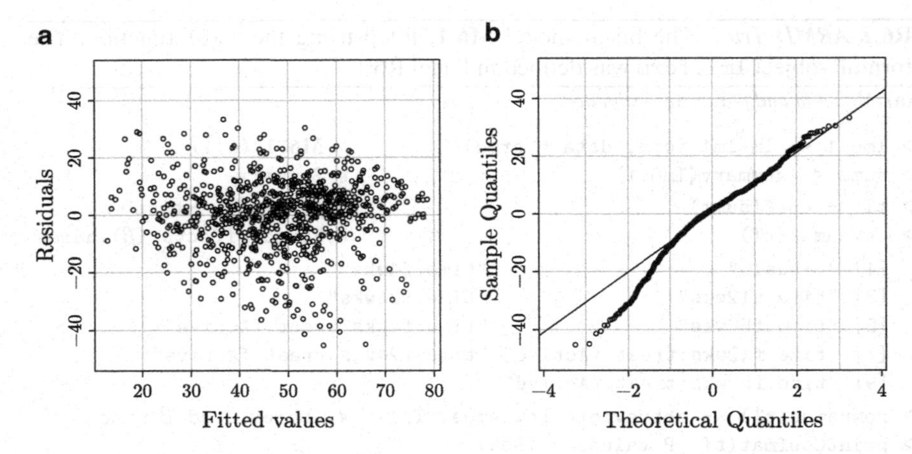

Fig. 6.1 *ARMD Trial*: Raw residuals for model **M6.1** (**a**) Residuals *versus* fitted values (**b**) Normal Q-Q plot

by the `anova()` function, are by default for the sequential approach (Sect. 4.7.1). Thus, for example, the t-test for `visual0` assumes that `visual0`, `time.f` and `time.f:treat.f`, are included in the alternative model, while the F-test assumes that no other terms besides `visual0` are present.

Note that the p value for the F-test for `time.f:treat.f` indicates a statistically significant result of the test at the 5% significance level. This suggests the presence of a time-varying treatment effect. As mentioned earlier, the negative point estimates for the interaction coefficients indicate a favorable placebo effect that increases over time. However, as we already mentioned, the model does not take into account the correlation between the visual acuity observations obtained from the same subject. It also does not take into account the heterogeneous variability present at different timepoints. Thus, it should not be used as a basis for inference.

The misspecification of model **M6.1** is reflected in Fig. 6.1 containing residual plots (Sect. 4.5.2). The scatterplot of raw residuals *versus* fitted values, presented in Fig. 6.1a, is obtained using the following traditional graphics commands:

```
> plot(fitted(lm6.1), resid(lm6.1))      # Fig. 6.1a
> abline(h = seq(-40, 40, by = 20), col = "grey")
> abline(v = seq( 10, 80, by = 10), col = "grey")
```

The generic functions `fitted()` and `resid()` (alias for `residuals()`) (Sect. 5.5) extract raw residuals (default) and fitted values, respectively, from the model-fit object `lm6.1`. The generic function `plot()` plots the residuals against the fitted values, i.e., against the estimated values of the linear predictor specified on the right-hand side of the model equation (6.1). The (vertical) width of the scatterplot clearly increases with increasing fitted values, which implies a nonconstant residual variance.

A normal Q–Q plot of the raw residuals in Fig. 6.1b is obtained by issuing the following commands:

```
> qqnorm(resid(lm6.1)); qqline(resid(lm6.1)) # Fig. 6.1b
```

The function `qqnorm()` creates the normal Q–Q plot, while the function `qqline()` adds a line passing through the first and third quantiles of the coordinates. The shape of the plot clearly deviates from a straight line. This may be an indication of a problem with the normality of the residuals. However, it may also be the effect of ignored heteroscedasticity and/or correlation of the visual acuity measurements. In any case, both the scatterplot in Fig. 6.1a and the Q–Q plot in Fig. 6.1b indicate problems with the fit of model **M6.1**.

6.4 Fitting a Linear Model Using the gls() Function

The `lm()` function fits LMs assuming independence between the observations. More flexible LMs, which allow for dependence between the observations, can be fitted using the `gls()` function. Of course, the function can also be used to fit the classical LM with independent observations, as described in Sect. 5.4. In Panel R6.3, we illustrate its use for this purpose by fitting model **M6.1**.

Toward this end, we first need to attach the **nlme** package. Model **M6.1** can then be fitted using the `gls()` function. The syntax is similar to that used for the `lm()` function. By default, `gls()` provides the REML estimates, discussed in Sect. 4.4.3. To obtain the ML estimates, the argument `method="ML"` should be used instead (see Sect. 5.4).

The results of the model fitted using the function `gls()` can be accessed in a variety of ways, as was explained in Sect. 5.5. The simplest way is to print the object itself or to use the `summary(fm6.1)` command (printouts not shown). By applying the function `intervals()`, as shown in Panel R6.3, confidence intervals can be constructed for all fixed-effects, $\boldsymbol{\beta}$, and parameter σ. Confidence intervals for the fixed-effects coefficients are obtained using the t-distribution with $n - p = 867 - 9 = 858$ degrees of freedom, while the interval for σ is constructed using the χ^2 distribution with 858 degrees of freedom, as discussed in Sect. 4.6.2.

The scatterplot and normal Q–Q plot of residuals can be obtained using the following traditional graphics commands:

```
> plot(predict(fm6.1), residuals(fm6.1))   # Same as Fig. 6.1a
> qqnorm(residuals(fm6.1))                 # Same as Fig. 6.1b
> qqline(residuals(fm6.1))
```

The resulting plots are not displayed; they are obviously identical to those presented in Figs. 6.1a and 6.1b.

The most interesting feature of the function `gls()` is its capability of fitting models with structured variance–covariance matrices for correlated observations.

R6.3 *ARMD Trial*: The 95% confidence intervals for fixed effects and residual standard deviation for the linear model **M6.1**, fitted using the `gls()` function. The formula-object `lm1.form` was defined in Panel R6.1

```
> require(nlme)                    # Attach nlme package
> fm6.1 <- gls(lm1.form,           # M6.1:(6.1)
+              data = armd)
> intervals(fm6.1)                 # 95% CI for β, σ
  Approximate 95% confidence intervals

   Coefficients:
                                  lower       est.     upper
  visual0                       0.77458   0.83037   0.88616
  time.f4wks                    4.26092   8.07531  11.88970
  time.f12wks                   3.27166   7.08066  10.88965
  time.f24wks                  -0.20332   3.63022   7.46376
  time.f52wks                  -5.65132  -1.74643   2.15846
  time.f4wks:treat.fActive     -5.54995  -2.35278   0.84440
  time.f12wks:treat.fActive    -6.93482  -3.70852  -0.48222
  time.f24wks:treat.fActive    -6.77400  -3.44915  -0.12430
  time.f52wks:treat.fActive    -7.96341  -4.47345  -0.98349
  attr(,"label")
  [1] "Coefficients:"

   Residual standard error:
   lower    est.  upper
  11.818 12.376 12.991
```

We will explore this feature in a more appropriate analysis of the armd data in Chaps. 9 and 12.

To conclude, we note that the results extracted from the model fits provided by objects `lm6.1` and `fm6.1` are identical, as observed in Table 6.1. This is due to the fact that for LMs with independent, homoscedastic errors, the REML estimates produced by `gls()` are exactly the same as the estimates obtained by OLS, which is used in `lm()` (see Sects. 4.4.1 and 4.4.3).

6.5 Chapter Summary

In this chapter, we illustrated the use of functions `lm()` and `gls()` to fit LMs with independent residual errors with homogeneous variance to data from the ARMD trial. In particular, we presented the main steps and tools related to the model-formula specification, creation and checking of the design matrix, extraction of results from the model-fit object, and investigation of the model fit. Similar steps and tools are used for more complex LMs that will be the focus of subsequent chapters. For fitting these models, the function `gls()` will be the primary instrument; its use for this purpose will be explored in more detail in the next chapters.

When presenting the diagnostic plots, we used traditional graphics tools. As an alternative, we could have used the tools from the **lattice** package. In fact, in subsequent chapters we will be using **lattice** more often.

Chapter 7
Linear Models with Heterogeneous Variance

7.1 Introduction

In Chap. 4, we formulated the classical LM for independent observations. The key assumptions underlying the model are that the observations are independent and normally distributed with a constant, i.e., homogeneous variance, and that the expected value of the observations can be expressed as a linear function of covariates.

We now relax the homoscedasticity assumption and allow for the observations to be heteroscedastic, i.e., to have different variances, while retaining the assumption that the observations are independent and normally distributed. We refer to this new class of models as LMs with heterogeneous variance.

In the presentation of the LMs with heterogeneous variance, we introduce important concepts of *variance function*, *WLS* estimation, *GLS* estimation, and *IRLS* estimation. These are general concepts, that are also important for more complex LMs, like those that will be described in Chaps. 10 and 13. The concepts in this chapter are introduced in a relatively simple framework, allowing for straightforward presentation.

The chapter is structured as follows. In Sects. 7.2 and 7.3, we describe the specification of the LMs with heterogeneous variance and, in particular, the use of variance functions. Sections 7.4–7.7 present estimation methods, model diagnostics, inferential tools, and model reduction and selection methods, respectively. Section 7.8 is devoted to a special class of models in which residual variance depends on the mean value. In Sect. 7.9, a summary of the chapter is offered.

In our presentation of the theoretical concepts underlying the LMs with heterogeneous variance, we focus on these that are implemented in R. The details of the implementation are discussed in Chap. 8.

A. Gałecki and T. Burzykowski, *Linear Mixed-Effects Models Using R: A Step-by-Step Approach*, Springer Texts in Statistics, DOI 10.1007/978-1-4614-3900-4_7,

7.2 Model Specification

Recall that in the classical LM with homogeneous variance, defined by (4.1)–(4.2) in Sect. 4.2, the variance $\text{Var}(y_i)$ of observation y_i of the dependent variable, displayed in (4.5), is equal to σ^2. We now relax the constant variance assumption and assume that

$$\text{Var}(y_i) = \sigma_i^2. \tag{7.1}$$

Therefore, we formulate an LM with heterogeneous variance by assuming that

$$y_i = x_i^{(1)}\beta_1 + \cdots + x_i^{(p)}\beta_p + \varepsilon_i \equiv \mathbf{x}_i'\boldsymbol{\beta} + \varepsilon_i \tag{7.2}$$

and

$$\varepsilon_i \sim \mathcal{N}(0, \sigma_i^2), \tag{7.3}$$

where ε_i are independent, i.e., ε_i is independent of $\varepsilon_{i'}$ for $i \neq i'$. Note that the fixed part of the LM, specified in (7.2), is exactly the same as that of the classical LM with homogeneous variance defined in Sect. 4.2. The only differences between the two models are different assumptions about residuals and their variance in (7.3), as compared to (4.2).

Similarly to (4.4), the model with heterogeneous variance, defined in (7.2)–(7.3), implies that

$$\text{E}(y_i) \equiv \mu_i = \mathbf{x}_i'\boldsymbol{\beta}. \tag{7.4}$$

It is important to note that the model contains in total $n+p$ parameters, including n parameters σ_i and p parameters $\boldsymbol{\beta}$. This is more than n, the number of observations. Therefore, the model is not identifiable. It may become identifiable, however, if we impose additional constraints on the residual variances $\sigma_1^2, \ldots, \sigma_n^2$.

One simple way to impose such constraints is to assume *known* variance weights. This case is described in Sect. 7.2.1. Another more general way is to represent variances more parsimoniously as a function of a small set of parameters. This can be accomplished by employing a *variance function*. The concept is introduced in Sect. 7.2.2.

7.2.1 Known Variance Weights

The simplest way to introduce heteroscedasticity and, at the same time, to reduce the number of variance parameters in model (7.2)–(7.3), is to assume that the variance of ε_i is equal to a *known* proportion of one (unknown) parameter σ^2. More specifically, we may associate with every observation i a known constant $w_i > 0$ and assume that $\text{Var}(\varepsilon_i) = \text{Var}(y_i) = \sigma^2/w_i$.

An LM with known weights is then specified as (7.2) with

$$\varepsilon_i \sim \mathcal{N}(0, \sigma^2/w_i). \tag{7.5}$$

Constants w_i are called "true" weights. The higher the weight for a given observation, the lower the variance, i.e., the more precisely recorded the value of y_i. However, in real-life applications, weights w_i are rarely known. Typically, given lack of knowledge about "true" weights, we assume that $w_i = 1$ for all observations, which means that, in fact, we assume the classical LM with homogeneous variance, defined by (4.1) and (4.2). In Sect. 7.4.1, we demonstrate that the model with known variance weights can be transformed back to an LM with homogeneous variance, defined by (4.1) and (4.2).

7.2.2 *Variance Function*

A more general and flexible way to introduce variance heterogeneity is by means of a *variance function* (Carroll and Ruppert 1988). More specifically, consider a function

$$\lambda(\boldsymbol{\delta}, \mu, \mathbf{v}),$$

which assumes positive values and is continuous and differentiable with respect to $\boldsymbol{\delta}$ for all legitimate values of $\boldsymbol{\delta}$. Note that μ is a scalar and $\boldsymbol{\delta}$ and $\mathbf{v}$ can be vectors.

We can then assume that the variance of the residual errors, i.e., $\mathrm{Var}(\varepsilon_i)$, is expressed as follows:

$$\mathrm{Var}(\varepsilon_i) = \sigma^2 \lambda^2(\boldsymbol{\delta}, \mu_i; \mathbf{v}_i), \tag{7.6}$$

where μ_i is defined in (7.4), σ is a scale parameter, $\mathbf{v}_i$ is a vector of (known) covariates defining the variance function for observation i, while the vector $\boldsymbol{\delta}$ contains a small set of variance parameters, common to all observations. Note that, because the function $\lambda(\cdot)$ on the right-hand side of (7.6) involves μ_i, it in fact depends on $\boldsymbol{\beta}$, too. However, we prefer to reflect this dependence in the notation indirectly by using μ_i, i.e., by pointing to the dependence of the variance of residual error on the mean value.

It is worth underscoring here that the parameter σ, used in (7.6), in general should be interpreted as a scale parameter. This is in contrast to the classical LM with homogeneous variance, defined in Sect. 4.2, in which σ can be interpreted as residual error standard deviation.

Note that, according to (7.6), $\lambda(\cdot)$ should, strictly speaking, be referred to as a function modeling standard deviation, not variance. However, the term variance function is commonly used when referring to $\lambda(\cdot)$, and we will follow this convention.

7.2.2.1 Specification at the Level of a Unit of Observation

For the ith observation, an LM with variance function $\lambda(\cdot)$, defined in (7.6), is specified by combining the mean structure, implied by (7.2), with the assumption that

$$\varepsilon_i \sim \mathcal{N}(0, \sigma^2 \lambda_i^2), \tag{7.7}$$

where

$$\lambda_i \equiv \lambda(\boldsymbol{\delta}, \mu_i; \mathbf{v}_i). \tag{7.8}$$

By using the variance function $\lambda(\cdot)$, we parsimoniously represent the variance σ_i^2 of ε_i, used in (7.3), as

$$\sigma_i^2 = \sigma^2 \lambda_i^2, \tag{7.9}$$

where σ^2 is an unknown scalar parameter and λ_i, as defined in (7.8), depends directly on the unknown variance parameters $\boldsymbol{\delta}$ and indirectly on $\boldsymbol{\beta}$ through μ_i.

For example, if we assume that $\lambda(\mu_i) = \mu_i$, from (7.8) it follows that $\sigma_i/\mu_i = \sigma$. Hence, in the context of this model, σ can be interpreted as a coefficient of variation.

Note that, upon taking $\lambda_i = \lambda(w_i) = 1/\sqrt{w_i}$, where w_i is an appropriately constructed variance covariate, the model specified by (7.2), (7.7), and (7.8) becomes equivalent to the model with known variance weights w_i, defined in Sect. 7.2.1.

7.2.2.2 Specification for All Data

Model equations (7.2), (7.7), and (7.8) apply to individual observations. They can be replaced by an equation accommodating all observations. Toward this end, we define a diagonal matrix $\boldsymbol{R}$:

$$\boldsymbol{R} \equiv \boldsymbol{\Lambda}\boldsymbol{\Lambda}, \tag{7.10}$$

where $\boldsymbol{\Lambda} = \text{diag}(\lambda_1, \ldots, \lambda_n)$ is a diagonal matrix, with elements defined by (7.8). By using (7.10), we can specify the model, defined by (7.2) and (7.7)–(7.8), as follows:

$$\mathbf{y} = \boldsymbol{X}\boldsymbol{\beta} + \boldsymbol{\varepsilon}, \tag{7.11}$$

where

$$\boldsymbol{\varepsilon} \sim \mathcal{N}(\mathbf{0}, \mathcal{R}), \quad \mathcal{R} = \sigma^2 \boldsymbol{R}, \tag{7.12}$$

and $\mathbf{y}$, $\boldsymbol{X}$, $\boldsymbol{\beta}$, and $\boldsymbol{\varepsilon}$ are defined as in Sect. 4.2.2. Note that (7.10)–(7.12) apply to all n observations simultaneously. This specification will become especially useful for models with correlated residual errors that will be introduced in Chap. 10.

7.3 Details of the Model Specification

In this section, we provide more details about the specification of the LMs with heterogeneous variance defined in Sect. 7.2. In particular, in Sect. 7.3.1, we provide examples of variance functions, while in Sect. 7.3.2, we discuss the issue of the unique parameterization of the functions.

7.3.1 *Groups of Variance Functions*

In this section, we present selected examples of the variance function $\lambda(\cdot)$, defined in (7.6). When referring to variance functions, we use names, such as varFixed$(\cdot)$ or varIdent$(\cdot)$, borrowed from the R syntax, which will be explained in Chap. 8.

For the sake of simplicity, variance functions $\lambda(\cdot)$ can be classified into the following four groups:

1. Known weights, $\lambda(\cdot) = \lambda(\mathbf{v})$
2. Variance functions depending on $\boldsymbol{\delta}$ but not on μ, i.e., $\lambda(\cdot) = \lambda(\boldsymbol{\delta}; \mathbf{v})$
3. Variance functions depending on $\boldsymbol{\delta}$ and μ, i.e., $\lambda(\cdot) = \lambda(\boldsymbol{\delta}, \mu; \mathbf{v})$
4. Variance functions depending on μ but not on $\boldsymbol{\delta}$, i.e., $\lambda(\cdot) = \lambda(\mu; \mathbf{v})$

In what follows, we will symbolically refer to groups 2–4 as <δ>-, <δ, μ>-, and <μ>-group, respectively.

Specification of an LM with heterogeneous variance, presented in Sect. 7.2, is very general and encompasses all four groups of variance functions. In other words, the use of a variance function from any of the aforementioned groups does not pose difficulties in terms of the model specification. However, in models involving variance functions from groups <δ, μ> or <μ>, the parameters $\boldsymbol{\beta}$ are shared by the mean and variance structures. For this reason, these models, referred to as *mean-variance* models, require different estimation approaches and inference techniques, as compared to the models involving known weights or variance functions from the <δ>-group. Consequently, the mean-variance models are discussed separately in Sect. 7.8.

Table 7.1 shows the information about the tables and sections, in which the reader can find examples of a particular group of variance functions and the description of the corresponding estimation method.

The simplest example of a variance function is varFixed$(v_i) = \sqrt{v_i}$. It belongs to the first group of variance functions, as it assigns fixed weights $w_i = 1/v_i$, determined by the values v_i $(v_i > 0)$ of a known variance covariate. Note that the function varFixed$(\cdot)$ depends on neither $\boldsymbol{\delta}$ nor μ_i. It simply implies an LM with known weights, described in Sect. 7.2.1.

Before discussing variance functions from groups <δ>, <δ, μ>, and <μ>, we note that they allow for multiple strata. To reflect this in the notation, we assume that the observations y_i are split into several strata, indexed by s $(s = 1, \ldots, S)$,

Table 7.1 A summary of the parts of Chap. 7 that contain the information about particular groups of variance functions and the corresponding estimation methods

Group	Arguments		Examples	Estimation algorithm	Section
	$\boldsymbol{\delta}$	μ_i			
Known weights	−	−	varFixed(·)	WLS	7.4.1
<δ>	+	−	Table 7.2	ML/REML	7.4.2
<δ, μ>	+	+	Table 7.3	ML/REML-based GLS	7.8.1.1
<μ>	−	+	Table 7.4	IRLS	7.8.1.2

Table 7.2 Examples of variance functions from the <δ>-group[a]

Function $\lambda(\cdot)$	λ_i	Description
varPower($\boldsymbol{\delta}; v_i, s_i$)	$\lvert v_i\rvert^{\delta_{s_i}}$	Power of a variance covariate v_i
varExp($\boldsymbol{\delta}; v_i, s_i$)	$\exp(v_i \delta_{s_i})$	Exponent of a variance covariate
varConstPower($\boldsymbol{\delta}; v_i, s_i$)	$\delta_{1,s_i} + \lvert v_i\rvert^{\delta_{2,s_i}}$	Constant plus power variance function $\delta_{1,s_i} > 0$
varIdent($\boldsymbol{\delta}; s_i$)	δ_{s_i}	Different variances per stratum $\delta_1 \equiv 1$, $\delta_s > 0$ for $s \neq 1$

[a]Function names used in the first column correspond to the names used in the package **nlme**

Table 7.3 Examples of variance functions from the <δ, μ>-group[a]

Function $\lambda(\cdot)$	λ_i	Description
varPower($\boldsymbol{\delta}, \mu_i; s_i$)	$\lvert \mu_i\rvert^{\delta_{s_i}}$	Power of $\lvert \mu_i \rvert$
varExp($\boldsymbol{\delta}, \mu_i; s_i$)	$\exp(\mu_i \delta_{s_i})$	Exponent of μ_i
varConstPower($\boldsymbol{\delta}, \mu_i; s_i$)	$\delta_{1,s_i} + \lvert \mu_i\rvert^{\delta_{2,s_i}}$	Constant plus power variance function $\delta_{1,s_i} > 0$

[a]Function names used in the first column correspond to the names used in the package **nlme**

with stratum-specific (not observation-specific) variance parameters $\boldsymbol{\delta}_s$. Further, we denote by s_i the stratum to which the ith observation belongs. Similarly to s, the index variable s_i assumes values $1, \ldots, S$.

The variance functions, presented in Table 7.2, belong to the <δ>-group. That is, they depend on a variance covariate, i.e., v_i, and on parameters $\boldsymbol{\delta} = (\delta_1, \ldots, \delta_S)$, but not on μ_i. Thus, we can refer to them as the mean-independent variance functions. Note that the function varIdent(·) is defined only for multiple strata.

In Table 7.3, we present selected examples of variance functions from the <δ, μ>-group. These functions are mean-dependent, as they depend on μ_i and on $\boldsymbol{\delta} = (\delta_1, \ldots, \delta_S)$.

Note that, by assuming a particular form of some of the variance functions shown in Table 7.3, we can obtain a function from the <μ>-group. This happens, for instance, for the varPower($\boldsymbol{\delta}, \mu_i$) or varExp($\boldsymbol{\delta}, \mu_i$), if we assume $\delta_{s_i} \equiv 1$ (or any other constant). Examples of functions from the <μ>-group are given in Table 7.4.

Table 7.4 Examples of variance functions from the <μ>-group[a]

Function $\lambda(\cdot)$	λ_i	Description
varPower$(\mu_i; s_i, \boldsymbol{\delta})$	$\lvert\mu_i\rvert^{\delta_{s_i}}$	Power of $\lvert\mu_i\rvert$, δ_{s_i} known
varExp$(\mu_i; s_i, \boldsymbol{\delta})$	$\exp(\mu_i \delta_{s_i})$	Exponent of μ_i, δ_{s_i} known
varConstPower$(\mu_i; s_i, \boldsymbol{\delta})$	$\delta_{1,s_i} + \lvert\mu_i\rvert^{\delta_{2,s_i}}$	Constant plus power variance function, $\delta_{1,s_i} > 0$, δ_{1,s_i} and δ_{2,s_i} known

[a]Function names used in the first column correspond to the names used in the package **nlme**

Variance functions allow for the modeling of many patterns of heteroscedasticity. For example, by using varPower$(\mu_i; s_i, \boldsymbol{\delta})$ from the <μ>-group with $\delta_{s_i} \equiv 1$, we obtain $\lambda(\mu_i) = \mu_i$ or, equivalently, $\sigma_i = \sigma\mu_i$. Thus, we get a model with a constant coefficient of variation σ.

In Sect. 8.2, we will demonstrate how the various variance functions are represented in the **nlme** package.

7.3.2 Aliasing in Variance Parameters

Note that, in the definition of the varIdent$(\cdot)$ variance function in Table 7.2, the constraint $\delta_1 \equiv 1$ was introduced. Without the constraint, the LM with heterogeneous variance, specified by (7.2), (7.7), and (7.8), and the varIdent$(\cdot)$ variance function with more than one stratum, would not be identifiable. This is because a fixed set of variances $(\sigma_1^2, \ldots, \sigma_S^2)$ can be represented as

$$(\sigma_1^2, \ldots, \sigma_S^2) = \sigma^2(\delta_1^2, \ldots, \delta_S^2) \tag{7.13}$$

upon defining

$$(\delta_1, \ldots, \delta_S) \equiv \left\{\frac{\sigma_1}{\sigma}, \ldots, \frac{\sigma_S}{\sigma}\right\}.$$

However, (7.13) can be equivalently represented, for $\sigma' \neq \sigma$, as

$$(\sigma_1^2, \ldots, \sigma_S^2) = (\sigma')^2\left\{(\delta_1')^2, \ldots, (\delta_S')^2\right\},$$

where

$$(\delta_1', \ldots, \delta_S') \equiv \left\{\frac{\sigma_1}{\sigma'}, \ldots, \frac{\sigma_S}{\sigma'}\right\}.$$

It follows that the representation (7.13) is not unique. To make it unique, constraints need to be imposed. A possible constraint is to assume, as it was done in Table 7.2 and as it is done by default in R, that $\delta_1 \equiv 1$. Under this constraint, the representation (7.13) holds uniquely with $\sigma^2 \equiv \sigma_1^2$ and

$$\boldsymbol{\delta} = (\delta_1, \ldots, \delta_S) \equiv \left\{1, \frac{\sigma_2}{\sigma_1}, \ldots, \frac{\sigma_S}{\sigma_1}\right\}. \tag{7.14}$$

In this way, the varIdent($\cdot$) variance function can be parameterized to define an LM with different residual variances for different strata:

$$\mathrm{Var}(\varepsilon_{s_i}) = \sigma^2_{s_i} = \sigma^2 \delta^2_{s_i}. \tag{7.15}$$

Note that, in this case, σ can be interpreted as residual standard deviation in the stratum $s = 1$.

7.4 Estimation

The parameters of the models specified in Sects. 7.2.1 and 7.2.2 can be estimated by using various approaches. Those depend, for example, on the form of the variance function, given in (7.6). More specifically, and as alluded to when presenting Table 7.1, different estimation methods are used for models specified by using variance functions from different groups.

In Sect. 7.4.1, we present the methods used to estimate the parameters of the model with known weights, which was specified in Sect. 7.2.1. Section 7.4.2 presents estimation methods for models defined by using a variance function from the <δ>-group. Discussion of the estimation approaches for the mean-variance models, i.e., models employing variance functions from the <δ, μ>- or <μ>-groups, is deferred until Sect. 7.8.1.2. Section 7.4.3 discusses an alternative parameterization of the variance function in LMs with heterogeneous variance, which is more suitable for numerical optimization. Finally, the uncertainty in parameter estimates is addressed in Sect. 7.4.4.

It is worth noting that (7.9) and (7.12), used in the specification of the LM with heterogeneous variance, imply a somewhat special role for σ^2, as compared to the $\boldsymbol{\delta}$ parameters. Namely, σ can be thought of as a scale parameter. On the other hand, the parameters $\boldsymbol{\delta}$ provide information about the relative magnitude of variation for different observations. Consequently, in some estimation approaches, more focus is given to the estimation of $\boldsymbol{\beta}$ and $\boldsymbol{\delta}$. Such approaches are actually used in R, and we will therefore primarily focus on these. More details about alternative approaches are available, for instance, in Verbeke and Molenberghs (2000).

7.4.1 *Weighted Least Squares*

From an estimation point of view, models with known variance weights (Sect. 7.2.1), which involve a variance function from the first group, do not impose any additional computational difficulties, as compared to the classical LM, described in Sect. 4.2. This is because, by multiplying both sides of (7.1) by $w_i^{1/2}$, we transform the known-weights LM back to the framework of the LM with homogeneous error variance, introduced in Sect. 4.2. More specifically, the transformed model can be written as

$$w_i^{1/2} y_i = w_i^{1/2} x_i^{(1)} \beta_1 + \ldots + w_i^{1/2} x^{(p)} \beta_p + w_i^{1/2} \varepsilon_i. \tag{7.16}$$

Note that, in the transformed model, the linearity with respect to $\boldsymbol{\beta}$ and independence of residual errors are maintained. Moreover, the variance of the transformed residual error is constant, i.e., $\text{Var}(w_i^{1/2}\varepsilon_i) = \sigma^2$.

Consequently, the estimates of $\boldsymbol{\beta}$ are obtained by minimization, with respect to $\boldsymbol{\beta}$, of a weighted residual sum of squares:

$$\sum_{i=1}^{n} w_i (y_i - \mathbf{x}_i' \boldsymbol{\beta})^2. \tag{7.17}$$

Explicit formulae for estimators of $\boldsymbol{\beta}$ and σ^2, built upon (4.12) and (4.13), are as follows:

$$\widehat{\boldsymbol{\beta}}_{\text{WLS}} \equiv \left(\sum_{i=1}^{n} w_i \mathbf{x}_i \mathbf{x}_i' \right)^{-1} \sum_{i=1}^{n} w_i \mathbf{x}_i y_i, \tag{7.18}$$

$$\widehat{\sigma}^2_{\text{WLS}} \equiv \sum_{i=1}^{n} w_i (y_i - \mathbf{x}_i' \widehat{\boldsymbol{\beta}}_{\text{WLS}})^2 / (n-p). \tag{7.19}$$

These are called WLS estimators.

7.4.2 Likelihood Optimization

In this section, we consider the model, defined by (7.2), (7.7), and (7.8), with the variance function $\lambda(\cdot)$ belonging to the <δ>-group, i.e.,

$$\lambda_i = \lambda(\boldsymbol{\delta}; \mathbf{v}_i). \tag{7.20}$$

Note that, compared to the general definition of $\lambda(\cdot)$, given by (7.6), we consider variance functions that depend on the vector of variance parameters $\boldsymbol{\delta}$ and on the vector of (known) covariates $\mathbf{v}_i$, but *not* on the expected value μ_i. Examples of such functions were given in Table 7.2.

7.4.2.1 Maximum-Likelihood Estimation

In this section, we first introduce the full log-likelihood function. We then consider the log-likelihood obtained by profiling out the $\boldsymbol{\beta}$ parameters, followed by profiling out σ^2.

Log-Likelihood for $\boldsymbol{\beta}$, σ^2, and $\boldsymbol{\delta}$

The log-likelihood function for the model, specified in (7.2), (7.7), and (7.8), is given by:

$$\ell_{\text{Full}}(\boldsymbol{\beta}, \sigma^2, \boldsymbol{\delta}) = -\frac{n}{2}\log(\sigma^2) - \frac{1}{2}\sum_{i=1}^{n}\log(\lambda_i^2) - \frac{1}{2\sigma^2}\sum_{i=1}^{n}\lambda_i^{-2}(y_i - \mathbf{x}_i'\boldsymbol{\beta})^2. \tag{7.21}$$

Note that $\ell_{\text{Full}}(\boldsymbol{\beta}, \sigma^2, \boldsymbol{\delta})$ depends on $\boldsymbol{\delta}$ through λ_i, defined in (7.20). In the special case of $\lambda_i \equiv 1$, the log-likelihood (7.21) becomes equivalent to the log-likelihood (4.15) for the classical LM.

Estimates of parameters $\boldsymbol{\beta}$, σ^2, and $\boldsymbol{\delta}$ can be obtained by simultaneously maximizing the log-likelihood function with respect to these parameters. In general, however, this is a numerically complex task, which requires finding an optimum in a multidimensional parameter space. This task can be simplified by employing the so-called profile likelihood technique.

Profiling Likelihood

Profiling of a likelihood function can be done in a variety of ways. Here, we will follow the profiling approach implemented in the `gls()` function of the **nlme** package. That is, we first profile out the $\boldsymbol{\beta}$ parameters, and then, we profile out σ^2.

More specifically, assume that $\boldsymbol{\delta}$ in (7.20) is *known*. Then, maximizing (7.21) with respect to $\boldsymbol{\beta}$ for every value of $\boldsymbol{\delta}$ leads to the following functional relationship between the optimum value $\widehat{\boldsymbol{\beta}}$ and $\boldsymbol{\delta}$:

$$\widehat{\boldsymbol{\beta}}(\boldsymbol{\delta}) \equiv \left(\sum_{i=1}^{n}\lambda_i^{-2}\mathbf{x}_i\mathbf{x}_i'\right)^{-1}\sum_{i=1}^{n}\lambda_i^{-2}\mathbf{x}_i y_i. \tag{7.22}$$

By plugging (7.22) into (7.21), we obtain the following log-profile-likelihood function:

$$\begin{aligned}\ell^*_{\text{ML}}(\sigma^2, \boldsymbol{\delta}) &\equiv \ell_{\text{Full}}(\widehat{\boldsymbol{\beta}}(\boldsymbol{\delta}), \sigma^2, \boldsymbol{\delta}) \\ &= -\frac{n}{2}\log(\sigma^2) - \frac{1}{2}\sum_{i=1}^{n}\log(\lambda_i^2) - \frac{1}{2\sigma^2}\sum_{i=1}^{n}\lambda_i^{-2}r_i^2, \end{aligned}\tag{7.23}$$

where

$$r_i \equiv r_i(\boldsymbol{\delta}) = y_i - \mathbf{x}_i'\widehat{\boldsymbol{\beta}}(\boldsymbol{\delta})$$
$$= y_i - \mathbf{x}_i'\left(\sum_{i=1}^{n}\lambda_i^{-2}\mathbf{x}_i\mathbf{x}_i'\right)^{-1}\sum_{i=1}^{n}\mathbf{x}_i\lambda_i^{-2}y_i, \tag{7.24}$$

and λ_i, defined by (7.20), depends on $\boldsymbol{\delta}$. Note that we use "*" in (7.23) to indicate that $\ell^*_{\text{ML}}(\sigma^2, \boldsymbol{\delta})$ is a log-profile-likelihood function. The advantage of using the function is that it does not depend on $\boldsymbol{\beta}$. Thus, optimization of the function is performed in a parameter space of a lower dimension.

Maximization of $\ell^*_{\text{ML}}(\sigma^2, \boldsymbol{\delta})$ with respect to σ^2 for every known value of $\boldsymbol{\delta}$ leads to the following functional relationship between the optimum value $\widehat{\sigma}^2$ and $\boldsymbol{\delta}$:

$$\widehat{\sigma}^2_{\text{ML}}(\boldsymbol{\delta}) \equiv \sum_{i=1}^{n}\lambda_i^{-2}r_i^2/n, \tag{7.25}$$

where $r_i \equiv r_i(\boldsymbol{\delta})$ are defined in (7.24).

Replacing σ^2 in (7.23) with the expression on the right-hand side of (7.25) yields a log-profile-likelihood function for $\boldsymbol{\delta}$:

$$\ell^*_{\text{ML}}(\boldsymbol{\delta}) \equiv \ell^*_{\text{ML}}(\widehat{\sigma}^2_{\text{ML}}(\boldsymbol{\delta}), \boldsymbol{\delta})$$
$$= -\frac{n}{2}\log(\widehat{\sigma}^2_{\text{ML}}) - \frac{1}{2}\sum_{i=1}^{n}\log(\lambda_i^2) - \frac{n}{2}, \tag{7.26}$$

where $\widehat{\sigma}^2_{\text{ML}} \equiv \widehat{\sigma}^2_{\text{ML}}(\boldsymbol{\delta})$.

The log-profile-likelihood function $\ell^*_{\text{ML}}(\boldsymbol{\delta})$, defined in (7.26), depends on $\boldsymbol{\delta}$, but does *not* depend on $\boldsymbol{\beta}$ nor on σ^2. Therefore, maximization of the function is much easier than the maximization of (7.21) over potentially many more parameters. Compared to $\ell^*_{\text{ML}}(\sigma^2, \boldsymbol{\delta})$ in (7.23), the function $\ell^*_{\text{ML}}(\boldsymbol{\delta})$ in (7.26) depends on one parameter less.

By maximizing $\ell^*_{\text{ML}}(\boldsymbol{\delta})$ with respect to $\boldsymbol{\delta}$, we obtain an estimator $\widehat{\boldsymbol{\delta}}_{\text{ML}}$ of $\boldsymbol{\delta}$. Whether or not the estimate can be presented in a closed-form expression depends on the chosen variance function $\lambda(\cdot)$, which defines λ_i. Plugging $\widehat{\boldsymbol{\delta}}_{\text{ML}}$ into (7.22) and (7.25) yields estimators $\widehat{\boldsymbol{\beta}}_{\text{ML}}$ and $\widehat{\sigma}^2_{\text{ML}}$ of $\boldsymbol{\beta}$ and σ^2, respectively:

$$\widehat{\boldsymbol{\beta}}_{\text{ML}} \equiv \widehat{\boldsymbol{\beta}}(\widehat{\boldsymbol{\delta}}_{\text{ML}}) = \left(\sum_{i=1}^{n}\widehat{\lambda}_i^{-2}\mathbf{x}_i\mathbf{x}_i'\right)^{-1}\sum_{i=1}^{n}\widehat{\lambda}_i^{-2}\mathbf{x}_iy_i, \tag{7.27}$$

$$\widehat{\sigma}^2_{\text{ML}} \equiv \widehat{\sigma}^2_{\text{ML}}(\widehat{\boldsymbol{\delta}}_{\text{ML}}) = \sum_{i=1}^{n}\widehat{\lambda}_i^{-2}\widehat{r}_i^2/n, \tag{7.28}$$

where $\widehat{\lambda}_i \equiv \lambda(\widehat{\boldsymbol{\delta}}_{\text{ML}}; \mathbf{v}_i)$ and $\widehat{r}_i \equiv r_i(\widehat{\boldsymbol{\delta}}_{\text{ML}})$ is defined in (7.24).

Similarly to the observation made in Sect. 4.4.2, the maximum-likelihood estimator $\widehat{\sigma}^2_{\text{ML}}$ of σ^2, obtained from the maximization of (7.26), is biased. In fact, the same comment applies to $\widehat{\boldsymbol{\delta}}_{\text{ML}}$. This is because neither of the two estimators adjusts for the uncertainty in the estimation of $\boldsymbol{\beta}$. For this reason, and similar to the case of the classical LM described in Sect. 4.4.2, σ^2 and $\boldsymbol{\delta}$ are preferably estimated by using the REML method. This is especially important for a small sample size. We will now describe this estimation approach.

7.4.2.2 Restricted Maximum-Likelihood Estimation

The idea of the REML estimation for the models, defined by (7.2), (7.7), and (7.8), and a variance function belonging to the $<\delta>$-group, is similar to the one used in the case of the classical LM for independent observations (Sect. 4.4.3). That is, to obtain unbiased estimates of σ^2 and $\boldsymbol{\delta}$, we should use an estimation approach that is orthogonal to the estimation of $\boldsymbol{\beta}$. This can be done by considering the likelihood function of a set of $n-p$ independent contrasts of $\mathbf{y}$ (Verbeke and Molenberghs 2000, p. 43–46). The resulting log-restricted-likelihood function is given by

$$\ell_{\text{REML}}(\sigma^2,\boldsymbol{\delta}) \equiv -\frac{n-p}{2}\log(\sigma^2) - \frac{1}{2}\sum_{i=1}^{n}\log(\lambda_i^2) - \frac{1}{2\sigma^2}\sum_{i=1}^{n}\lambda_i^{-2}r_i^2$$
$$-\frac{1}{2}\log\left[\det\left(\sum_{i=1}^{n}\lambda_i^{-2}\mathbf{x}_i\mathbf{x}_i'\right)\right], \tag{7.29}$$

with $\det(\mathbf{A})$ denoting the determinant of matrix $\mathbf{A}$ and r_i defined in (7.24).

We may profile out σ^2 from $\ell_{\text{REML}}(\cdot)$ by observing that, for a known value of $\boldsymbol{\delta}$, the function is maximized by

$$\widehat{\sigma}^2_{\text{REML}}(\boldsymbol{\delta}) \equiv \sum_{i=1}^{n}\lambda_i^{-2}r_i^2/(n-p). \tag{7.30}$$

By plugging (7.30) into (7.29), we obtain a log-profile-restricted-likelihood function that depends only on $\boldsymbol{\delta}$:

$$\ell^*_{\text{REML}}(\boldsymbol{\delta}) \equiv \ell_{\text{REML}}(\widehat{\sigma}^2_{\text{REML}}(\boldsymbol{\delta}),\boldsymbol{\delta}). \tag{7.31}$$

By maximization of (7.31) with respect to $\boldsymbol{\delta}$, we obtain an estimator $\widehat{\boldsymbol{\delta}}_{\text{REML}}$ of $\boldsymbol{\delta}$. Note that the resulting $\widehat{\boldsymbol{\delta}}_{\text{REML}}$ is also used in (7.22) to yield an estimator $\widehat{\boldsymbol{\beta}}_{\text{REML}}$ of $\boldsymbol{\beta}$.

It is worth noting that the log-restricted-likelihood function $\ell_{\text{REML}}(\cdot)$ is not designed to obtain estimates of $\boldsymbol{\beta}$. Toward this end, the formula (7.22), obtained for the ML estimation, is used instead. It is important to stress that, although the same formula, (7.22), is used to obtain the estimator $\widehat{\boldsymbol{\beta}}_{\text{REML}}$ of $\boldsymbol{\beta}$, the estimator is different from $\widehat{\boldsymbol{\beta}}_{\text{ML}}$. This is because the ML estimator $\widehat{\boldsymbol{\beta}}_{\text{ML}}$ results from the use of the ML estimator $\widehat{\boldsymbol{\delta}}_{\text{ML}}$ of $\boldsymbol{\delta}$, which is obtained from the maximization of (7.26), and which differs from the REML estimator $\widehat{\boldsymbol{\delta}}_{\text{REML}}$, obtained from maximizing (7.31).

7.4.3 Constrained Versus Unconstrained Parameterization of the Variance Parameters

For some variance functions like, e.g., varIdent($\cdot$) (Sect. 7.3.1), the parameters $\boldsymbol{\delta}$ are constrained to be positive. This complicates the issue of finding their estimates, as it leads to a constrained optimization problem for log-likelihood functions like, e.g., (7.31).

To overcome the problem, the optimization is performed by using an alternative, unconstrained parameterization (Pinheiro and Bates 1996). More specifically, the components δ_s of $\boldsymbol{\delta}$ are expressed as $\delta_s \equiv e^{\delta_s^*}$. Subsequently, $\delta_s^* = \log(\delta_s)$ are used as the parameters of the variance function and of the optimized log-likelihood function. Note that δ_s^* are no longer bound to be positive, which simplifies the optimization task.

Similarly, if optimization over $\sigma^2 > 0$ is required, e.g., in (7.23) and (7.29), the parameter can be replaced by its logarithm.

7.4.4 Uncertainty in Parameter Estimation

The variance-covariance matrix of $\widehat{\boldsymbol{\beta}}$ is estimated by

$$\widehat{\text{Var}}(\widehat{\boldsymbol{\beta}}) \equiv \widehat{\sigma}^2 \left(\sum_{i=1}^{n} \widehat{\lambda}_i^{-2} \mathbf{x}_i \mathbf{x}_i' \right)^{-1} = \widehat{\sigma}^2 \left(\boldsymbol{X}' \widehat{\boldsymbol{\Lambda}}^{-2} \boldsymbol{X} \right)^{-1}, \tag{7.32}$$

where $\widehat{\lambda}_i \equiv \lambda(\widehat{\boldsymbol{\delta}}; \mathbf{v}_i)$ and $\widehat{\boldsymbol{\Lambda}}$ is a diagonal matrix with $\widehat{\lambda}_i$ on the diagonal. Formula (7.32) is similar to (4.22), obtained for the classical LM. Note that $\widehat{\sigma}^2$, $\widehat{\boldsymbol{\delta}}$, and, consequently, $\widehat{\lambda}_i$ depend on the estimation method, i.e., whether WLS, ML, or REML is used.

It is worth noting that Rothenberg (1984) showed that, for models defined by (7.2), (7.7), and (7.8), with the variance function $\lambda(\cdot)$ belonging to the <δ>-group, the variance-covariance matrix of $\widehat{\boldsymbol{\beta}}$ depends, up to the second-order approximation, on the precision of the estimation of $\boldsymbol{\delta}$. Thus, especially for small sample sizes, the standard errors, obtained from (7.32), may be too small. It also follows that correct specification of the variance function may improve the precision of estimation of $\boldsymbol{\beta}$ (Sect. 7.8.2).

To assess the uncertainty in the estimates of parameters σ^2 and $\boldsymbol{\delta}$, one could adopt several approaches related to the estimation techniques described in Sect. 7.4. We focus on those that are implemented in R. Consider the two log-restricted-likelihood functions $\ell_{\text{REML}}(\cdot)$ and $\ell^*_{\text{REML}}(\cdot)$, defined by (7.29) and (7.31), respectively. The latter includes one parameter less, namely, σ, and can be used to obtain an estimator of $\boldsymbol{\delta}$. This is actually the solution used for this purpose in the `gls()` function of the **nlme** package. The variance-covariance matrix of $\widehat{\boldsymbol{\delta}}$ could also be obtained from the inverse of the observed Fisher information matrix (Sect. 4.6.1) of the log-profile-likelihood function $\ell^*_{\text{REML}}(\cdot)$. However, the drawback of using this approach is that it does not allow for the estimation of the variance of $\widehat{\sigma}^2$ nor the covariances between the estimates of σ^2 and of $\boldsymbol{\delta}$. To overcome this shortcoming, the variance-covariance matrix of $\widehat{\sigma}^2$ and $\widehat{\boldsymbol{\delta}}$ can be estimated by using the inverse of the observed Fisher information matrix of the log-likelihood $\ell_{\text{REML}}(\cdot)$, which depends on $\boldsymbol{\delta}$ *and* σ^2. This is also the approach adopted by the `gls()` function from the **nlme** package.

It should be stressed that, because the methods of the assessment of the uncertainty in the estimates of the parameters σ^2 and $\boldsymbol{\delta}$ described above are likelihood-based, they require a correct specification of the model, including the specification of the mean and variance structures and the normality assumption (Sect. 7.8.2).

7.5 Model Diagnostics

In the case of the heterogeneous variance model, defined by (7.2) and (7.7)–(7.8), with a variance function belonging to the <δ>-group, the diagnostic tools, described in Sect. 4.5, need to be modified. In particular, because of heteroscedasticity, neither the raw nor scaled residuals, presented in Tables 4.1 and 4.2, can be expected to exhibit a constant-variability scatter when plotted against predicted values. Nevertheless, with some care, the plots can be used to search for systematic patterns that might suggest problems with the linearity of effects of covariates, with outlying observations or may allow to detect patterns in residual variance heterogeneity.

To check for constant-variability and outlying observations, Pearson residuals are more useful. They are obtained by appropriately scaling of the raw residuals, as described in Sect. 7.5.1.

7.5.1 Pearson Residuals

In Chap. 4, we considered scaling residuals by dividing raw residuals by estimates of σ (Table 4.1). Another set of scaled residuals, displayed in Table 4.2, involved an additional adjustment based on the hat matrix (Sect. 4.5.1). As already mentioned, the use of these residuals for LMs with heterogeneous variance is somewhat limited.

Table 7.5 Examples of Pearson residuals for linear models with heterogeneous variance

Pearson residual	R naming convention	Mathematical formula[c]
Standardized by $\sqrt{\text{Var}(y_i)}$		$\widehat{\varepsilon}_i/\sqrt{\text{Var}(y_i)}$
Internally studentized[a]	Standardized	$\widehat{\varepsilon}_i/\sqrt{\widehat{\text{Var}}(y_i)}$
Externally studentized[b]	Studentized	$\widehat{\varepsilon}_i/\sqrt{\widehat{\text{Var}}(y_{(-i)})}$

[a] $\widehat{\text{Var}}(y_i)$ is an estimate of $\text{Var}(y_i)$ based on all observations
[b] $\widehat{\text{Var}}(y_{(-i)})$ is an estimate of $\text{Var}(y_i)$ after excluding the ith observation
[c] $\widehat{\varepsilon}_i = \mathbf{y}_i - \mathbf{x}_i'\widehat{\boldsymbol{\beta}}$

A different set of scaled residuals, also helpful in the context of LMs with heterogeneous variance, is obtained by dividing the raw residuals by the estimated standard deviation of the dependent variable, $[\widehat{\text{Var}}(y_i)]^{1/2}$. The resulting residuals are called *internally studentized* or, using the R convention, *standardized Pearson residuals* and are presented in Table 7.5. We will simply refer to them as Pearson residuals. Their main advantage is that they are less heteroscedastic, as compared to the raw residuals. However, the heteroscedasticity, related to the heterogeneity of $\text{Var}(y_i)$, is not completely removed. Moreover, the correlation between Pearson residuals, similar to that induced by the hat matrix (4.23) for the raw residuals in the classical LM (Sect. 4.5.1), is not removed either.

It appears sensible to generalize the hat matrix to LMs with heterogeneous variance. Consequently, we could try to adjust Pearson residuals in a similar way to that used for raw residuals (Table 4.2). However, this approach is not supported by the `gls()` function from the **nlme** package used to fit models with heterogenous variance and therefore we do not consider it further.

7.5.2 *Influence Diagnostics*

Influence diagnostics for LMs with heterogeneous variance resemble, to a large extent, those presented in Sect. 4.5.3 for LMs with homogeneous variance. The primary difference is that models with heterogeneous variance have an additional set of variance function parameters, namely, the parameters stored in vector $\boldsymbol{\delta}$. Thus, we should try to first investigate the combined influence of a given observation on the estimates of *all* parameters, including $\boldsymbol{\beta}$, $\boldsymbol{\delta}$, and σ. As a consequence, the diagnostics based on the likelihood displacement should be performed first. Toward this end, the generic likelihood-displacement definition (4.27) is used, with $\widehat{\Theta} \equiv (\widehat{\boldsymbol{\beta}}', \widehat{\boldsymbol{\delta}}', \widehat{\sigma}^2)'$ and the log-likelihood given in (7.21).

After identifying influential observations by using the likelihood displacement, we may try to narrow down their impact. For example, we may use Cook's distance, similar to that given in (4.26), to determine whether a particular observation affects estimation of $\boldsymbol{\beta}$. In addition, it may also be useful to apply Cook's distance to $\boldsymbol{\delta}$ as well.

7.6 Inference

In this section, we discuss the issue of testing hypotheses about parameters of the model, specified by (7.2), (7.7) and (7.8). Again, we focus our discussion on models which involve a variance function from the <δ>-group.

In particular, in Sect. 7.6.1, we briefly discuss the use of tests of statistical significance, while in Sect. 7.6.2, we consider the construction of confidence intervals.

7.6.1 Tests of Statistical Significance

Inference for the LM with heterogeneous variances typically focuses on the fixed-effects parameters $\boldsymbol{\beta}$. In particular, linear hypotheses of the form defined by (4.30) in Sect. 4.6 are of primary interest, so we describe them first. Afterward we consider testing hypotheses about the variance parameters.

7.6.1.1 Tests for Fixed Effects

The LM with known weights, presented in Sect. 7.2.1, does not assume homogeneous variance, so, strictly speaking, it does not meet the assumptions of the classical LM, specified in Sect. 4.2. However, following the representation (7.16) as a weighted LM, statistical inference for LMs with known weights can be performed within the classical LM framework, described in Sect. 4.6.

For the LM, defined by (7.2), (7.7), and (7.8), with the variance function $\lambda(\cdot)$ from the <δ>-group, linear hypotheses about $\boldsymbol{\beta}$ can be tested by using the F-test given by (4.36) in Sect. 4.6.1. The employed variance-covariance matrix of $\widehat{\boldsymbol{\beta}}$ is computed from (7.32). Note, however, that the distribution of the test under the null hypothesis is no longer a central F-distribution with rank($\boldsymbol{L}$) numerator and $n-p$ denominator degrees of freedom. This is because the test ignores the uncertainty related to the estimation of the $\boldsymbol{\delta}$ parameters. It appears, though, that the true distribution of the test statistic can still be *approximated* by a central F-distribution with the numerator degrees of freedom equal to rank($\mathbf{L}$). The number of denominator degrees of freedom needs to be determined from the data. For this purpose, several methods are available. These include, for example, a so-called Satterthwaite-type approximation (Satterthwaite 1941; Fai and Cornelius 1996) and the method suggested by Kenward and Roger (1997). However, the function `gls()`, available in R for fitting LMs with heterogeneous variance, ignores uncertainty related to estimation of $\boldsymbol{\delta}$ parameters and simply uses the central F-distribution with rank($\mathbf{L}$) numerator and $n-p$ denominator degrees of freedom. Thus, we will not discuss the issue of the approximation of the degrees of freedom, in spite of its importance. For further details, we refer to Verbeke and Molenberghs (2000). The issue of the choice of the degrees of freedom for the approximate F-test carries forward and applies to the models considered in Chaps. 10 and 13.

Alternatively, the LR test (Sect. 4.6.1), based on the ML estimation, can be used to test linear hypotheses pertaining to the $\boldsymbol{\beta}$ parameters. Typically, a χ^2-distribution with rank($\mathbf{L}$) degrees of freedom is used as the null distribution for the evaluation of the results of the test. It has to be stressed that, in contrast to the ML-based LR test, the REML-based LR test cannot be used to test linear hypotheses about $\boldsymbol{\beta}$. This is due to the fact that the last term in (7.29), which defined the log-restricted-likelihood, potentially depends on the parameterization of the fixed effects. Thus, calculating the test statistics based on models with different mean structures may imply using design matrices employing different parameterizations for the nested models and, consequently, comparing different log-restricted-likelihood functions.

7.6.1.2 Tests for Variance Parameters

Sometimes it is of interest to test a hypothesis about the variance parameters $\boldsymbol{\delta}$. In particular, the hypothesis implying equality of variances for some groups of observations is of interest. Such a hypothesis can be formulated by specifying equality constraints on the elements of $\boldsymbol{\delta}$.

Toward this end, the LR test, mentioned in Sect. 4.6.1, is used. More precisely, the test statistic is constructed based on the two nested models estimated with and without taking into account the constraints. Asymptotically, the null distribution of the test is approximately a χ^2 distribution with the number of degrees of freedom equal to the difference in the number of variance parameters between the null and alternative models. Three important comments are in order, though. First, the test should be based on the REML, because the ML estimates of $\boldsymbol{\delta}$ are biased, especially for small sample size. Second, the models involved should have exactly the same mean structure. This is because log-restricted-likelihoods are only fully comparable for LMs with the same mean structure. Third, it is required that under the null hypothesis the variance function parameters do not lie on the boundary of the parameter space.

As an example of testing a hypothesis about variance parameters, consider a model with $\text{Var}(y_i)$ varying across S strata, i.e., with the residual variance defined with the use of the varIdent($\cdot$) variance function (Sect. 7.3.1). Thus, $\boldsymbol{\delta}$ is given by (7.14). One might be interested in testing whether a homoscedastic variance structure might be appropriate. The corresponding null hypothesis would specify that $\boldsymbol{\delta} = (1, \ldots, 1)'$. The LR test would be calculated based on a homoscedastic null model with $\text{Var}(y_i) \equiv \sigma^2$ and $\boldsymbol{\delta} = (1, \ldots, 1)'$ and a heteroscedastic alternative model with $\boldsymbol{\delta}$ given by (7.14). The resulting test statistic would have asymptotically a χ^2 distribution with $S - 1$ degrees of freedom.

We may also think of the second situation, in which under the null hypothesis the parameter $\boldsymbol{\delta}$ indeed lies on the boundary space. This may occur, for example, if we test whether the parameter δ_1 in the varConstPower($\cdot$) function is equal to zero. In this case, the LR test statistic under null does not have a χ^2 distribution (Shapiro 1985).

It is worth noting that, in practice, both for testing fixed effects and variance parameters, the null distribution of the test statistics is merely *approximated* by a theoretical distribution. Thus, an important alternative approach to evaluation of the test results is to simulate the null distribution of the test statistic.

7.6.2 Confidence Intervals for Parameters

Confidence intervals for individual components of the parameter vector $\boldsymbol{\beta}$ can be constructed based on a t-distribution used as an approximate distribution for the t-test statistic (Sect. 4.6). The comments related the choice of the number of degrees of freedom for the F-test (see Sect. 7.6.1) apply to the t-test statistic, too.

Confidence intervals for the variance parameters $\boldsymbol{\delta}$ are somewhat more difficult to obtain. The main issue is that, for some variance functions like, for example, varIdent(·) (Sect. 7.3.1), the elements of $\boldsymbol{\delta}$ are *constrained* to being positive. A solution to this problem implemented in R is similar to the one used to overcome the constrained optimization problem, i.e., by considering the logarithmic transformation of the parameters (Sect. 7.4.3). The construction of confidence intervals is then based on using a normal-distribution approximation to the distribution of the ML or REML estimators of the transformed/unconstrained parameters.

For instance, consider the model defined by (7.2), (7.7), and (7.8), with the residual variance specified by the varIdent(·) variance function (Sect. 7.3.1). In this case, the variance parameters are σ and $\boldsymbol{\delta}$, where the latter is given in (7.14). Note that σ and the components of $\boldsymbol{\delta}$ are constrained to be positive. By assuming a normal-distribution approximation to the distribution of the ML- or REML-based estimator of the logarithm of δ_s, the $(1-\alpha/2)100\%$ confidence interval for δ_s is given by

$$\exp\left[\widehat{\log\delta_s} \pm z_{1-\alpha/2}\sqrt{\{\mathbf{I}^{-1}\}_{\delta_s\delta_s}}\right], \tag{7.33}$$

where $\{\mathbf{I}^{-1}\}_{\delta_s\delta_s}$ is the diagonal element, corresponding to $\log\delta_s$, of the inverse of the observed Fisher information matrix (Sect. 4.6.1) of the log-likelihood (7.21) or log-restricted-likelihood (7.29), while $z_{1-\alpha/2}$ is the $(1-\alpha/2)100$-th percentile of the standard normal distribution (Pinheiro and Bates 2000, Sect. 2.4.3). In a similar way, by assuming a normal-distribution approximation to the distribution of the ML- or REML-based estimator of the logarithm of $\widehat{\sigma}^2$, a confidence interval for σ^2 can be obtained.

7.7 Model Reduction and Selection

In Sect. 7.6.1, we described statistical tests for fixed effects and variance parameters. In addition to testing research hypotheses, they are also the most commonly used tool for model reduction. Similarly to the classical LM, in the context of the LM with heterogeneous variance defined by a variance function from the <δ>-group (Sect. 7.2), the need to discriminate between nonnested models may arise. This can happen, for example, when two models with different mean structures and variance functions are considered as suitable candidates to be fitted to the same dataset. In such a case, the use of the LR test is not valid. A solution is the use of information criteria, described in Sect. 4.7.2.

When dealing with selection of the best LM with heterogeneous variance, we need to keep in mind that, in contrast to the classical LM, there are two sets of parameters, namely, $\boldsymbol{\beta}$ and $\boldsymbol{\delta}$. Consequently, we use the same model reduction and selection tools, but in the process we often alternate between reducing the set of the fixed effects and finding the optimal variance structure. The idea of alternating between two sets of parameters carries over to more complex models, including LMMs. An explanation of this issue in the context of LMMs can be found in Verbeke and Molenberghs (2000, Fig. 9.1).

7.8 Mean-Variance Models

In this section, we present the estimation approaches and other theoretical issues related to the mean-variance models. Recall that, by mean-variance models, we consider LMs with heterogeneous residual variance, which is specified by a variance function depending on mean value μ_i, i.e., by a variance function from the <δ, μ>- or <μ>-group (see Tables 7.3 and 7.4, respectively). It is worth mentioning that the concept of mean-variance models carries over to other LMs, including LMs for correlated data and LMMs, which will be discussed in Chaps. 10 and 13, respectively.

7.8.1 Estimation

In this section, we consider estimation of the parameters of the model, defined by (7.2), (7.7), and (7.8), with the variance function $\lambda(\cdot)$ depending on μ_i. First, we present estimation using GLS for models involving variance functions from the <δ, μ>-group. Next, we discuss estimation using IRLS, which is applicable to models involving variance functions from the <μ>-group.

7.8.1.1 Estimation Using Generalized Least Squares

First, we consider a model with the residual variance defined by a variance function belonging to the <δ, μ>-group:

$$\lambda_i = \lambda(\boldsymbol{\delta}, \mu_i; \mathbf{v}_i). \tag{7.34}$$

Thus, we consider variance functions that, for any value of the vector of (known) covariates $\mathbf{v}_i$, depend on the vector of variance parameters $\boldsymbol{\delta}$ and on the expected value $\mu_i \equiv \mathbf{x}_i'\boldsymbol{\beta}$. Examples of such functions were given in Table 7.3.

Estimation of such models could, in principle, be based on the maximization of log-likelihood (7.21) over σ, $\boldsymbol{\beta}$, and $\boldsymbol{\delta}$. However, besides the numerical complexity, the joint maximization of the log-likelihood encounters an additional problem, because the parameters $\boldsymbol{\beta}$ are shared by the mean and variance structures, through μ_i. This is the rationale for using GLS instead of ML. In the remainder of this section, we briefly summarize the GLS estimation. A more detailed exposition can be found, for instance, in Davidian and Giltinan (1995).

Maximum-Likelihood-Based Generalized Least Squares

As an introduction to the GLS estimation, we consider the following objective function:

$$\ell^*_{\text{PL}}(\sigma^2, \boldsymbol{\delta}; \boldsymbol{\beta}) \equiv \ell_{\text{Full}}(\boldsymbol{\beta}, \sigma^2, \boldsymbol{\delta}), \tag{7.35}$$

derived from the full log-likelihood (7.21). Note that we essentially assume $\boldsymbol{\beta}$ to be known. Consequently, for every value of $\boldsymbol{\beta}$, the function $\ell^*_{\text{PL}}(\cdot)$ has two arguments, i.e., σ^2 and $\boldsymbol{\delta}$. Note that, for $\boldsymbol{\beta} = \widehat{\boldsymbol{\beta}}(\boldsymbol{\delta})$, given in (7.22), $\ell^*_{\text{PL}}(\cdot)$ is equivalent to $\ell^*_{\text{ML}}(\cdot)$, defined by (7.23).

By investigating (7.21), we note that optimization of $\ell^*_{\text{PL}}(\sigma^2, \boldsymbol{\delta}; \boldsymbol{\beta})$ with respect to σ and $\boldsymbol{\delta}$ is equivalent to the optimization of the full log-likelihood for a sensibly defined *working* LM with heterogeneous variance. More specifically, the working model is defined by (7.2), (7.7), and (7.8), with *all* terms used in the mean structure absorbed into an offset (Sect. 4.3). For this reason, $\ell^*_{\text{PL}}(\cdot)$ is referred to as the *pseudo-likelihood* function (Carroll and Ruppert 1988), hence the subscript "PL" in the notation.

In the next step, we profile out σ from $\ell^*_{\text{PL}}(\cdot)$ in a similar way as we did it for $\ell^*_{\text{ML}}(\cdot)$, defined by (7.23). To this end, we note that maximization of $\ell^*_{\text{PL}}(\sigma^2, \boldsymbol{\delta}; \boldsymbol{\beta})$ with respect to σ^2 for every known value of $\boldsymbol{\delta}$ leads to the following functional relationship between $\widehat{\sigma}^2$ and $\boldsymbol{\delta}$:

$$\widehat{\sigma}^2_{\text{PL}}(\boldsymbol{\delta}; \boldsymbol{\beta}) \equiv \sum_{i=1}^{n} \lambda_i^{-2} r_i^2 / n, \tag{7.36}$$

where $r_i \equiv y_i - \mathbf{x}_i'\boldsymbol{\beta}$. Note that, in contrast to $r_i(\cdot)$ defined in (7.24), this time r_i does not depend on $\boldsymbol{\delta}$ because $\boldsymbol{\beta}$ is considered to be known. Replacing σ^2 in (7.35) with the expression on the right-hand side of (7.36) yields a pseudo-log-profile-likelihood function for $\boldsymbol{\delta}$:

$$\ell^*_{\text{PL}}(\boldsymbol{\delta}; \boldsymbol{\beta}) \equiv \ell^*_{\text{PL}}(\widehat{\sigma}^2_{\text{PL}}(\boldsymbol{\delta}), \boldsymbol{\delta}; \boldsymbol{\beta}). \tag{7.37}$$

It is important to point out that, given that $\boldsymbol{\beta}$ is fixed, the corresponding variance function, derived from (7.34), no longer depends on μ_i. Consequently, the variance function belongs to the <δ>-group and, therefore, the likelihood-based estimation methods, presented in Sect. 7.4.2, can be used.

Based on the considerations related to (7.35), the GLS algorithm proceeds as follows:

1. Assume initial values $\widehat{\boldsymbol{\beta}}^{(0)}$ of $\boldsymbol{\beta}$ and $\widehat{\boldsymbol{\delta}}^{(0)}$ of $\boldsymbol{\delta}$ and set the iteration counter $k = 0$.
2. Increase k by 1.
3. *Use* $\widehat{\beta}_i^{(k-1)}$ *to (re)define the variance function* $\lambda^{(k)}(\cdot)$.

   Calculate $\widehat{\mu}_i^{(k)} \equiv \mathbf{x}_i'\widehat{\boldsymbol{\beta}}^{(k-1)}$. (Re)define the variance function $\lambda^{(k)}(\cdot)$ as $\lambda^{(k)}(\boldsymbol{\delta};\widehat{\mu}_i^{(k)},\mathbf{v}_i) \equiv \lambda(\boldsymbol{\delta},\widehat{\mu}_i^{(k)};\mathbf{v}_i)$, where $\lambda(\cdot)$ is defined by (7.34). Note that $\lambda^{(k)}(\cdot)$ is from the <δ>-group of variance functions.
4. *Keep* $\widehat{\boldsymbol{\beta}}^{(k-1)}$ *fixed and optimize* $\ell^*_{PL}(\boldsymbol{\delta})$ *to find* $\widehat{\boldsymbol{\delta}}^{(k)}$.

   Use the function $\lambda^{(k)}(\cdot)$ in defining the pseudo-likelihood function, $\ell^*_{\text{PL}}(\boldsymbol{\delta};\widehat{\boldsymbol{\beta}}^{(k-1)})$, specified by (7.37). Optimize this function with respect to $\boldsymbol{\delta}$ to obtain the new estimate $\widehat{\boldsymbol{\delta}}^{(k)}$ of $\boldsymbol{\delta}$. Values of $\widehat{\boldsymbol{\delta}}^{(k-1)}$ can be used as the initial values for the optimization.
5. *Fix* $\widehat{\boldsymbol{\delta}}^{(k)}$ *and use WLS to find* $\widehat{\boldsymbol{\beta}}^{(k)}$.

   Use $\widehat{\boldsymbol{\delta}}^{(k)}$ to compute the values $\widehat{\lambda}_i^{(k)} \equiv \lambda(\widehat{\boldsymbol{\delta}}^{(k)},\widehat{\mu}_i^{(k)};\mathbf{v}_i)$. Using $w_i \equiv \left[\widehat{\lambda}_i^{(k)}\right]^{-2}$ as weights, compute the WLS estimate $\widehat{\boldsymbol{\beta}}^{(k)}$ of $\boldsymbol{\beta}$ by using (7.18).
6. Iterate between steps 2–5 until convergence or until a pre-determined number of iterations k.
7. Compute the final estimate of σ^2 by plugging the obtained estimates of $\boldsymbol{\delta}$ and $\boldsymbol{\beta}$ into (7.36).

The algorithm described above is called the *generalized least squares pseudo-likelihood* (GLS-PL) estimation. Note that it involves two iterative loops: an external and an internal one. The internal loop is related to the optimization of the function (7.37) in the step 4. The external loop is related to the repeated execution of the steps 2–5.

In general, the ML-based GLS-PL algorithm does not have to produce an ML estimator. Similarly to the observation made for the estimator $\widehat{\sigma}^2_{\text{ML}}$, given by (7.28), the estimator of σ^2, obtained from the GLS-PL algorithm, is likely to be biased. In fact, the same comment applies to the estimator of $\boldsymbol{\delta}$. For this reason, σ^2 and $\boldsymbol{\delta}$ are preferably estimated by using the GLS estimation based on REML. We will now briefly describe this estimation approach.

Restricted-Maximum-Likelihood-Based Generalized Least Squares

This estimation approach is based on a doubly iterative algorithm similar to the ML-based GLS-PL algorithm, described in the previous section. Compared to the latter, the algorithm is modified as follows:

- In step 4, instead of maximizing (7.37), a corresponding REML-based log-likelihood function (not shown) is maximized with respect to $\boldsymbol{\delta}$.
- In step 7, to obtain the final estimate of σ^2, we use a modified version of (7.36), with the denominator n replaced by $n - p$.

It is expected that, by using the REML-based estimators of $\boldsymbol{\delta}$ and σ^2 in both modifications described above, the resulting final estimators of the parameters are less biased, as compared to the estimators obtained by the ML-based GLS-PL algorithm.

7.8.1.2 Estimation Using Iteratively Re-weighted Least Squares

In this section, we consider the model, defined by (7.2), (7.7), and (7.8), with the variance function $\lambda(\cdot)$ belonging to the <μ>-group:

$$\lambda_i = \lambda(\mu_i; \mathbf{v}_i). \tag{7.38}$$

Thus, we consider variance functions that depend on the vector of the expected values $\mu_i = \mathbf{x}_i'\boldsymbol{\beta}$ and on (known) covariates stored in $\mathbf{v}_i$. Examples of such functions were given in Table 7.4.

Note that, because the variance functions from the <μ>-group do not depend on $\boldsymbol{\delta}$, the estimation problem simplifies, because we only need to estimate $\boldsymbol{\beta}$ and σ^2. However, the dependence of $\lambda(\cdot)$ on $\boldsymbol{\beta}$ implies the need for an iterative procedure to find an estimator for the latter parameters. More specifically, $\boldsymbol{\beta}$ can be estimated by the following IRLS algorithm:

1. Assume initial values $\widehat{\boldsymbol{\beta}}^{(0)}$ of $\boldsymbol{\beta}$ and set the iteration counter $k = 0$.
2. Increase k by 1.
3. Calculate $\widehat{\mu}_i^{(k)} \equiv \mathbf{x}_i'\widehat{\boldsymbol{\beta}}^{(k-1)}$ and $\widehat{\lambda}_i^{(k)} \equiv \lambda(\widehat{\mu}_i^{(k)}; \mathbf{v}_i)$, where $\lambda(\cdot)$ is defined by (7.38).
4. *No optimization needed.*
5. Compute the WLS estimate $\widehat{\boldsymbol{\beta}}^{(k)}$ of $\boldsymbol{\beta}$ by using (7.18) and $w_i \equiv \left[\widehat{\lambda}_i^{(k)}\right]^{-2}$.
6. Iterate between steps 2 and 5 until convergence or until a pre-determined number of iterations k.
7. Compute the final estimate of σ^2 by plugging the obtained estimate of $\boldsymbol{\beta}$ in (7.36).

The algorithm can be viewed as a simplified version of the ML-based GLS-PL algorithm outlined in Sect. 7.8.1.1. The main simplification occurs in step 4, in which the internal loop, necessary in the GLS-PL algorithm to update the current values of $\boldsymbol{\delta}$ estimates, has been dropped.

Note that in the last step of the IRLS algorithm, the REML-based GLS estimator of σ^2 can be used.

7.8.2 *Model Diagnostics and Inference*

Carroll and Ruppert (1988) and Davidian and Giltinan (1995) show that the estimators of $\boldsymbol{\beta}$, obtained by any of the algorithms described in Sect. 7.8.1.1, are asymptotically normally distributed with a variance-covariance matrix, which can be estimated as in (7.32), but with $\widehat{\lambda}_i \equiv \lambda(\widehat{\mu}_i, \widehat{\boldsymbol{\delta}}; \mathbf{v}_i)$ or $\widehat{\lambda}_i \equiv \lambda(\widehat{\mu}_i)$. However, standard errors, based on the estimated matrix, may need to be treated with caution. This is because the estimate does not take into account the uncertainty related to the use of estimates of $\boldsymbol{\beta}$ and $\boldsymbol{\delta}$ to compute the weights $w_i \equiv (\widehat{\lambda}_i)^{-2}$ in the step 5 of the GLS-PL and IRLS algorithms. In fact, by using the second-order approximation, it can be shown (Rothenberg 1984; Carroll and Ruppert 1988) that the variance-covariance matrix of $\widehat{\boldsymbol{\beta}}$ depends on the precision of the estimation of $\boldsymbol{\delta}$. Thus, especially for small sample sizes, the standard errors, obtained from (7.32), may be too small.

Keeping this issue in mind, linear hypotheses about $\boldsymbol{\beta}$ can be tested by using the F-test (4.36), along the lines discussed in Sect. 7.6 for the case of LMs specified with the use of variance functions from the <δ>-group. Confidence intervals for the elements of $\boldsymbol{\beta}$ can be constructed by the method mentioned in Sect. 7.6.2.

The use of LR tests for testing hypotheses about $\boldsymbol{\beta}$ is complicated by the fact that $\boldsymbol{\beta}$ is not estimated based on a likelihood neither in the GLS-PL nor in the IRLS algorithm. For instance, the ML-version of the GLS-PL estimate of $\boldsymbol{\beta}$ is obtained by using the log-profile-pseudo-likelihood (7.37). Nevertheless, an LR test for $\boldsymbol{\beta}$ can be constructed by considering the core part of log-likelihood (7.21) with the estimated weights λ_i and σ^2:

$$-\frac{1}{2\widehat{\sigma}^2}\sum_{i=1}^{n}(\widehat{\lambda}_i)^{-2}(y_i - \mathbf{x}_i'\boldsymbol{\beta})^2, \tag{7.39}$$

with $\widehat{\lambda}_i \equiv \lambda(\widehat{\mu}_i, \widehat{\boldsymbol{\delta}}; \mathbf{v}_i)$ or $\widehat{\lambda}_i \equiv \lambda(\widehat{\mu}_i)$, and estimates $\widehat{\boldsymbol{\beta}}$, $\widehat{\boldsymbol{\delta}}$, and $\widehat{\sigma}^2$ obtained under the alternative model. Comparison of the values of (7.39) for the null and alternative models, along the lines described in Sect. 4.6.1, yields an LR test for $\boldsymbol{\beta}$.

As mentioned at the beginning of this section, the precision of estimates of $\boldsymbol{\beta}$, obtained by the algorithms presented in Sect. 7.8.1, depends on the precision of estimation of $\boldsymbol{\delta}$. From this point of view, correct specification of the residual variance structure is desirable, as it would increase the efficiency of the estimation of $\boldsymbol{\beta}$ and yield valid estimates of $\boldsymbol{\delta}$ and σ^2. However, the PL-GLS and IRLS algorithms provide valid estimates of $\boldsymbol{\beta}$, as long as the mean structure of the model is correctly chosen. Thus, the algorithms can be also be applied even with a working, misspecified variance structure.

Inference on $\boldsymbol{\delta}$ and σ^2, based on the estimates produced by the PL-GLS and IRLS procedures, is complicated by several factors. First, any misspecification of the variance structure of the data may lead to biased estimates of $\boldsymbol{\delta}$and σ^2. However, correct specification of the structure is often difficult. Second, the variability of the estimates depends, in a complicated form, on the variability of the estimate of $\boldsymbol{\beta}$ and on the true (unknown) third- and higher-order moments of the data. Consequently, the estimation of the variance-covariance matrix of the estimates of $\boldsymbol{\delta}$ and σ^2 is difficult. For all these reasons, inference on $\boldsymbol{\delta}$ and σ^2, in the context of the mean-variance models, should be treated with caution and we will not discuss it further. Interested readers are referred to, e.g., the monograph by Carroll and Ruppert (1988, Chap. 3) for more details.

Model diagnostics for mean-variance models can be based on the Pearson residuals (Table 7.5). The comments related to the incomplete removal of the heteroscedasticity and correlation between the Pearson residuals, given in Sect. 7.5.1, apply here as well.

7.9 Chapter Summary

In this chapter, we briefly reviewed the formulation of an LM for independent observations with heterogeneous variance. For brevity, in the presentation, we tried to use as much as possible the theory introduced in Chap. 4 and to focus mainly on the concepts essential for addressing the variance heterogeneity. Thus, particularly when describing the specification of the model in Sects. 7.2 and 7.3, we focused on the concept of variance function.

The classification of variance functions into four groups (Sect. 7.3.1) has important implications in terms of the choice of estimation methods and properties of the resulting estimates. In Sect. 7.4, we described the estimation methods for the models involving known weights or variance functions from the <δ>-group, which do not depend on the mean value. An important point was the modification of the estimation methods, presented in Sect. 4.4 for the classical LM, to allow for the estimation of the variance function parameters. To a large extent, similar estimation methods will be used for more complex models in the later chapters. In Sect. 7.5, we offered a review of the diagnostic methods, while in Sects. 7.6 and 7.7, we described the inferential tools available for models involving variance functions that do not depend on the mean value. As in Sect. 7.4, we focused on the adaptation of the methods developed for the classical LM to the case of independent observations with heterogeneous variance. In this respect, it is worth noting that, as compared to the classical LM, the F-distribution of the F-test statistic for the LMs with heterogeneous variance is only approximate, not exact.

In Sect. 7.8, we discussed the mean-variance models, i.e., the models involving variance functions from the $<\delta,\mu>$- and $<\mu>$-groups, which do depend on the mean value. In particular, in Sect. 7.8.1, we briefly summarized the estimation approaches for such models, which typically involve iterative algorithms. On the other hand, in Sect. 7.8.2, we reviewed the issues related to the inference based on the mean-variance models.

The use of variance function will be instrumental in formulating other models, including LMMs in Chap. 13. In this respect, it is worth noting that the $<\delta>$-group of variance functions is the most common choice for LMs, especially for LMMs. The $<\delta,\mu>$- and $<\mu>$-groups of variance functions are less frequently applied, because their use implies that the variance of the dependent variable is related to fixed effects $\boldsymbol{\beta}$, which complicates both the model estimation and inference.

Chapter 8
Fitting Linear Models with Heterogeneous Variance: The `gls()` Function

8.1 Introduction

In Chap. 7, we introduced several concepts related to the LM for independent, normally distributed observations with heterogeneous variance. Compared to the classical LM (Chap. 4), the formulation of the model included a new component, namely, the variance function, which is used to take into account heteroscedasticity of the dependent variable.

In the current chapter, we review the tools available in R for fitting LMs for independent, normally distributed observations with heterogeneous variance. Sections 8.2 and 8.3 present the details of the implementation of variance functions in R. In Sect. 8.4, we briefly discuss the use of the `gls()` function from the **nlme** package, which is the primary tool to fit LMs for independent observations with heterogeneous variance. Finally, Sect. 8.5 explains how the details of the estimated form of the model can be accessed.

8.2 Variance-Function Representation: The *varFunc* Class

An important component needed in the context of an LM with heterogeneous variance, is the variance function, defined in Sect. 7.2.2. Several examples of variance functions were given in Tables 7.1–7.4. In this section, we provide the details of the implementation of the functions in the **nlme** package and illustrate them with examples.

A. Gałecki and T. Burzykowski, *Linear Mixed-Effects Models Using R: A Step-by-Step Approach*, Springer Texts in Statistics, DOI 10.1007/978-1-4614-3900-4_8,

8.2.1 Variance-Function Constructors

The **nlme** package provides *constructor functions* designed to create specialized objects, representing different variance functions. Examples of the constructor functions and the class of the returned object are presented in Table 8.1.

Each created object belongs to the corresponding class, named after the constructor function. For example, the `varIdent()` constructor function is used to create objects of class *varIdent*, representing the variance function defined in (7.15), which assigns different variances to different strata. Note that the objects inherit from the *varFunc* class. A list of variance functions available in **nlme** can be obtained from the R help system by issuing the `?varClasses` command (after loading the package).

The constructor functions allow exploring the features of the corresponding variance functions. They also allow the choosing of user-defined initial values for the function parameters. The constructors are primarily used to specify variance functions, with the help of the `weight` argument, for the model-fitting functions `gls()`, `lme()`, and `nlme()`.

8.2.1.1 Arguments of the Variance-Function Constructors

For the `varIdent()`, `varExp()`, and `varPower()` constructor functions, there are three (optional) arguments available: `value`, `form`, and `fixed`. The argument `value` is a numeric vector or a list of numeric values, which specifies the values of the variance parameter vector $\boldsymbol{\delta}$, as defined in (7.6). The default value is 0, implying a constant variance. The argument `form` provides a one-sided formula, which indicates the vector of variance covariates $\mathbf{v}_i$ and, if required, a stratification variable that defines the strata $s = 1, \ldots, S$ for $\boldsymbol{\delta}$ (Sect. 7.3.1). The default value of the argument is `~1`, implying a constant variance with no strata. Finally, `fixed` can be used to fix the values of chosen (possibly, all) variance parameters in the numerical optimization in the modeling functions. Toward this end, we should provide a named numeric vector with the values, or a named list indicating the strata, for which the parameters are to be kept constant. The argument defaults to `NULL`, corresponding to no fixed coefficients.

For the constructor-function `varConstPower()`, the argument `value` is replaced by arguments `const` and `power`. The arguments are numeric vectors or lists of numeric values, which specify the coefficients for the constant and the power terms, respectively (Sect. 7.3.1).

The constructor-function `varFixed()` uses only one argument, `value`, which is a one-sided formula of the form `~v`. It specifies the variance covariate `v`, which induces a fixed (known) variance structure.

More information on the arguments of the variance-function constructors can be obtained by issuing command `?varClasses`.

Table 8.1 *R syntax*: Examples of constructor functions and classes representing variance functions in the package **nlme**. Created objects inherit from the *varFunc* class

Constructor function	Class of the returned object	Description
`varFixed()`	*varFixed*	Fixed weights, determined by a variance covariate
`varIdent()`	*varIdent*	Different variances per stratum
`varExp()`	*varExp*	Exponential of a variance covariate
`varPower()`	*varPower*	Power of a variance covariate
`varConstPower()`	*varConstPower*	Constant plus power of a variance covariate
`varComb()`	*varComb*	Combination of variance functions

8.2.2 Initialization of Objects of Class varFunc

Typically, the next step, after defining an object representing a variance function, is to evaluate the variance structure in the context of a given dataset. The process is referred to as *initialization*. It is conducted using the generic `Initialize()` function. The function accepts two arguments: `object` and `data`. Based on the class of the object specified in the first argument, the function dispatches an appropriate method, which initializes the object using the data provided in the second argument. More information on the process of initialization of a *varFunc*-class object can be obtained by issuing the command `?Initialize.varFunc`. In what follows, we illustrate the initialization of an object of class *varIdent*.

8.2.2.1 Illustration: Using the `varIdent()` *Variance Function*

We continue with the varIdent($\cdot$) variance function, given in (7.15), as an example. Panel R8.1 shows the construction and initialization of an object of class *varIdent*. Objects `val`, `fix`, and `frm` are used to specify the values of the arguments `value`, `fixed`, and `form`, respectively, in the definition of the object `vf0`. The formula `form` specifies that variance depends on the value of the variance covariate, i.e., the factor `time.f`. Vectors `val` and `fix`, taken together, define the initial values of the variance-function coefficients, i.e., the three ratios of standard deviation (SD) of the residual error for weeks 12, 24, and 52, relative to the reference value at week 4. More specifically, the initial value of SD at 12 weeks is specified as a half of that at 4 weeks. On the other hand, the value of SD at 24 weeks and 52 weeks is taken as twice and three times as high, respectively, as the value at 4 weeks. The use of the argument `fixed=fix` implies that the value of the coefficient corresponding to the variance for week 52 will not change during any optimization steps in modeling routines. Finally, the `Initialize()` command initializes the object `vf0` and stores the result in `vf0i`. Initialization is performed by evaluating the `vf0` object for the variance covariate, i.e., the factor `time.f` from the `armd` dataset.

R8.1 *R syntax*: Definition and initialization of an object of class *varIdent*

```
> (val <- c("12wks" = 0.5, "24wks" = 2))  # δ1≡1, δ2=0.5, δ3=2
 12wks 24wks
   0.5   2.0
> (fix <- c("52wks" = 3))                  # δ4=3 (fixed)
 52wks
     3
> frm  <- formula(~1|time.f)         # time.f is a stratifying factor
> (vf0 <-
+    varIdent(value = val,           # Var. function object defined...
+             fixed = fix,
+             form  = frm))
 Variance function structure of class varIdent with no parameters, ...
> (vf0i <- Initialize(vf0, armd))            # ... and initialized
 Variance function structure of class varIdent representing
  4wks 12wks 24wks 52wks
   1.0   0.5   2.0   3.0
```

8.3 Inspecting and Modifying Objects of Class *varFunc*

In Sect. 7.4.3, we mentioned that, for numerical optimization purposes, the use of an alternative, unconstrained parameterization of the variance function may be advantageous. The information about the values of the coefficients, which correspond to the different possible parameterizations, can be extracted from an appropriate, initialized *varFunc*-class object.

The primary tool to extract or modify coefficients from any object is the generic `coef()` function. It has specific methods for different classes. For example, to extract coefficients from an object of class *varIdent*, we use the method `coef.varIdent()`, dispatched by invoking the function `coef()`.

The primary arguments of the `coef.varIdent()` method are `object`, `unconstrained`, and `allCoef`. The argument `object` indicates an object of class *varIdent*. The value of the logical argument `unconstrained` specifies the type of the parameterization applied to the coefficients of the variance function. More specifically, the coefficients (parameters) can be presented on a constrained or unconstrained scale (Sect. 7.4.3). The value of the logical argument `allCoef` indicates whether all coefficients or only those, which were not designated to be fixed in numerical optimization routines, are to be returned. Note that similar methods and arguments apply to other classes of variance functions, such as *varPower*, *varExp*, etc.

Panel R8.2 demonstrates how to extract and to modify coefficients of a variance function. For illustration purposes, we continue with the object `vf0i`, which represents the initialized variance-function `varIdent`, defined in Panel R8.1. In Panels R8.2a and R8.2b, we illustrate the results of the use of all four combinations of the possible values of arguments `unconstrained` and `allCoef` of the `coef()` function.

The syntax displayed in Panel R8.2a returns the coefficients expressed on the natural/constrained scale. The code shown in Panel R8.2b displays the coefficients on the unconstrained scale. Note that the `coef()` function, when applied to the object `vf0i` with the arguments `unconstrained=FALSE` and `allCoef=TRUE` returns all four coefficients, including the one corresponding to 4 weeks, on the original, constrained scale. In Sect. 7.3.2, it was explained that the parameter corresponding to 4 weeks was set to 1 to avoid aliasing with the scale parameter σ. By default, the function `coef()` uses the arguments `unconstrained = TRUE` and `allCoef = FALSE` and returns on the unconstrained scale only those variance-function parameters, which are allowed to vary in optimization routines.

R8.2 *R syntax*: Extracting and assigning coefficients to an initialized object of class *varIdent*. The object `vf0i` was created in Panel R8.1

(a) *Coefficients on the natural/constrained scale* ($\delta_1 \equiv 1$, δ_2, δ_3, δ_4)

```
> coef(vf0i, unconstrained = FALSE, allCoef = TRUE) # All δ coefs
  4wks 12wks 24wks 52wks
   1.0   0.5   2.0   3.0
> coef(vf0i, unconstrained = FALSE, allCoef = FALSE)# Varying only
 12wks 24wks
   0.5   2.0
```

(b) *Coefficients on the unconstrained scale* ($\delta_2^*, \delta_3^*, \delta_4^*$)

```
> coef(vf0i, unconstrained = TRUE, allCoef = TRUE)  # All δ* coefs
 [1] -0.69315  0.69315  1.09861
> coef(vf0i, unconstrained = TRUE, allCoef = FALSE) # Varying (default)
 [1] -0.69315  0.69315
> coef(vf0i) <- c(-0.6, 0.7)                        # New coefs assigned
> coef(vf0i, allCoef = TRUE)                        # All coefs printed
 [1] -0.6000  0.7000  1.0986
```

At the bottom of Panel R8.2b, we illustrate how to assign new values of the variance-function coefficients for `gls()` function in Sect. 8.4. This might be useful if we want, for example, to use different initial values of the coefficients. In Panel R8.2b, we assign new values to the two coefficients, which are not fixed and which correspond to the levels of the factor `time.f` corresponding to 12 and 24 weeks. Toward this end, we apply the *replacement* function `coef()<-`. The used syntax is of the form `coef(object) <- value`. Note that *`value`* is a numeric vector with the replacement values for the coefficients associated with the variance-function object. The vector has to have the same length as coef(object), and its elements have to be given in the unconstrained form. For objects of class *varIdent*, coefficients on the unconstrained scale are obtained by taking the natural logarithm of the corresponding constrained parameters (Sect. 7.4.3). Moreover, the object has to be initialized before any new values can be assigned to its coefficients. See Panel R8.1 to confirm that the object vf0i is indeed initialized.

In Panel R8.3, we illustrate how to extract information about an initialized *varFunc* object. Toward this end, we use the object vf0i, with coefficients modified in Panel R8.2b, as an example. Using the generic function summary(), we obtain a description of the variance-function structure, stored in the object. The use of the functions formula() and getCovariate() allows extracting, respectively, the formula and the name of the variance covariate, which were used to define the structure. By applying the function getGroupsFomula(), we obtain the formula used to define the strata. In our case, the strata were defined by the levels of the factor time.f. Using the generic function getGroups(), we extract the information about the groups attribute of the object vf0i and store it in the object stratum. By applying the functions length() and unique(), we obtain the information about the number of observations and the unique values of the stratifying factor, respectively. With the help of the varWeights()[3:6] command, we obtain reciprocals of the weights for observations 3 to 6 in the armd data frame. In our case, these are the observations for subject #2, for whom visual acuity measurements at all four timepoints are available. Finally, by applying the function logLik(), we obtain the sum of the logarithms of the weights, which is the contribution of the variance-function structure, represented by the object vf0i, to the log-likelihood (7.21).

Note that syntax similar to that used in Panels R8.1–R8.3 can be applied to objects inheriting from any other *varFunc* class. It should also be mentioned that the generic update() function can also be used to modify objects of class *varFunc*. Finally, in Table 8.2, we present additional ways to extract elements of an initialized object of this class.

8.4 Using the gls() Function to Fit Linear Models with Heterogeneous Variance

LMs for independent observations with heterogeneous variance can be fitted in R using the function gls() from the **nlme** package. The function allows for the use of both known weights and variance functions. It implements the ML and REML estimation methods (Sect. 7.4.2), as well as GLS (Sect. 7.8.1.1).

The basic use and arguments of the gls() were briefly described in Sect. 5.4. The mean structure of the model is defined by the argument model. It specifies

R8.3 *R syntax*: Extracting information from a *varIdent*-class object. The object vf0i was created in Panel R8.1

```
> summary(vf0i)                    # Summary
  Variance function:
   Structure: Different standard deviations per stratum
   Formula: ~1 | time.f
   Parameter estimates:
     4wks    12wks    24wks    52wks
  1.00000 0.54881 2.01375 3.00000
> formula(vf0i)                    # Variance function formula
  ~1 | time.f
> getCovariate(vf0i)               # Variance covariate
  NULL
> getGroupsFormula(vf0i)           # Formula for variance strata
  ~time.f
  <environment: 0x00000000086a3500>
> length(stratum <-                # Length of stratum indicator
+          getGroups(vf0i))
  [1] 867
> unique(stratum)                  # Unique strata
  [1] "4wks"  "12wks" "24wks" "52wks"
> stratum[1:6]                     # First six observations
  [1] "4wks"  "12wks" "4wks"  "12wks" "24wks" "52wks"
> varWeights(vf0i)[3:6]            # Variance weights 1/λ_i:(7.8)
     4wks    12wks    24wks    52wks
  1.00000 1.82212 0.49659 0.33333
> logLik(vf0i)                     # Contribution to the log-likelihood
  'log Lik.' -227.83 (df=2)
```

a two-sided model formula, which defines the dependent variable and the design matrix (Sects. 5.2 and 5.3). Arguments `data`, `subset`, and `na.action` are used to create the model frame necessary to evaluate the model formula. The default estimation method of the `gls()` function is REML, defined by (7.31) in Sect. 7.4.2.

The argument which allows one to specify the variance function is `weights`. When specifying it, we generally use an object of class *varFunc*, which defines the variance function and, at the same time, provides the initial values for the likelihood-optimization routine. Thus, a typical use of the argument is of the form `weights=`*`varFunc`*`(form=`*`formula`*`)`, where *`varFunc`* is a variance-function constructor (Table 8.1), while *`formula`* is a one-sided-formula object, necessary to define the variance covariate(s) and strata (Sect. 8.2).

Alternatively, `weights` can be given directly as a one-sided formula. In this case, the formula is used as the argument of the `varFixed()` function, corresponding to fixed variance weights. The default value of the `weights` argument is `NULL`, which implies an LM with homoscedastic residual errors. Note that the use of the argument weights also adds variance covariates to the model frame. Argument weights can prove useful to specify user-defined values of the variance parameters.

An important optional argument of the gls() function is control. It contains a list of components used to define various options controlling the execution of the estimation algorithm. The auxiliary function glsControl() returns the default list of options and can be used to efficiently modify them. The arguments of the glsControl() function include maxIter and msMaxIter, which are used to limit the number of external and internal iterations in the GLS algorithm (Sect. 7.8.1.1). To obtain the full list of arguments, the args(glsControl) or ?glsControl commands can be used.

As a result of fitting a model with the use of the gls() function, an object of class *gls*, representing the fit of the model, is created. A description of the components of the object can be obtained by issuing the ?glsObject command.

8.5 Extracting Information From a Model-fit Object of Class *gls*

To extract the results from an object of class *gls*, typically created by the gls() function, generic functions such as print(), summary(), predict(), etc., can be used. Additional functions and syntax useful to extract information about mean structure and scale parameter σ are presented in Sect. 5.5 and Table 5.5.

In Table 8.2, we present selected functions and methods to extract the results pertaining to the variance structure of a fitted model. We assume that the model-fit results are stored in a hypothetical object gls.fit. In Table 8.2a, we demonstrate how to extract selected results directly from gls.fit. First, we obtain the applied form of the gls()-function call and store it in the object cl. Subsequently, the form of the weights argument is obtained by extracting the cl$weights component of the cl object. Confidence intervals (CIs) for the constrained variance-function coefficients are obtained by applying the intervals() function, with the argument which="var-cov", to the model-fit object. The intervals are constructed by transforming the CIs for the corresponding unconstrained coefficients (Sect. 7.6.2). Pearson residuals (Sect. 7.5.1) are obtained by applying the resid() function, with the argument type="pearson", to the model-fit object.

By extracting and storing the modelStruct component of the model-fit object in the object mSt, we get access to the estimated variance structure of the model. Details of the estimated form of the variance function can be obtained by extracting the varStruct component of the mSt object and saving it in the object vF of the *varFunc* class.

In Table 8.2b, we illustrate how to extract various components of the variance-function structure stored in object vF. For instance, the application of the summary() function provides a description of the variance function, together with

Table 8.2 *R syntax*: Extracting components of the variance structure contained in a hypothetical object `gls.fit` of *gls* class, representing a fit of a linear model with heterogeneous variance obtained using the `gls()` function. To find out how to extract other results from the `gls.fit` object, refer to Table 5.5

(**a**) Extracting results directly from the object `gls.fit` of class *gls*

Model-fit component to be extracted	Syntax
`gls()`-call	`(cl <- getCall(gls.fit))`
`weights` argument	`cl$weights`
95% CI for $\boldsymbol{\delta}$	`intervals(gls.fit, which="var-cov")$varStruct`
Pearson residuals	`resid(gls.fit, type="pearson")`
Var-cov structure	`mSt <- gls.fit$modelStruct`
Variance function	`vF <- mSt$varStruct`

(**b**) Extracting results from an auxiliary object `vF` of class *varFunc*

Variance structure component to be extracted	Syntax
Summary	`summary(vF)`
Variance-function formula	`formula(vF)`
Variance-function covariate	`getCovariate(vF)`
Formula for stratification	`getGroupsFormula(vF)`
Stratification variable	`getGroups(vF)`
$\widehat{\boldsymbol{\delta}}^*$ (unconstrained)	`coef(vF)`
$\widehat{\boldsymbol{\delta}}$ (constrained)	`coef(vF, unconstrained=FALSE)`
Contribution to LogLik	`logLik(vF)`
Variance weights $(1/\widehat{\lambda}_i)$	`varWeights(vF)`

the variance function coefficients on the original, constrained scale. On the other hand, the formula and the covariate, used in the specification of the variance-function, are obtained using the `formula()` and `getCovariate()` functions, respectively.

Results of hypothesis tests of the fixed effects and variance parameters (Sect. 7.6.1) for the fitted LM for independent, heteroscedastic observations can be accessed by the methods mentioned in Sect. 5.6. In particular, the `anova()` function uses likelihoods for nested models to provide the LR tests. Note that, when testing hypotheses about variance parameters is of interest, the REML-based LR test should be used, especially in the smaller datasets, and the null and alternative models should have the same mean structure (Sect. 7.6.1). It follows that the model-fit objects, used in the `anova()`-function call, should be obtained by applying the default `method=REML` argument of the `gls()` function.

Alternatively, `anova()` provides the ML-based Akaike's and Bayesian information criteria (Sect. 4.7.2) for all of the models for which model-fit objects are specified in the function call. The criteria can be used to choose the best-fitting models from a set of nonnested models having the same fixed part, but with different variance structures.

8.6 Chapter Summary

In this chapter, we presented the tools available in R to fit LMs for independent, heteroscedastic observations. From this point of view, the important issue was the implementation of the concept of the variance function. We described the tools available for this purpose in Sect. 8.2. In particular, the variance-function constructors and the initialization of the variance-function structure in the context of given data were presented. In Sect. 8.3, we reviewed different ways to extract information from an initialized object of class *varFunc*. In Sect. 8.4, we discussed the use of the `gls()` function to fit these models. The way, in which the variance functions are implemented in the `gls()` function, is similar to their implementation in the `lme()` function, which is used to fit LMMs. Thus, the presentation of this particular aspect of the `gls()` function will be useful also for the latter models.

Finally, we note that, occasionally, it may be of interest to define a new class representing a user-defined variance function. In this case, it is recommended to explore, in the first instance, the information related to an already defined standard class like, e.g., *varPower*. The source code of the `varPower()` constructor function is obtained by simply typing `varPower` at the command prompt of the command window in R. The methods for the class *varPower* are identified by issuing the command `methods(class=varPower)`. Users may define their own variance functions by writing appropriate constructor functions, similar to `varPower()`. Such functions should return objects of a class with the same name and with attributes similar to those of already defined classes, such as the *varPower* class. In addition, at least three methods, i.e., `coef`, `coef<-`, and `Initialize` have to be written for the newly defined class.

Chapter 9
ARMD Trial: Linear Model with Heterogeneous Variance

9.1 Introduction

In this chapter, we continue with the analysis of the visual acuity measurements collected in the ARMD trial. For illustrative purposes, in Chap. 6 we considered LMs with independent, homoscedastic residual errors. In the current chapter, we allow for heterogeneous variance, but we keep the assumption of independence. Note that the results of the exploratory analysis indicate that the assumption is most likely incorrect. Thus, the models constructed in the present chapter are considered mainly for purposes of illustration of the concept of the variance function (Sect. 7.2.2) and of the R tools that implement the concept (Sect. 8.2). Models that accommodate correlated observations will be considered in Chaps. 12 and 16.

The chapter is structured as follows. In Sect. 9.2, we specify an LM with independent, heteroscedastic residual errors for the visual acuity measurements from the ARMD dataset, and we fit the model to the data with the help of the `gls()` function. In Sect. 9.3, we consider an alternative set of models, which use the power variance function for a more parsimonious representation of the variance structure of the data. The results of fitting of the models are presented in Sect. 9.3.1. Section 9.3.2 briefly discusses goodness-of-fit issues. A summary of chapter's contents is provided in Sect. 9.4.

9.2 A Linear Model with Independent Residual Errors and Heterogeneous Variance

In Sect. 6.2, we specified model **M6.1** with independent residual errors and homogeneous variance. For the reader's convenience, we display the model equation below:

$$\text{VISUAL}_{it} = \beta_{0t} + \beta_1 \times \text{VISUAL0}_i + \beta_{2t} \times \text{TREAT}_i + \varepsilon_{it}. \tag{9.1}$$

A. Gałecki and T. Burzykowski, *Linear Mixed-Effects Models Using R: A Step-by-Step Approach*, Springer Texts in Statistics, DOI 10.1007/978-1-4614-3900-4_9,

Note that i indexes patients and t indexes measurement times: $t = 1, 2, 3$, and 4 for 4, 12, 24, and 52 weeks, respectively. Additionally, it was assumed that the residuals ε_{it} were normally distributed with a constant variance:

$$\varepsilon_{it} \sim N(0, \sigma^2).$$

Note, however, that the exploratory analysis of the ARMD data (Sect. 3.2.2) indicated that the variances of the visual acuity measurements, obtained at different timepoints, differed. Thus, in this section, we relax the assumption about the homogeneity of residual variance. More specifically, we consider a model having the same mean structure as the one defined in (9.1), but we allow for heteroscedasticity:

$$\varepsilon_{it} \sim N(0, \sigma_t^2). \tag{9.2}$$

Thus, variances $\sigma_t^2 \geq 0$ $(t = 1,2,3,4)$ are allowed to be different for each timepoint. We refer to this new model, defined by (9.1) and (9.2), as model **M9.1**.

Following (7.15), we can re-parameterize the variances as follows:

$$\text{Var}(\text{VISUAL}_{it}) \equiv \sigma_t^2 = \begin{cases} \sigma^2 & \text{for t} = 1 \text{ (4 weeks)}, \\ \sigma^2 \delta_2^2 & \text{for t} = 2 \text{ (12 weeks)}, \\ \sigma^2 \delta_3^2 & \text{for t} = 3 \text{ (24 weeks)}, \\ \sigma^2 \delta_4^2 & \text{for t} = 4 \text{ (52 weeks)}, \end{cases} \tag{9.3}$$

where $\delta_2 \equiv \sigma_2/\sigma_1$, $\delta_3 \equiv \sigma_3/\sigma_1$, and $\delta_4 \equiv \sigma_4/\sigma_1$. Thus, parameters δ_2, δ_3, and δ_4, are the ratios of standard deviations (SDs) of visual acuity measurements for weeks 12, 24, and 52, relative to SD at week 4 (the reference level). Note that, according to (9.3), the scale parameter σ can be interpreted as SD at 4 weeks.

9.2.1 Fitting the Model Using the `gls()` *Function*

As shown in Panel R9.1 to fit model **M9.1**, we use the `gls()` function from the **nlme** package. Panel R9.1a includes the details of the function call. To define the mean structure of the model, we use the formula `lm1.form`, the same as the one specified in Panel R6.1 for model **M6.1**. To allow for heterogeneous variance, we set the `weights` argument to an object of the *varIdent* class, created with the help of the `varIdent()` constructor function. The *varIdent* class represents a variance structure with different variances for different strata (Table 7.2). In our case, the strata are defined by the levels of the factor `time.f`, which is indicated using the formula `~1|time.f` in the `form` argument of the `varIdent()` constructor function. That is, we allow for different variances of visual acuity measurements at different timepoints. Note that, by default, the `gls()` function uses the REML estimation (Sects. 7.4.2 and 8.4).

R9.1 *ARMD Trial*: Estimates and confidence intervals for timepoint-specific variance for model **M9.1**

(a) *Selected results for the model with timepoint-specific variances*

```
> lm1.form <-                                    # See also R6.1
+    formula(visual ~ -1 + visual0 + time.f + treat.f:time.f)
> fm9.1 <-                                       # M9.1
+    gls(lm1.form,
+        weights = varIdent(form = ~1|time.f), # Var. function; <δ>-group
+        data = armd)
> summary(fm9.1)
  Generalized least squares fit by REML
    Model: visual ~ visual0 + time.f + time.f:treat.f - 1
    Data: armd
       AIC    BIC  logLik
    6740.3 6802.1 -3357.1
  ...      [snip]
> fm9.1$modelStruct$varStruct                    # (9.3): δ̂1≡1, δ̂2, δ̂3, δ̂4
  Variance function structure of class varIdent representing
    4wks  12wks  24wks  52wks
  1.0000 1.3976 1.6643 1.8809
> (intervals(fm9.1, which = "var-cov"))          # 95% CI for δ2,δ3,δ4, & σ
  Approximate 95% confidence intervals

   Variance function:
          lower   est.  upper
  12wks 1.2269 1.3976 1.5921
  24wks 1.4576 1.6643 1.9004
  52wks 1.6409 1.8809 2.1559
  attr(,"label")
  [1] "Variance function:"

   Residual standard error:
   lower   est.  upper
  7.5190 8.2441 9.0391
```

(b) *REML-based LR test of homoscedasticity. The object* `fm6.1` *was created in Panel R6.3*

```
> anova(fm6.1, fm9.1)                            # M6.1 ⊂ M9.1
        Model df    AIC    BIC  logLik   Test L.Ratio p-value
  fm6.1     1 10 6821.6 6869.2 -3400.8
  fm9.1     2 13 6740.3 6802.1 -3357.1 1 vs 2  87.326  <.0001
```

We store the resulting model fit in the object `fm9.1`. The results of the fitted model can be printed out using the `summary()` function (Sect. 8.5). The printout is long and we only present a short part of it in Panel R9.1a; more details are presented in Table 9.1 later in this chapter. By referring to the `varStruct` component of the `modelStruct` component of the model-fit object (Table 8.2a), we display the estimated values of the $\boldsymbol{\delta}$ variance-function coefficients. The estimates indicate an increasing variability of visual acuity measurements in time. This is consistent with the results of exploratory analysis shown, e.g., in Panel R3.6.

By using the `intervals()` function (Table 8.2a), we obtain the estimates and 95% confidence intervals for the parameters $\boldsymbol{\delta}$ and σ. The confidence intervals are computed using the methods described in Sect. 7.6.2. The standard deviation at week 4 is estimated to be equal to 8.24. The 95% confidence intervals for the variance-function coefficients slightly overlap, but suggest timepoint-specific variances.

To formally test the hypothesis that the variances are timepoint specific, we apply, in Panel R9.1b, the `anova()` function. The LR-test statistic is calculated based on the likelihood for model **M6.1**, which assumed homoscedasticity, and that for model **M9.1**, which assumes heteroscedasticity. The results of the fit of the former model were stored in the object `fm6.1` (Panel R6.3). Note that both models differ only by their variance structure and were fitted using the `gls()` function with the default estimation method, i.e., REML. Moreover, model **M6.1** is nested within **M9.1**, because the former can be obtained from the latter by specifying $\sigma_1^2 = \sigma_2^2 = \sigma_3^2 = \sigma_4^2$. Thus, we can use the LR test (Sect. 4.6.1) to test the null hypothesis of homoscedasticity. The result of the test, based on the restricted-likelihood function (Sect. 7.4.2), is statistically significant ($p < 0.0001$). Consequently, we conclude that the data provide evidence for heterogeneous variances of visual acuity measurements at different measurement occasions. In further modeling, we will therefore assume heteroscedasticity.

9.3 Linear Models with the varPower(·) Variance-Function

In this section, for illustration purposes, we consider models with more parsimonious parameterizations of the variance structure. Toward this end, we use the varPower(·) variance function (Sect. 7.3.1).

The mean structure for all models introduced in this section is the same as the one specified in (9.1). The models differ with respect to the assumed form of the variances σ_t^2 of the residual errors ε_{it}.

Model **M9.2** specifies that

$$\sigma_{it} = \sigma\lambda_{it} = \sigma\lambda(\delta; \text{TIME}_{it}) = \sigma(\text{TIME}_{it})^{\delta}, \tag{9.4}$$

where $\text{TIME}_{it} = 4$, 12, 24, and 52 weeks for $t = 1$, 2, 3, and 4, respectively. The function $\lambda(\cdot)$, used in (9.4), is an example of the varPower(·) variance function from the <δ>-group (Table 7.2), with $\boldsymbol{\delta} \equiv \delta$, variance covariate TIME_{it}, and no strata. It specifies that the variance is a power function of the time (in weeks), at which the visual acuity measurement was taken.

Model **M9.3** assumes that the power coefficient depends on treatment:

$$\sigma_{it} = \sigma\lambda_{it} = \sigma\lambda\{(\delta_1, \delta_2); \text{TIME}_{it}\} = \begin{cases} \sigma(\text{TIME}_{it})^{\delta_1} & \text{for Active,} \\ \sigma(\text{TIME}_{it})^{\delta_2} & \text{for Placebo.} \end{cases} \tag{9.5}$$

Similarly to (9.4), the function $\lambda(\cdot)$ is the varPower(·) variance function from the <δ>-group (Table 7.2), with variance covariate TIME_{it} and $\boldsymbol{\delta} = (\delta_1, \delta_2)'$ for the two strata that correspond to the treatment groups.

Model **M9.4** specifies that the variances are a power function of the mean value:

$$\sigma_{it} = \sigma\lambda_{it} = \sigma\lambda(\mu_{it}, \delta) = \sigma(\mu_{it})^{\delta}, \tag{9.6}$$

where $\mu_{it} \equiv \beta_{0t} + \beta_1 \times \text{VISUAL0}_i + \beta_{2t} \times \text{TREAT}_i$ is the predicted (mean) value of VISUAL_{it}, as implied by (9.1). The function $\lambda(\cdot)$, used in (9.6), is an example of the varPower(·) variance function from the <δ, μ>-group (Table 7.3), with $\boldsymbol{\delta} \equiv \delta$ and no strata.

Finally, model **M9.5** assumes a constant coefficient of variation, i.e., it assumes that SDs of the visual acuity measurements are proportional to the mean values:

$$\sigma_{it} = \sigma\lambda_{it} = \sigma\lambda(\mu_{it}) = \sigma\mu_{it}. \tag{9.7}$$

The function $\lambda(\cdot)$, used in (9.7), is similar to the one used in (9.6), but with $\boldsymbol{\delta} \equiv 1$. Thus, it is an example of the varPower(·) variance function from the <μ>-group (Table 7.4). Note that, according to (9.7), $\sigma_{it}/\mu_{it} = \sigma$, i.e., the scale parameter can be interpreted as a coefficient of variation, constant for all timepoints.

9.3.1 Fitting the Models Using the `gls()` *Function*

In Panel R9.2, we fit the models, **M9.2–M9.5**, which have the same mean structure as model **M9.1**, but employ the variance functions from different groups. In addition, we illustrate model selection using the REML-based LR tests and AIC.

The syntax for fitting the models is given in Panel R9.2a. All models are fitted using the generic function `update()` to modify the `weights` argument, as compared to the function call used for model **M9.1** in Panel R9.1. In the argument, an appropriate form of the `varPower()` constructor-function is used. Note that, for models **M9.2** and **M9.3**, the formula specified in the `form` argument of the `varPower()` function refers to the continuous-time variable `time` rather than to the factor `time.f`. For model **M9.4**, the `varPower()`-function call does not use any arguments. This default syntax is equivalent to specifying `varPower(form=~fitted(.))`, which means that the fitted values $\widehat{\mu}_i$ are used as the variance covariate. Finally, the use of the argument `fixed=1` for model **M9.5** implies that, in the absence of any stratifying variable, the power coefficient δ is fixed at 1.

Models **M9.2** and **M9.3** are fitted using the REML estimation, suitable for variance functions belonging to the <δ>-group (Sect. 7.4.2). Model **M9.4** employing variance function from the <δ, μ>-group is fitted using the REML-based GLS (Sect. 7.8.1.1). The model **M9.5** involving variance function from <μ>-group is fitted using the REML-based IRLS (Sect. 7.8.1.2).

R9.2 *ARMD Trial*: Model selection using the REML-based LRT and AIC for models **M9.1–M9.5**. The model-fit object fm9.1 was created in Panel R9.1a

(a) *Models with various variance functions*

```
> fm9.2 <-                                          # M9.2 ← M9.1
+   update(fm9.1,
+          weights = varPower(form = ~time)) # (9.4), <δ>-group
> fm9.3 <-                                          # M9.3 ← M9.1
+   update(fm9.1,                                   # (9.5), strata=treat.f
+          weights = varPower(form = ~time|treat.f))
> fm9.4 <-                                          # M9.4 ← M9.1
+  update(fm9.1, weights = varPower())              # (9.6), <δ,μ>-group
> fm9.5 <-                                          # M9.5 ← M9.1
+  update(fm9.1,
+         weights = varPower(fixed = 1))            # (9.7), <μ>-group
```

(b) *Test of the variance structure: equal power of time for the two treatments*

```
> anova(fm9.2, fm9.3)                               # M9.2 ⊂ M9.3
      Model df    AIC    BIC  logLik   Test  L.Ratio p-value
fm9.2     1 11 6738.1 6790.4 -3358.1
fm9.3     2 12 6740.1 6797.2 -3358.1 1 vs 2 0.015529  0.9008
```

(c) *Test of the variance structure: power of time vs. timepoint-specific variances*

```
> anova(fm9.2, fm9.1)                               # M9.2 ⊂ M9.1
      Model df    AIC    BIC  logLik   Test L.Ratio p-value
fm9.2     1 11 6738.1 6790.4 -3358.1
fm9.1     2 13 6740.3 6802.1 -3357.1 1 vs 2   1.832  0.4001
```

(d) *Test of the variance structure: power of the mean value equal to 1*

```
> anova(fm9.5, fm9.4)                               # M9.5 ⊂ M9.4
      Model df    AIC    BIC  logLik   Test L.Ratio p-value
fm9.5     1 10 6965.9 7013.5 -3473.0
fm9.4     2 11 6823.1 6875.4 -3400.6 1 vs 2  144.84  <.0001
```

(e) *AIC for models* **M9.1– M9.5**

```
> AIC(fm9.1, fm9.2, fm9.3,                          # Nonnested models
+     fm9.4, fm9.5)                                 # Smaller AIC ≈ better fit
      df    AIC
fm9.1 13 6740.3
fm9.2 11 6738.1
fm9.3 12 6740.1
fm9.4 11 6823.1
fm9.5 10 6965.9
```

Note that, in Panel R9.2, we do not present parameter estimates of models **M9.2–M9.5**. Instead, we summarize the results later in this chapter in Tables 9.1 and 9.2.

9.3.1.1 Inference and Model Selection

First, we note that model **M9.2** is nested both in **M9.1** and in **M9.3**. Using the LR test based on models **M9.2** and **M9.3** allows testing the hypothesis that the power variance function for the two treatment groups is actually the same. More formally, we test the null hypothesis $\delta_1 = \delta_2$, against the alternative $\delta_1 \neq \delta_2$, with δ_1 and δ_2 defined in (9.5). The LR test is conducted in Panel R9.2b with the help of the `anova()` function. Note that the models have the same mean structure and that the test is based on the restricted likelihood, as required when the LR test is applied to verify hypotheses about variance-function parameters (Sects. 4.7.2 and 7.6.1). The result of the test is statistically not significant ($p = 0.9$). It indicates that a common-power variance function of the TIME covariate can be used for both treatment groups.

We may now ask the question whether the common-power variance function can be used as a more parsimonious representation of the variance structure of the data. To answer the question, we use the LR test based on the nested models **M9.1** and **M9.2**. Thus, we test the null hypothesis

$$\sigma_1^2 = 4^\delta \sigma^2, \quad \sigma_2^2 = 12^\delta \sigma^2, \quad \sigma_3^2 = 24^\delta \sigma^2, \quad \text{and} \quad \sigma_4^2 = \sigma^2 52^\delta$$

against the alternative

$$\sigma_1^2 \neq 4^\delta \sigma^2 \quad \text{or} \quad \sigma_2^2 \neq 12^\delta \sigma^2 \quad \text{or} \quad \sigma_3^2 \neq 24^\delta \sigma^2 \quad \text{or} \quad \sigma_4^2 \neq \sigma^2 52^\delta,$$

with the parameters defined in (9.4). Again, the test is based on the restricted likelihood. The result of the test displayed in Panel R9.2c is statistically not significant ($p = 0.4$). It suggests that the fit of model **M9.2**, measured by the value of the restricted log-likelihood, is not statistically significantly worse than the fit of model **M9.1**. Hence, model **M9.2**, which specifies that the variance is a power function of the time (in weeks), offers an adequate description of the variance structure of the data.

In Panel R9.2d, we use the REML-based LR test carried out using the likelihoods for models **M9.4** and **M9.5**. The test verifies the null hypothesis, implied by model **M9.5**, that, if we assume a variance function in the form of a power function of the mean value, the power coefficient is equal to 1. The result of the test is statistically significant ($p < 0.0001$). Thus, it allows rejecting the null hypothesis. Note that, given that models **M9.4** and **M9.5** are both mean-variance models, the inference on the implied variance structure, based on the result of the LR test, may need to be treated with caution (Sect. 7.8.2).

The results presented in Panel R9.2a–d indicate that, among models **M9.1–M9.3**, model **M9.2** seems to offer an adequate description of variance structure of the data. On the other hand, of models **M9.4** and **M9.5**, the former is more appropriate. Thus, it would be of interest to decide which of the models **M9.2** or **M9.4** fits the data better. However, the models are not nested, so we cannot compare them with the use of the LR test. Toward this aim, we need to apply the information criteria (Sect. 4.7.2).

In Panel R9.2e, we provide the values of AIC (Sect. 4.7.2) for models **M9.1–M9.5**. The results suggest that model **M9.2** offers a better fit to the data than model **M9.4**. In fact, AIC for the former is the smallest, as compared to the other four models. Thus, based on the information criterion, model **M9.2** offers the most adequate description of the data.

In Panel R9.3, we show how to extract the information about the estimated form of the variance function from the two objects representing model fits. Specifically, we extract corresponding information for objects `fm9.2` and `fm9.3` representing fitted models **M9.2** and **M9.3**, respectively. The applied extractor functions were discussed in Sect. 8.5. Note that model **M9.3**, as it has just been mentioned, is not the best model in a statistical sense, but it nicely illustrates several features related to the use of variance functions like, e.g., the use of a strata for the variance parameters, which are not present in the other considered models.

The estimated value of the power coefficient for model **M9.2** is very close to the estimated treatment-specific coefficients for model **M9.3**. This is in agreement with the result of the LR test, which was presented in Panel R9.2b. The estimated power coefficient for the variance function and σ of model **M9.2** indicate that the variance of visual acuity measurements increases according to the following relationship:

$$\text{Var}(\text{VISUAL}_{it}) = \sigma_t^2 \approx (6 \times \text{TIME}_{it}^{0.25})^2 = 36 \times \sqrt{\text{TIME}_{it}},$$

where TIME_{it} is the week at which the t-th measurement was taken.

Tables 9.1 and 9.2 display the REML-based estimates, obtained using the `gls()` function to fit models **M9.1– M9.5**. There are virtually no differences in the estimates of the fixed-effects coefficients for models **M9.1–M9.3**. In this respect, model **M9.5** is the most distinct one. The estimates of the timepoint-specific treatment effects are remarkably consistent among the five models, though. All the models suggest an increasing, negative effect of the "active" treatment compared to placebo.

The estimated standard errors of the fixed-effects coefficients vary more noticeably between all the models. This is related to the differences in the assumed residual-variance structure; as it was noted in Sect. 7.8.2, the precision of estimates of $\boldsymbol{\beta}$ depends on the (correct) specification of the structure.

Table 9.1 *ARMD Trial*: The REML parameter estimates[a], obtained using the `gls()` function, for models with different variance functions from the <δ>-group

	Parameter	`fm9.1`	`fm9.2`	`fm9.3`
Model label		**M9.1**	**M9.2**	**M9.3**
Log-REML value		−3357.15	−3358.06	−3358.06
AIC		6740.29	6738.13	6740.11
Fixed effects:				
Visual acuity at t = 0	β_1	0.86(0.03)	0.86(0.03)	0.86(0.03)
Time (4wks)	β_{01}	6.27(1.60)	6.28(1.62)	6.28(1.61)
Time (12wks)	β_{02}	5.28(1.76)	5.29(1.75)	5.29(1.75)
Time (24wks)	β_{03}	1.84(1.91)	1.84(1.89)	1.84(1.88)
Time (52wks)	β_{04}	−3.56(2.07)	−3.56(2.12)	−3.55(2.12)
Tm(4wks):Trt(Actv)	β_{21}	−2.33(1.09)	−2.33(1.12)	−2.33(1.11)
Tm(12wks):Trt(Actv)	β_{22}	−3.69(1.53)	−3.69(1.48)	−3.69(1.48)
Tm(24wks):Trt(Actv)	β_{23}	−3.43(1.88)	−3.43(1.82)	−3.43(1.82)
Tm(52wks):Trt(Actv)	β_{24}	−4.44(2.23)	−4.44(2.32)	−4.44(2.32)
Variance functions:				
a. varIdent(TIME): (9.3)				
12 *vs.* 4 wks	δ_2	1.40(1.23,1.59)		
24 *vs.* 4 wks	δ_3	1.66(1.46,1.90)		
52 *vs.* 4 wks	δ_4	1.88(1.64,2.16)		
b. Power of time: (9.4)	δ		0.25(0.20,0.30)	
c. Power of time: (9.5)				
Active	δ_1			0.25(0.20,0.31)
Placebo	δ_2			0.25(0.20,0.30)
Scale	σ	8.24(7.52,9.04)	5.97(5.15,6.93)	5.97(5.15,6.93)

[a]Approximate SE for fixed effects and 95% CI for covariance parameters are included in parentheses

Table 9.2 *ARMD Trial*: The REML parameter estimates[a], obtained using the `gls()` function, for the mean-variance models

	Parameter	`fm9.4`	`fm9.5`
Model label		**M9.4**	**M9.5**
Log-REML value		−3400.55	−3472.97
AIC		6823.10	6965.94
Fixed effects:			
Visual acuity at t = 0	β_1	0.83(0.03)	0.76(0.03)
Time (4wks)	β_{01}	7.87(1.96)	10.79(1.81)
Time (12wks)	β_{02}	6.84(1.96)	10.62(1.80)
Time (24wks)	β_{03}	3.38(1.97)	7.00(1.72)
Time (52wks)	β_{04}	−2.11(2.02)	3.92(1.67)
Tm(4wks):Trt(Actv)	β_{21}	−2.36(1.62)	−2.44(1.78)
Tm(12wks):Trt(Actv)	β_{22}	−3.69(1.64)	−4.35(1.74)
Tm(24wks):Trt(Actv)	β_{23}	−3.46(1.70)	−3.11(1.67)
Tm(52wks):Trt(Actv)	β_{24}	−4.43(1.79)	−5.53(1.54)
Variance function:			
power (μ^{δ}) : (9.6)	δ	−0.06(−0.21,0.09)	1
Scale	σ	15.51(8.74,27.50)	0.29(0.28,0.30)

[a]Approximate SE for fixed effects and 95% CI for covariance parameters are included in parentheses

R9.3 *ARMD Trial*: Extracting information about the variance functions for models **M9.2** and **M9.3**

(a) *Model* **M9.2***: Power-of-time variance function*

```
> mSt2 <- fm9.2$modelStruct                  # Model structure
> vF2 <- mSt2$varStruct                      # Variance function:(9.4)
> summary(vF2)                               # Summary: δ̂.
  Variance function:
   Structure: Power of variance covariate
   Formula: ~time
   Parameter estimates:
    power
  0.25193
> summary(fm9.2)$sigma                       # σ̂
  [1] 5.9749
```

(b) *Model* **M9.3***: Power-of-time with treatment-specific coefficients*

```
> mSt3 <- fm9.3$modelStruct                  # Model structure
> vF3  <- mSt3$varStruct                     # Variance function:(9.5)
> summary(vF3)                               # Summary: δ̂1, δ̂2
  Variance function:
   Structure: Power of variance covariate, different strata
   Formula: ~time | treat.f
   Parameter estimates:
   Active Placebo
  0.25325 0.25113
> coef(vF3)                                  # δ̂1, δ̂2
   Active Placebo
  0.25325 0.25113
> formula(vF3)                               # Variance function formula
  ~time | treat.f
> varWeights(vF3)[3:10]                      # Weights for two subjects
   Active  Active  Active  Active Placebo Placebo Placebo Placebo
  0.70393 0.53297 0.44716 0.36764 0.70600 0.53578 0.45019 0.70600
```

9.3.2 Model-Fit Evaluation

Although the AIC values, presented in Panel R9.2e, suggest that model **M9.2** with the varPower$(\cdot)$ variance function is the best-fitting one, we know that the model does not offer a proper description of the data, because it ignores the within-subject correlation between the visual acuity measurements. For illustrative purposes, we will assess the fit of the model using residual plots.

R9.4 *ARMD Trial*: Residual plots for model **M9.2**. The model-fit object `fm9.2` was created in Panel R9.2a

(a) *Raw residuals*

```
> library(lattice)
> plot(fm9.2,                                       # Fig. 9.1a
+      resid(., type = "response") ~ fitted(.)) # Raw vs. fitted
> plot(fm9.2,                                       # Raw vs. time (not shown)
+      resid(., type = "response") ~ time)       # (See Fig. 9.1a)
> bwplot(resid(fm9.2) ~ time.f,                   # Fig. 9.1b
+         pch = "|", data = armd)                   # Raw vs. time.f.
```

(b) *Pearson residuals*

```
> plot(fm9.2,                                       # Fig. 9.1c
+      resid(., type = "pearson" ) ~ fitted(.)) # Pearson vs. fitted
> plot(fm9.2,                                       #  vs. time (not shown)
+      resid(., type = "pearson") ~ time)        # (See Fig. 9.1c)
> bwplot(                                           # Fig. 9.1d
+   resid(fm9.2, type = "pearson") ~ time.f,      # Pearson vs. time.f
+   pch = "|", data = armd)
```

(c) *Scale-location plots*

```
> plot(fm9.2,                                   # Fig. 9.2a
+      sqrt(abs(resid(., type = "response"))) ~ fitted(.),
+      type = c("p", "smooth"))
> plot(fm9.2,                                   # Fig. 9.2b
+      sqrt(abs(resid(., type = "pearson"))) ~ fitted(.),
+      type = c("p", "smooth"))
```

The R code in Panel R9.4 constructs several plots of raw residuals (Sect. 7.5) for model **M9.2**. In particular, in Panel R9.4a scatterplots of the residuals *versus* fitted values and *versus* the time covariate are created with the help of the `plot()` function. The first of the plots is shown in Fig. 9.1a. It displays an asymmetric pattern, with large positive (negative) residuals present mainly for small (large) fitted values.

To evaluate the distribution of the raw residuals, we use the function `bwplot()` from the package **lattice** (Sect. 3.2.2) to create a box-and-whiskers plot of the residuals for each timepoint. The resulting graph is shown in Fig. 9.1b. The box-and-whiskers plots clearly show an increasing variance of the residuals.

Note that, in Panel R9.4a, we create a draft of the graph presented in Fig. 9.1b. We do not show the details on how to enhance it by providing labels for the horizontal axis, because a suitable syntax can be inferred from Panel R3.4.

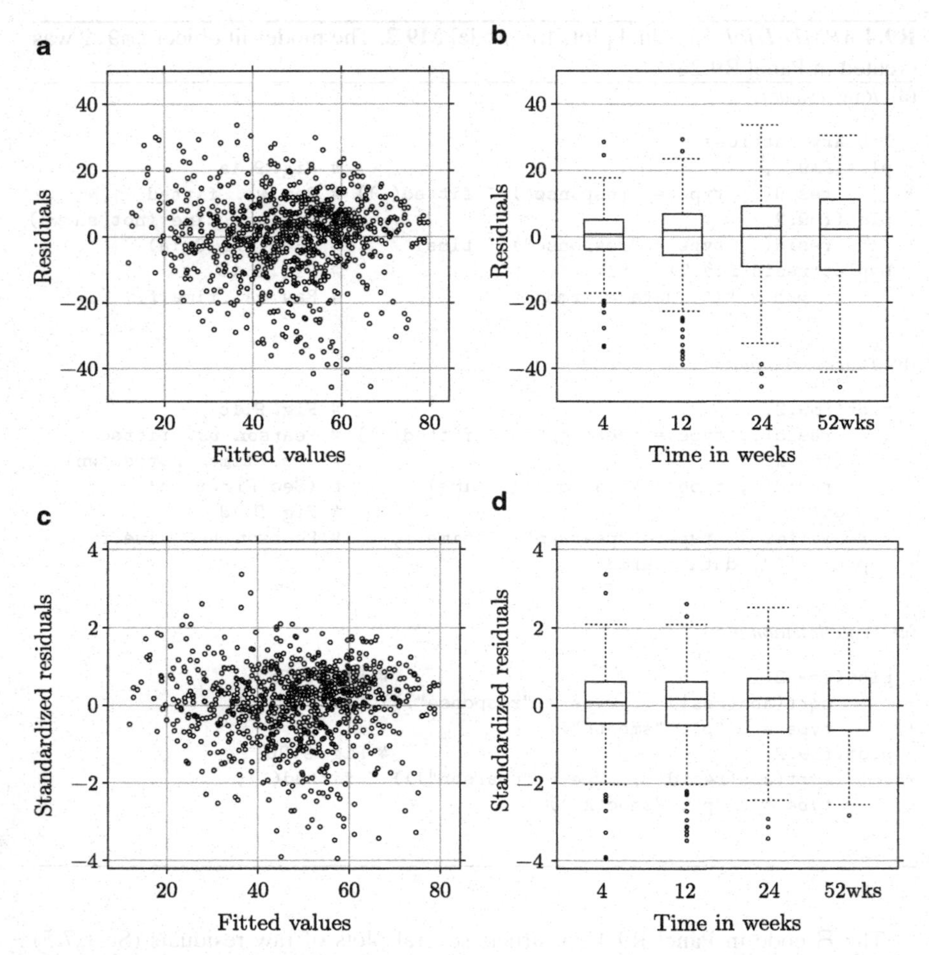

Fig. 9.1 *ARMD Trial*: Residual plots for model **M9.2** (**a**) Raw *versus* fitted (**b**) Raw *versus* time (**c**) Pearson *versus* fitted (**d**) Pearson *versus* time

In Panel R9.4b, we create corresponding plots of Pearson residuals (Sect. 7.5.1). The scatterplot of the residuals *versus* fitted values is shown in Fig. 9.1c. Similarly to the plot of the raw residuals, it displays an asymmetric pattern. The box-and-whiskers plots of the Pearson residuals for each timepoint are shown in Fig. 9.1d. The plots illustrate the effect of scaling: the variance of the residuals is virtually constant.

In Panel R9.4c, we construct the scale-location plots for the raw and Pearson residuals. These are the scatterplots of the square-root transformation of the absolute value of the residuals *versus* fitted values. The plots allow for detection of patterns in the residual variance. The plots, constructed in Panel R9.4c, include a smooth curve, which facilitates a visual assessment of a trend.

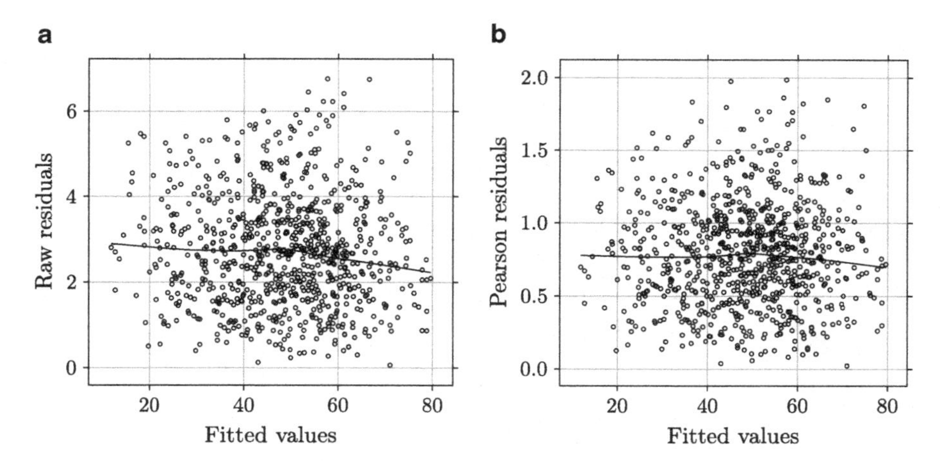

Fig. 9.2 *ARMD Trial*: Scale-location plots for model **M9.2** based on (**a**) Raw and (**b**) Pearson residuals

The plot for the raw residuals, shown in Fig. 9.2a, suggests a dependence between the residual variance and the mean value. However, this may be an artifact of the heteroscedasticity of the raw residuals, which was observed in Fig. 9.1b. Thus, it might be better to look at the scale-location plot for the Pearson residuals. The plot is shown in Fig. 9.2b; it does not indicate any clear trend in the residual variance.

Figure 9.3 presents a scatterplot matrix of the Pearson residuals for all four measurement occasions. The figure was constructed using the `splom()` function for the data for 188 subjects with all four postrandomization visual acuity measurements. The 95% confidence ellipses were added using the `ellipse()` function from the **ellipse** package. For brevity, we do not show the R code for creating the figure. The scatterplots clearly show a violation of the assumption of the independence of observations: residuals for different measurement occasions are correlated. The correlation coefficient decreases with the increasing distance between the timepoints. Of course, some caution is needed in interpreting the strength of correlation, because the estimated residuals are correlated even if the independence assumption holds (Sect. 4.5.1).

9.4 Chapter Summary

In this chapter, we considered an LM for independent observations with heterogeneous variance. We illustrated its application using the ARMD dataset. Strictly speaking, the model is not suitable for the analysis of this dataset, as it ignores the dependence of visual acuity measurements obtained for the same individual. Thus, the presented results should mainly be treated as an illustration of the important theoretical concepts and R software tools available for this type of models.

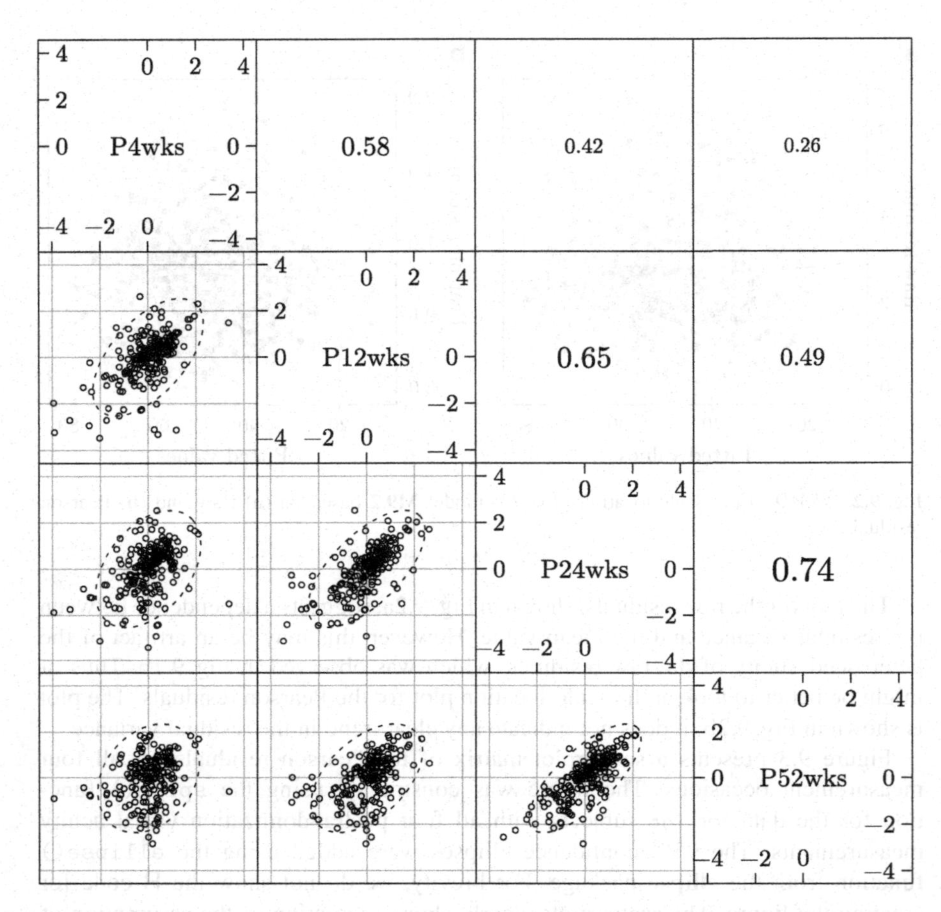

Fig. 9.3 *ARMD Trial*: Scatterplot matrix of the Pearson residuals for model **M9.2** (complete cases only, $n = 188$; correlation coefficients above the diagonal)

In particular, we focused on the concept of the variance function, which allows addressing heteroscedasticity. Variance functions are also useful in formulation of LMMs, which are discussed in later chapters. Thus, from this point of view, a good understanding of the concept and its implementation in R is important.

We considered several models, constructed with the help of the varIdent(·) (Sect. 9.2) and varPower(·) (Sect. 9.3) variance functions. Table 9.3 summarizes the models. The mean structure of all of the models was defined in (9.1). Model **M9.1** assumed different variances of visual acuity measurements taken at different timepoints. Model **M9.2** specified that the variances changed proportionally to the power of the number of weeks at which the measurements were taken. Model **M9.3** postulated that the power coefficient was different for different treatment groups. Models **M9.4** and **M9.5** assumed that the variances depended on a power function

Table 9.3 *ARMD Trial*: Summary of the models defined in Chap. 9. The mean structure for all models was defined in (9.1) and can be expressed using formula: `visual ~ -1 + time.f + treat.f:time.f`

Model	Section	R-syntax	R-object	Variance function (eq.)/group
M9.1	9.2	R9.1	`fm9.1`	Time-specific variance (9.2, 9.3)/<δ>-group
M9.2	9.3	R9.2	`fm9.2`	varPower(·) for TIME (9.4)/<δ>-group
M9.3	9.3	R9.2	`fm9.3`	varPower(·) for TIME (9.5)/<δ>-group; stratified by `treat.f`
M9.4	9.3	R9.2	`fm9.4`	varPower(·) for μ (9.6)/<δ, μ>-group
M9.5	9.3	R9.2	`fm9.5`	varPower(·) for μ (9.6)/<μ>-group, $\delta = 1$

of the mean value of the measurements: the former model assumed the power coefficient was unknown, while the latter assumed that the power coefficient was equal to 1.

All the models were fitted to the data by applying the `gls()` function with an appropriate value of the `weights` argument. In Sects. 9.2 and 9.3.1, we illustrated how the information on the fitted model can be extracted from the model-fit object of class *gls* and how to select the best model. Additionally, in Sect. 9.3.2, we reviewed the goodness-of-fit model **M9.2** the best fitting, according to AIC. As expected, symptoms of violation of the independence assumption were detected. In the next chapter, we will discuss a class of models that takes correlation of observations into account.

Part III
Linear Fixed-Effects Models for Correlated Data

Chapter 10
Linear Model with Fixed Effects and Correlated Errors

10.1 Introduction

The essential assumption for the LMs considered in Part II of the book was that the observations collected during the study were independent of each other. This assumption is restrictive in studies which use sampling designs that lead to correlated data. Such data result, for example, from studies collecting measures over time, i.e., in a longitudinal fashion; in designs which involve clustering or grouping, e.g., cluster-randomization clinical trials; in studies collecting spatially correlated data, etc. Note that, in contrast to Part II, for such designs, the distinction between sampling units (e.g., subjects in a longitudinal study) and analysis units (e.g., time-specific measurements) is important.

In Part III of the book, we consider a class of more general LMs that allow relaxing the assumptions of independence and variance homogeneity. We will refer to these models as LMs with fixed effects and correlated residual errors for grouped data, or simply as LMs for correlated data. The models can be viewed as an example of *population-averaged* models, i.e., models in which the parameters are interpreted as quantifying effects of covariates on the marginal mean value of the dependent variable for the *entire* population.

The goal of the current chapter is to describe the fundamental concepts of the theory of LMs for correlated data. In particular, we introduce the important notion of a *correlation structure*. It is a general concept, which is also applicable to LMMs that will be described in Chap. 13. The concept is introduced in this chapter in a relatively simple framework, allowing for a better exposition. By combining correlation structures with variance functions (Sect. 7.3), flexible forms of variance-covariance matrices can be specified for an LM for correlated data.

This chapter is structured as follows. In Sects. 10.2–10.6, we review the theory underlying the LMs for correlated data. In particular, in Sects. 10.2 and 10.3 we describe the specification of the models the use of correlation structures. Sections 10.4–10.6 present estimation methods, model diagnostics, and model

A. Gałecki and T. Burzykowski, *Linear Mixed-Effects Models Using R: A Step-by-Step Approach*, Springer Texts in Statistics, DOI 10.1007/978-1-4614-3900-4_10,

reduction and selection methods, respectively. Section 10.7 is devoted to the models in which variance depends on the mean value. In Sect. 10.8, a summary of the chapter is offered.

In our explanations, we refer to the material presented in Chaps. 4 and 7, indicating the necessary modifications. We focus on the concepts and methods that are implemented in R. The details of the corresponding R syntax will be discussed in Chap. 11.

10.2 Model Specification

In this section, we specify an LM with fixed effects and correlated residual errors for grouped data with hierarchical structure introduced in Chap. 2. For ease of exposition, we focus initially on data with a single-level of grouping, with N groups (levels of a grouping factor) indexed by i ($i = 1,\ldots,N$) and n_i observations per group indexed by j ($j = 1,\ldots,n_i$).

We build on (7.10)–(7.12) for the LM for independent observations with heterogeneous variance, presented in Chap. 7. More concretely, we assume that, for group i, the model for a continuous dependent variable $\mathbf{y}_i$ is expressed as

$$\mathbf{y}_i = \boldsymbol{X}_i\boldsymbol{\beta} + \boldsymbol{\varepsilon}_i, \tag{10.1}$$

where

$$\mathbf{y}_i \equiv \begin{pmatrix} y_{i1} \\ \vdots \\ y_{ij} \\ \vdots \\ y_{in_i} \end{pmatrix}, \quad \boldsymbol{\varepsilon}_i \equiv \begin{pmatrix} \varepsilon_{i1} \\ \vdots \\ \varepsilon_{ij} \\ \vdots \\ \varepsilon_{in_i} \end{pmatrix}, \tag{10.2}$$

$$\boldsymbol{X}_i \equiv \begin{pmatrix} x_{i1}^{(1)} & x_{i1}^{(2)} & \ldots & x_{i1}^{(p)} \\ \vdots & \vdots & \ddots & \vdots \\ x_{in_i}^{(1)} & x_{in_i}^{(2)} & \ldots & x_{in_i}^{(p)} \end{pmatrix} \equiv \left(\mathbf{x}_i^{(1)}\ \mathbf{x}_i^{(2)}\ \ldots\ \mathbf{x}_i^{(p)} \right), \tag{10.3}$$

$\boldsymbol{\beta}$ is defined in (4.3), $\boldsymbol{X}_i$ is a design matrix for the i-th group, and the vector of the within-group residual errors, $\boldsymbol{\varepsilon}_i$, is assumed to have a multivariate normal distribution. More specifically,

$$\boldsymbol{\varepsilon}_i \sim \mathcal{N}_{n_i}(\mathbf{0}, \boldsymbol{\mathcal{R}}_i), \tag{10.4}$$

where, for the variance-covariance matrix $\mathcal{R}_i$, a representation similar to (7.12) is assumed, that is,

$$\mathcal{R}_i = \sigma^2 \boldsymbol{R}_i, \tag{10.5}$$

with σ^2 denoting an unknown scalar parameter. Finally, we assume that vectors of residual errors for different groups are independent, i.e., $\boldsymbol{\varepsilon}_i$ is independent of $\boldsymbol{\varepsilon}_{i'}$ for $i \neq i'$.

It is straightforward to note that the mean and variance of $\mathbf{y}_i$ are given as follows:

$$\mathrm{E}(y_{ij}) \equiv \mu_{ij} = \mathbf{x}_{ij}'\boldsymbol{\beta}, \tag{10.6}$$

$$\mathrm{Var}(\mathbf{y}_i) = \sigma^2 \boldsymbol{R}_i. \tag{10.7}$$

The formulation of models described above allows for data with more than one level of grouping. Multiple levels of grouping would be reflected by introducing factors, related to the different group levels, into the design matrix $\boldsymbol{X}_i$, and by assuming a particular form of the variance-covariance matrix $\mathcal{R}_i$. However, to deal with multiple levels of grouping, only a small modification of the R syntax, used for a setting with a single level of grouping, is required. Thus, the latter setting is the most important one from an R-syntax point of view. Hence, in the remainder of this chapter, we will focus on models for the data with a single level of grouping.

10.3 Details of Model Specification

It is important to note that the LM with correlated errors, specified by (10.1)–(10.5), is not identifiable in its most general form. This is because of nonuniqueness of the representation (10.5) and because the model potentially involves too many unknown parameters related to the variance-covariance matrix of the residual errors $\boldsymbol{\varepsilon}_i$. The issue is similar to the one described in Sect. 7.2 for the LM with heterogeneous variance.

The model (10.1)–(10.5) may become identifiable, however, if we impose additional constraints on the residual variance-covariance matrices $\mathcal{R}_i$. A solution is to represent the matrices as functions of a small number of parameters. An approach, implemented, e.g., in the function `gls()` of the **nlme** package, is based on the fact that $\boldsymbol{R}_i$, defined in (10.5), can be decomposed as

$$\boldsymbol{R}_i = \boldsymbol{\Lambda}_i \boldsymbol{C}_i \boldsymbol{\Lambda}_i, \tag{10.8}$$

where $\boldsymbol{\Lambda}_i$ is a diagonal matrix with nonnegative diagonal elements, and $\boldsymbol{C}_i$ is a correlation matrix. Note the similarity of the decomposition, described in (10.8), to that specified in (7.10) for the LM with independent, heteroscedastic observations. By using $\boldsymbol{\Lambda}_i$ in (10.8), we allow for heteroscedasticity of observations within group i, while by employing the correlation matrix $\boldsymbol{C}_i$, we allow for correlation between the observations within the group.

By employing disjoint sets of parameters for $\boldsymbol{C}_i$ and $\boldsymbol{\Lambda}_i$, we use the decomposition (10.8) to model $\boldsymbol{R}_i$. More concretely, we assume that the diagonal elements of the diagonal matrix $\boldsymbol{\Lambda}_i$ are, in general, expressed as

$$\{\boldsymbol{\Lambda}_i\}_{j,j} \equiv \lambda_{i_{j,j}} = \lambda(\mu_{ij}, \boldsymbol{\delta}; \mathbf{v}_{ij}), \tag{10.9}$$

where $\lambda(\cdot)$ is a variance function (Sects. 7.2.2 and 7.3.1).

Similarly to (7.6), $\boldsymbol{\delta}$ is a vector of variance parameters, and $\mathbf{v}_{ij}$ is a vector of (known) variance covariates. Moreover, we assume that the matrix $\boldsymbol{C}_i$ is specified using a set of parameters $\boldsymbol{\varrho}$, which will be defined in Sect. 10.3.2. Thus, formally speaking, (10.8) should be written as

$$\boldsymbol{R}_i(\mu_{ij}, \boldsymbol{\theta}_R; \mathbf{v}_{ij}) = \boldsymbol{\Lambda}_i(\mu_{ij}, \boldsymbol{\delta}; \mathbf{v}_{ij})\boldsymbol{C}_i(\boldsymbol{\varrho})\boldsymbol{\Lambda}_i(\mu_{ij}, \boldsymbol{\delta}; \mathbf{v}_{ij}) \tag{10.10}$$

where $\boldsymbol{\theta}_R \equiv (\boldsymbol{\delta}', \boldsymbol{\varrho}')'$. However, to simplify notation, we will often suppress the use of $\boldsymbol{\theta}_R$, μ_{ij}, and $\mathbf{v}_{ij}$ in formulae, unless specified otherwise.

The classical LM, specified in Sect. 4.2, is obtained as a special case of model (10.1)–(10.5), with $\boldsymbol{R}_i$ given by (10.10), upon assuming that $n_i = 1$ and that $\boldsymbol{R}_i = 1$ for all i. Independence and homoscedasticity of the residual errors, $\boldsymbol{\varepsilon}_j$, then follows from the normality assumption (10.4). Also, the LMs for independent, heteroscedastic observations, specified in Sect. 7.2, can be seen as a special case of model (10.1)–(10.5), with $\boldsymbol{R}_i$ given by (10.10), if we assume that $n_i = 1$ and that $\boldsymbol{R}_i = \lambda_i^2$, where λ_i is defined in (7.8).

It should be noted that, by employing separate sets of parameters in (10.10), namely, $\boldsymbol{\delta}$ for $\boldsymbol{\Lambda}_i$ and $\boldsymbol{\varrho}$ for $\boldsymbol{C}_i$, additional constraints are imposed on the structure of the matrix $\mathcal{R}_i$, as defined in (10.5). For example, variance-covariance matrices composed by a variance function and correlation matrix sharing some of the parameters are not allowed under this framework.

In what follows, in Sects. 10.3.1 and 10.3.2, we review the use of the variance and correlation functions used in the decomposition (10.10).

10.3.1 Variance Structure

Similarly to the case of the LM for independent observations with heterogeneous variance, specified in Sect. 7.2, the elements of the matrix $\boldsymbol{\Lambda}_i$, given in (10.10), are defined using the variance function (Sects. 7.2.2 and 7.3.1). For data with a single level of grouping, the variance function definition, presented in (7.6), is modified by the use of double indices i and j, so that the variance of the residual errors is written as

$$\text{Var}(\varepsilon_{ij}) = \sigma^2\lambda^2(\mu_{ij}, \boldsymbol{\delta}; \mathbf{v}_{ij}), \tag{10.11}$$

where μ_{ij} is the mean value, given in (10.6), $\mathbf{v}_{ij}$ is a vector of (known) variance covariates, $\boldsymbol{\delta}$ is a vector of covariance parameters, and $\lambda(\cdot)$ is a continuous function with respect to $\boldsymbol{\delta}$.

The decomposition (10.10) allows for the use of both mean-independent and mean-dependent variance functions (Table 7.1). However, as mentioned in, e.g., Sects. 7.4 and 7.8, the application of variance functions that depend on the mean value requires the use of more advanced estimation and inferential approaches. For this reason, in the next sections, we will mainly concentrate on the use of variance functions from the <δ>-group, which do not depend on the mean value (Table 7.2). The use of mean-dependent functions from the <δ, μ>- and <δ>-groups (Tables 7.3 and 7.4, respectively) will be discussed in Sect. 10.7.

Note that, for variance functions that do not depend on the mean value, (10.11) simplifies to

$$\mathrm{Var}(\varepsilon_{ij}) = \sigma^2 \lambda^2(\boldsymbol{\delta}; \mathbf{v}_{ij}). \tag{10.12}$$

10.3.2 Correlation Structure

In this section, we present selected examples of structures for the correlation matrix $\boldsymbol{C}_i$, defined in (10.10). Following the convention used in R, the matrix $\boldsymbol{C}_i$ is specified by assuming that the correlation coefficient between two residual errors, ε_{ij} and $\varepsilon_{ij'}$, corresponding to two observations from the same group i, is given by

$$\mathrm{Corr}(\varepsilon_{ij}, \varepsilon_{ij'}) = h[d(\mathbf{t}_{ij}, \mathbf{t}_{ij'}), \boldsymbol{\varrho}], \tag{10.13}$$

where $\boldsymbol{\varrho}$ is a vector of correlation parameters, $d(\mathbf{t}_{ij}, \mathbf{t}_{ij'})$ is a distance function of vectors of position variables $\mathbf{t}_{ij}$ and $\mathbf{t}_{ij'}$ corresponding to, respectively, ε_{ij} and $\varepsilon_{ij'}$, and $h(\cdot,\cdot)$ is a continuous function with respect to $\boldsymbol{\varrho}$, such that it takes values between -1 and 1, and $h(0, \boldsymbol{\varrho}) \equiv 1$.

By assuming various distances and correlation functions, a variety of correlation structures can be obtained. In what follows, we limit our discussion to the structures, which are implemented in R. When referring to them, we will use names borrowed from the **nlme** package. The correlation structures include:

corCompSymm	a compound-symmetry structure corresponding to uniform correlation.
corAR1	corresponding to an autoregressive process of order 1.
corARMA	corresponding to an autoregressive moving average (ARMA) process.
corCAR1	corresponding to a continuous-time autoregressive process.
corSymm	a general correlation matrix.
corExp	exponential spatial correlation.
corGaus	Gaussian spatial correlation.
corLin	linear spatial correlation.
corRatio	rational quadratic spatial correlation.
corSpher	spherical spatial correlation.

Table 10.1 Examples of serial and spatial correlation structures

Correlation structure	Function $h(.,.)$	Comment
Serial	(Auto)correlation function	
corCompSymm[a]	$h(k,\varrho) \equiv \varrho$	$k = 1,2,\ldots;\ \lvert\varrho\rvert < 1$
corAR1	$h(k,\varrho) \equiv \varrho^k$	$k = 0,1,\ldots;\ \lvert\varrho\rvert < 1$
corCAR1	$h(s,\varrho) \equiv \varrho^s$	$s \geq 0;\ \varrho \geq 0$
corSymm	$h(d(j,j'),\varrho) \equiv \varrho_{j,j'}$	$j < j';\ \lvert\varrho_{jj'}\rvert < 1$
Spatial	Correlation function	
corExp	$h(s,\varrho) \equiv e^{-s/\varrho}$	$s \geq 0;\ \varrho > 0$
corGaus	$h(s,\varrho) \equiv e^{-(s/\varrho)^2}$	$s \geq 0;\ \varrho > 0$
corLin	$h(s,\varrho) \equiv (1 - s/\varrho)\mathrm{I}(s < \varrho)$	$s \geq 0;\ \varrho > 0$
corRatio	$h(s,\varrho) \equiv 1 - (s/\varrho)^2/\{1 + (s/\varrho)^2\}$	$s \geq 0;\ \varrho > 0$
corSpher	$h(s,\varrho) \equiv [1 - 1.5(s/\varrho) + 0.5(s/\varrho)^3]\mathrm{I}(s < \varrho)$	$s \geq 0;\ \varrho > 0$

[a]The names of the structures follow the convention used in the **nlme** package

The correlation functions $h(\cdot,\cdot)$, corresponding to the structures listed above (except for an ARMA process, which is excluded for brevity, but explained in a more detail later in this section), are described in Table 10.1.

The correlation structures can be classified into two main groups:

1. "Serial" structures (*corCompSymm*, *corAR1*, *corARMA*, *corCAR1*, *corSymm*).
2. "Spatial" structures (*corExp*, *corGaus*, *corLin*, *corRatio*, *corSpher*).

The reason for using quotation marks in the names of the groups of correlations structures is that, in principle, the split follows the convention used in R and does not necessarily reflect the properties of these structures in their most general form. In what follows, however, we will use the naming convention proposed above.

The first group, listed above, corresponds to the correlation structures which are defined in the context of time-series or longitudinal data. The second group corresponds to correlation structures which are defined in the context of spatial data. We will now review the properties of the two groups of correlation structures, with a focus on the aspects relevant for their implementation in R.

10.3.3 Serial Correlation Structures

For the *corCompSymm*, *corAR1*, *corARMA*, and *corSymm* correlation structures, it is assumed that $\mathbf{t}_{ij}$ are simply positive integer scalars, i.e., $\mathbf{t}_{ij} \equiv j$, describing the position of observation in a time-series/longitudinal sequence. For *corCAR1*, the actual value of measurement time is actually used. For *corCompSymm*, *corAR1*, and *corARMA*, the distance function is simplified even further by assuming that the function depends on the *time lag*, i.e., the absolute difference, k, of the two position indices: $k = |j - j'|$. For these correlation structures, the function $h(\cdot,\cdot)$

simply depends on k and ϱ. Note that, for time-series data, the function is often called an *autocorrelation function*.

For instance, the simplest serial correlation structure, *compound-symmetry* (*corCompSymm*), assumes a constant correlation between all within-group residual errors. This means that

$$\text{Corr}(\varepsilon_{ij}, \varepsilon_{ij'}) = \varrho, \tag{10.14}$$

which corresponds to (10.13) upon defining, for $j \neq j'$ and $k = 1, 2, \ldots,$

$$h(k, \varrho) \equiv \varrho. \tag{10.15}$$

A more advanced example of a serial correlation structure, *corARMA*, is obtained from an ARMA process. The process corresponds to longitudinal observations, for which a current observation can be expressed as a sum of (1) a linear combination of, say p, previous observations; (2) a linear combination of, say q, mean-zero, independent and identically distributed residual random errors from previous observations; and (3) a mean-zero, independent residual random error for the current measurement. The structure is described by $p+q$ parameters. Unlike the correlation structures shown in Table 10.1, the (auto)correlation function of an ARMA process cannot be expressed by a simple, closed-form expression, but it is defined by a recursive relation (Box et al. 1994). More details about the ARMA structure can be found in, e.g., Jones (1993) and Pinheiro and Bates (2000, Sect. 5.3.1).

10.3.4 Spatial Correlation Structures

The second group of correlation structures, which includes *corExp*, *corGaus*, *corLin*, *corRatio*, and *corSpher*, corresponds to structures that are defined in the context of spatial data. For these structures, it is allowed that $\mathbf{t}_{ij}$ are genuine two- or more dimensional real-number vectors. Note, however, that a unidimensional vector can also be used, which allows the application of the structures to time-series/longitudinal data.

For instance, the exponential correlation structure, *corExp*, is given by

$$\text{Corr}(\varepsilon_{ij}, \varepsilon_{ij'}) = e^{-s_{ij,ij'}/\varrho}, \tag{10.16}$$

where $s_{ij,ij'} = d(\mathbf{t}_{ij}, \mathbf{t}_{ij'})$ is a real number equal to the distance between the two position vectors $\mathbf{t}_{ij}$ and $\mathbf{t}_{ij'}$ corresponding to observations j and j', respectively, from the same group i. The corresponding function h is defined as

$$h(s, \varrho) \equiv e^{-s/\varrho}. \tag{10.17}$$

Note that, to explicitly define the spatial correlation structures, in addition to defining correlation functions, given in Table 10.1, we should also provide a distance function. There are several possibilities here. The most natural choice is the Euclidean distance, i.e., the square root of the sum, over all dimensions, of the squares of distances. Other possible distance functions include the "maximum" (or Tchebyshev) metric, i.e., the maximum, over all dimensions, of the absolute differences; and Manhattan (or "city block", "taxicab") distance, i.e., the sum, over all dimensions, of the absolute differences. Note that these three choices correspond to the L_2, L_∞ (Cantrell 2000), and L_1 metrics, respectively. In Sect. 11.4.2, we demonstrate examples of using these distance functions.

It is worth noting that in the spatial correlation literature, the parameter ϱ, used in Table 10.1 for the spatial structures, is referred to as *range*. The reader may want to verify that all spatial correlation functions $h(s,\varrho)$, presented in Table 10.1, are continuous and monotonically nonincreasing with respect to s at $s = 0$. This characteristic reflects a commonly observed feature of the data that observations being further apart are correlated to a lesser degree.

As already mentioned, the value of $h(0,\varrho)$ is equal to 1. This requirement can be relaxed by including the so-called *nugget effect*, an abrupt change in correlation at small distances (discontinuity at zero), which can be defined by the condition that $h(s,\varrho)$ tends to $1-\varrho_0$, with $\varrho_0 \in (0,1)$, when s tends to 0. In other words, a discontinuity at $s = 0$ can be allowed for. Consequently, a correlation function $h_{\varrho_0}(\cdot,\cdot)$ containing a nugget effect can be obtained from any continuous spatial correlation function $h(\cdot,\cdot)$ by defining

$$h_{\varrho_0}(s,\varrho) \equiv \begin{cases} (1-\varrho_0)h(s,\varrho) & \text{if} \quad s > 0, \\ 1 & \text{if} \quad s = 0. \end{cases} \tag{10.18}$$

Instead of the correlation function, spatial correlation structures are often represented by the *semivariogram function* or simply *semivariogram* (Cressie 1991). For the cases considered in this book, the semivariogram function can be defined as the complement of the correlation function, that is,

$$\gamma(s,\varrho) \equiv 1 - h(s,\varrho). \tag{10.19}$$

Similarly to (10.18), the nugget effect can be included in the semivariogram by defining

$$\gamma_{\varrho_0}(s,\varrho) \equiv \begin{cases} \varrho_0 + (1-\varrho_0)\gamma(s,\varrho) & \text{if} \quad s > 0, \\ 0 & \text{if} \quad s = 0. \end{cases} \tag{10.20}$$

Consequently, $\gamma(s,\varrho)$ tends to ϱ_0, with $\varrho_0 \in (0,1)$, when s tends to 0.

Figure 10.1 presents an example of semivariogram and correlation functions for the exponential correlation structure with the range $\varrho = 1$ and nugget $\varrho_0 = 0.2$.

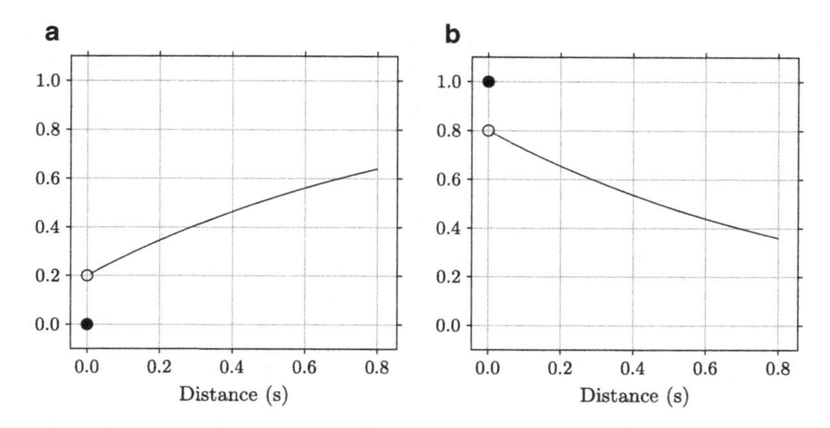

Fig. 10.1 Semivariogram (**a**) and correlation (**b**) functions for *corExp* structure with the range $\varrho = 1$ and nugget $\varrho_0 = 0.2$

10.4 Estimation

The main goal of fitting model (10.1)–(10.5) to the data is to obtain estimates of parameters $\boldsymbol{\beta}$, σ^2, and $\boldsymbol{\theta}_R$. In Sects. 10.4.1 and 10.4.2, we present the methods to estimate the parameters. As in the case of the LM for independent observations with heterogeneous variance, the choice of the estimation method depends on the form of the variance function. Thus, in this section, we focus on the estimation approaches for simpler models defined with the use of variance functions from the <δ>-group, which do not depend on the mean value (see Table 7.2). The use of mean-dependent functions will be discussed in Sect. 10.7.

In our presentation, we will refer to the description of the methods presented in Sect. 7.4 and shortly describe their modifications necessary for the application to the model (10.1)–(10.5).

Additionally, in Sect. 10.4.3, we address the issue of the most suitable, from a numerical optimization point of view, parameterization of the model (10.1)–(10.5), while in Sect. 10.4.4, we discuss the assessment of the uncertainty of the estimators of the parameters.

10.4.1 Weighted Least Squares

In this section, we consider the simple case of the model (10.1)–(10.5), with matrices $\boldsymbol{R}_i$ *known* for all groups. Similarly to models with known variance weights, presented in Sect. 7.2.1, the model with known matrices $\boldsymbol{R}_i$ does not pose any additional computational difficulties, as compared to the classical LM introduced in Sect. 4.2. This is because, by multiplying $\boldsymbol{X}_i$ and $\mathbf{y}_i$ in (10.1) on the left-hand side by $\boldsymbol{W}_i^{1/2} \equiv \boldsymbol{R}_i^{-1/2}$, where $\boldsymbol{R}_i^{-1/2}$ is the upper-triangular Cholesky decomposition of $\boldsymbol{R}_i^{-1}$, i.e., $\boldsymbol{R}_i^{-1} = (\boldsymbol{R}_i^{-1/2})'\boldsymbol{R}_i^{-1/2}$, we transform the model with correlated residual errors back to an LM with independent, homoscedastic errors. More specifically, the transformed model can be written as:

$$\boldsymbol{W}_i^{1/2}\mathbf{y}_i = \boldsymbol{W}_i^{1/2}\boldsymbol{X}_i\boldsymbol{\beta} + \boldsymbol{W}_i^{1/2}\boldsymbol{\varepsilon}_i. \tag{10.21}$$

Note that, in the transformed model, the linearity with respect to $\boldsymbol{\beta}$ is maintained. Moreover, the variance-covariance matrix of the transformed residual error vector is

$$\text{Var}(\boldsymbol{W}_i^{1/2}\boldsymbol{\varepsilon}_i) = \boldsymbol{W}_i^{1/2}\text{Var}(\boldsymbol{\varepsilon}_i)\left(\boldsymbol{W}_i^{1/2}\right)' = \boldsymbol{R}_i^{-1/2}(\sigma^2\boldsymbol{R}_i)\left(\boldsymbol{R}_i^{-1/2}\right)' = \sigma^2\boldsymbol{I}_{n_i}.$$

The estimates of $\boldsymbol{\beta}$ are obtained by the minimization, with respect to $\boldsymbol{\beta}$, of a weighted residual sum of squares

$$\sum_{i=1}^{n}(\mathbf{y}_i - \boldsymbol{X}_i\boldsymbol{\beta})'\boldsymbol{W}_i(\mathbf{y}_i - \boldsymbol{X}_i\boldsymbol{\beta}), \tag{10.22}$$

which is an extension of (7.17). Explicit formulae for WLS estimators for $\boldsymbol{\beta}$ and σ^2, built upon (7.18) and (7.19), are as follows:

$$\widehat{\boldsymbol{\beta}}_{\text{WLS}} \equiv \left(\sum_{i=1}^{N}\boldsymbol{X}_i'\boldsymbol{W}_i\boldsymbol{X}_i\right)^{-1}\sum_{i=1}^{N}\boldsymbol{X}_i'\boldsymbol{W}_i\mathbf{y}_i, \tag{10.23}$$

$$\widehat{\sigma}^2_{\text{WLS}} \equiv \frac{1}{n-p}\sum_{i=1}^{N}\left(y_i - \boldsymbol{X}_i\widehat{\boldsymbol{\beta}}_{\text{WLS}}\right)'\boldsymbol{W}_i\left(y_i - \boldsymbol{X}_i\widehat{\boldsymbol{\beta}}_{\text{WLS}}\right), \tag{10.24}$$

where $\boldsymbol{W}_i \equiv \boldsymbol{R}_i^{-1}$ and $n = \sum_{i=1}^{N} n_i$.

10.4.2 Likelihood-Based Estimation

When the variance function does not depend on μ_{ij}, i.e., if it belongs to the <δ>-group of variance functions (Sect. 7.3.1 and Table 7.2), the ML or REML estimation are used, along the lines described in Sect. 7.4.2. In particular, the full log-likelihood function for the model (10.1)–(10.5) is given by

$$\ell_{\text{Full}}(\boldsymbol{\beta}, \sigma^2, \boldsymbol{\theta}_R) \equiv -\frac{n}{2}\log(\sigma^2) - \frac{1}{2}\sum_{i=1}^{N}\log[\det(\boldsymbol{R}_i)]$$
$$-\frac{1}{2\sigma^2}\sum_{i=1}^{N}(\mathbf{y}_i - \boldsymbol{X}_i\boldsymbol{\beta})'\boldsymbol{R}_i^{-1}(\mathbf{y}_i - \boldsymbol{X}_i\boldsymbol{\beta}). \tag{10.25}$$

Note that $\ell_{\text{Full}}(.)$ depends on $\boldsymbol{\theta}_R$ through $\boldsymbol{R}_i \equiv \boldsymbol{R}_i(\boldsymbol{\theta}_R)$. Estimates of the parameters $\boldsymbol{\beta}$, σ^2, and $\boldsymbol{\theta}_R$ can be obtained by a simultaneous maximization of the log-likelihood function with respect to these parameters. However, this is a numerically complex task. An alternative is to consider profiling out $\boldsymbol{\beta}$ from (10.25). Toward this aim, assuming that $\boldsymbol{\theta}_R$ is known, (10.25) is maximized with respect to $\boldsymbol{\beta}$ and σ^2 (see Sect. 7.4.2). This leads to the following expressions for estimators of these parameters, expressed as functions of $\boldsymbol{\theta}_R$:

$$\widehat{\boldsymbol{\beta}}(\boldsymbol{\theta}_R) \equiv \left(\sum_{i=1}^{N} \boldsymbol{X}_i'\boldsymbol{R}_i^{-1}\boldsymbol{X}_i\right)^{-1} \sum_{i=1}^{N} \boldsymbol{X}_i'\boldsymbol{R}_i^{-1}\mathbf{y}_i, \tag{10.26}$$

$$\widehat{\sigma}^2(\boldsymbol{\theta}_R) \equiv \sum_{i=1}^{N} \mathbf{r}_i'\boldsymbol{R}_i^{-1}\mathbf{r}_i/n, \tag{10.27}$$

where $\mathbf{r}_i \equiv \mathbf{r}_i(\boldsymbol{\theta}_R) = \mathbf{y}_i - \boldsymbol{X}_i\widehat{\boldsymbol{\beta}}(\boldsymbol{\theta}_R)$ and $\boldsymbol{R}_i \equiv \boldsymbol{R}_i(\boldsymbol{\theta}_R)$. The expressions correspond to (7.22) and (7.25), presented for the LM with heterogeneous variance in Sect. 7.4.2.

Plugging (10.26) back into (10.25) yields a log-profile-likelihood function, which depends on σ^2 and $\boldsymbol{\theta}_R$:

$$\ell^*_{\text{ML}}(\sigma^2, \boldsymbol{\theta}_R) \equiv \ell_{\text{Full}}(\widehat{\boldsymbol{\beta}}(\boldsymbol{\theta}_R), \sigma^2, \boldsymbol{\theta}_R). \tag{10.28}$$

Maximization of (10.28) over σ^2 yields the estimator given in (10.27). Plugging the estimator into (10.28) yields a log-profile-likelihood function, which depends only on $\boldsymbol{\theta}_R$:

$$\ell^*_{\text{ML}}(\boldsymbol{\theta}_R) \equiv \ell_{\text{Full}}(\widehat{\boldsymbol{\beta}}(\boldsymbol{\theta}_R), \widehat{\sigma}^2(\boldsymbol{\theta}_R), \boldsymbol{\theta}_R). \tag{10.29}$$

By maximizing the function, we obtain the ML estimator $\widehat{\boldsymbol{\theta}}_R$ of $\boldsymbol{\theta}_R$. Note that, in general, the estimator cannot be presented in a closed form. Plugging $\widehat{\boldsymbol{\theta}}_R$ into (10.26) and (10.27) yields the ML estimators of $\boldsymbol{\beta}$ and σ^2, respectively.

For reasons similar to those mentioned in Sects. 4.4.2 and 7.4.2, the ML estimator of σ^2, obtained from (10.27) with $\widehat{\boldsymbol{\theta}}_R$ replacing $\boldsymbol{\theta}_R$, is biased. In fact, the same comment applies to $\widehat{\boldsymbol{\theta}}_R$. Therefore, σ^2 and $\boldsymbol{\theta}_R$ are often estimated by maximizing the following log-restricted-likelihood function (see Sect. 7.4.2):

$$\begin{aligned}\ell^*_{\text{REML}}(\sigma^2, \boldsymbol{\theta}_R) \equiv\ & \ell_{\text{Full}}(\widehat{\boldsymbol{\beta}}(\boldsymbol{\theta}_R), \sigma^2, \boldsymbol{\theta}_R) + \frac{p}{2}\log(\sigma^2) \\ & -\frac{1}{2}\log\left[\det\left(\sum_{i=1}^{N} \boldsymbol{X}_i'\boldsymbol{R}_i^{-1}\boldsymbol{X}_i\right)\right],\end{aligned} \tag{10.30}$$

where $\widehat{\boldsymbol{\beta}}(\boldsymbol{\theta}_R)$ is specified in (10.26).

The parameter σ^2 can also be profiled out from the log-likelihood function (10.30). That is, σ^2 is expressed using the following formula:

$$\widehat{\sigma}^2(\boldsymbol{\theta}_R) \equiv \sum_{i=1}^{N} \mathbf{r}_i' \boldsymbol{R}_i^{-1} \mathbf{r}_i / (n-p), \tag{10.31}$$

which results from the maximization of (10.30) over σ^2. The $\mathbf{r}_i$ is specified in (10.27). The expression on the right-hand side of (10.31) is then plugged into (10.30), what results in an objective function that depends only on $\boldsymbol{\theta}_R$:

$$\ell^*_{\text{REML}}(\boldsymbol{\theta}_R) \equiv \ell^*_{\text{REML}}(\widehat{\sigma}^2(\boldsymbol{\theta}_R), \boldsymbol{\theta}_R). \tag{10.32}$$

The estimator of $\boldsymbol{\theta}_R$, obtained as a result of maximization of the log-profile-likelihood function (10.32), is then plugged into (10.26) and (10.31) to provide the REML estimates of $\boldsymbol{\beta}$ and σ^2.

10.4.3 Constrained Versus Unconstrained Parameterization of the Variance-Covariance Matrix

One of the important issues related to finding the maximum of functions like, e.g., (10.32), is the fact that the solution should lead to a symmetric and positive-definite matrix $\boldsymbol{R}_i$. From a numerical point of view, such a constrained optimization problem is difficult to solve. Note that a similar issue applies to the LM for independent observations with heterogeneous variance and it was addressed in Sect. 7.4.3. A possible solution is to parameterize $\boldsymbol{R}_i$ in such a way that the optimization problem becomes unconstrained.

Toward this end, we consider the decomposition given by (10.10). In the decomposition, $\boldsymbol{\Lambda}_i$ is a diagonal matrix with diagonal elements expressed using a variance function, which depends on parameters $\boldsymbol{\delta}$. The parameters are unconstrained or constrained to be positive (see Sect. 7.3.1). In the latter case, the solution is to express the components δ_s of $\boldsymbol{\delta}$ as $\delta_s = e^{\log(\delta_s)}$ and use $\delta_s^* \equiv \log(\delta_s)$ as the parameters for the variance function.

The second component of the decomposition, given by (10.10), is the correlation matrix $\boldsymbol{C}_i$. Thus, it is constrained to be symmetric and positive-definite.

To deal with this constraint in the optimization algorithm, a transformation of the elements of the matrix $\boldsymbol{C}_i$ can be sought, such that the transformed elements are unconstrained and that they ensure the positive-definiteness of the matrix. The transformation may depend on the form of the correlation matrix.

For instance, for an $n_i \times n_i$ matrix $\boldsymbol{C}_i$, corresponding to the autoregressive-of-order-1 correlation structure (see Table 10.1), application of Fisher's z-transform to ϱ,

$$\varrho^* \equiv \log \frac{1+\varrho}{1-\varrho}, \tag{10.33}$$

allows expressing the elements of the matrix $\boldsymbol{C}_i$ in terms of an unconstrained parameter ϱ^*. Note that, sometimes, a factor 0.5 is used in front of the transform, but this factor is immaterial for our purposes. At the same time, the back-transformation

$$\varrho = \frac{e^{\varrho^*} - 1}{e^{\varrho^*} + 1} \tag{10.34}$$

guarantees that $\varrho \in (-1, 1)$ and that the matrix $\boldsymbol{C}_i$ is positive-definite.

On the other hand, to guarantee the positive-definiteness of an $n_i \times n_i$ correlation matrix, corresponding to the compound-symmetry correlation structure (see Table 10.1), its eigenvalues, equal to $1 + (n_i - 1)\varrho$ and $1 - \varrho$, need to be positive. By transforming the parameter ϱ by the following, modified Fisher's z-transform:

$$\varrho^* \equiv \log \frac{\frac{1}{n^*-1} + \varrho}{1-\varrho}, \tag{10.35}$$

where $n^* \equiv \max_i(n_i)$ (to allow for varying n_i), we express the elements of the matrix $\boldsymbol{C}_i$ in terms of an unconstrained parameter ϱ^*. Again, the back-transformation of ϱ^* to ϱ guarantees that matrix $\boldsymbol{C}_i$ is positive-definite.

A solution for a general correlation matrix $\boldsymbol{C}_i$ is to consider its Cholesky decomposition $\boldsymbol{C}_i = \boldsymbol{U}_i'\boldsymbol{U}_i$, while representing the upper-triangular matrix $\boldsymbol{U}_i$ in terms of *spherical coordinates* (Pinheiro and Bates 1996). Toward this end, the diagonal and above-diagonal elements of $\boldsymbol{U}_i$ are represented as follows:

$$u_{kl} \equiv \begin{cases} 1, & \text{for } k = l = 1, \\ \cos(u^*_{l-1,1}), & \text{for } 1 = k < l, \\ \left\{\prod_{j=1}^{k-1} \sin(u^*_{l-1,j})\right\} \cos(u^*_{l-1,k}), & \text{for } 1 < k < l, \\ \prod_{j=1}^{l-1} \sin(u^*_{l-1,j}), & \text{for } 1 < k = l, \end{cases}$$

where u^*_{kl} ($k = 1, \ldots, n_i - 1$, $l = 1, \ldots, k$) are the spherical coordinates. To ensure uniqueness of the parameterization, we need to assume that $u^*_{kl} \in (0, \pi)$. An unconstrained set of parameters ϱ^*_{kl} is obtained by transforming the coordinates as follows:

$$\varrho^*_{kl} \equiv \log \frac{u^*_{kl}}{\pi - u^*_{kl}}. \tag{10.36}$$

This allows expressing the parameters $\boldsymbol{\varrho}$, defining $\boldsymbol{C}_i$, as a function of the parameters ϱ^*_{kl}. The latter form a set of unconstrained parameters that can be used for numerical optimization purposes.

An additional advantage of the use of the spherical coordinates is that $\cos(u^*_{k-1,1}) = \varrho_{1k}$. Thus, by permuting rows and columns of $\boldsymbol{C}_i$, confidence intervals for the parameters $\boldsymbol{\varrho}$ can be easily obtained from the intervals for $u^*_{k-1,1}$ or, equivalently, for $\varrho^*_{k-1,1}$.

For the spatial correlation structures, displayed in Table 10.1, a common constraint for the parameter ϱ is $\varrho > 0$. If present, the nugget, ϱ_0, is restricted to lie within the unit interval, i.e., $\varrho_0 \in (0,1)$. Consequently, for numerical optimization purposes, ϱ is replaced by its logarithm, while ϱ_0 is transformed using the logit transformation:

$$\varrho_0^* \equiv \log \frac{\varrho_0}{1-\varrho_0}. \tag{10.37}$$

Note that we presented unconstrained parameterizations for selected correlation structures. The transformations, which preserve the positive-definiteness of the matrix $\boldsymbol{C}_i$ are relatively simple for the compound-symmetry and the autoregressive-of-order-1 structures, while it is more complex for a general structure. We need to keep in mind that for some correlation structures, especially for those described by multiple parameters, there is no guarantee that such transformations exist. Finally, to assure unconstrained optimization for σ^2, the parameter is replaced by its logarithm.

10.4.4 Uncertainty in Parameter Estimation

The variance-covariance matrix of $\widehat{\boldsymbol{\beta}}$ is estimated by

$$\widehat{\text{Var}}(\widehat{\boldsymbol{\beta}}) \equiv \widehat{\sigma}^2 \left(\sum_{i=1}^{N} \boldsymbol{X}_i' \widehat{\boldsymbol{R}}_i^{-1} \boldsymbol{X}_i \right)^{-1}, \tag{10.38}$$

where $\widehat{\sigma}^2$ and $\widehat{\boldsymbol{R}}_i$ are estimated by one of the methods described in Sects. 10.4.1 and 10.4.2.

The variance-covariance matrix of $\widehat{\sigma}^2$ and $\widehat{\boldsymbol{\theta}}_R$ can be estimated in various ways. As indicated in Sect. 7.4.4, a possible solution, which is implemented in the `gls()` function from the **nlme** package in R, is to use the inverse of the negative Hessian of the log-likelihood (10.28) or (10.30), evaluated at the estimated values of σ^2 and $\boldsymbol{\theta}_R$, i.e., the inverse of the observed Fisher information matrix (Sect. 4.6.1).

10.5 Model Diagnostics

As it was the case for the LMs for independent observations (see Sects. 4.5 and 7.5), after fitting an LM for correlated data, and before making any inferences based on it, it is important to check whether the model assumptions are fulfilled. Toward this end, tools similar to those described in Sects. 4.5 and 7.5 are used. In this section, we review the tools and their modifications required for the LMs for correlated data.

10.5.1 Residual Diagnostics

For checking the fit of an LM for correlated data, residual plots are used. The raw residuals are defined as $\widehat{\boldsymbol{\varepsilon}}_i \equiv \mathbf{y}_i - \boldsymbol{X}_i\widehat{\boldsymbol{\beta}}$. The use of these residuals requires caution, because the model (10.1)–(10.5) allows for heteroscedasticity. Thus, the comments regarding residual diagnostics for the LM for independent observation with heterogeneous variance, provided in Sect. 7.5, apply here as well. Consequently, Pearson residuals (see Sect. 7.5.1) are more useful for checking for, e.g., outlying observations.

Pearson residuals have variance approximately equal to 1. Their within-group correlations approximate the elements of the correlation matrix $\boldsymbol{C}_i$. Therefore, the residuals are well suited to investigate whether an appropriate correlation structure was used in the model. Toward this end, we estimate the semivariogram function, defined in (10.19), using the squared differences between pairs of residuals (Cressie 1991; Pinheiro and Bates 2000, Sect. 5.3.2):

$$\frac{1}{2N(s)}\sum_{i=1}^{N}\sum_{d(\mathbf{t}_{ij},\mathbf{t}_{ij'})=s}(r_{ij}-r_{ij'})^2, \tag{10.39}$$

where $N(s)$ is the number of pairs of observations separated by a distance of s units. An estimator, which is more robust with respect to the presence of outliers, is given in Cressie and Hawkins (1980) by

$$\frac{1}{0.457+0.494/N(s)}\left(\frac{1}{2N(s)}\sum_{i=1}^{N}\sum_{d(\mathbf{t}_{ij},\mathbf{t}_{ij'})=s}|r_{ij}-r_{ij'}|^{1/2}\right)^4. \tag{10.40}$$

Note that, as compared to (10.39), it uses the square-root differences between pairs of residuals.

A potential complication, when interpreting Pearson residuals, is the fact that the model (10.1)–(10.5) also allows for correlation between residual errors. Thus, for example, even if we studentize the residuals (Sect. 4.6.1), the overall Q-Q plots, based on all estimated residuals, are not appropriate for checking the normality of the residual random error.

A possible solution is to obtain approximately independent residuals using the transformation of the residuals based on the Cholesky decomposition of the matrix $\boldsymbol{R}_i$ (see Sect. 4.5.2). That is, to use the transformed residuals $(\sigma\boldsymbol{U}_i')^{-1}\widehat{\boldsymbol{\varepsilon}}_i$, where the upper-triangular matrix $\boldsymbol{U}_i$ is obtained from the Cholesky decomposition of the matrix $\boldsymbol{R}_i$, i.e., $\boldsymbol{R}_i = \boldsymbol{U}_i'\boldsymbol{U}_i$ (Schabenberger 2004). The vector of the transformed residuals should be approximately normally distributed with mean $\mathbf{0}$ and variance-covariance matrix equal to an identity matrix. That is, the elements of the vector should be uncorrelated and follow the standard normal distribution. Note that, in the **nlme** package, these transformed residuals are referred to as *normalized residuals*. We will use this term in the remainder of our book.

10.5.2 Influence Diagnostics

Influence diagnostics for LMs for correlated data uses similar tools that were presented in Sect. 4.5.3 for LMs with homogeneous variance. To investigate the combined influence of a given observation on the estimates of *all* model parameters, the likelihood displacement is used. Toward this end, the defining equation (4.27) is modified by specifying $\widehat{\Theta} \equiv (\widehat{\boldsymbol{\beta}}', \widehat{\boldsymbol{\theta}}'_R, \widehat{\sigma}^2)'$ and using the log-likelihood function given in (10.25).

After identifying influential observations using likelihood displacement, Cook's distance, similar to that given in (4.26), may be used to determine whether a particular observation affects estimation of $\boldsymbol{\beta}$.

10.6 Inference and Model Selection

Inference for the LMs with correlated errors, specified in (10.1)–(10.5), focuses on the fixed-effect parameters $\boldsymbol{\beta}$ and/or the variance-covariance parameters $\boldsymbol{\theta}_R$. In this section, we focus on the inference for models defined with the use of variance functions from the <δ>-group, which do not depend on the mean value (see Table 7.2). Inference for the mean-variance models will be discussed in Sect. 10.7.

When testing hypotheses about the parameters $\boldsymbol{\beta}$, the methods and issues described in Sect. 7.6.1 for the case of the LM for independent observations with heterogeneous variance, apply as well. In particular, linear hypotheses of the form defined by (4.30) may be tested using the F-test, given by (4.36). The variance-covariance matrix of $\widehat{\boldsymbol{\beta}}$ is computed using (10.38). In general, similarly to the case discussed in Sect. 7.6.1, the null distribution of the test statistics is not a central F distribution with p and $n-p$ degrees of freedom. Instead, the approximate test is performed using the central F-distribution with sensibly approximated numerator and denominator degrees of freedom. In R, the **lme()** function uses a crude approximation with rank($\boldsymbol{L}$) numerator and $n-p$ denominator degrees of freedom. An alternative approach is to use an approximate LR test (Sect. 7.6.1) or a simulation technique.

Inference about $\boldsymbol{\theta}_R$ is based, in general, on two approaches. One is based on the use of the LR test and the other is based on the use of information criteria.

The first approach is applied along the lines similar to those described in Sect. 7.6.1 for the case of the LM for independent observations with heterogeneous variance. The comments about the use of the REML in the construction of the LR test apply here as well.

The second approach, based on the information criteria, is used when the hypothesis about $\boldsymbol{\theta}_R$ cannot be expressed in the way that it would lead to two nested models. In this case, we can use the information criteria like AIC or BIC (Sect. 4.7.2) to select the model that seems to better fit the data.

The information criteria can be also used for the more general problem of model selection, i.e., for discrimination between nonnested models, which differ both in the variance-covariance and the mean structures. In this case, the criteria are applied in a way similar to the one described in Sect. 7.7.

Obviously, irrespectively of the approach chosen for the model reduction or selection, the fit of the final model should be formally checked using the residual diagnostic methods, described in Sect. 10.5.

Confidence intervals for individual components of the parameter vector $\boldsymbol{\beta}$ can be constructed based on the t-test statistic (Sect. 4.6.2).

Confidence intervals for the variance-covariance parameters σ and $\boldsymbol{\delta}$ can be obtained by considering a normal-distribution approximation to the distribution of the ML- or REML-based estimator of a transformation of the parameters (Sect. 7.6.2).

A similar idea can be applied to construct confidence intervals for the correlation parameters $\boldsymbol{\varrho}$. More specifically, for the component ϱ_s of $\boldsymbol{\varrho}$, we can consider the unconstrained parameter ϱ_s^*, obtained using Fisher's z-transform, given in (10.33), and apply the normal-distribution approximation to the distribution of the ML or REML estimate of ϱ_s^*. The resulting $(1-\alpha/2)100\%$ confidence interval for ϱ_s is given by

$$\left[\frac{\exp\left[\widehat{\varrho_s^*} - V_{\varrho_s^*}(\alpha)\right]-1}{\exp\left[\widehat{\varrho_s^*} - V_{\varrho_s^*}(\alpha)\right]+1}, \frac{\exp\left[\widehat{\varrho_s^*} + V_{\varrho_s^*}(\alpha)\right]-1}{\exp\left[\widehat{\varrho_s^*} + V_{\varrho_s^*}(\alpha)\right]+1}\right], \tag{10.41}$$

with

$$V_{\varrho_s^*}(\alpha) \equiv z_{1-\alpha/2}\sqrt{\{\mathbf{I}^{-1}\}_{\varrho_s^*\varrho_s^*}},$$

where $\{\mathbf{I}^{-1}\}_{\varrho_s^*\varrho_s^*}$ is the diagonal element of the inverse of the observed Fisher information matrix based on the log-(profile)-likelihood or log-(profile)-restricted-likelihood corresponding to ϱ_s^*, and $z_{1-\alpha/2}$ is the $(1-\alpha/2)100$-th percentile of the standard normal distribution (Pinheiro and Bates 2000, Sect. 2.4.3).

Note that the use of Fisher's z-transform, given in (10.33), does not guarantee, in general, that the matrix, constructed from the back-transformed parameters $\boldsymbol{\varrho}$, will be positive-definite. Thus, while the application of Fisher's z-transform for the purpose of construction of confidence intervals for the parameters $\boldsymbol{\varrho}$ is justified, for the optimization purposes the use of the transformations, described in Sect. 10.4.3, is required.

10.7 Mean-Variance Models

The concept of mean-variance models was introduced in Sect. 7.8 in the context of the LMs for independent observations with heterogeneous variance. It carries naturally over to LMs with fixed effects for correlated data.

For the mean-variance models, the decomposition (10.10) involves mean-dependent variance functions from <δ, μ> and <μ>-groups (see Tables 7.3 and 7.4, respectively). Thus, in particular, the residual error variance is given by (10.11) and it depends on the fixed-effects parameters $\boldsymbol{\beta}$. This dependence complicates the estimation not only of $\boldsymbol{\beta}$, but also of σ^2 and $\boldsymbol{\theta}_R$.

First, let us consider the case when the variance function depends on μ_{ij} *and* $\boldsymbol{\theta}_R$, i.e., if it belongs to the <δ, μ>-group of variance functions (see Sect. 7.3.1 and Table 7.3). Then, the estimates of the parameters $\boldsymbol{\beta}$, σ^2, and $\boldsymbol{\theta}_R$ can be obtained using the GLS approaches similar to those described in Sect. 7.8.1.

In particular, we start with the following pseudo-likelihood objective function:

$$\ell^*_{\text{PL}}(\sigma^2, \boldsymbol{\theta}_R; \boldsymbol{\beta}) \equiv \ell_{\text{Full}}(\boldsymbol{\beta}, \sigma^2, \boldsymbol{\theta}_R), \tag{10.42}$$

derived from the full log-likelihood (10.25) by assuming $\boldsymbol{\beta}$ to be known. Consequently, for every value of $\boldsymbol{\beta}$, the function $\ell^*_{\text{PL}}(\cdot)$ has two arguments, i.e., σ^2 and $\boldsymbol{\theta}_R$. Next, we profile out σ from $\ell^*_{\text{PL}}(\cdot)$ in a similar way as we did it in Sect. 7.8.1. Toward this end, we use the following functional relationship between $\widehat{\sigma}^2$, which maximizes (10.42) for a fixed $\boldsymbol{\theta}_R$, and $\boldsymbol{\theta}_R$:

$$\widehat{\sigma}^2_{\text{PL}}(\boldsymbol{\theta}_R) \equiv \sum_{i=1}^{N} \mathbf{r}_i' \boldsymbol{R}_i^{-1} \mathbf{r}_i / n, \tag{10.43}$$

where $\mathbf{r}_i \equiv \mathbf{y}_i - \boldsymbol{X}_i \boldsymbol{\beta}$ and $\boldsymbol{R}_i \equiv \boldsymbol{R}_i(\boldsymbol{\theta}_R)$. Replacing σ^2 in (10.42) with the expression on the right-hand side of (10.43) yields a log-pseudo-profile-likelihood function for $\boldsymbol{\theta}_R$:

$$\ell^*_{\text{PL}}(\boldsymbol{\theta}_R; \boldsymbol{\beta}) \equiv \ell^*_{\text{PL}}(\widehat{\sigma}^2_{\text{PL}}(\boldsymbol{\theta}_R), \boldsymbol{\theta}_R; \boldsymbol{\beta}) \tag{10.44}$$

Then, the following algorithm, similar to the one described in Sect. 7.8.1, is used to estimate $\boldsymbol{\beta}$ and $\boldsymbol{\theta}_R$:

1. Assume initial values $\widehat{\boldsymbol{\beta}}^{(0)}$ of $\boldsymbol{\beta}$, $\widehat{\boldsymbol{\theta}}_R^{(0)}$ of $\boldsymbol{\theta}_R$, and set the iteration counter $k = 0$.
2. Increase k by 1.
3. *Use* $\widehat{\boldsymbol{\beta}}_i^{(k-1)}$ *to (re)define variance function* $\lambda^{(k)}(\boldsymbol{\delta})$.

   Calculate $\widehat{\boldsymbol{\mu}}_i^{(k)} \equiv \boldsymbol{X}_i \widehat{\boldsymbol{\beta}}^{(k-1)}$. (Re)define variance function $\lambda^{(k)}(\boldsymbol{\delta}; \widehat{\mu}_{ij}^{(k)}, \mathbf{v}_{ij}) \equiv \lambda(\widehat{\mu}_{ij}^{(k)}, \boldsymbol{\delta}; \mathbf{v}_{ij})$ from the <δ>-group, where $\lambda(\cdot)$ is defined by (10.12).
4. *Keep* $\widehat{\boldsymbol{\beta}}^{(k-1)}$ *fixed and optimize* $\ell^*_{\text{PL}}(\boldsymbol{\theta}_R; \widehat{\boldsymbol{\beta}}^{(k-1)})$ *to find* $\widehat{\boldsymbol{\theta}}_R^{(k)}$.

   Use the function $\lambda^{(k)}(\cdot)$ in defining the log-pseudo-likelihood function, $\ell^*_{\text{PL}}(\boldsymbol{\theta}_R; \widehat{\boldsymbol{\beta}}^{(k-1)})$, specified by (10.44). Optimize this function with respect to $\boldsymbol{\theta}_R$

to obtain the new estimate $\widehat{\boldsymbol{\theta}}_R^{(k)}$. Values of $\widehat{\boldsymbol{\theta}}_R^{(k-1)}$ can be used as initial values for the optimization.

5. *Fix* $\widehat{\boldsymbol{\theta}}_R^{(k)}$ *and use WLS to find* $\boldsymbol{\beta}^{(k)}$.

   Use $\widehat{\boldsymbol{\theta}}_R^{(k)}$ to derive $\boldsymbol{R}_i^{(k)}$ and to compute the WLS estimate $\widehat{\boldsymbol{\beta}}^{(k)}$ of $\boldsymbol{\beta}$ using (10.26).
6. Iterate between steps 2 and 5 until convergence or until a predetermined number of iterations k.
7. Compute the final, ML-based estimate of σ^2, by plugging the obtained estimates of $\boldsymbol{\theta}_R$ and $\boldsymbol{\beta}$ into (10.43).

Note that the aforementioned construction can also be applied while starting from the log-restricted-likelihood, given in (10.30). With modifications similar to those described in Sect. 7.8.1, it will result in an algorithm leading to REML-based estimates of the parameters.

If the variance function depends on μ_{ij}, i.e., if it belongs to the <μ>-group of variance functions (see Sect. 7.3.1 and Table 7.4), estimates of the parameters $\boldsymbol{\beta}$, σ^2, and $\boldsymbol{\theta}_R$ can be obtained using an IRLS procedure similar to the one described in Sect. 7.8.1.

The issues related to the inference for the mean-variance models, defined in the context of the model (10.1)–(10.5), are similar to those mentioned in Sect. 7.8.2. In particular, provided that the mean structure of the model is correctly specified, misspecification of the variance-covariance structure does not bias the point estimate of $\boldsymbol{\beta}$, but decreases its efficiency. More specifically, if we denote by $\mathcal{R}_i$ the model-based variance-covariance matrix of $\mathbf{y}_i$, then it can be shown (see, e.g., Davidian and Giltinan 1995 or Verbeke and Molenberghs 2000) that

$$\text{Var}(\widehat{\boldsymbol{\beta}}) = \left(\sum_{i=1}^{N} \boldsymbol{X}_i' \mathcal{R}_i^{-1} \boldsymbol{X}_i\right)^{-1} \sum_{i=1}^{N} \boldsymbol{X}_i' \boldsymbol{A}_i \boldsymbol{X}_i \left(\sum_{i=1}^{N} \boldsymbol{X}_i' \mathcal{R}_i^{-1} \boldsymbol{X}_i\right)^{-1}, \tag{10.45}$$

where

$$\boldsymbol{A}_i \equiv \mathcal{R}_i^{-1} \text{Var}(\mathbf{y}_i) \mathcal{R}_i^{-1}. \tag{10.46}$$

If $\mathcal{R}_i = \text{Var}(\mathbf{y}_i)$, i.e., if the variance-covariance structure is correctly specified, then $\boldsymbol{A}_i = \mathcal{R}_i^{-1}$ and (10.45) reduces to

$$\text{Var}(\widehat{\boldsymbol{\beta}}) = \left(\sum_{i=1}^{N} \boldsymbol{X}_i' \mathcal{R}_i^{-1} \boldsymbol{X}_i\right)^{-1}, \tag{10.47}$$

and can be estimated as in (10.38). However, if $\mathcal{R}_i \neq \text{Var}(\mathbf{y}_i)$, then (10.45) implies a loss of efficiency of $\widehat{\boldsymbol{\beta}}$. Moreover, it indicates that the estimator (10.38) underestimates the true variance-covariance matrix of $\widehat{\boldsymbol{\beta}}$. A corrected estimator, based on formula (10.45), can be constructed. We do not discuss the construction here; interested readers can find more information on this topic in the monographs by, e.g., Davidian and Giltinan (1995) or Verbeke and Molenberghs (2000).

Inference on parameters $\boldsymbol{\theta}_R$ and σ^2, similarly to the case mentioned in Sect. 7.8.2, is difficult due to the need for correct specification of the form of the variance-covariance structure and due to a complex dependence of the parameters of the asymptotic distribution of the estimates of $\boldsymbol{\theta}_R$ and σ^2 on, e.g., true third- and higher-order moments of data. For this reason, we do not discuss it further.

10.8 Chapter Summary

In this chapter, we reviewed the formulation of an LM with fixed effects and correlated residual errors, applicable to grouped data. This class of models is an example of *population-average* models. To the extent possible, we used the concepts and theory introduced in Chaps. 4 and 7 in the context of LMs for independent observations with homogeneous and heterogeneous variance, respectively. Compared to those models, the new component, used in the model formulation, was the correlation structure, described in Sect. 10.3.2. It is an important component of the model, as it allows taking into account in the analysis the dependence of observations made within the same group. This concept will also be used in the formulation of LMMs in Chap. 13.

Estimation methods for LMs for correlated data, which used mean-independent variance functions, were discussed in Sect. 10.4. From the discussion, it should be clear that they are based on the similar approaches that are used for LMs for independent observations. It is also worth mentioning that the log-likelihood functions, described in Sect. 10.4.2, play an important role in the construction of the estimation approaches for LMMs.

In Sects. 10.5 and 10.6, we offered a review of the diagnostic and model reduction/selection methods, respectively, which are available for LMs for correlated data and mean-independent variance. Essentially, the methods are based on the concepts similar to those used in the case of the LM for independent observations with heterogeneous variance.

In Sect. 10.7, we discussed the mean-variance models, i.e., models involving variance functions from the <δ, μ>- and <μ>-groups (see Tables 7.3 and 7.4, respectively), which do depend on the mean value. The estimation methods and the inferential issues are very similar to those presented in Sect. 7.8. They will also appear when discussing the formulation of LMMs.

Note that, in the context of LMs for correlated data, the grouping and, consequently, correlation of the data was primarily reflected in the correlation structure used in the modeling. It is possible to imagine a situation, where the levels of data hierarchy can be used in defining various sources of variability of the data, e.g., between- and within-groups. Such an approach is used in LMMs, which will be discussed in Chap. 13.

Chapter 11
Fitting Linear Models with Fixed Effects and Correlated Errors: The `gls()` Function

11.1 Introduction

In Chap. 10, we summarized the main concepts underlying the construction of the LM with fixed effects and correlated residual errors for normally distributed, grouped data. An important component of the model is the correlation function, which is used to take into account the correlation between the observations belonging to the same group.

In this chapter, we review the tools available in R for fitting LMs for correlated data. The primary tool to fit the models is the `gls()` function from the **nlme** package. In Sects. 11.2 and 11.3, we present the details of the representation of correlation functions and how to extract related information. Section 11.4 contains a few examples of the correlation functions that are available in R. In Sect. 11.5, we explain how the details of the estimated correlation structure of an LM for correlated data can be extracted from a model-fit object created with the use of the `gls()` function. Note that additional information about the syntax and the use of the function `gls()` has already been provided in Sects. 5.4, 5.5, 8.4, and 8.5. Finally, Sect. 11.7 includes a summary of the contents of the chapter.

11.2 Correlation-Structure Representation: The *corStruct* Class

An important component, needed in the context of the LM model for correlated data, is the correlation structure for residual errors, defined in Sect. 10.3.2. In this section, we provide details about the implementation of correlation structures in the form of objects inheriting from the *corStruct* class implemented in the **nlme** package.

A. Gałecki and T. Burzykowski, *Linear Mixed-Effects Models Using R: A Step-by-Step Approach*, Springer Texts in Statistics, DOI 10.1007/978-1-4614-3900-4_11,

11.2.1 Correlation-Structure Constructor Functions

Correlation structure was defined in (10.13); several examples were given in Table 10.1. The package **nlme** provides several constructor functions designed to create specialized objects representing different correlation structures. Each created object belongs to the class named after the constructor function. For example, the constructor function `corCompSymm()` creates objects of class *corCompSymm*. The objects represent the compound-symmetry correlation structure, defined in (10.15). Note that all of these objects also inherit from the *corStruct* class. Of course, this applies to objects created by other constructor functions as well. A list of correlation structures available in the package **nlme** can be obtained from R's help system by issuing the `?corClasses` command.

Correlation-structure constructors are primarily used to specify correlation structures, with the help of the `correlation` argument, for the model-fitting functions `gls()`, and `lme()`. They also allow exploring the details of correlation structures, to choose user-defined initial values, or to fix values of correlation parameters in the numerical optimization procedures.

11.2.1.1 Arguments of the Correlation-Structure Constructor Functions

For the serial correlation functions (see Table 10.1), similar to the case of variance functions (see Sect. 8.2), three arguments are available in R: `value`, `form`, and `fixed`. The first one specifies the values of the correlation-parameter vector $\boldsymbol{\varrho}$. The second one provides a one-sided formula that defines the indices j (10.13) by specifying a position variable, and, optionally, a grouping factor. Note that observations in different groups are assumed to be uncorrelated. The default value of the `form` argument is `~1`, which amounts to using the order of the observations in the data as a position variable, without any grouping. Finally, `fixed=TRUE` can be used to fix all values of the correlation parameters in the numerical optimization in the modeling functions. Note that the default value is `fixed=FALSE`.

For the spatial correlation structures (see Table 10.1), apart from the `value`, `form`, and `fixed` arguments, two additional arguments are available: `nugget` and `metric`.

Note that, for these correlation structures, the argument `form` is a one-sided formula of the form `~S1+···+Sp|g`, where `S1` through `Sp` are spatial position variables and g, optionally, is a grouping factor. When a grouping factor is present, the correlation structure is assumed to apply only to the observations sharing the same level of a grouping factor; in contrast, observations with different levels are assumed to be uncorrelated. It is worth mentioning that the spatial position variables can be unidimensional, what allows to apply the "spatial" structures also to, e.g., longitudinal data.

If `nugget=FALSE`, which is the default, no nugget effect is assumed (see (10.18) and (10.20)). In that case, `value` should have only one element, indicating the (positive) value of the "range" parameter ϱ. If `nugget=TRUE`, the argument `value` can contain one or two elements, with the first indicating the range (constrained to be a positive value) and the second providing the nugget effect (a value between zero and one). The default is `value=numeric(0)`, a numeric vector of length 0, which results in the assignment, upon the initialization of a *corStruct* object, of the range equal to the 90% of the minimum between-pairs distance and of the nugget effect equal to 0.1.

The argument `metric` is an optional character string. It can be used to specify the distance metric, i.e., the function $d(s,\varrho)$, defined in (10.13). Three options are currently available: `metric="euclidean"` for the Euclidean metric, `metric="maximum"` for the maximum metric, and `metric="manhattan"` for the Manhattan metric. The definitions of these metrics have been provided in Sect. 10.3.2.

Initialization of Objects of Class corStruct

After an object, which inherited from the *corStruct* class, has been defined using an appropriate constructor function, it is then typically evaluated in the context of a given data set. This process, called initialization, was already shortly described in Sect. 8.2 for the *varFunc*-class objects. The main tool is the generic `Initialize()` function. We show examples of initialization of objects of class *corStruct* in Panels R11.1 and R11.3–R11.5 later in this chapter.

11.3 Inspecting and Modifying Objects of Class *corStruct*

In this section, we describe the functions and methods that allow extracting information about initialized objects of *corStruct* class. In particular, in Sect. 11.3.1, we discuss the use of the `coef()` generic function to extract and modify the coefficients of such objects. In Sect. 11.3.2, we present the application of the `Variogram()` function to obtain the semivariogram. Section 11.3.3 describes the use of the `corMatrix()` function to display the form of the correlation matrix corresponding to the object.

11.3.1 Coefficients of Correlation Structures

In Sect. 10.4.3, we mentioned that, e.g., for numerical optimization purposes, the use of an alternative, unconstrained parameterization of the correlation structure may be of interest. The information about values of the correlation coefficients, which correspond to the different possible parameterizations, can be extracted from an appropriate, initialized *corStruct* object.

Similar to the case of variance functions (Sect. 8.3), the primary tool to extract or modify coefficients of a correlation-structure object is the generic `coef()` function. For instance, to obtain coefficients from an object of class *corAR1*, the method `coef.corAR1()` is dispatched.

The primary arguments of the `coef.corStruct` method are `object` and `unconstrained`. The argument `object` indicates an object inheriting from the particular *corStruct* class. The value of the logical argument `unconstrained` specifies the type of the parameterization applied to the coefficients of the correlation structure. More specifically, the coefficients (parameters) can be presented on a constrained or unconstrained scale (Sect. 7.4.3). In the first case, the elements of the vector $\boldsymbol{\varrho}$ are provided. For instance, for the *corAR1* class, it is the value of the scalar parameter ϱ (see Table 10.1). In the second case, the values of the unconstrained transformations of parameters $\boldsymbol{\varrho}$, are returned. For instance, for the *corAR1* class, it is the value of the parameter ϱ^* corresponding to Fisher's z-transform of ϱ, as defined in (10.33). On the other hand, for the general *corSymm* class, the values of the transformed spherical coordinates of the Cholesky decomposition of the correlation matrix are returned, as defined in (10.36).

Coefficients of an initialized *corStruct* object can be modified with the use of the "`coef<-`" function. Toward this end, the syntax `coef(`*`object`*`)<-`*`value`* is used, where *`object`* is the initialized *corStruct* object and *`value`* is a vector of values to be assigned to the coefficients of the object. Note that the vector has to have the appropriate length, corresponding to the length of the `coef(`*`object`*`)` vector. Moreover, the values have to be given in the *unconstrained* form. Thus, for instance, to replace coefficients of an initialized *corAR1* object, we use the syntax `coef(`*`object`*`)<-`*`value`*, where *`value`* is a scalar resulting from applying Fisher's z-transform (10.33) to the correlation coefficient ϱ, which we want to use as the argument for the new form of the *corAR1* structure object.

11.3.2 Semivariogram

The semivariogram and correlation function were introduced in Sect. 10.3.2. The values of these functions can be computed with the use of the generic function `Variogram()` that can be applied to objects inheriting from the *corSpatial* class. Note that the function can be also applied to objects of other classes, including, for instance, *gls* and *lme*. Its arguments depend on the class of the object. The information about the arguments used for the *corSpatial* class is obtained by issuing the command `?Variogram.corSpatial`.

The `Variogram()` function is available for several *corSpatial* classes, including *corExp*, *corGaus*, *corLin*, *corRatio*, and *corSpher*. For objects of these classes, the main arguments of the function are `object`, `distance`, `sig2`, and `length.out`. The argument `object` specifies an initialized object of the specific *corStruct* class. The distances, at which the semivariogram is to be computed, are specified optionally by providing a numeric vector to the `distance` argument. The optional sig2 argument is used to provide a numeric value for the process variance, which, by default, is equal to 1. Finally, length.out is an optional integer that defines the length of the sequence of distances, when distance = NULL. By default, length.out = 50.

R11.1 *R syntax*: Semivariogram and correlation function plots for *corExp*

```
> tx <- c(0, 10^-2, 0.8)              # Auxilary vector
> cX <-                               # corExp object defined
+   corExp(value = c(1, 0.2),         # range ϱ:(10.16), nugget ϱ0:(10.18)
+          form = ~tx,
+          nugget = TRUE)             # Nugget defined
> Dtx <- data.frame(tx)
> (cXi <-                             # corExp object initialized
+   Initialize(cX, data = Dtx))
  Correlation structure of class corExp representing
   range nugget
     1.0    0.2
> (getCovariate(cXi))                 # tx diffs: 2-1, 3-1, 3-2
  [1] 0.01 0.80 0.79
> Vrg <- Variogram(cXi)               # Semi-variogram created ...
> plot(Vrg, smooth = FALSE,           # ... and plotted. Fig. 10.1a
+       type = "l")
> corFunDt <-                         # Data for correlation function
+    data.frame(dist = Vrg$dist,
+               corF = 1 - Vrg$variog)
> plot(corFunDt,                      # Corr function plotted with ...
+       type = "l", ylim = c(0,1)) # ... traditional graphics ...
> xyplot(corF ~ dist,                 # ... and xyplot().  Fig. 10.1b
+    data = corFunDt, type = "l")
```

Panel R11.1 presents the use of the Variogram() function to obtain a semivariogram for the *corExp* correlation structure (Table 10.1). The object cX of class *corExp* is defined with the range $\varrho = 1$ and the nugget effect $\varrho_0 = 0.2$ (see (10.18)). The values of the parameters are specified using the value argument of the corExp() constructor function. Additionally, a one-dimensional position variable tx, indicated in the argument form, is used. The elements of the vector tx are chosen in such a way that their differences cover the desired range of values from values very close to zero up to 0.8. The correlation structure is initialized using the actual numeric values of the variable tx and stored in the object cXi. The resulting correlation matrix can be printed with the use of the function corMatrix(); for brevity, we do not include the result in Panel R11.1. Using the function Variogram(), the semivariogram is calculated and stored in the object Vrg. The object is subsequently used to plot the (theoretical) semivariogram.

Panel R11.1 presents also two methods to plot the correlation function. Both methods use a data frame, created from the `dist` and `variog` components of the `Vrg` object. We do not show the resulting plots. However, a similar syntax was used to obtain plots of the semivariogram and correlation function shown in panels (a) and (b) of Fig. 10.1, respectively.

11.3.3 The `corMatrix()` *Function*

To obtain the correlation matrix represented by an initialized object of *corStruct* class, the generic function `corMatrix()` is used. The arguments of the function are `object`, `covariate`, and `corr`. The argument `object` is an object of class *corStruct*, for which we want to obtain information about. If the object is initialized, `corMatrix()` returns, by default, the correlation matrix or a list of correlation matrices, depending on whether the correlation structure was initialized for a single or multiple covariate vectors. If the object is not initialized, the argument `covariate` is used to provide a covariate vector (matrix), or a list of covariate vectors (matrices), at which the values of the correlation matrix are to be evaluated. The argument `corr` is a logical value. By default, `corr=TRUE` and indicates that the function should return the correlation matrix or a list of correlation matrices, represented by the *corStruct*-class object. If `corr=FALSE`, the function returns a transpose of the inverse of the square root of the correlation matrix (or a list of such matrices). That is, if $\boldsymbol{C} = \boldsymbol{U}'\boldsymbol{U}$ is a correlation matrix, the use of `corr=FALSE` yields $(\boldsymbol{U}^{-1})'$. The examples of using `corMatrix()` function are given in the next chapter in Panels R11.3–R11.5.

11.4 Illustration of Correlation Structures

In the next three sections, we illustrate the use of constructor functions, initialization, and extracting information for various correlation structures. The first two structures, represented by the *corCompSymm* and *corAR1* classes, are examples of serial correlation structures. The last structure, *corExp*, is an example of a spatial correlation structure.

For illustration purposes, in Panel R11.2, we generate a hypothetical data frame `df` containing two subjects. The first subject has four consecutive observations, indicated by the values of the variable `occ`. The observations are made at different locations in a two-dimensional space, with the coordinates given by the `loc1` and `loc2` position variables. The second subject has only three observations, with the observation for the third occasion missing. Note that, for the two subjects, the coordinates of the observations made at the same occasion differ.

R11.2 *R syntax*: Hypothetical data to illustrate various correlation structures

```
> subj <- rep(1:2, each = 4)                  # Two subjects
> occ  <- rep(1:4, 2)                         # Four observations each
> loc1 <- rep(c(0, 0.2, 0.4, 0.8), 2)         # First coordinate
> loc2 <-                                     # Second coordinate
+    c(0, 0.2, 0.4, 0.8, 0, 0.1, 0.2, 0.4)
> df0  <-                                     # Hypothetical data frame
+    data.frame(subj, occ, loc1, loc2)
> (df  <-                                     # Occ = 3 for subj.2 deleted
+    subset(df0, subj != 2 | occ != 3))
   subj occ loc1 loc2
 1    1   1  0.0  0.0
 2    1   2  0.2  0.2
 3    1   3  0.4  0.4
 4    1   4  0.8  0.8
 5    2   1  0.0  0.0
 6    2   2  0.2  0.1
 8    2   4  0.8  0.4
```

11.4.1 Compound Symmetry: The corCompSymm Class

We continue with the compound-symmetry correlation structure, defined in (10.15), as an example. Panel R11.3 shows the use and initialization of an object of class *corCompSymm*.

The `corCompSymm(value=0.3, form=~1|subj)` command specifies the compound-symmetry function with a constant correlation of 0.3. It is worth mentioning that the syntax could have been abbreviated to `corCompSymm(0.3, ~1|subj)`.

The `Initialize()` function initializes the *corCompSymm* object for the hypothetical data set. As a result, the same correlation structure, though with different dimensions, is obtained for both subjects. Note that, had we provided the position variable by specifying `corCompSymm(0.3, ~occ|subj))`, the result would not have changed, because, by default, `corCompSymm()` ignores any position variable. This is understandable, because the compound-symmetry correlation structure assumes a constant correlation coefficient between any two observations.

The command `coef(cs)` provides the coefficients of the initialized object `cs` in the unconstrained form. Note that the obtained value of the coefficients is equal to $\log(1/3+0.3) - \log(1-0.3)$, corresponding to the modified Fisher's z-transform (10.35) of $\varrho = 0.3$. The use of the `getCovariate(cs)` command allows to obtain the position vectors for both subjects included in our hypothetical data set. By applying the `corMatrix()` function to the object `cs`, we obtain the correlation matrices, defined for the subjects. Note that, by default, the function prints out the correlation matrices, which is equivalent to the use of the `corr=TRUE` argument.

R11.3 *R syntax*: Defining and initializing an object of class *corCompSymm*. The data frame df was defined in Panel R11.2

```
> cs <-                                       # Object defined...
+     corCompSymm(value = 0.3, form = ~1|subj)
> cs <- Initialize(cs, df)                    # ... initialized
> coef(cs, unconstrained = FALSE)             # Constrained coefficient
  Rho
  0.3
> coef(cs)                       # Unconstrained = log((1/3+.3)/(1-.3))
  [1] -0.10008
> getCovariate(cs)                            # Positions in series
  $`1`
  [1] 1 2 3 4

  $`2`
  [1] 1 2 3
> corMatrix(cs)                               # Corr. matrix displayed
  $`1`
       [,1] [,2] [,3] [,4]
  [1,]  1.0  0.3  0.3  0.3
  [2,]  0.3  1.0  0.3  0.3
  [3,]  0.3  0.3  1.0  0.3
  [4,]  0.3  0.3  0.3  1.0

  $`2`
       [,1] [,2] [,3]
  [1,]  1.0  0.3  0.3
  [2,]  0.3  1.0  0.3
  [3,]  0.3  0.3  1.0
```

11.4.2 Autoregressive Structure of Order 1: The corAR1 Class

The object cs1 in Panel R11.4 represents an uninitialized *corAR1* correlation structure. By applying to the object the function coef(), with the unconstrained= FALSE argument, we obtain the value of the defining correlation coefficient $\varrho = 0.3$ (see Table 10.1). Using the argument unconstrained=TRUE, we obtain the coefficient on the unconstrained scale, i.e., the value of $\log(1.3/0.7)$, resulting from Fisher's z-transform (10.33) of $\varrho = 0.3$. The object cs1i represents a corAR1 structure, which has been initialized for the data frame df2 that contains the covariate vector tx. The application of the function corMatrix() to the initialized object prints out the corresponding correlation matrix. The same result is obtained, as shown in Panel R11.4, by applying the function corMatrix(), with the argument covariate = tx, to the uninitialized object cs1. In this case, the correlation structure, defined in cs1, is evaluated at the covariate vector tx.

In Panel R11.4, the object chL contains the coefficients of the transpose of the inverse of the square root of the correlation matrix (Sect. 10.4.3), corresponding to cs1i. The last command in Panel R11.4 illustrates the back-transformation leading to the correlation matrix (Sect. 11.3.3).

R11.4 Extracting and assigning coefficients to an object of class *corAR1*

```
> cs1 <- corAR1(0.3, form = ~tx)    # Uninitialized corAR1 struct
> coef(cs1, unconstrained = FALSE) # Constrained coefficient
  Phi
  0.3
> coef(cs1)                        # Unconstrained = log((1+.3)/(1-.3))
  [1] 0.61904
> tx  <- 1:4                       # A covariate with values 1, 2, 3, 4
> corMatrix(cs1, covariate = tx)   # Corr(R_i) of uninitialized object
        [,1] [,2] [,3]  [,4]
  [1,] 1.000 0.30 0.09 0.027
  [2,] 0.300 1.00 0.30 0.090
  [3,] 0.090 0.30 1.00 0.300
  [4,] 0.027 0.09 0.30 1.000
> df2 <- data.frame(tx)            # An auxiliary data frame
> cs1i <-                          # Initialized corAR1 object
+    Initialize(cs1, data = df2)
> corMatrix(cs1i)                  # corAR1 matrix displayed
        [,1] [,2] [,3]  [,4]
  [1,] 1.000 0.30 0.09 0.027
  [2,] 0.300 1.00 0.30 0.090
  [3,] 0.090 0.30 1.00 0.300
  [4,] 0.027 0.09 0.30 1.000
> (chL <-                          # Cholesky factor L=(U')^-1
+    corMatrix(cs1i, corr = FALSE))
           [,1]     [,2]     [,3]   [,4]
  [1,]  1.00000  0.00000  0.00000 0.0000
  [2,] -0.31449  1.04828  0.00000 0.0000
  [3,]  0.00000 -0.31449  1.04828 0.0000
  [4,]  0.00000  0.00000 -0.31449 1.0483
  attr(,"logDet")
  [1] 0.14147
> solve(t(chL) %*% chL)            # Back to Corr(R_i) = U'U =(L'L)^-1
        [,1] [,2] [,3]  [,4]
  [1,] 1.000 0.30 0.09 0.027
  [2,] 0.300 1.00 0.30 0.090
  [3,] 0.090 0.30 1.00 0.300
  [4,] 0.027 0.09 0.30 1.000
```

A word of caution is worth issuing with regard to the use of serial correlation classes other than *corCompSymm*. For these classes, specifying the `form=~1|g` argument for the appropriate constructor function indicates the use of the order of the observations in the group as the position index. When data are balanced, i.e., when all subjects have got all measurements, or when they reveal monotone missingness patterns (Sect. 3.2.1), this will work fine. However, if, for some subjects, intermittent measurements are missing, the use of the observation order can result in the wrong correlation structure. Such a case is illustrated in Panel R11.5 for the `corAR1` correlation structure. In the part (a) of the panel, the first `corAR1()` statement defines the object `car` of class *corAR1*, with the parameter $\varrho = 0.3$, and with the order of observations within a subject used as the position variable. Consequently, after initializing the object `car` using the data from the `df` data frame, the correlation matrix for the second subject contains the value $\varrho^2 = 0.09$ as the correlation coefficient between the first and third observation. However, these observations were actually made at the first and *fourth* occasion, respectively, so the correct value is $\varrho^3 = 0.027$. To correctly specify this value, the `occ` variable should be used as the position variable using the `form= occ | subj` argument, as shown in the first `corAR1()` statement of the part (b) of Panel R11.5.

Note that, for data with measurement timepoints common to all subjects, this caution is required only for nonmonotone missing data patterns. Nevertheless, in case of the constructor functions for serial correlation classes other than *corCompSymm*, it is prudent to always use a position variable, which reflects the proper positions of the observations in a sequence for each group (subject), in the `form` argument.

11.4.3 Exponential Structure: The corExp Class

In Panel R11.6, we illustrate the definition and initialization of an object of class *corExp*.

The hypothetical data frame `df`, specified in Panel R11.2, is used for illustration. The first `corExp()` statement, in Panel R11.6a, defines the *corExp*-class object `ceE`, which represents the exponential correlation structure, given by (10.16). By default, the Euclidean distance between the position vectors, specified by the position variables `loc1` and `loc2`, is used. The `Initialize()` statement initializes the object and computes the correlation structure coefficients using the data from the data frame `df`. The resulting correlation matrices for both subjects are displayed with the use of the `corMatrix()` statement. Note that the matrices differ, because the spatial coordinates of the measurements differ for the subjects.

In Panel R11.6b, the distance function is changed to the Manhattan metric (Sect. 10.3.2). Toward this end, the `metric="man"` argument is used in the call to the `corExp()` constructor function. The resulting correlation matrix, displayed only for the first subject, is different from the one obtained using the Euclidean distance.

R11.5 *R syntax*: Defining and initializing an object of class *corAR1*. The data frame df was defined in Panel R11.2

(a) *Not a recommended syntax*

```
> car <-                              # Not-recommended syntax ...
+    corAR1(value = 0.3, form = ~1|subj)
> carI <- Initialize(car, df)         # corAR1 class object initialized
> getCovariate(carI)     # Position=order of observations for a subject
  $`1`
  [1] 1 2 3 4

  $`2`
  [1] 1 2 3
> corMatrix(carI)[[1]]    # Correct matrix for the 1st subject
        [,1] [,2] [,3]  [,4]
  [1,] 1.000 0.30 0.09 0.027
  [2,] 0.300 1.00 0.30 0.090
  [3,] 0.090 0.30 1.00 0.300
  [4,] 0.027 0.09 0.30 1.000
> corMatrix(carI)[[2]]    # Incorrect matrix for the 2nd subject
       [,1] [,2] [,3]
  [1,] 1.00  0.3 0.09
  [2,] 0.30  1.0 0.30
  [3,] 0.09  0.3 1.00
```

(b) *Recommended syntax*

```
> car1 <- corAR1(value = 0.3, form = ~occ|subj)   # Recommended syntax
> car1 <- Initialize(car1, df)        # corAR1 classs object initialized
> getCovariate(car1)      # Correct positions based on the occ variable
  $`1`
  [1] 1 2 3 4

  $`2`
  [1] 1 2 4
> corMatrix(car1)[[2]]    # Correct matrix for the 2nd subject
        [,1] [,2]  [,3]
  [1,] 1.000 0.30 0.027
  [2,] 0.300 1.00 0.090
  [3,] 0.027 0.09 1.000
```

R11.6 *R syntax*: Defining and initializing an object of class *corExp*. The data frame `df` was defined in Panel R11.2

(a) *Euclidean metric*

```
> ceE <- corExp(value=1, form= ~loc1 + loc2 | subj)# Euclidean metric
> ceE <- Initialize(ceE, df)
> corMatrix(ceE)             # List with corr matrices for both subjects
$`1`
        [,1]    [,2]    [,3]    [,4]
[1,] 1.00000 0.75364 0.56797 0.32259
[2,] 0.75364 1.00000 0.75364 0.42804
[3,] 0.56797 0.75364 1.00000 0.56797
[4,] 0.32259 0.42804 0.56797 1.00000

$`2`
        [,1]    [,2]    [,3]
[1,] 1.00000 0.79963 0.40884
[2,] 0.79963 1.00000 0.51129
[3,] 0.40884 0.51129 1.00000
```

(b) *Manhattan metric*

```
> ceM <-                                             # Manhattan metric
+    corExp(1, ~ loc1 + loc2 | subj, metric = "man")
> ceM <- Initialize(ceM, df)
> corMatrix(ceM)[[1]]                   # Corr matrix for the 1st subject
        [,1]    [,2]    [,3]    [,4]
[1,] 1.00000 0.67032 0.44933 0.20190
[2,] 0.67032 1.00000 0.67032 0.30119
[3,] 0.44933 0.67032 1.00000 0.44933
[4,] 0.20190 0.30119 0.44933 1.00000
```

(c) *Nugget effect*

```
> ceEn <-                               # nugget = 0.2
+    corExp(c(1, 0.2), ~ loc1 + loc2 | subj, nugget = TRUE)
> ceEn <- Initialize(ceEn, df)
> coef(ceEn, unconstrained=FALSE)  # Constrained ϱ, ϱ0
  range nugget
    1.0    0.2
> corMatrix(ceEn)[[1]]                  # Corr matrix for the 1st subject
        [,1]    [,2]    [,3]    [,4]
[1,] 1.00000 0.60291 0.45438 0.25807
[2,] 0.60291 1.00000 0.60291 0.34244
[3,] 0.45438 0.60291 1.00000 0.45438
[4,] 0.25807 0.34244 0.45438 1.00000
```

Finally, in Panel R11.6c, a nugget effect is used (Sect. 10.3.2). Toward this end, the `nugget=TRUE` argument is specified in the `corExp()` constructor-function call, together with the value of the effect equal to 0.2, given as the second element of the `value` argument.

11.5 Using the `gls()` Function

The function most frequently used in R to fit LMs for correlated data is the `gls()` function from the **nlme** package. It allows fitting models, defined by (10.1)–(10.5), with various forms for the variance-covariance matrix, $\mathcal{R}_i$, of the within-group residual errors.

The main arguments of the function `gls()`, i.e., `model`, `data`, `subset`, `na.action`, and `method` were introduced in Sect. 5.4 in the context of LMs for independent observations with homogeneous variance. In Sect. 8.4, we described an additional argument, namely, `weights`, which allows specifying the variance function for LMs for independent observations with heterogeneous variances, which were introduced in Chap. 7. We illustrated the use of these arguments in Chaps. 6 and 9. Note that all these arguments play essentially the same role and have the same syntax for the models introduced in this chapter.

In the context of LMs for correlated data, the additional important argument of the `gls()` function is `correlation`. The argument specifies an object that inherits from the *corStruct* class, which defines the correlation structure. Thus, a typical use of the argument is of the form `correlation=` *`corStruct`*`(form=`*`formula`*`)`, where *`corStruct`* is a correlation-structure constructor function (Table 10.1), while *`formula`* is a one-sided formula (Sect. 11.2), which indicates the position and grouping variables used in defining the correlation structure. The default value of the argument is `correlation=NULL`, which implies uncorrelated residual errors. This argument can prove useful when user defined initial values need to be assigned to a vector of $\boldsymbol{\theta}_R$ parameters.

Note that the information about the grouping of the data, relevant in the context of the models considered in this chapter, can be introduced into a `gls()`-function call in two ways. The preferred, transparent way is by specifying a formula (Sect. 11.2), indicating the grouping factors, in the correlation-structure constructor function used in the `correlation` argument. In this way, the grouping of the data can be directly inferred from the definition of the model. An alternative is to use an object of *groupedData* class in the `data` argument. As mentioned in Sect. 2.6, the *groupedData* class has some limitations. Also, in this way, the assumed grouping of the data is not reflected by any means in the definition of the model. Therefore, the use of the *groupedData* objects is not recommended.

11.6 Extracting Information from a Model-Fit Object of Class *gls*

To extract the results from a *gls* model-fit object, generic functions such as `print()`, `summary()`and `predict()`, can be used (Sect. 5.5 and Table 5.5). The methods to extract the results pertaining to the variance structure were described in Sect. 8.5 and Table 8.2.

In Table 11.1, we present selected functions and methods to extract the results pertaining to the correlation structure of a fitted LM for correlated data. They are very similar to the methods used to obtain the details of the fitted variance structure (Table 8.2); the difference lies mainly in the use of different components of the model-fit object. As in Sect. 8.5, we assume that the model fit results are stored in a hypothetical object `gls.fit`. In Table 11.1a, we demonstrate how to extract selected results directly from `gls.fit`. First, we obtain the applied form of the `gls()`-function call and store it in the object `cl`. Subsequently, the applied form of the `correlation` argument is obtained by extracting the `cl$correlation` component of the `cl` object. Confidence intervals (CIs) for the constrained correlation-function coefficients are obtained by extracting the `corStruct` component of the object resulting from the application of the `intervals()` function, with the argument `which="var-cov"`, to the model-fit object. The intervals are constructed by transforming CIs for the corresponding unconstrained coefficients (Sect. 10.6).

Table 11.1 *R syntax*: extracting results pertaining to the correlation structure from a hypothetical object `gls.fit` of class *gls*, representing a fit of a linear model for correlated data

(a) Extracting results directly from the `gls.fit` model-fit object	
Model fit component to be extracted	Syntax
`gls()`-call	`(cl <- getCall(gls.fit))`
`correlation=` argument	`cl$correlation`
95% CI for $\boldsymbol{\varrho}$	`intervals(gls.fit, which = "var-cov")$corStruct`
$\widehat{\mathcal{R}}_i$ matrices	`getVarCov(gls.fit)`
Normalized residuals	`resid(gls.fit, type = "normalized")`
Var-cov structure	`mSt <- gls.fit$modelStruct`
Correlation structure (CorSt)	`cSt <- mSt$corStruct`
(b) Extracting results from an auxiliary object `cSt` of *corStruct* class	
Correlation structure component to be extracted	Syntax
Summary	`summary(cSt)`
CorSt formula	`formula(cSt)`
CorSt covariate	`getCovariate(cSt)`
$\widehat{\boldsymbol{\varrho}}^*$ (unconstrained)	`coef(cSt)`
$\widehat{\boldsymbol{\varrho}}$ (constrained)	`coef(cSt, unconstrained = FALSE)`
Contribution to log-likelihood	`logLik(cSt)`
$\widehat{\boldsymbol{C}}_i$ matrices	`corMatrix(cSt)`
Log-determinant	`logDet(cSt)`

The estimated form of the $\mathcal{R}_i$ matrices (Sect. 10.2) is obtained by applying the `getVarCov()` function. The normalized residuals (Sect. 10.5) are extracted by applying the `resid()` function, with the argument `type="normalized"`, to the model-fit object.

By extracting and storing, in the object `mSt`, the `modelStruct` component of the model-fit object, we get access to the estimated variance-covariance structure of the model. In particular, details on the estimated form of the correlation structure are contained in the `corStruct` component of the *modelStruct* object. We extract and store the component in the object `cSt`. In Table 11.1b, we illustrate how to extract various elements of the fitted form of the correlation structure. For instance, the application of the `summary()` function to the object `cSt` provides a description of the correlation function, together with the estimates of the coefficients on the original, constrained scale. On the other hand, the formula and the covariate, used in the specification of the correlation function, are obtained using the `formula()` and `getCovariate()` functions, respectively. The correlation-function coefficients on the unconstrained and constrained scale are obtained by applying the generic `coef()` function with the argument `unconstrained` set to `TRUE` (default) and `FALSE`, respectively (Sect. 11.3.1).

Compared to Table 8.2, Table 11.1 includes two additional functions. The function `corMatrix()` gives, by default, the estimated correlation matrices for all groups in the analyzed data (Sect. 11.3.3). The function `logDet()`, on the other hand, extracts the sum of the logarithms of the determinants of the square roots of the correlation matrices.

As it was the case for the LM for independent, heteroscedastic observations (Sect. 8.5), results of the hypothesis tests based on the fitted LM for correlated data (Sect. 10.6) are accessed by the `anova()` generic function. Note that, as it was mentioned in Sect. 10.6, the results of the tests are assessed using the central F-distribution with the number of the denominator degrees of freedom equal to $n-p$. The function `anova()` also provides the information criteria which can be used to choose the best-fitting models from a set of nonnested models with different variance-covariance structures.

11.7 Chapter Summary

In this chapter, we presented the tools available in R to fit LMs for correlated data. The important issue was the implementation of the concept of the correlation structure. In Sect. 11.2, we described the relevant R tools. In particular, the correlation-structure constructor functions and their arguments were presented in Sects. 11.2.1 and 11.2.1.1, respectively. In Sect. 11.3, we described the tools available for inspection and modification of objects of class *corStruct*. Several examples of correlation structures were presented in Sect. 11.4.

The main tool to fit LMs for correlated data in R is the `gls()` function. Its main arguments were already described in Sects. 5.4 and 8.4. In Sect. 11.5, we introduced another argument, `correlation`. It allows modeling of the within-group correlation between residual errors. In the argument, the correlation-structure constructor functions, described in Sect. 11.2, are used. Note that a similar argument is used in the `lme()` function, which is applied to fit LMMs. Thus, the presentation of this particular feature of the `gls()` function will be relevant also for the latter models.

Finally, in Sect. 11.6, we discussed the tools used to extract the information about the estimated form of the within-group correlation structure from a *gls* model-fit object. The tools used for extracting information about other aspects of the fitted model were already presented in Sects. 5.5 (Table 5.5) and 8.5 (Table 8.2).

We note that, occasionally, it may be of interest to define a new correlation structure. In such a situation, as was mentioned for the case of variance functions (Sect. 8.6), it is recommended to explore, in the first instance, the information related to an already defined standard class, e.g., *corCompSymm*. Users may define their own correlation structures by writing appropriate constructor functions, similar to `corCompSymm()`. Such functions should return objects of a class with the same name as the constructor function and with attributes similar to those of already defined classes, like *corCompSymm*. In addition, at least three methods, namely, `coef()`, `coef<-`, and `corMatrix()`, have to be written for a newly created constructor function.

In the next chapter, we illustrate the use of the models and R syntax, introduced in this chapter, by applying them to the ARMD case study.

Chapter 12
ARMD Trial: Modeling Correlated Errors for Visual Acuity

12.1 Introduction

In Chap. 9, we analyzed the ARMD data using a model that assumed independence between repeated measurements of visual acuity for an individual patient. Given the longitudinal study design, in which sampling units, i.e., subjects, are *different* from the units of analysis, i.e., visual acuity measurements, the assumption of independence typically does not hold. This consideration is confirmed by the results of the exploratory analysis presented in Sect. 3.2, which indicate that the assumption is clearly not fulfilled. Thus, in the current chapter, we further modify our analysis to account for the correlation between the repeated measurements. In particular, we use LMs with fixed effects and correlated residual errors, defined in Sect. 10.2.

We begin by an analysis of a model that assumes a constant correlation between the visual acuity measurements (Sect. 12.3). In Sect. 12.4, we extend the analysis by allowing the correlation to differ depending on the timepoints, at which the measurements were taken. Finally, in Sect. 12.5, we allow for models with an unconstrained, general variance-covariance structure. Model-fit diagnostics are presented in Sect. 12.6. Model reduction and inference are discussed in Sect. 12.7. Section 12.8 offers a summary of the analyses conducted in the chapter.

12.2 The Model with Heteroscedastic, Independent Residual Errors Revisited

In Chap. 9, we analyzed the ARMD data using several LMs of the following form:

$$\text{VISUAL}_{it} = \beta_{0t} + \beta_1 \times \text{VISUAL0}_i + \beta_{2t} \times \text{TREAT}_i + \varepsilon_{it}, \tag{12.1}$$

where residual errors ε_{it} were assumed to be normally distributed with mean 0 but where different variance functions were used to model the variance heterogeneity.

A. Gałecki and T. Burzykowski, *Linear Mixed-Effects Models Using R: A Step-by-Step Approach*, Springer Texts in Statistics, DOI 10.1007/978-1-4614-3900-4_12,

Model **M9.3**, introduced in Sect. 9.3, with a variance function in the form of a power function of time, showed the best fit among the models considered in Chap. 9. Note that, by using the notation introduced in Sect. 10.2, the 4×4 diagonal variance-covariance matrix $\mathcal{R}_i$ for a subject with all four post-randomization visual acuity observations can be represented as

$$\mathcal{R}_i = \sigma^2 \boldsymbol{\Lambda}_i \mathbf{C}_i \boldsymbol{\Lambda}_i, \tag{12.2}$$

with $\mathbf{C}_i \equiv \mathbf{I}_4$, a 4×4 identity matrix, and Λ_i defined as

$$\boldsymbol{\Lambda}_i \equiv \begin{pmatrix} (\text{TIME}_{i1})^\delta & 0 & 0 & 0 \\ 0 & (\text{TIME}_{i2})^\delta & 0 & 0 \\ 0 & 0 & (\text{TIME}_{i3})^\delta & 0 \\ 0 & 0 & 0 & (\text{TIME}_{i1})^\delta \end{pmatrix}, \tag{12.3}$$

where TIME_{it} is the week at which the t-th measurement was taken. Diagonal elements of $\boldsymbol{\Lambda}_i$ were defined by the varPower($\cdot$) variance function, specified in (9.4).

According to model **M9.3**, specified by (12.1)–(12.3), the variances of the visual acuity measurements at different timepoints are allowed to be different, but the measurements are assumed to be independent. In view of the results of the exploratory analysis, presented in Sect. 3.2, the latter assumption is not correct.

In this chapter, we modify the model, so that the visual acuity measurements, obtained for the same individual, are allowed to be correlated. We consider initially models with the same mean structure as the one defined in (12.1), but with different variance-covariance (correlation) structures.

12.2.1 Empirical Semivariogram

In choosing an appropriate correlation structure, the empirical semivariogram, defined in Sect. 10.5 and employed in Panel R12.1, can be helpful.

It is worth noting that, in the ARMD trial, time differences for all possible six pairs of timepoints are different. Therefore, as shown in Panel R12.1a, we can estimate the semivariogram to calculate correlation coefficients between Pearson residuals for every pair of timepoints, separately. The residuals are computed based on the estimated form of model **M9.3**, stored in the object `fm9.2`. For example, the correlation between the residuals at 4 and 12 weeks is equal to $0.59 = 1 - 0.41$. A graphical representation of the estimated semivariogram plotted *versus* the time distance is presented in Fig. 12.1a. The semivariogram was obtained using the `Variogram()` function with the numeric `time` variable used in the `form` argument (Sect. 11.3.2).

Note that, when applying the function to a model-fit object of class *gls*, many additional arguments, above those mentioned in Sect. 11.3.2, are available. In particular, the argument `robust` can be used. If `robust=TRUE`, the semivariogram is

R12.1 *ARMD Trial*: Empirical semivariograms for Pearson residuals for model **M9.3**. The model-fit object `fm9.2` was created in Panel R9.2

(a) *Per time difference*

```
> (Vg1 <- Variogram(fm9.2, form = ~ time | subject))
    variog dist n.pairs
  1 0.41144    8     224
  2 0.35739   12     214
  3 0.54202   20     212
  4 0.23819   28     190
  5 0.46203   40     194
  6 0.61401   48     193
> plot(Vg1, smooth = FALSE, xlab = "Time difference")      # Fig. 12.1a
```

(b) *Per time lag*

```
> (Vg2 <- Variogram(fm9.2, form = ~tp | subject))
    variog dist n.pairs
  1 0.34060    1     628
  2 0.50380    2     406
  3 0.61401    3     193
> plot(Vg2, smooth = FALSE, xlab = "Time Lag")             # Fig. 12.1b
```

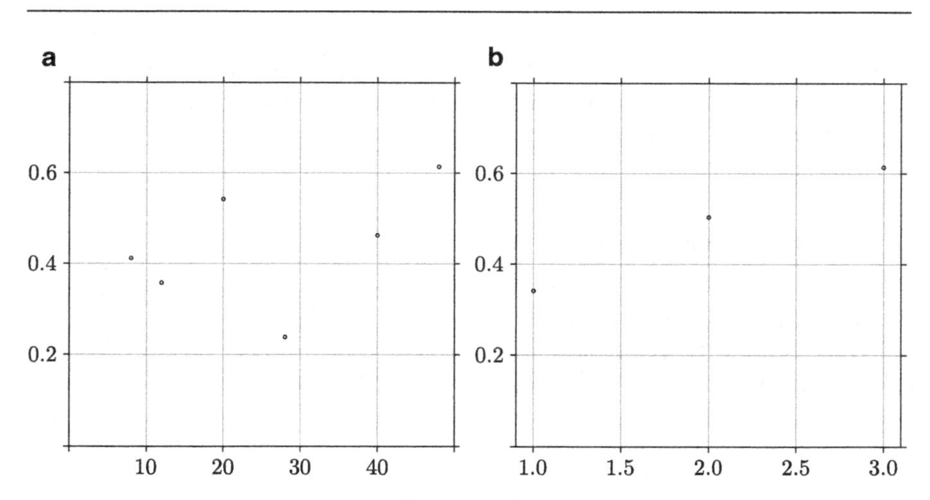

Fig. 12.1 *ARMD Trial*: Empirical semivariograms for model **M9.2** (**a**) Per time-difference (**b**) Per time-lag

estimated using the robust estimator, given in (10.40). By default, `robust=FALSE`, and the semivariogram function is estimated using the estimator defined in (10.39). This is the case for the `Variogram()`-function call used in Panel R12.1a.

An alternative form of the variogram is obtained in Panel R12.1b using the time lag, i.e., the absolute difference between two position indices. This variant was estimated using the `Variogram()` function with the position variable `tp` used in the `form` argument. The variable indicates the position of a measurement in the sequence of planned measurements and is equal to 1 for week 4, 2 for week 12, 3 for week 24, and 4 for week 52 measurements (see Sect. 2.2.2). As there are at most four post-randomization observations per subject, there are only three possible time-lag values. The corresponding plot is shown in Fig. 12.1b. Note that, because there are only a few time-difference and time-lag values, smoothing of the plots will not help in their interpretation. For this reason, we set the `smooth` argument in the `plot()`-function calls, shown in Panel R12.1, to `FALSE`. The plots of the semivariograms, presented in Fig. 12.1, suggest that the correlation decreases with the time difference/lag. Thus, a correlation structure like, e.g., a compound symmetry, will most likely not fit the data well. A more appropriate structure might be, e.g., an autoregressive process of order 1 (see Table 10.1). Nevertheless, in the next section, for illustrative purposes, we consider a model with a compound-symmetry correlation structure.

12.3 A Linear Model with a Compound-Symmetry Correlation Structure

With the use of the function `gls()`, we can fit a modified variant of model **M9.3**, which takes into account the correlation between the repeated measurements of visual acuity measurements. In a first attempt, for illustrative purposes, we allow for a constant correlation between the measurements.

12.3.1 Model Specification

We assume that the mean structure of the model, specified in this section and labeled as **M12.1**, is the same as the one implied by (12.1). On the other hand, we assume that the marginal variance-covariance matrix is of the form defined in (12.2), with $\boldsymbol{\Lambda}_i$ given in (12.3) and

$$\mathbf{C}_i \equiv \begin{pmatrix} 1 & \varrho & \varrho & \varrho \\ \varrho & 1 & \varrho & \varrho \\ \varrho & \varrho & 1 & \varrho \\ \varrho & \varrho & \varrho & 1 \end{pmatrix}. \qquad (12.4)$$

As a result, we obtain a compound-symmetry correlation structure and a heterogeneous compound-symmetry variance-covariance structure.

R12.2 *ARMD Trial*: Model **M12.1** with a compound-symmetry correlation structure

(a) *Fitting model* **M12.1**

```
> lm1.form <-                                    # (12.1)
+    formula(visual ~ -1 + visual0 + time.f + treat.f:time.f )
> fm12.1 <-                                      # M12.1
+    gls(lm1.form, weights = varPower(form = ~time),
+        correlation = corCompSymm(form = ~1|subject),
+        data = armd)
```

(b) *95% CIs for the variance-covariance parameters*

```
> intervals(fm12.1, which = "var-cov")     # CIs for ϱ:(12.4), δ:(12.3), σ
  Approximate 95% confidence intervals

   Correlation structure:
         lower    est.   upper
  Rho 0.50447 0.57326 0.63664
  attr(,"label")
  [1] "Correlation structure:"

   Variance function:
           lower    est.   upper
  power 0.21581 0.25982 0.30382
  attr(,"label")
  [1] "Variance function:"

   Residual standard error:
   lower   est.  upper
  5.2357 5.9815 6.8336
```

12.3.2 Syntax and Results

The decomposition (12.2) allows us to model the variance-covariance matrix $\mathcal{R}_i$ using the *varPower* class of variance functions and *corCompSymm* class of correlation functions, described in Sects. 8.2 and 11.4.1, respectively. In Panel R12.2, we apply the `gls()` function to fit model **M12.1**.

More specifically, in Panel R12.2a, we use the `weights` argument combined with the `varPower()` constructor function and the `correlation` argument combined with the `corCompSymm()` constructor function. The `correlation = corCompSymm(form = ~1|subject)` argument indicates that we use the same correlation coefficient for different observations for each level of the `subject` factor. That is, we allow for a constant correlation of visual acuity measurements made at different timepoints for the same patient. Note that we do not explicitly specify the `method` argument. Thus, the default value, i.e., `method="REML"`, is used (Sect. 5.4). This is because, at this point, we focus on the estimation of the variance-covariance structure of the data.

Table 12.1 *ARMD Trial*: The REML-based parameter estimates[a] for models with various correlation structures fitted using the `gls()` function

	Parameter	`fm12.1`	`fm12.2`	`fm12.3`
Model label		**M12.1**	**M12.2**	**M12.3**
Log-REML value		−3216.46	−3186.46	−3176.60
Fixed effects				
Visual acuity at t = 0	β_1	0.92(0.04)	0.88(0.04)	0.89(0.04)
Time (4wks)	β_{01}	2.98(2.18)	5.23(2.22)	4.73(2.14)
Time (12wks)	β_{02}	1.94(2.29)	4.17(2.31)	3.67(2.25)
Time (24wks)	β_{03}	−1.78(2.40)	0.51(2.41)	0.00(2.37)
Time (52wks)	β_{04}	−6.98(2.58)	−4.85(2.56)	−5.33(2.56)
Tm(4wks):Trt(Actv)	β_{21}	−2.26(1.13)	−2.28(1.15)	−2.29(1.10)
Tm(12wks):Trt(Actv)	β_{22}	−3.59(1.51)	−3.60(1.49)	−3.60(1.49)
Tm(24wks):Trt(Actv)	β_{23}	−2.90(1.84)	−3.13(1.78)	−3.13(1.82)
Tm(52wks):Trt(Actv)	β_{24}	−5.33(2.31)	−4.74(2.21)	−4.95(2.31)
Variance function				
power (TIME^{δ})	δ	0.26(0.22,0.30)	0.23(0.18,0.28)	0.27(0.22,0.32)
Correlation structure				
C-S (12.4)	ϱ_{CS}	0.57(0.50,0.64)		
AR1 (12.5)	ϱ_{AR1}		0.66(0.60,0.70)	
General (12.6)				
cor(1,2)	ϱ_{12}			0.58(0.49,0.66)
cor(1,3)	ϱ_{13}			0.45(0.33,0.55)
cor(1,4)	ϱ_{14}			0.30(0.15,0.44)
cor(2,3)	ϱ_{23}			0.65(0.57,0.72)
cor(2,4)	ϱ_{24}			0.53(0.42,0.63)
cor(3,4)	ϱ_{34}			0.77(0.70,0.82)
Scale	σ	5.98(5.24,6.83)	6.36(5.50,7.34)	5.74(4.95,6.65)

[a] Approximate SE for fixed effects and 95% CI for covariance parameters are included in paratheses

Results of the fit of the model, stored in the object `fm12.1`, can be accessed using the `summary()` command. The output is extensive and we do not display it here; the detailed results are shown in Table 12.1. Instead, in Panel R12.2b, we present the approximate 95% confidence intervals (CIs) for the variance-covariance parameters of the model using the `intervals()` function. The output includes the point estimates and lower and upper limits of the CIs for ϱ, for the power coefficient δ of the power variance function, and for σ. Underlying computations employ the formulae introduced in Sects. 7.6.2 and 10.6.

The results indicate that, according to the assumed compound-symmetry correlation structure, the correlation coefficient of any two visual acuity measurements obtained for the same patient ϱ is equal to 0.573. The 95% CI for the correlation coefficient confirms that there is a nonnegligible correlation between the visual acuity measurements, as noted earlier. The scale parameter σ is estimated to be equal to 5.98. The estimated power coefficient of the variance function, 0.260, is very close to the value of 0.252 obtained for model **M9.3** fitted in Panel R9.3. It indicates an increasing variability of the measurements over time.

R12.3 *ARMD Trial*: Estimated variance-covariance structure of the fitted model **M12.1**. The model-fit objects `fm9.2` and `fm12.1` were created in Panels 9.2 and R12.2, respectively

(a) *The marginal variance-covariance structure*

```
> fm12.1vcov <-                                    # R̂_i
+    getVarCov(fm12.1, individual = "2")
> nms <- c("4wks", "12wks", "24wks", "52wks")
> dnms <- list(nms, nms)                           # Dimnames created
> dimnames(fm12.1vcov) <- dnms                     # Dimnames assigned
> print(fm12.1vcov)
 Marginal variance covariance matrix
         4wks   12wks   24wks   52wks
 4wks  73.531  56.077  67.143  82.081
 12wks 56.077 130.140  89.323 109.200
 24wks 67.143  89.323 186.560 130.740
 52wks 82.081 109.200 130.740 278.810
   Standard Deviations: 8.575 11.408 13.659 16.698
> print(cov2cor(fm12.1vcov),                       # Ĉ_i: (12.4)
+       corr = TRUE, stdevs = FALSE)
 Marginal correlation matrix
          4wks   12wks   24wks   52wks
 4wks  1.00000 0.57326 0.57326 0.57326
 12wks 0.57326 1.00000 0.57326 0.57326
 24wks 0.57326 0.57326 1.00000 0.57326
 52wks 0.57326 0.57326 0.57326 1.00000
```

(b) *Test of independence vs. compound-symmetry correlation structure*

```
> anova(fm9.2, fm12.1)                      # M9.2 ⊂ M12.1
        Model df    AIC    BIC  logLik   Test L.Ratio p value
 fm9.2      1 11 6738.1 6790.4 -3358.1
 fm12.1     2 12 6456.9 6514.0 -3216.5 1 vs 2  283.21  <.0001
```

In Panel R12.3, we display the estimated variance-covariance and correlation matrices. In addition, we test the hypothesis about the need of compound-symmetry correlation in the model. Specifically, in Panel R12.3a, we use the `getVarCov()` function (Sect. 11.6) to obtain an estimate of the variance-covariance matrix. The argument `individual="2"` indicates that we request the matrix for subject "2", for whom all four post-randomization measurements are available. To simplify the printout, we modify the default names of the rows and columns of the matrix. The resulting correlation structure is obtained by transforming the variance-covariance matrix into a correlation matrix with the use of the `cov2cor()` function (Sect. 3.2.3). Note that, when printing the latter matrix, we use arguments `corr=TRUE` and `stdevs=FALSE`. The first argument chooses the format of the printout suitable for a correlation matrix, while the second one suppresses the display of the standard deviations, which are irrelevant for a correlation matrix.

At this point, we might want to test whether allowing for the correlation of the dependent variable in the model is indeed important. Toward this end, in Panel R12.3b, we use the LR test employing the restricted likelihoods for models stored in objects `fm9.2` and `fm12.1` using the `anova()` function. The test is based on the restricted likelihoods, because we use it for a comparison of models with the same mean structures (see Sects. 4.7.2 and 7.6.1).

In this case, the `anova()` command subtracts the value of $-2\times$(restricted log-likelihood) of the more general model **M12.1**, the fit of which is stored in the object `fm12.1`, from the corresponding value of the nested model **M9.3**, the fit of which is stored in the object `fm9.2`. The difference is referred to the appropriate χ^2 distribution. The result of the LR test is clearly statistically significant, indicating the importance of the adjustment for the correlation in modeling the data.

Given the results of the exploratory analysis (Sect. 3.2) and the shape of the semivariograms, presented in Fig. 12.1, a compound-symmetry correlation structure is most likely not suitable to describe the ARMD data. Therefore, in the next section, we consider an autoregressive process of order 1 (see Table 10.1), which might be more appropriate.

12.4 Heteroscedastic Autoregressive Residual Errors

In this section, we consider the use of an autoregressive process of order 1 (see Table 10.1) that allows for an unequal correlation of the visual acuity measurements taken at different occasions. Note that the mean structure of the newly defined model is the same as the one implied by (12.1).

12.4.1 Model Specification

Instead of assuming a constant correlation, we might assume that the correlation decreases for measurements obtained at more distant timepoints. Although visual acuity was not measured at equally spaced intervals, it does make sense, pragmatically speaking, to consider the use of the autoregressive correlation structure with lag 1 (see Sect. 10.3.2). The structure implies that two observations separated by t time units are correlated with a correlation coefficient equal to ϱ^t, where ϱ is the lag–1 correlation. Thus, we assume that the variance-covariance matrix is represented by (12.2), with $\boldsymbol{\Lambda}_i$ given by (12.3) and

$$\mathbf{C}_i \equiv \begin{pmatrix} 1 & \varrho & \varrho^2 & \varrho^3 \\ \varrho & 1 & \varrho & \varrho^2 \\ \varrho^2 & \varrho & 1 & \varrho \\ \varrho^3 & \varrho^2 & \varrho & 1 \end{pmatrix}. \tag{12.5}$$

We label the newly defined model as **M12.2**.

R12.4 *ARMD Trial*: Model **M12.2** with an AR(1) correlation structure. The model-fit object fm9.2 was created in Panel R9.2

(a) *Fitting model* **M12.2**

```
> fm12.2 <-                                   # M12.2 ← M9.2
+    update(fm9.2,                            # (12.5)
+           correlation = corAR1(form = ~tp|subject),
+           data = armd)
```

(b) *95% CIs for variance-covariance parameters*

```
> intervals(fm12.2, which = "var-cov") # CIs for ϱ:(12.5), δ:(12.3), σ
  Approximate 95% confidence intervals

   Correlation structure:
          lower    est.   upper
  Phi1 0.60398 0.65731 0.70478
  attr(,"label")
  [1] "Correlation structure:"

   Variance function:
          lower    est.   upper
  power 0.1832 0.23119 0.27918
  attr(,"label")
  [1] "Variance function:"

   Residual standard error:
   lower   est.  upper
  5.5036 6.3563 7.3411
```

12.4.2 Syntax and Results

In Panel R12.4, we fit model **M12.2** and display 95% CIs for variance-covariance parameters. Note that, in Panel R12.4a, rather than using a new call to the gls() function, we use the update() function to modify only the aspect of interest, i.e., the correlation structure, of the model represented by the object fm9.2. In particular, we apply the autoregressive correlation structure, defined in (12.5), using the corAR1() constructor function to modify the value of the correlation argument.

For the visual acuity data, we might simply assume that two adjacent measurements (e.g., those at weeks 4 and 12, or at weeks 12 and 24, etc.) are correlated with a correlation coefficient equal to ϱ, say, while measurements separated by one intermittent observation (e.g., those at weeks 4 and 24, or at weeks 12 and 52), are correlated with a correlation coefficient of ϱ^2 and so on. This would suggest the use of the call `corAR1(form = ~1 | subject)`, i.e., the use of the order of the observations in the data as a covariate. However, as was noted in Sect. 3.2.1, in the dataset there are 8 patients with a nonmonotone missing data pattern. For these patients, the use of the `~1 | subject` formula is not correct, as the order of visual acuity measurements does not lead to a proper correlation assignment. This issue was explained in Sect. 11.4.2. To avoid the problem, we need to use a position variable that indicates, for each visual acuity measurement, the proper rank (position) of the particular measurement in the planned sequence of measurements. The data frame `armd` contains such a variable, named `tp` (Sect. 2.2.2). The variable takes values equal to 1 for week 4, 2 for week 12, 3 for week 24, and 4 for week 52 measurements. We should use the variable in the formula of the `corAR1()` function. That is, we should apply the syntax `corAR1(~tp | subject)`, as shown in Panel R12.4a.

The results of the fit of the model can be accessed using the `summary(fm12.2)` call. The output is extensive and we do not present it here. Instead, in Panel R12.4b, we show the approximate 95% CIs for the REML-based estimates of the parameters of the $\mathcal{R}_i$ and $\mathbf{C}_i$ matrices. The estimates of all parameters of model **M12.2** are displayed in Table 12.1.

In Panel R12.5, we continue with the presentation of the results for Model **M12.2**. The estimated correlation matrix, shown in Panel R12.5a, suggests that the correlation coefficient for the visual acuity measurements adjacent in time (e.g., those at weeks 4 and 12, or at weeks 12 and 24, etc.) is equal to $\varrho = 0.66$. This value is higher than 0.57 obtained for the compound-symmetry structure (see Panel R12.2a). On the other hand, the measurements separated by, e.g., one intermittent observation (e.g., those at weeks 4 and 24), are correlated with the correlation coefficient equal to $\varrho^2 = 0.43$, which is lower than 0.57.

Model **M12.1**, defined by (12.1) and (12.3)–(12.4), and represented by the object `fm12.1`, is not nested within model **M12.2**, represented by the object `fm12.2`. Therefore, we cannot compare them directly using the LR test. We can nevertheless use the `anova()` function to compare their information criteria (Sect. 4.7.2). Note that, alternatively, the `AIC()` function could be used. Results are shown in Panel R12.5b. The smaller the value of AIC, the better is the fit of a model. For the compound-symmetry model **M12.1**, AIC is equal to 6456.9, while for the autoregressive model **M12.2**, it is equal to 6396.9. Thus, we conclude that the model with the autoregressive correlation structure provides a better description of the data, in line with the results of the exploratory analysis and of the considerations based on the empirical semivariogram (Sect. 12.2).

Although the autoregressive correlation structure allows for differences in the values of correlation coefficients, it assumes a particular form of these differences, specified in (12.5). Of course, this assumption can be incorrect. Therefore, in the next section, we consider a fully general correlation structure, which does not impose any constraints on the possible differences.

R12.5 *ARMD Trial*: Estimated variance-covariance structure of the fitted model **M12.2** with an AR(1) correlation structure. Objects `fm12.1`, `dnms`, and `fm12.2` were created in Panels R12.2, R12.3, and R12.4, respectively

(a) *The marginal variance-covariance structure*

```
> fm12.2vcov <-
+     getVarCov(fm12.2, individual = "2")
> dimnames(fm12.2vcov) <- dnms
> fm12.2vcov                                   # R_i matrix
  Marginal variance covariance matrix
          4wks   12wks   24wks   52wks
  4wks  76.698  64.992  50.144  39.411
  12wks 64.992 127.470  98.346  77.296
  24wks 50.144  98.346 175.620 138.030
  52wks 39.411  77.296 138.030 251.100
    Standard Deviations: 8.7578 11.29 13.252 15.846
> fm12.2cor <- cov2cor(fm12.2vcov)
> print(fm12.2cor, digits = 2,                 # C_i: (12.5)
+       corr = TRUE, stdevs = FALSE)
  Marginal correlation matrix
        4wks 12wks 24wks 52wks
  4wks  1.00  0.66  0.43  0.28
  12wks 0.66  1.00  0.66  0.43
  24wks 0.43  0.66  1.00  0.66
  52wks 0.28  0.43  0.66  1.00
```

(b) *Compound-symmetry vs. autoregressive correlation (nonnested models)*

```
> anova(fm12.1, fm12.2)                        # M12.1 vs. M12.2
         Model df    AIC  BIC  logLik
  fm12.1     1 12 6456.9 6514 -3216.5
  fm12.2     2 12 6396.9 6454 -3186.5
```

12.5 General Correlation Matrix for Residual Errors

In this section, we present a model that allows for a fully general correlation structure. The mean structure of the model is defined by (12.1).

12.5.1 Model Specification

The model is obtained by assuming that the variance-covariance matrix $\mathcal{R}_i$ is defined by (12.2), with $\boldsymbol{\Lambda}_i$ given by (12.3) and

$$\mathbf{C}_i \equiv \begin{pmatrix} 1 & \varrho_{12} & \varrho_{13} & \varrho_{14} \\ \varrho_{12} & 1 & \varrho_{23} & \varrho_{24} \\ \varrho_{13} & \varrho_{23} & 1 & \varrho_{34} \\ \varrho_{14} & \varrho_{24} & \varrho_{34} & 1 \end{pmatrix}. \tag{12.6}$$

Note that the matrix $\mathbf{C}_i$ specifies a completely general correlation structure, with (potentially) different correlation coefficients for different pairs of measurements. We will refer to the model, defined in this section, as **M12.3**.

12.5.2 Syntax and Results

In Panel R12.6, we fit model **M12.3** and extract approximate 95% CIs for the variance-covariance parameters. First, in Panel R12.6a, we update the model-fit object `fm12.2` by changing the value of the `correlation` argument to `corAR1(form = ~1 | subject)`. The `corSymm()` constructor function (Sect. 11.2.1) specifies general (unconstrained) correlations (Sect. 10.3.2) between the visual acuity measurements for a `subject`.

In Panel R12.6b, we present the approximate 95% CIs for the variance-covariance parameters of the model. Panel R12.7 displays the variance-covariance and correlation matrices. We observe that, according to model **M12.3**, the correlation decreases for visual acuity measurements more distant in time, as it was seen for the autoregressive correlation model **M12.2**. The last column of Table 12.1 displays the REML-based estimates of all of the model parameters.

In Panel R12.8, we test hypotheses about the variance-covariance structure pertaining to model **M12.3**. Because model **M12.2** is nested within model **M12.3**, we use the LR test based on the two models. Toward this end, we apply the `anova()` function. The result is shown in Panel R12.8a. Note that both models have the same mean structure and that the test is based on REML, because the objects `fm12.2` and `fm12.3` were obtained using the default estimation method (REML) of the `gls()` function. Thus, the test is constructed in the form suitable for comparison of models with different variance-covariance structures (Sects. 4.7.2 and 7.6.1).

The result of the test is significant at the 5% significance level. It indicates that model **M12.3** provides a better fit than model **M12.2**.

It might be of interest to check whether a model with the general correlation structure, defined by (12.6), and with the most general variance function, which allows arbitrary (positive) variances of the visual acuity measurements made at different timepoints, could offer a better fit than model **M12.3**, which uses the variance function equal to the power function of the time covariate. Such a model can be obtained using the varIdent($\cdot$) variance function with timepoint-specific variance parameters (see Table 7.2 in Sect. 7.3.1). In Panel R12.8b, we fit the model to the ARMD data using the REML estimation method. Toward this end, we apply

R12.6 *ARMD Trial*: Model **M12.3** with a general correlation structure. The model-fit object fm12.2 was created in Panel R12.4

(a) *Fitting model* **M12.3**

```
> fm12.3 <-                              # M12.3 ← M12.2
+     update(fm12.2, correlation = corSymm(form = ~tp|subject),
+            data = armd)
```

(b) *95% CIs for variance-covariance parameters*

```
> intervals(fm12.3,                      # 95% CIs for ϱ:(12.6), δ:(12.3), σ
+            which = "var-cov")
  Approximate 95% confidence intervals

   Correlation structure:
               lower    est.   upper
  cor(1,2) 0.48963 0.58206 0.66155
  cor(1,3) 0.33240 0.44820 0.55068
  cor(1,4) 0.15182 0.30062 0.43610
  cor(2,3) 0.57117 0.65122 0.71900
  cor(2,4) 0.41930 0.53096 0.62680
  cor(3,4) 0.69847 0.76578 0.81966
  attr(,"label")
  [1] "Correlation structure:"

   Variance function:
            lower    est.   upper
  power 0.21908 0.27126 0.32345
  attr(,"label")
  [1] "Variance function:"

   Residual standard error:
   lower   est.  upper
  4.9539 5.7379 6.6460
```

the update() function to the object fm12.3 with the weights argument set to weights = varIdent(form=~1|time.f). The results are stored in the model-fit object fmA.vc. We then apply the anova() function to that object and the object fm12.3. From the output we observe that the AIC value for the model, corresponding to fmA.vc, is equal to 6389.4. The value is larger then the value of 6387.2, obtained for the object fm12.3, which corresponds to model **M12.3**. Moreover, the result of the REML-based LR test, which is based on the two models, is not statistically significant ($p = 0.40$). These results indicate that, as compared to the model with the general variance and correlation structures, the simpler model **M12.3** provides an adequate summary of the data.

R12.7 *ARMD Trial*: Estimated variance-covariance structure of the fitted model **M12.3** with a general variance-covariance structure. The model-fit object `fm12.3` was created in Panel R12.6

```
> fm12.3vcov <-                                        # R_i hat
+     getVarCov(fm12.3, individual = "2")
> dimnames(fm12.3vcov) <- dnms
> fm12.3vcov
  Marginal variance covariance matrix
          4wks    12wks    24wks    52wks
  4wks  69.846  54.769  50.897  42.105
  12wks 54.769 126.760  99.627 100.190
  24wks 50.897  99.627 184.630 174.380
  52wks 42.105 100.190 174.380 280.860
    Standard Deviations: 8.3574 11.259 13.588 16.759
> fm12.3cor <- cov2cor(fm12.3vcov)                     # C_i hat: (12.6)
> print(fm12.3cor, corr = TRUE, stdevs = FALSE)
  Marginal correlation matrix
           4wks    12wks    24wks    52wks
  4wks  1.00000 0.58206 0.44820 0.30062
  12wks 0.58206 1.00000 0.65122 0.53096
  24wks 0.44820 0.65122 1.00000 0.76578
  52wks 0.30062 0.53096 0.76578 1.00000
```

R12.8 *ARMD Trial*: Tests of hypotheses about the variance-covariance parameters of model **M12.3**. Model-fit objects `fm12.2` and `fm12.3` were created in Panels R12.4 and R12.6, respectively

(a) *Autoregressive of order 1 vs. a general correlation structure*

```
> anova(fm12.2, fm12.3)                      # M12.2 ⊂ M12.3
         Model df    AIC  BIC  logLik   Test L.Ratio p-value
  fm12.2     1 12 6396.9 6454 -3186.5
  fm12.3     2 17 6387.2 6468 -3176.6 1 vs 2  19.711  0.0014
```

(b) *Power-of-time variance function vs. timepoint-specific variances*

```
> fmA.vc    <-                               # Alternative model
+    update(fm12.3, weights = varIdent(form = ~1|time.f))
> anova(fm12.3, fmA.vc)                      # M12.3 ⊂ alternative
         Model df    AIC    BIC  logLik   Test L.Ratio p-value
  fm12.3     1 17 6387.2 6468.0 -3176.6
  fmA.vc     2 19 6389.4 6479.7 -3175.7 1 vs 2  1.8432  0.3979
```

In Sect. 12.6, we evaluate the goodness of fit of model **M12.3** in more detail.

At this point it might be worthwhile to compare the results, presented in Table 9.1 for model **M9.3**, with the results for models **M12.1–M12.3**, which are shown in Table 12.1. The mean structure of all of the models is exactly the same; they differ with respect to the variance-covariance structure. When comparing the estimates of the fixed-effects coefficients from Tables 9.1 and 12.1, two observations can be made. First, there are some differences in the values of the point estimates between the two tables. They are most pronounced for the estimates of the main effects of time. However, given the precision of the estimates, the differences are not dramatic. The second observation is related to the precision of the estimates. The estimated standard errors, presented in Table 12.1, are, in general, larger than the corresponding values for model **M9.3**, displayed in Table 9.1. This implies that accounting for the correlation between the visual acuity measurements led to a loss in the precision of estimation of the mean structure parameters. The loss can be explained by the fact that a set of, n say, correlated observations contains less information than a corresponding set of n independent observations. Thus, when the correlation is taken into account in a model, larger standard errors of the mean structure parameters can be expected. Note, however, that these estimates of the true standard deviations are better, i.e., less biased than the estimates obtained for a model, which assumes independence of observations. Consequently, by basing the inference on the former estimates, a better control of the Type I error probability should be obtained.

12.6 Model-Fit Diagnostics

In this section, we evaluate the goodness of fit of model **M12.3**. In particular, in Panel R12.9, we consider syntax for plots of raw (Sect. 12.6.1), Pearson (Sect. 12.6.2), and normalized (Sect. 12.6.3) residuals to investigate various aspects of the fit of the model to the ARMD data.

12.6.1 Scatterplots of Raw Residuals

To assess the goodness of fit of model **M12.3**, we first look at the scatterplot of the raw residuals (Sect. 4.5.1) for each timepoint and each treatment group. To enhance the interpretation of the graph, we superimpose a box-and-whiskers plot over each scatterplot. Toward this end, in Panel R12.9a, we use the function `bwplot()` from the package **lattice** (Sect. 3.2.2). Note that we precede the use of the `bwplot()` function with a definition of an auxiliary panel function. The latter combines a one-dimensional scatterplot (stripplot) with a box-and-whiskers plot and adds a grid of horizontal lines aligned with the axis labels. The function is then used in the `panel` argument of the `bwplot()` function. Note that, in the first argument of

R12.9 *ARMD Trial*: Residual plots for model **M12.3**. The model-fit object `fm12.3` was created in Panel R12.6

(a) *Plots (and boxplots) of raw residuals*

```
> panel.bwxplot0 <-
+   function(x,y, subscripts, ...)
+ {
+         panel.grid(h = -1)
+         panel.stripplot(x, y, col = "grey", ...)
+         panel.bwplot(x, y, pch = "|", ...)
+         }
> bwplot(resid(fm12.3) ~ time.f | treat.f,              # Fig. 12.2
+      panel = panel.bwxplot0,
+      ylab = "Residuals", data = armd)
```

(b) *Plots of Pearson residuals vs. fitted values*

```
> plot(fm12.3)                                          # Fig. 12.3a
> plot(fm12.3,                                          # Fig. 12.3b
+      resid(., type = "p") ~ fitted(.) | time.f)
> stdres.plot <-
+   plot(fm12.3, resid(., type = "p") ~ jitter(time) | treat.f,
+      id = 0.01, adj = c(-0.3, 0.5 ), grid = FALSE)
> plot(update(stdres.plot,                              # Fig. 12.4
+             xlim = c(-5,59), ylim = c(-4.9, 4.9), grid = "h"))
```

(c) *Plots (and boxplots) of normalized residuals*

```
> bwplot(                                               # Fig. 12.7
+    resid(fm12.3, type = "n") ~ time.f | treat.f,
+    panel = panel.bwxplot,                             # User defined panel
                                                        (not shown)
+    data = armd)
> qqnorm(fm12.3, ~resid(., type = "n") | time.f)        # Fig. 12.8
```

`bwplot()`, we use a formula requesting a plot of raw residuals *versus* the levels of the `time.f` factor, separately for the levels of the `treat.f` factor. The residuals are extracted from the model-fit object `fm12.3` by applying the `resid()` function (Sect. 5.5).

The resulting graph is shown in Fig. 12.2. The box-and-whiskers plots clearly show an increasing variance of the residuals with timepoint. This reflects the heteroscedasticity, already noted in, e.g., Sect. 6.3 or 9.3.2.

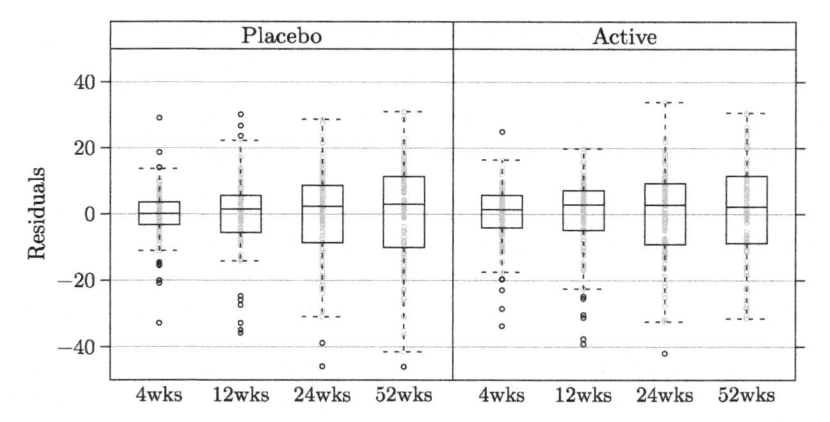

Fig. 12.2 *ARMD Trial*: Stripplots (and box-and-whiskers plots) of raw residuals for each time-point and treatment group for model **M12.3**

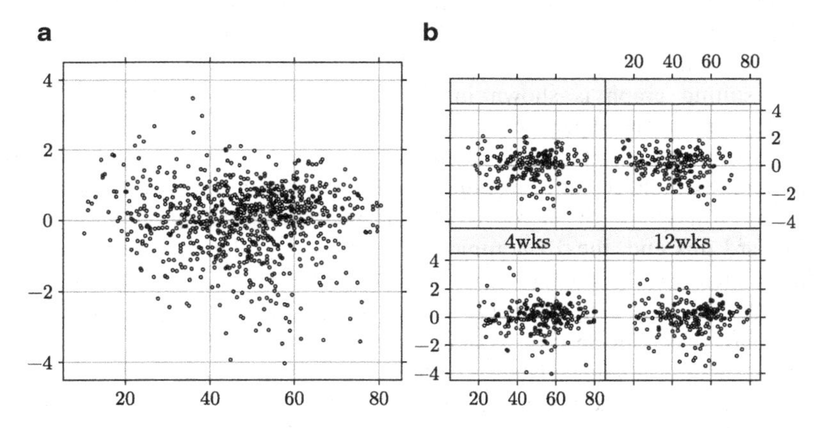

Fig. 12.3 *ARMD Trial*: Scatterplots of Pearson residuals *versus* fitted values for model **M12.3** (**a**) default plot (**b**) by `time.f`

12.6.2 Scatterplots of Pearson Residuals

In this section, we turn our attention to Pearson residuals. As described in Sect. 7.5.1, they are obtained from the raw residuals by dividing the latter by an estimate of the appropriate residual standard deviation. Hence, they should be more homoscedastic, and their scatterplots should be easier to interpret. However, because the residuals are correlated within groups, some degree of caution is required when interpreting the plots.

In Panel R12.9b, we first apply the generic `plot()` function to the model-fit object `fm12.3`, which represents the estimated form of model **M12.3**. We use the default setting of the function arguments. As a result, we obtain a default scatterplot of Pearson residuals *versus* the fitted values, shown in Fig. 12.3a. Unfortunately, in the scatterplot, the residuals for different timepoints are plotted together. As a result, due to the correlation of the residuals corresponding to the measurements obtained for the same patient at different timepoints, the plot reveals a pattern, with a few large, positive residuals in the upper-left part and a few negative ones in the lower-right part.

A more informative plot can be constructed by noting that the residuals for each timepoint should be (approximately) uncorrelated. Therefore, it is more appropriate to present them separately for each timepoint. Toward this end, we use the second `plot()`-function call, shown in Panel R12.9b. In the call, we use a formula indicating explicitly that we require a separate plot of the standardized residuals *versus* fitted values for each level of the `time.f` factor. Note that, in the formula, the `type="p"` argument is specified in the `resid()` function. The argument indicates the use of Pearson residuals (Sect. 5.5). Moreover, on the right-hand side of the formula, we use the function `fitted()` (Sect. 5.5) to extract the fitted values from the model-fit object. Note that, instead of indicating the name of the object, we use `.` (dot). This shortened syntax implies that the fitted values are to be extracted from the object `fm12.3`, the name of which is provided in the first argument of the `plot()`-function call.

The resulting graph is shown in Fig. 12.3b. The four scatterplots show a somewhat more balanced pattern.

If we assume that the residuals should approximately follow a standard normal distribution, we might consider absolute values greater than, e.g., the 95th percentile as outlying. It might be of interest to identify the corresponding observations in a plot. Toward this end, the `id` argument can be used in the appropriate `plot()`-function statement. This is done in the third `plot()`-function call, shown in Panel R12.9b. In the call, we apply the argument `id=0.01`. This indicates that the residuals larger, in absolute value, than the 99th percentile of the standard normal distribution should be labeled in the plot by the number of the corresponding observation from the ARMD data frame. To avoid cluttering of the labels, on the left-hand side of the formula specified in the `plot()`-function call, we apply the function `jitter()` to the variable `time`. The function adds a small amount of noise to the variable. As a result, the overlap of the labels should be reduced. We also use the argument `adj=c(-0.3,0.5)` to move the labels to the right (horizontally) and the center (vertically) of the plotted symbol. The resulting plot is stored in the object `stdres.plot`. Subsequently, we update the object by adding suitable limits for the two axes and a grid of horizontal lines.

The resulting graph is shown in Fig. 12.4. It presents the scatterplots of Pearson residuals grouped by timepoint and treatment. The number of residuals larger, in absolute value, than the 99th percentile of the standard normal distribution, is not excessive, given the total number of observations.

As mentioned at the beginning of this section, the main issue in the interpretation of Pearson residuals is the fact that they are correlated. Figure 12.5 presents the scatterplot of the residuals against time for model **M12.3** separately for each treatment group. In the plot, residuals for a few randomly selected individuals have been connected by lines. The plot illustrates the correlation between the residuals obtained for the same individual. For instance, in the panel for the `Active` treatment

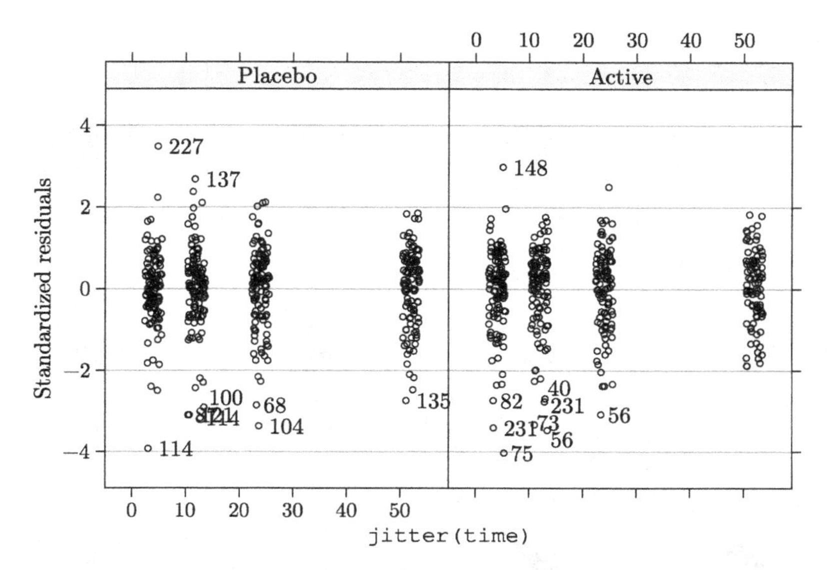

Fig. 12.4 *ARMD Trial*: Scatterplots of Pearson residuals *versus* time per treatment group for model **M12.3**. Points are jittered along time axis

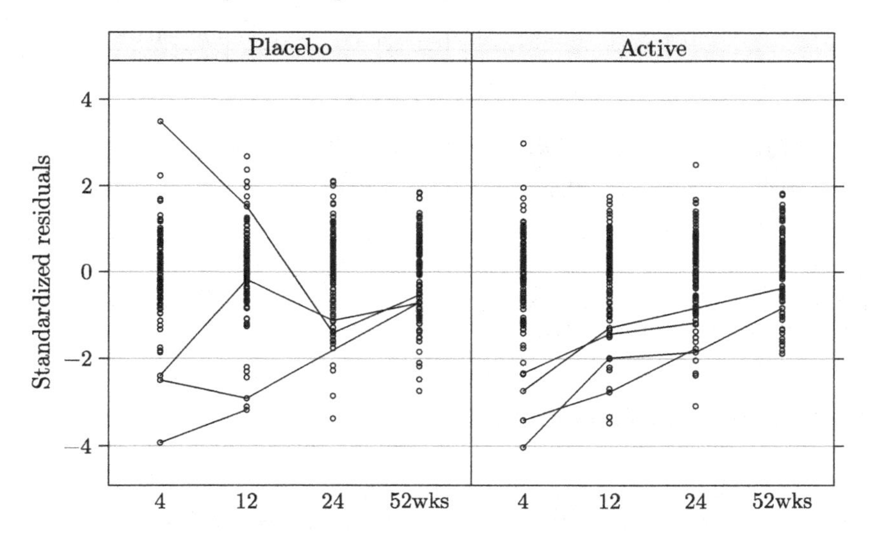

Fig. 12.5 *ARMD Trial*: Scatterplots of Pearson residuals *versus* time per treatment group for model **M12.3**. Residuals for selected subjects are connected with lines

group, the residuals obtained for each of the selected individuals tend to have negative values. Note that, for brevity's sake, we do not show the R syntax necessary to create the graph.

To remove the correlation between Pearson residuals, we may use the normalized residuals (Sect. 10.5). Their application is considered in the next section.

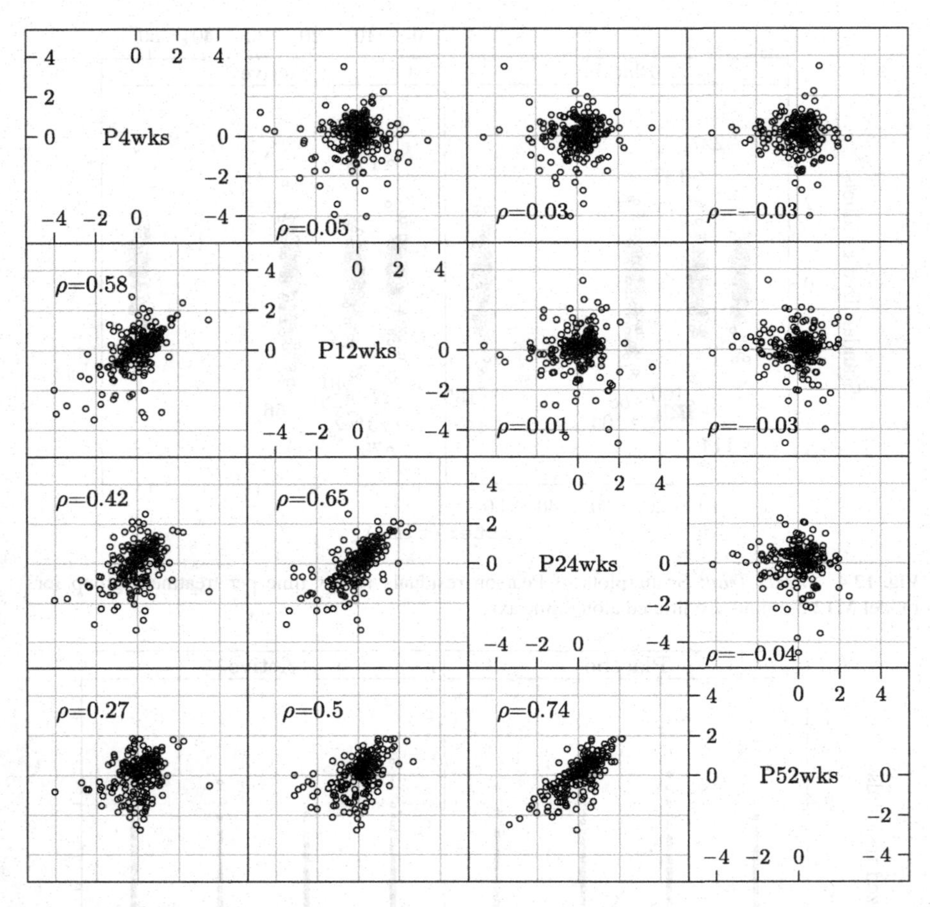

Fig. 12.6 *ARMD Trial*: Scatterplot matrix of Pearson (below the diagonal) and normalized (above the diagonal) residuals for model **M12.3**

12.6.3 Normalized Residuals

Normalized residuals are obtained from a transformation of the raw residuals based on the Cholesky decomposition of the residual variance-covariance matrix (Sect. 10.5). Ideally, the residuals should become uncorrelated.

Figure 12.6 shows the scatterplots of Pearson residuals (below the diagonal) and the normalized residuals (above the diagonal) for all pairs of timepoints for model **M12.3**. To conserve space, we do not show the R syntax used to create the figure. The scatterplots of Pearson residuals show a correlation between the residuals corresponding to different timepoints. On the other hand, the plots for the normalized residuals clearly illustrate that the residuals are (approximately) uncorrelated.

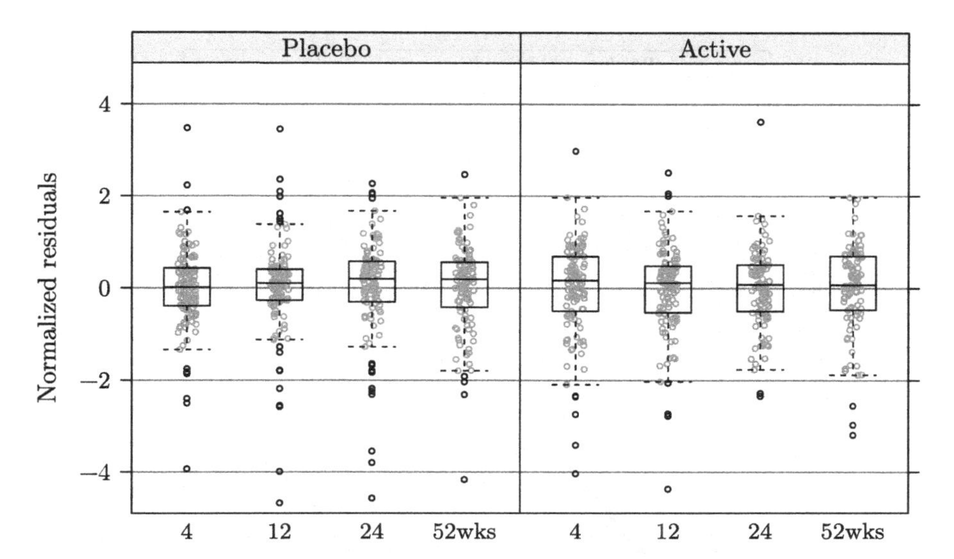

Fig. 12.7 *ARMD Trial*: Stripplots (and box-and-whiskers plots) of normalized residuals for each timepoint and treatment group for model **M12.3**. Points between whiskers are jittered along time axis to reduce an overlap

Figure 12.7 presents stripplots of the normalized residuals grouped by timepoint and treatment. To enhance the interpretation, box-and-whiskers plots were superimposed on the stripplots. In Panel R12.9c, we present the basic form of the `bwplot()`-function call that could be used to create a graph similar to the one shown in Fig. 12.7. The call is very similar to the one used in Panel R12.9a. The main difference is the use of the `type="n"` argument in the `resid()` function, which extracts the normalized residuals (Sect. 5.5). Note that the graph, presented in Fig. 12.7, was created by a modified version of the syntax from Panel R12.9c. To conserve space, we do not show the modification.

As compared to Fig. 12.4, the plot in Fig. 12.7 shows a few more extreme residuals with negative values, smaller than −4. Nevertheless, the number of residuals with an absolute value larger than, e.g., 2 is about the same.

To check the normality assumption, we may want to inspect the normal Q-Q plot of the normalized residuals. In Panel R12.9c, we present a suitable `qqnorm()`-function call. When applied to model-fit objects of *gls* class, the function is typically called by using a syntax of the form `qqnorm(`*`gls.fit, form`*`)`, where *`gls.fit`* is a model-fit object for class *gls* and *`form`* is a one-sided formula, which specifies the desired type of plot. In the formula, any variable from the data frame used to produce the model-fit object can be referred to. Separate graphs for the levels of a grouping factor *`g`* are obtained by specifying `|` *`g`* in the formula (as it is done for `time.f` in the syntax presented in Panel R12.9c). The expression to the left of the `|` operator must be equal to a vector of residuals. The default formula is `~resid (., type = "p")`, which corresponds to a normal Q-Q plot of Pearson residuals.

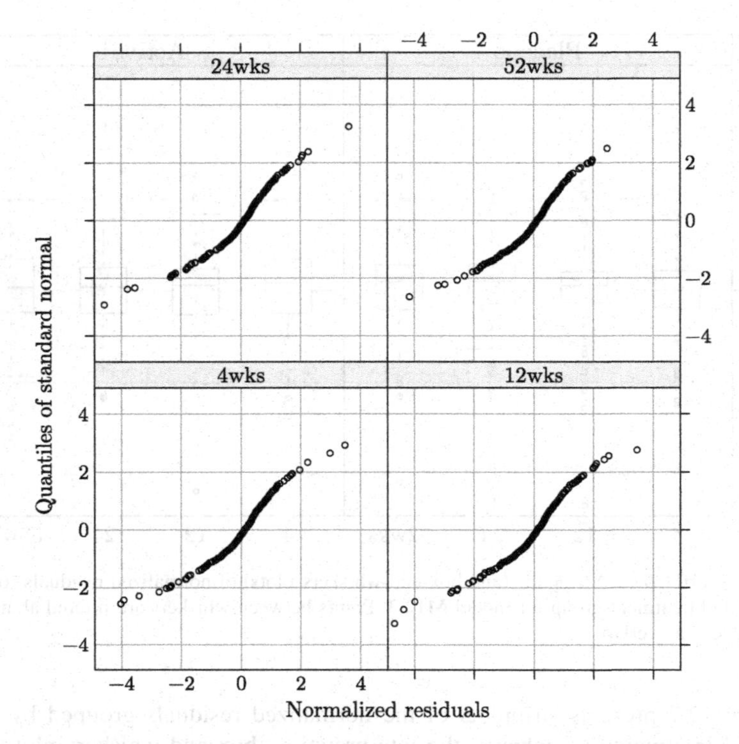

Fig. 12.8 *ARMD Trial*: Normal Q-Q plots of normalized residuals for the fitted model **M12.3**

In the formula used in the `qqnorm()`-function call in Panel R12.9c, we require a plot of the normalized residuals by applying the argument `type="n"` of the `resid()` function (Sect. 11.6).

The Q-Q plot, corresponding to the `qqnorm()`-function call from Panel R12.9c, is presented in Fig. 12.8. Although normalized residuals should be approximately uncorrelated, we graph separate Q-Q plots per timepoint to remove the influence of any residual correlation. The patterns shown in Fig. 12.8 appear to be reasonably close to straight lines. Thus, the normality assumption seems to be plausible for the ARMD data.

12.7 Inference About the Mean Structure

In Sects. 12.3–12.5, we focused on modeling the variance-covariance structure of the visual acuity data. In particular, we kept the same mean structure, defined in (12.1), for all models considered in those sections, while changing the form of the variance-covariance structure. In Sect. 12.6, we presented results that suggested that model **M12.3**, defined by (12.1)–(12.3) and (12.6), provided a reasonable fit to the visual acuity data. Thus, we use it as a basis for inference about the mean structure parameters. In particular, we perform tests of hypotheses about the fixed effects (Sect. 10.6). Toward this end, in Panel R12.10, we use the `anova()` function. Note the use of the `update()` function to refit model **M12.3** using the ML estimation.

R12.10 *ARMD Trial*: Sequential F-tests for fixed effects of model **M12.3**. The model-fit object `fm12.3` was created in Panel R12.6

```
> anova(update(fm12.3, method = "ML"))          # M12.3
  Denom. DF: 858
                   numDF F-value p-value
  visual0              1  9867.8  <.0001
  time.f               4    26.6  <.0001
  time.f:treat.f       4     2.0  0.0869
```

The results, displayed in Panel R12.10, correspond to the F-tests (4.36) of the linear hypotheses (4.30) for parameters β_{0t}, β_1, and β_{2t} of (12.1). In particular, the linear hypotheses specify that the parameters are equal to zero. For instance, the test for the `time.f:treat.f` interaction pertains to the null hypothesis

$$H_0:\ \beta_{21}=\beta_{22}=\beta_{23}=\beta_{24}=0,$$

which can be expressed in the form $\boldsymbol{L\beta}=\mathbf{0}$ by specifying

$$\boldsymbol{\beta}=(\beta_{01},\beta_{02},\beta_{03},\beta_{04},\beta_1,\beta_{11},\beta_{12},\beta_{13},\beta_{14})'$$

and

$$\boldsymbol{L}=(\mathbf{0}_{4\times 5}\ \mathbf{I}_4),$$

where $\mathbf{0}_{4\times 5}$ is a 4×5 matrix of zeroes and $\mathbf{I}_4$ is the 4×4 identity matrix. As it was mentioned in Sect. 5.6, by default, the `anova()` function provides results of the sequential tests. Thus, for instance, the test for `time.f`, presented in Panel R12.10, is obtained by comparing the alternative model, which contains the intercept, `visual0`, and `time.f` as covariates, with the null model, which contains only the intercept and `visual0`.

The nonsignificant, at the 5% significance level, result of the test for the `time.f:treat.f` interaction suggests that a simpler, constant treatment effect might be plausible. We might also want to consider a mean linear trend of visual acuity measurements. To check these hypotheses, in the next section, we consider a series of alternative models with appropriate mean structures. In all these models, we modify (simplify) the mean structure, while fixing the variance-covariance structure at the one defined by the power variance-function (12.3) and the general correlation matrix (12.6).

12.7.1 Models with the General Correlation Structure and Power Variance Function

In the process of simplifying the mean structure of model **M12.3**, we consider models labeled as **M12.3a**, **M12.4**, and **M12.5**. All these models have the same variance-covariance structure, defined by (12.2), (12.3) and (12.6). They differ with respect to mean structures, as described below.

Model **M12.3a** is given by

$$\begin{aligned}\text{VISUAL}_{it} &= \beta_0 + \beta_{0t} + \beta_1 \times \text{VISUAL0}_i + \beta_2 \times \text{TREAT}_i \\ &\quad + \beta_{2t} \times \text{TREAT}_i + \varepsilon_{it}.\end{aligned} \tag{12.7}$$

Compared to (12.1), the mean structure, implied by (12.7), contains two additional terms, which involve parameters β_0 and β_2. The parameters can be thought of as the "overall" mean and treatment effects, respectively. Note that they are aliased with β_{0t} for $t = 1,\ldots,4$, and with β_{2t} for $t = 1,\ldots,4$, respectively. To account for the aliasing, we constrain β_{01} and β_{21} to zero. Although models **M12.3** and **M12.3a** are equivalent, we use different labels to indicate the different parameterizations. The parameterization, introduced in (12.7), is applied to allow for a comparison of the parameter estimates of model **M12.3** with those of models **M12.4** and **M12.5**, which will be defined next.

Model **M12.4** is defined as

$$\begin{aligned}\text{VISUAL}_{it} &= \beta_0 + \beta_1 \times \text{VISUAL0}_i + \beta_2 \times \text{TIME}_t \\ &\quad + \beta_3 \times \text{TREAT}_i + \beta_4 \times \text{TIME}_t \times \text{TREAT}_i + \varepsilon_{it}.\end{aligned} \tag{12.8}$$

In contrast to (12.1) and (12.7), the mean structure, specified in (12.8), assumes a linear effect of the continuous TIME variable.

Finally, model **M12.5** is obtained by removing the interaction between time and treatment from (12.8):

$$\begin{aligned}\text{VISUAL}_{it} &= \beta_0 + \beta_1 \times \text{VISUAL0}_i + \beta_2 \times \text{TIME}_t \\ &\quad + \beta_3 \times \text{TREAT}_i + \varepsilon_{it}.\end{aligned} \tag{12.9}$$

In the next section, the models are fitted to the ARMD data and compared against each other to assess which of them provides the most parsimonious representation of the mean structure of the data.

12.7.2 Syntax and Results

In Panel R12.11, we fit models **M12.3a**, **M12.4**, and **M12.5** to the ARMD dataset using the `gls()` function. We compare the model fits using the LR tests. As we focus on testing hypotheses about fixed effects, for reasons explained at the end of Sect. 4.7.2, we use the ML estimation (Sect. 10.4.2).

R12.11 *ARMD Trial*: Likelihood ratio and sequential F-tests of hypotheses about the fixed effects for models **M12.3a**, **M12.4**, and **M12.5**. The model-fit object `fm12.3` was created in Panel R12.6

(a) *Models needed for LR test*

```
> lm1a.form <- formula (visual ~ visual0 + time.f          # (12.7)
+      + treat.f + time.f:treat.f)
> fm12.3a <- update(fm12.3, lm1a.form,         # M12.3a ← M12.3
+      method="ML", data=armd)
> lm2.form <- formula(visual ~ visual0 + time              # (12.8)
+      + treat.f + treat.f:time)
> fm12.4 <- update(fm12.3, lm2.form,           # M12.4 ← M12.3
+      method="ML", data=armd)
> lm3.form <-  update(lm2.form, . ~ . - treat.f:time)      # (12.9)
> fm12.5 <- update(fm12.3, lm3.form,           # M12.5 ← M12.3
+      method="ML", data=armd)
```

(b) *LR test for the mean linear time trend and interaction term*

```
> anova(fm12.3a, fm12.4, fm12.5)       # M12.3a ⊃ M12.4 ⊃ M12.5
          Model df    AIC    BIC  logLik   Test L.Ratio p-value
  fm12.3a     1 17 6395.7 6476.7 -3180.9
  fm12.4      2 13 6389.6 6451.5 -3181.8 1 vs 2  1.8367  0.7658
  fm12.5      3 12 6389.0 6446.2 -3182.5 2 vs 3  1.4129  0.2346
```

(c) *Sequential-approach F-tests for terms in model* **M12.5**

```
> anova(fm12.5)
  Denom. DF: 863
              numDF F-value p-value
  (Intercept)     1  9374.1  <.0001
  visual0         1   613.0  <.0001
  time            1   104.3  <.0001
  treat.f         1     6.0  0.0146
```

More specifically, in Panel R12.11a, we first fit model **M12.3a** by modifying the formula `lm1.form` (see Sect. 12.3.2) to allow for an overall intercept. We then update the object `fm12.3` with the modified formula `lm1a.form` and with the `method="ML"` argument.

Then, to fit model **M12.4**, we modify the LM formula using the continuous `time` variable. Subsequently, we update the object `fm12.3` with the modified formula `lm2.form` and with the `method="ML"` argument.

Finally, to fit model **M12.5**, we remove the interaction between the linear effect of time and treatment from the LM formula, and we update `fm12.3` with the modified formula `lm3.form` and with the `method="ML"` argument. Note that we use the abbreviated syntax `.~.-treat.f:time` to remove the `treat.f:time` interaction from formula `lm2.form` (Sect. 5.2.1).

In Panel R12.11b, we use the `anova()` function to compare the models. As a result, we obtain two LR tests. One compares the likelihood of model **M12.3a** with that of **M12.4** and tests the hypothesis that a mean linear time trend of the visual acuity measurements can be assumed. The other test compares the likelihood of model **M12.4** with that of model **M12.5** and tests the hypothesis that a constant treatment effect, in a model with a linear time trend, can be assumed.

The results of both tests are statistically not significant at the 5% significance level. Thus, they indicate that model **M12.3a**, which is equivalent to **M12.3**, can be simplified by assuming a mean linear trend of visual acuity measurements in time and a constant treatment effect. That is, model **M12.3** can be simplified to model **M12.5**.

The fit of model **M12.5**, defined by (12.3), (12.6), and (12.9), does not change much, as compared to model **M12.3**: the scatterplot and the normal Q-Q plot of residuals (not shown) are comparable to those obtained for model **M12.3** (Sect. 12.6). The result of the F-test for the effect of treatment, β_2, in the simplified model **M12.5** is shown in Panel R12.11c. The test is obtained by applying the `anova()` function to the model-fit object `fm12.5`. The result of the test is statistically significant at the 5% significance level ($p = 0.015$).

Table 12.2 shows the details of the models fitted in Panel R12.11. The results of model **M12.5**, stored in the object `fm12.5`, are displayed in the last column of the table. The parameter estimates suggest a statistically significant negative treatment effect, as noted earlier. The effects of the baseline visual acuity measurement and of the measurement time are also statistically significant. They suggest that post-randomization visual acuity measurements are higher by 0.89 for each unit increase in the baseline measurement, and that they decrease linearly by 0.23 with each week. The estimated variance-covariance and correlation matrices, displayed in Panel R12.12, imply that the residual variance increases with time, while the correlation between the measurements decreases when the observations are made at more distant timepoints (in terms of the order, not time distance).

12.8 Chapter Summary

In this chapter, we considered the LM with fixed effects for correlated data that was defined, in general terms, in Sect. 10.2. The model allows taking into account the correlation between the observations belonging to the same group.

R12.12 *ARMD Trial*: The estimated variance-covariance and correlation matrices for the model **M12.5**. The model-fit object fm12.5 was created in Panel R12.11

```
> fm12.5vcov <- getVarCov(fm12.5,              # R_i
+       individual="2")
> dimnames(fm12.5vcov) <- dnms                 # Dimnames assigned
> fm12.5vcov
  Marginal variance covariance matrix
          4wks    12wks   24wks   52wks
  4wks  68.990  53.905  50.102  41.127
  12wks 53.905 125.520  98.433  98.942
  24wks 50.102  98.433 183.100 172.940
  52wks 41.127  98.942 172.940 279.010
    Standard Deviations: 8.306 11.203 13.532 16.704
> fm12.5cor <- cov2cor(fm12.5vcov)             # C_i: (12.6)
> print(fm12.5cor, corr=TRUE, stdevs=FALSE)
  Marginal correlation matrix
           4wks    12wks   24wks   52wks
  4wks  1.00000 0.57928 0.44578 0.29643
  12wks 0.57928 1.00000 0.64930 0.52872
  24wks 0.44578 0.64930 1.00000 0.76514
  52wks 0.29643 0.52872 0.76514 1.00000
```

We illustrated the features of the model by applying it to the ARMD dataset. We considered several models, constructed with the help of the varPower($\cdot$) (Sects. 7.3.1 and 9.3) variance function and various correlation structures. Table 12.3 provides information about the models defined in this chapter. The final model was model **M12.5**. Its mean structure was defined by (12.9), while the variance-covariance structure was defined by the power variance function (12.3) and the general correlation structure (12.6). The model accounted for the correlation between the visual acuity measurements obtained for the same individual. Given that it provided an acceptable fit to the data, it could be used for inference about the mean and variance-covariance structure.

In the process of arriving at the form of the final model, we used the results presented in Chap. 9. We assumed that the variances of the visual acuity measurements were adequately described by the variance function defined as a power function of the measurement time. We then fixed the mean structure at (12.1) and built a series of models (see Table 12.1) with increasingly more complex correlation structures: a compound symmetry (Sect. 12.3), an autoregressive process of order 1 (Sect. 12.4), and a general correlation structure (Sect. 12.5). The latter gave the best fit to the data, according to the results of the LR test and AIC (Sect. 12.5.2). Given the adequate fit of the corresponding model **M12.3** (Sect. 12.6), we fixed the resulting variance-covariance function and considered a series of models (see Table 12.2) with more parsimonious mean structures than the one given by (12.1). With the help of the LR

Table 12.2 *ARMD Trial*: The ML-based parameter estimates[a] for models with various mean structures and the general correlation matrix fitted using the `gls()` function

	Parameter	fm12.3a	fm12.4	fm12.5
Model label		**M12.3a**	**M12.4**	**M12.5**
Log-ML value		−3180.87	−3181.78	−3182.49
Fixed effects				
Intercept	β_0	0.77(2.22)	5.61(2.14)	5.84(2.13)
Visual acuity at t=0	β_1	0.89(0.04)	0.89(0.04)	0.89(0.04)
Time (in weeks)	β_2		−0.21(0.03)	−0.23(0.02)
Trt(Actv *vs.* Plcb)	β_3	−3.49(1.38)	−2.17(1.12)	−2.61(1.06)
Tm × Treat(Actv)	β_4		−0.05(0.05)	
Time poly(3)				
Linear	β_{02}	−7.83(1.16)		
Quadratic	β_{03}	0.29(0.66)		
Cubic	β_{04}	0.74(0.59)		
Time poly(3) × Treat(Actv)				
Linear:Trt(Actv)	β_{22}	−1.74(1.69)		
Quadratic × Treat(Actv)	β_{23}	−0.06(0.96)		
Cubic × Treat(Actv)	β_{24}	−0.81(0.85)		
Variance function				
Power (TIME$^{\delta}$)	δ	0.27(0.22,0.32)	0.27(0.22,0.32)	0.27(0.22,0.32)
Correlation structure				
General (12.6)				
cor(1,2)	ϱ_{12}	0.58(0.49,0.66)	0.58(0.49,0.66)	0.58(0.49,0.66)
cor(1,3)	ϱ_{13}	0.45(0.33,0.55)	0.45(0.33,0.55)	0.45(0.33,0.55)
cor(1,4)	ϱ_{14}	0.30(0.15,0.43)	0.30(0.15,0.43)	0.30(0.15,0.43)
cor(2,3)	ϱ_{23}	0.65(0.57,0.72)	0.65(0.57,0.72)	0.65(0.57,0.72)
cor(2,4)	ϱ_{24}	0.53(0.42,0.63)	0.53(0.42,0.63)	0.53(0.42,0.62)
cor(3,4)	ϱ_{34}	0.77(0.70,0.82)	0.77(0.70,0.82)	0.77(0.70,0.82)
Scale	σ	5.70(4.92,6.60)	5.70(4.92,6.60)	5.69(4.92,6.59)

[a] Approximate SE for fixed effects and 95% CI for covariance parameters are included in parentheses

Table 12.3 *ARMD Trial*: Summary of the models defined in Chap. 12. Variance function defined as a power function of time specified using syntax `varPower(form ~ time)` is assumed for all models

Model	Section	Syntax	R Object	Mean (eq.)	CorStruct (eq.)
REML estimation					
M12.1	12.3	R12.2	`fm12.1`	(12.1)	*corCompSymm* (12.4)
M12.2	12.4	R12.4	`fm12.2`	(12.1)	*corAR1* (12.5)
M12.3[a]	12.5	R12.6	`fm12.3`	(12.1)	*corSymm* (12.6)
ML estimation					
M12.3a[a]	12.7.1	R12.11	`fm12.3a`	(12.7)	*corSymm* (12.6)
M12.4	12.7.1	R12.11	`fm12.4`	(12.8)	*corSymm* (12.6)
M12.5	12.7.1	R12.11	`fm12.5`	(12.9)	*corSymm* (12.6)

[a] Models **M12.3** and **M12.3a** are equivalent but expressed using different parameterizations.

test we found that the mean structure, defined by (12.9), provided an adequate description of the data (Sect. 12.7.2). This led us to the adoption of model **M12.5** as the final model.

It should be stressed that the strategy described above was adopted for illustrative purposes only. In practice, we should start building the model using the most general mean and variance-covariance structures. Then, while keeping the mean structure fixed, we might consider simplifying the variance-covariance structure. If a more parsimonious structure with a satisfactory fit to the data has been found, we could consider simplifying the mean structure. After arriving at the final model, we should check whether its fit is satisfactory. If it is the case, the model could serve as a basis for inference.

The models considered in this chapter used the four post-randomization visual acuity measurements as the correlated observations of the dependent variable. This approach offers some advantages like, e.g., the possibility of an explicit adjustment for the baseline visual acuity value through the inclusion of an appropriate covariate in the mean structure. On the other hand, within this approach, the adjustment for the possible imbalance in visual acuity at baseline between the two treatment groups is more difficult. An alternative could be to consider modeling of all the measurements, including those obtained at baseline. In this approach, adjusting for an imbalance at baseline could be accomplished by, e.g., the inclusion, in the mean structure of the model, of an interaction term between the treatment indicator and the time factor variable. On the other hand, the adjustment for the effect of the baseline visual acuity on the post-randomization measurements would be only implicit through the variance-covariance structure of the measurements. Also, the interpretation of the treatment effects would require a bit more caution, because the effects of interest would be those at the post-randomization occasions and not the one at baseline. Given these considerations, we decided to focus in the chapter on the models for the post-randomization measurements.

Yet another approach to the analysis of the visual acuity measurements from the ARMD dataset can be proposed by recognizing that the data have a hierarchical structure. That is, the measurements (analysis units) are grouped within the individuals (sampling units). The variance-covariance structure of the measurements can then be seen as resulting from the variability arising at the two levels of hierarchy. A modeling approach that looks at the data from such a hierarchical point of view uses LMMs. In the next part of the book we focus on these models.

Part IV
Linear Mixed-Effects Models

Chapter 13
Linear Mixed-Effects Model

13.1 Introduction

In Chap. 10, we presented models with fixed effects for correlated data. They are examples of population-averaged models, because their mean-structure parameters can be interpreted as effects of covariates on the mean value of the dependent variable in the entire population. The association between the observations in a dataset was a result of a grouping of the observations sharing the same level of a grouping factor(s). An example of grouped data is longitudinal data, with multiple measurements collected over time for an individual. This is an example of data with a single level of grouping (Sect. 10.2): measurements are grouped at the level of an individual. Such a hierarchy is present in the ARMD data (Sect. 2.2 and Chaps. 6, 9, and 12), with multiple visual acuity measurements available for individual patients. Another example of data with a single level of grouping is meta-analysis data, with patients grouped within clinical trials.

An example of data with a multilevel hierarchy is student's scores. Scores, e.g., for the same course across several years, are grouped for a student, students are grouped into classes, classes into schools, schools within districts, etc. Consequently, the total variability of the scores can be seen as resulting from their variability within students, between students, between classes within the same school, between schools in the same district, etc. Such a structure is present in the data for the Instructional Improvement Study, presented in Sect. 2.4.

Note that, because of grouping, a complex association structure of the observed data can be anticipated. For instance, we can expect correlation not only between the scores for an individual student, but also between scores from different students from the same class or between scores for different students from the same school. As argued in Part III of the book, the correlations should be taken into account in the analysis of the data.

In this chapter, we consider the analysis of continuous, hierarchical data using a different class of models, namely, LMMs. They allow taking into account the correlation of observations contained in a dataset. Moreover, they allow us to

A. Gałecki and T. Burzykowski, *Linear Mixed-Effects Models Using R: A Step-by-Step Approach*, Springer Texts in Statistics, DOI 10.1007/978-1-4614-3900-4_13,

effectively partition overall variation of the dependent variable into components corresponding to different levels of data hierarchy. The models are examples of *subject-specific* models, because they include subject-specific coefficients.

In this chapter, we describe the specification of LMMs for hierarchical data. We build upon the concepts introduced in Parts II and III of the book. We provide only essential theoretical information, linked to the concepts and methods used in R. For a more detailed exposition of the theory of LMMs, the reader is referred to the monographs by, e.g., Searle et al. (1992), Davidian and Giltinan (1995), Vonesh and Chinchilli (1997), Pinheiro and Bates (2000), Verbeke and Molenberghs (2000), Demidenko (2004), Fitzmaurice et al. (2004), or West et al. (2007).

This chapter is structured as follows. In Sects. 13.2–13.4, we describe the formulation of the model. Sections 13.5–13.7 are devoted to, respectively, the estimation approaches, diagnostic tools, and inferential methods used for the LMMs, in which the (conditional) residual variance-covariance matrix is independent of the mean value. This is the most common type of LMMs used in practice. In Sect. 13.8, we focus on the LMMs, in which the (conditional) residual variance-covariance matrix depends of the mean value. Section 13.9 summarizes the contents of this chapter and offers some general concluding comments.

In our presentation, we focus on the formulation and methods for LMMs applicable to data with a single level of grouping, with N groups indexed by $i = 1, \ldots, N$, each containing n_i observations. The extension of the formulation to multilevel grouped data is presented at the end of Sect. 13.2 and in Sect. 13.8.

13.2 The Classical Linear Mixed-Effects Model

In this section, we describe specification of the classical LMMs in their general form. Essentially, the formulation corresponds to the one proposed in the classical paper by Laird and Ware (1982).

In particular, in Sect. 13.2.1, we provide model specification at a particular level of grouping factor, while in Sect. 13.2.2 we describe a specification for all data. Extension of the classical LMM is presented in Sect. 13.3. More detailed aspects of the classical and extended model specification are discussed in Sect. 13.4.

13.2.1 Specification at a Level of a Grouping Factor

For hierarchical data with a single level of grouping, we can formulate the classical LMM at a given level of a grouping factor as follows:

$$\mathbf{y}_i = \boldsymbol{X}_i\boldsymbol{\beta} + \boldsymbol{Z}_i\mathbf{b}_i + \boldsymbol{\varepsilon}_i, \tag{13.1}$$

where $\mathbf{y}_i$, $\boldsymbol{X}_i$, $\boldsymbol{\beta}$, and $\boldsymbol{\varepsilon}_i$ are the vector of continuous responses, the design matrix, and the vector of residual errors for group i, specified in (10.2) and (10.3), respectively, while $\boldsymbol{Z}_i$ and $\mathbf{b}_i$ are the matrix of covariates and the corresponding vector of random effects:

$$\boldsymbol{Z}_i \equiv \begin{pmatrix} z_{i1}^{(1)} & z_{i1}^{(2)} & \dots & z_{i1}^{(q)} \\ \vdots & \vdots & \ddots & \vdots \\ z_{in_i}^{(1)} & z_{in_i}^{(2)} & \dots & z_{in_i}^{(q)} \end{pmatrix} = \left(\mathbf{z}_i^{(1)} \; \mathbf{z}_i^{(2)} \dots \mathbf{z}_i^{(q)} \right), \quad \mathbf{b}_i \equiv \begin{pmatrix} b_{i1} \\ \vdots \\ b_{iq} \end{pmatrix}. \tag{13.2}$$

Similar to the design matrix $\boldsymbol{X}_i$, the matrix $\boldsymbol{Z}_i$ contains known values of q covariates, with corresponding unobservable effects $\mathbf{b}_i$. Moreover,

$$\mathbf{b}_i \sim \mathcal{N}_q(\mathbf{0}, \mathcal{D}), \quad \boldsymbol{\varepsilon}_i \sim \mathcal{N}_{n_i}(\mathbf{0}, \mathcal{R}_i), \quad \text{with} \quad \mathbf{b}_i \perp \boldsymbol{\varepsilon}_i, \tag{13.3}$$

i.e., the residual errors $\boldsymbol{\varepsilon}_i$ for the same group are independent of the random effects $\mathbf{b}_i$. This particular assumption plays the key role in distinguishing a classical LMM from an extended LMM. In addition, we assume that vectors of random effects and residual errors for different groups are independent of each other, i.e., $\mathbf{b}_i$ is independent of $\boldsymbol{\varepsilon}_{i'}$ for $i \neq i'$.

We also specify that

$$\mathcal{D} = \sigma^2 \boldsymbol{D} \quad \text{and} \quad \mathcal{R}_i = \sigma^2 \boldsymbol{R}_i, \tag{13.4}$$

where σ^2 is an unknown scale parameter. In general, we will assume that $\mathcal{D}$ and $\mathcal{R}_i$ are positive-definite, unless stated otherwise.

The representation (13.4), in its general form, is not unique. To make it identifiable, similar to the case of the LM for correlated data (Sect. 10.3), we will specify the structure of the matrix $\boldsymbol{R}_i$ in terms of a set of parameters for a variance function and a correlation matrix (Sect. 13.4.2). The specification will imply constraints on $\boldsymbol{R}_i$ making (13.4) identifiable.

In addition to the fixed-effects parameters $\boldsymbol{\beta}$ for the covariates used in constructing the design matrix $\boldsymbol{X}_i$, model (13.1) includes two random components: the within-group residual errors $\boldsymbol{\varepsilon}_i$ and the random effects $\mathbf{b}_i$ for the covariates included in the matrix $\boldsymbol{Z}_i$. The presence of fixed and random effects of known variables gives rise to the name of the model.

In many cases, the (random) effects included in $\mathbf{b}_i$ have corresponding (fixed) effects, contained in $\boldsymbol{\beta}$. Consequently, the matrix $\boldsymbol{Z}_i$ is often created by selecting a subset of appropriate columns of the matrix $\boldsymbol{X}_i$. In such a situation, it is said that the corresponding fixed and random effects are "coupled."

Model (13.1)–(13.4) is commonly referred to as a *two-stage* model (Davidian and Giltinan 1995) or *two-level* model (West et al. 2007). However, some authors, e.g., Pinheiro and Bates (2000), call it a *single-level* LMM, because it applies to a data hierarchy defined by a single level of grouping. In what follows, we will use the latter terminology, as it is reflected in the nomenclature used in R.

The classical LMM, defined in (13.1)–(13.4), can be adapted to multilevel grouped data. For instance, a model for data with two levels of grouping, with observations grouped into N first-level groups (indexed by $i = 1,\ldots,N$), each with n_i second-level (sub-)groups (indexed by $j = 1,\ldots,n_i$) containing n_{ij} observations, can be written as

$$\mathbf{y}_{ij} = \boldsymbol{X}_{ij}\boldsymbol{\beta} + \boldsymbol{Z}_{1,ij}\mathbf{b}_i + \boldsymbol{Z}_{2,ij}\mathbf{b}_{ij} + \boldsymbol{\varepsilon}_{ij}, \tag{13.5}$$

with

$$\mathbf{b}_i \sim \mathcal{N}_{q_1}(\mathbf{0}, \mathcal{D}_1), \quad \mathbf{b}_{ij} \sim \mathcal{N}_{q_2}(\mathbf{0}, \mathcal{D}_2), \quad \text{and} \quad \boldsymbol{\varepsilon}_{ij} \sim \mathcal{N}_{n_{ij}}(\mathbf{0}, \mathcal{R}_{ij}),$$

where the random vectors $\mathbf{b}_i$, $\mathbf{b}_{ij}$, and $\boldsymbol{\varepsilon}_{ij}$ are independent of each other. In model (13.5), $\mathbf{b}_i$ are the random effects associated with the first-level groups, while $\mathbf{b}_{ij}$ are the random effects, independent of the first-level random effects, associated with the second-level groups. Design matrices $\boldsymbol{Z}_{1,ij}$ and $\boldsymbol{Z}_{2,ij}$ can, but do not have to be, identical. Following Pinheiro and Bates (2000), this model can be referred to as a *two-level* LMM.

13.2.2 Specification for All Data

In this section, we briefly describe the specification of the single-level LMM, given by (13.1)–(13.4), for all data. Generalization to multilevel LMMs is obvious, though notationally more complex.

Let $\mathbf{y} \equiv (\mathbf{y}_1', \mathbf{y}_2', \ldots, \mathbf{y}_N')'$ be the vector containing all $n = \sum_{i=1}^{N} n_i$ observed values of the dependent variable. Similarly, let $\mathbf{b} \equiv (\mathbf{b}_1', \mathbf{b}_2', \ldots, \mathbf{b}_N')'$ and $\boldsymbol{\varepsilon} \equiv (\boldsymbol{\varepsilon}_1', \boldsymbol{\varepsilon}_2', \ldots, \boldsymbol{\varepsilon}_N')'$ be the vectors containing all Nq random effects and n residual errors, respectively. Define matrices

$$\boldsymbol{X} \equiv \begin{bmatrix} \boldsymbol{X}_1 \\ \boldsymbol{X}_2 \\ \vdots \\ \boldsymbol{X}_N \end{bmatrix} \quad \text{and} \quad \boldsymbol{Z} \equiv \begin{bmatrix} \boldsymbol{Z}_1 & \mathbf{0} & \ldots & \mathbf{0} \\ \mathbf{0} & \boldsymbol{Z}_2 & \ldots & \mathbf{0} \\ \vdots & \vdots & \ddots & \vdots \\ \mathbf{0} & \mathbf{0} & \ldots & \boldsymbol{Z}_N \end{bmatrix}, \tag{13.6}$$

where $\mathbf{0}$ denotes a matrix with all elements equal to 0. Note that, to simplify notation, we do not indicate the dimensions of the matrices, as they can be deduced from (13.2). Overall, $\boldsymbol{X}$ is of dimension $n \times p$, while $\boldsymbol{Z}$ is of dimension $n \times Nq$.

Models (13.1)–(13.4) can then be written for *all* data as follows:

$$\mathbf{y} = \boldsymbol{X}\boldsymbol{\beta} + \mathbf{Z}\mathbf{b} + \boldsymbol{\varepsilon}, \tag{13.7}$$

with

$$\mathbf{b} \sim \mathcal{N}_{Nq}(\mathbf{0}, \sigma^2 \mathsf{D}) \quad \text{and} \quad \boldsymbol{\varepsilon} \sim \mathcal{N}_n(\mathbf{0}, \sigma^2 \boldsymbol{R}), \tag{13.8}$$

where

$$\mathsf{D} \equiv I_N \otimes D = \begin{bmatrix} D & 0 & \dots & 0 \\ 0 & D & \dots & 0 \\ \vdots & \vdots & \ddots & \vdots \\ 0 & 0 & \dots & D \end{bmatrix}, \quad R \equiv \begin{bmatrix} R_1 & 0 & \dots & 0 \\ 0 & R_2 & \dots & 0 \\ \vdots & \vdots & \ddots & \vdots \\ 0 & 0 & \dots & R_N \end{bmatrix}, \tag{13.9}$$

with $\otimes$ denoting the (right) Kronecker product.

It is worth noting that the particular, block-diagonal form of matrices Z, D, and R, given in (13.6), (13.8), and (13.9), respectively, results from the fact that the single-level LMM, defined by (13.1)–(13.4), assumes a particular hierarchy of data and random effects, as explicitly shown in (13.3). In particular, the model assumes that random effects for different groups, defined by levels of a particular factor, are independent. Informally, we can describe the hierarchy as generated by grouping factors, with one being *nested* within the other.

It is possible, however, to formulate random-effects models by using the representation (13.7) with non-block-diagonal matrices Z, D, and R. This is the case, for instance, of models with *crossed* random effects. We will describe this type of models in Chap. 15.

13.3 The Extended Linear Mixed-Effects Model

In some cases, the assumption that the residual errors ε_i are independent of the random effects $\mathbf{b}_i$, as specified in (13.3), may be too restrictive. For instance, as is done in the mean-variance models, we might postulate that the variance of the residual errors depends on the subject-specific mean value. If we relax the assumption, we obtain an extended LMM. The model is specified by using (13.1)–(13.2) and replacing (13.3) by

$$\mathbf{b}_i \sim \mathcal{N}_q(\mathbf{0}, \mathcal{D}), \quad \text{and} \quad \varepsilon_i \mid \mathbf{b}_i \sim \mathcal{N}_{n_i}(\mathbf{0}, \mathcal{R}_i), \tag{13.10}$$

with $\mathcal{D}$ and $\mathcal{R}_i$ decomposed further as in (13.4). We will refer to the above specification as a *hierarchical* specification.

Note that, if we assume that ε_i in (13.10) is independent of the random effects $\mathbf{b}_i$, then we obtain the classical LMM, specified by (13.1)–(13.4). Thus, the extended LMM allows for a more general modeling approach, as compared to the classical LMM.

A hierarchical specification of a *two-level*, extended LMM, corresponding to (13.5), would amount to assuming that

$$\mathbf{b}_i \sim \mathcal{N}_{q_1}(\mathbf{0}, \mathcal{D}_1), \quad \mathbf{b}_{ij} \mid \mathbf{b_i} \sim \mathcal{N}_{q_2}(\mathbf{0}, \mathcal{D}_2), \quad \text{and} \quad \varepsilon_{ij} \mid \mathbf{b}_i, \mathbf{b}_{ij} \sim \mathcal{N}_{n_{ij}}(\mathbf{0}, \mathcal{R}_{ij}).$$

13.4 Distributions Defined by the y and b Random Variables

Both the classical (Sect. 13.2) and extended (Sect. 13.3) LMMs introduce two continuous random variables $\mathbf{b}$ and $\mathbf{y}$. They are described by two probability density functions, which play essential role in defining LMMs. The first one is an unconditional distribution of (unobserved) random effects $\mathbf{b}$, defined by (13.8). The second one is a conditional distribution of the (random) dependent variable $\mathbf{y}$, assuming that random effects are known. In the next two sections, we provide a more detailed description of the two distributions, which completes the model specification for the classical and extended LMMs. In Sect. 13.4.3, we will introduce additional auxiliary distributions related to $\mathbf{y}$ and $\mathbf{b}$ random variables.

13.4.1 Unconditional Distribution of Random Effects

The unconditional distribution $f_b(\mathbf{b}_i)$ of the random effects $\mathbf{b}_i$, defined by (13.3), is a multivariate normal distribution with zero mean and variance-covariance matrix $\mathcal{D}$. Taking into account (13.4), we write

$$\mathcal{D}(\sigma^2, \boldsymbol{\theta}_D) = \sigma^2 \boldsymbol{D}(\boldsymbol{\theta}_D), \tag{13.11}$$

where $\boldsymbol{\theta}_D$ is a vector of parameters, which represent the (scaled by σ^2) variances and covariances of the elements of $\mathbf{b}_i$.

Note that, according to (13.11), the matrix $\boldsymbol{D}$, used to define the variance-covariance matrix of random effects $\mathbf{b}_i$, is parameterized using a vector of parameters $\boldsymbol{\theta}_D$. In many cases, it is assumed that any two elements of the vector $\mathbf{b}_i$ can be correlated and there are no restrictions imposed on the matrix $\mathcal{D}$, except that it is positive-definite and symmetric. In this case, $\mathcal{D}$ has a general structure of a positive-definite matrix, with $q(q+1)/2$ distinct elements corresponding to q variances and $q(q-1)/2$ covariances of the random effects included in $\mathbf{b}_i$. Consequently, $\boldsymbol{\theta}_D$ contains $q(q+1)/2$ distinct parameters. Although q is typically small, estimating all of the parameters may be difficult if, e.g., the sample size n is limited. In such a situation, a simplified structure of the matrix $\mathcal{D}$ can be chosen. For instance, a diagonal form can be assumed, which is equivalent to assuming that all elements of the vector $\mathbf{b}_i$ are independent. Plausibility of the assumption will depend on the data at hand. In this case, $\boldsymbol{\theta}_D$ contains only q distinct parameters.

13.4.2 Conditional Distribution of y Given the Random Effects

Note that, from (13.1) to (13.4), it follows that, for the classical LMMs, the conditional distribution, $f_{y|b}(\mathbf{y}_i|\mathbf{b}_i)$, of $\mathbf{y}_i$ given $\mathbf{b}_i$ is multivariate normal, with the mean and variance defined as:

$$\mathrm{E}(\mathbf{y}_i|\mathbf{b}_i) \equiv \boldsymbol{\mu}_i = \boldsymbol{X}_i\boldsymbol{\beta}+\boldsymbol{Z}_i\mathbf{b}_i \tag{13.12}$$

$$\mathrm{Var}(\mathbf{y}_i|\mathbf{b}_i) = \sigma^2\boldsymbol{R}_i, \tag{13.13}$$

with $\boldsymbol{\mu}_i \equiv (\mu_{i1},\ldots,\mu_{i,n_i})'$ and

$$\mathrm{E}(y_{ij}|\mathbf{b}_i) \equiv \mu_{ij} = \mathbf{x}_{ij}'\boldsymbol{\beta}+\mathbf{z}_{ij}'\mathbf{b}_i, \tag{13.14}$$

where $\mathbf{x}_{ij} \equiv (x_{ij}^{(1)},\ldots,x_{ij}^{(p)})'$ and $\mathbf{z}_{ij} \equiv (z_{ij}^{(1)},\ldots,z_{ij}^{(q)})'$ are column vectors, which contain the values of predictors X and Z for the j-th observation from the i-th group. Thus, *conditionally* on the (unknown) values of the random effects $\mathbf{b}_i$, the mean value of the dependent-variable vector $\mathbf{y}_i$ is defined by a linear combination of the vectors of the X- and Z-covariates included, as columns, in the group-specific design matrices $\boldsymbol{X}_i$ and $\boldsymbol{Z}_i$, corresponding to the fixed effects $\boldsymbol{\beta}$ and random effects $\mathbf{b}_i$, respectively. Moreover, the conditional variance-covariance matrix of $\mathbf{y}_i$ is equal to the variance-covariance matrix of the residual errors $\boldsymbol{\varepsilon}_i$.

In their most general form, LMMs are not identifiable, because of the nonuniqueness of the representation (13.4) and because they potentially contain too many unknown parameters (see similar comments in Sects. 7.2 and 10.3). To make them identifiable, similarly to the matrix $\mathcal{D}$, we can consider representing elements of $\mathcal{R}_i$ as functions of a limited set of parameters $\boldsymbol{\theta}_R$, distinct from $\boldsymbol{\theta}_D$.

For the matrix $\mathcal{R}_i$, similarly to the approach described in Sect. 10.3 and implemented in R, we could consider the decomposition, given by (10.10), and combine it with the use of variance functions (Sects. 7.2.2 and 7.3.1) and correlation structures (Sect. 10.3.2). Thus, $\mathcal{R}_i$ would become parsimoniously parameterized in terms of a set of parameters of a variance function and a correlation structure. In this way not only the number of parameters of the model would be reduced, but the representation (13.4) would become identifiable.

To allow for the use of variance functions from the <δ, μ>- and <μ>-groups (Sect. 7.3.1), we follow the hierarchical specification (13.10) and apply the decomposition (10.11) to the conditional distribution of $\boldsymbol{\varepsilon}_i$ given $\mathbf{b}_i$. Consequently, we can postulate that

$$\mathrm{Var}(\varepsilon_{ij} \mid \mathbf{b}_i) = \sigma^2\lambda^2(\mu_{ij},\boldsymbol{\delta};\mathbf{v}_{ij}), \tag{13.15}$$

with μ_{ij} defined in (13.14). It follows that, upon combining the use of the variance function with a correlation structure (Sect. 10.3), we can write that

$$\mathrm{Var}(\boldsymbol{\varepsilon}_i \mid \mathbf{b}_i) = \sigma^2\boldsymbol{R}_i(\boldsymbol{\mu}_i,\boldsymbol{\theta}_R;\mathbf{v}_i), \tag{13.16}$$

with $\boldsymbol{\theta}_R \equiv (\boldsymbol{\delta},\boldsymbol{\varrho})$, where $\boldsymbol{\delta}$ is a vector of variance parameters employed by the variance function $\lambda(\cdot)$, $\boldsymbol{\varrho}$ is a vector of parameters related to the chosen correlation structure for the matrix $\boldsymbol{R}_i$, and $\mathbf{v}_i \equiv (\mathbf{v}_{i1}',\ldots,\mathbf{v}_{i,n_i}')'$ is a vector of variance covariates for the observations from the ith group.

Equations (13.15) and (13.16) imply that, for models with mean-dependent variance functions from the <δ, μ>- and <μ>-groups (Sect. 7.3.1), $\boldsymbol{\varepsilon}_i$ depend on $\mathbf{b}_i$ through $\boldsymbol{\mu}_i$. This violates the assumption of the classical LMM and leads to the extended LMM.

Extended LMMs, defined with the use of variance functions from the <δ, μ>- and <μ>-groups, pose theoretical and computational difficulties. For this reason, in the current chapter we will mainly focus on the classical LMMs, defined by (13.1)–(13.4), i.e., for which the matrix $\boldsymbol{R}_i$ is specified with the use of a mean-independent variance function. Models defined with the use of mean-dependent variance functions, which we term mean-variance models (see also Sects. 7.8 and 10.7), will be treated separately in Sect. 13.8.

For the *mean-independent* functions, such as those from the <δ>-group (see Table 7.2), the definition (13.16) can be simplified as follows:

$$\mathrm{Var}(\boldsymbol{\varepsilon}_i \mid \mathbf{b}_i) = \mathrm{Var}(\boldsymbol{\varepsilon}_i) = \sigma^2 \boldsymbol{R}_i(\boldsymbol{\theta}_R; \mathbf{v}_i). \tag{13.17}$$

Note that (13.17) is concordant with the assumption that the residual errors $\boldsymbol{\varepsilon}_i$ are independent of the random effects $\mathbf{b}_i$. Consequently, the hierarchical model specification with mean-independent variance functions leads to the classical LMM, with $\mathcal{R}_i = \sigma^2 \boldsymbol{R}_i(\boldsymbol{\theta}_R; \mathbf{v}_i)$. Essentially, this is the LMM formulation developed by Laird and Ware (1982).

It is worth noting that the choice of the structure of matrices $\mathcal{D}$ and $\mathcal{R}_i$ or, equivalently, $\boldsymbol{D}$ and $\boldsymbol{R}_i$ has consequences for the form of the *marginal* variance-covariance matrix of vector $\mathbf{y}_i$, implied by model (13.1)–(13.4). This form will be discussed in Sect. 13.5.1.

13.4.3 Additional Distributions Defined by y and b

In this section, we introduce additional auxiliary distributions related to LMMs. They build on distributions defined in Sects. 13.4.1 and 13.4.2 and play important role in the various aspects of model fitting and checking model assumptions.

13.4.3.1 Joint Distribution of y and b

The joint distribution $f_{y,b}(\mathbf{y}_i, \mathbf{b}_i)$ of $\mathbf{y}$ and $\mathbf{b}$ for the classical LMMs can be specified by taking the product of the unconditional distribution of the random effects $\mathbf{b}$ and the conditional distribution of $\mathbf{y}$ defined in Sects. 13.4.1 and 13.4.2:

$$f_{y,b}(\mathbf{y}_i, \mathbf{b}_i) = f_{y|b}(\mathbf{y}_i \mid \mathbf{b}_i) f_b(\mathbf{b}_i).$$

Given that the component distributions, $f_b(\mathbf{b})$ and $f_{y|b}(\mathbf{y} \mid \mathbf{b}) f_b(\mathbf{b})$, are multivariate normal, the joint distribution is also normal. We refer to the joint distribution in Sect. 13.5.3.

13.4.3.2 Marginal Distribution of y

The marginal distribution $f_y(\mathbf{y}_i)$ of $\mathbf{y}_i$ is obtained by "integrating out" the random effects $\mathbf{b}_i$ from the joint distribution of $\mathbf{y}_i$ and $\mathbf{b}_i$. More specifically, we calculate the density of the marginal distribution of $\mathbf{y}_i$ as

$$f_y(\mathbf{y}_i) = \int f_{y,b}(\mathbf{y}_i, \mathbf{b}_i)\, \mathrm{d}\mathbf{b}_i = \int f_{y|b}(\mathbf{y}_i \mid \mathbf{b}_i) f_b(\mathbf{b}_i)\, \mathrm{d}\mathbf{b}, \tag{13.18}$$

where $f_{y,b}$ is the density of the joint distribution of $\mathbf{y}_i$ and $\mathbf{b}_i$, $f_{y|b}$ is the conditional distribution of $\mathbf{y}_i$ given $\mathbf{b}_i$, and f_b is the density of the unconditional distribution of $\mathbf{b}_i$. Given that $f_{y,b}$ and f_b are densities of multivariate normal distributions, the marginal distribution of $\mathbf{y}$ is also multivariate normal and it can be derived analytically. In fact, it is given in (13.26).

13.4.3.3 Posterior Distribution of b Given y Is Known

The distribution $f_b(\mathbf{b}_i)$ of random effects $\mathbf{b}_i$ defined in (13.3) does *not* depend on the observed values of $\mathbf{y}_i$. Therefore, in the Bayesian setting, it is referred to as *prior* distribution of $\mathbf{b}_i$. Assuming that the observed values of $\mathbf{y}_i$ are equal to $\mathbf{y}_i^{(obs)}$, the so-called *posterior* distribution of $\mathbf{b}_i$ conditional on $\mathbf{y}_i^{(obs)}$ can be calculated using the following general formula:

$$f_{b|y}(\mathbf{b}_i|\mathbf{y}_i) \equiv f_{b|y}(\mathbf{b}_i|\mathbf{y}_i = \mathbf{y}_i^{(obs)}) = \frac{f_{y|b}(\mathbf{y}_i \mid \mathbf{b}_i) f_b(\mathbf{b}_i)}{\int f_{y|b}(\mathbf{y}_i \mid \mathbf{b}_i) f_b(\mathbf{b}_i)\, \mathrm{d}\mathbf{b}}. \tag{13.19}$$

Assuming that the parameters $\boldsymbol{\beta}, \boldsymbol{\theta}$ are known, the *posterior* distribution $f_{b|y}(\mathbf{b}_i|\mathbf{y}_i)$ for the classical LMMs is multivariate normal. Based on the observed data, we often estimate this distribution using its (*posterior*) mean:

$$\widehat{\mathbf{b}}_i(\boldsymbol{\beta}, \boldsymbol{\theta}) \equiv \widehat{\mathbf{b}}_i = \boldsymbol{D}\boldsymbol{Z}_i'\boldsymbol{V}_i^{-1}(\mathbf{y}_i^{(obs)} - \boldsymbol{X}_i\boldsymbol{\beta}). \tag{13.20}$$

Since the posterior mean is a linear function of $\mathbf{y}_i$, the variance-covariance matrix of the $\widehat{\mathbf{b}}_i$ estimator is equal to

$$\mathrm{Var}(\widehat{\mathbf{b}}_i) = \sigma^2 \boldsymbol{D}\boldsymbol{Z}_i' \left\{ \boldsymbol{V}_i^{-1} - \boldsymbol{V}_i^{-1}\boldsymbol{X}_i \left(\sum_{i=1}^{N} \boldsymbol{X}_i'\boldsymbol{V}_i^{-1}\boldsymbol{X}_i \right)^{-1} \boldsymbol{X}_i'\boldsymbol{V}_i^{-1} \right\} \boldsymbol{Z}_i\boldsymbol{D}. \tag{13.21}$$

To make inference about random effects, we are often interested in assessing the variability of the $\widehat{\mathbf{b}}_i - \mathbf{b}_i$ difference. The following formula can be used:

$$\mathrm{Var}(\widehat{\mathbf{b}}_i - \mathbf{b}_i) = \mathcal{D} - \mathrm{Var}(\widehat{\mathbf{b}}_i). \tag{13.22}$$

It follows from the formula that, for any linear combination of random effects represented by the column vector $\boldsymbol{\lambda}$, the following inequality (see (7.7) in Verbeke and Molenberghs 2000) holds:

$$\text{Var}(\boldsymbol{\lambda}'\widehat{\mathbf{b}}_i) \leq \text{Var}(\boldsymbol{\lambda}'\mathbf{b}_i) = \boldsymbol{\lambda}'\boldsymbol{\mathcal{D}}\boldsymbol{\lambda}. \tag{13.23}$$

This inequality is one of many ways which illustrate "shrinkage" of the random effects toward the prior mean of $\mathbf{b}_i$, i.e., toward zero. We revisit this issue in Sect. 13.6.1. On a side, we note that, for the LMM defined by (13.1)–(13.4), the posterior mean (13.20) is also the mode of the density of the posterior distribution of $\mathbf{b}_i$, given $\mathbf{y}_i$. In fact, the use of the mode to predict the random effects can be applied to mixed-effects models in general, including GLMMs and NLMMs. However, for mixed-effects models other than the LMMs (13.1)–(13.4), the mode does not have, in general, to be equal to the posterior mean.

13.5 Estimation

In this section, we present methods to obtain a set of estimates of parameters $\boldsymbol{\beta}$, σ^2, $\boldsymbol{\theta}_D$, and $\boldsymbol{\theta}_R$ for the classical LMM, defined by (13.1)–(13.4). The case of the extended, mean-variance model will be discussed separately in Sect. 13.8.

In Sect. 13.5.1, we present the marginal model, implied by the classical LMM. The marginal model allows estimating the LMM using the methods presented in Sect. 10.4 for the LM with fixed effects and correlated residual errors. In Sect. 13.5.2, we briefly describe the necessary modifications of the methods. In particular, we focus on the approaches that are implemented in R. Section 13.5.4 briefly discusses the issue of the parameterization of the classical LMM, while in Sect. 13.5.5 we describe the methods to assess the uncertainty of the parameter estimates. To complete the description of the estimation approaches, in Sect. 13.5.6, we briefly discuss approaches alternative to those presented in Sect. 13.5.2.

13.5.1 The Marginal Model Implied by the Classical Linear Mixed-Effects Model

For the classical LMM, Equations (13.12)–(13.13) and (13.17) imply that the marginal mean and variance-covariance matrix of $\mathbf{y}_i$ are given as follows:

$$\text{E}(\mathbf{y}_i) = \boldsymbol{X}_i\boldsymbol{\beta}, \tag{13.24}$$

$$\begin{aligned} \text{Var}(\mathbf{y}_i) &\equiv \boldsymbol{\mathcal{V}}_i(\sigma^2, \boldsymbol{\theta}; \mathbf{v}_i) \\ &= \sigma^2 \boldsymbol{V}_i(\boldsymbol{\theta}; \mathbf{v}_i) = \sigma^2[\boldsymbol{Z}_i\boldsymbol{D}(\boldsymbol{\theta}_D)\boldsymbol{Z}_i' + \boldsymbol{R}_i(\boldsymbol{\theta}_R; \mathbf{v}_i)], \end{aligned} \tag{13.25}$$

where $\boldsymbol{\theta}' \equiv (\boldsymbol{\theta}_D', \boldsymbol{\theta}_R')'$. Note that, to simplify the notation, from now on, we will, in general, suppress the use of $\boldsymbol{\theta}$ and $\mathbf{v}_i$ in the formulae, in line with conventions used in Parts II and III of the book.

From (13.24) and (13.25), it follows that, marginally,

$$\mathbf{y}_i \sim \mathcal{N}_{n_i}(\boldsymbol{X}_i\boldsymbol{\beta}, \sigma^2\boldsymbol{Z}_i\boldsymbol{D}\boldsymbol{Z}_i' + \sigma^2\boldsymbol{R}_i). \tag{13.26}$$

The marginal mean value of the dependent variable vector $\mathbf{y}_i$, similarly to the linear model (10.1)–(10.5), is defined by a linear combination of the vectors of covariates included, as columns, in the group-specific design matrix $\boldsymbol{X}_i$, with parameters $\boldsymbol{\beta}$. Moreover, the variance-covariance matrix of $\mathbf{y}_i$ consists of two components. The first one, $\sigma^2\boldsymbol{Z}_i\boldsymbol{D}\boldsymbol{Z}_i'$, is contributed by the random effects $\mathbf{b}_i$. The second one, $\sigma^2\boldsymbol{R}_i$, is related to the residual errors $\boldsymbol{\varepsilon}_i$. Hence, strictly speaking, the model employing random effects, specified in (13.1)–(13.4), implies a marginal normal distribution, defined by (13.26), which is similar to distributions considered in Chap. 10 in the context of LMs for correlated data, but with the variance-covariance matrix of $\mathbf{y}_i$ of a very specific parametric form, given by (13.25).

It is worth observing that the marginal model, defined by (13.24)–(13.26), does not involve the random effects $\mathbf{b}_i$. Thus, the matrix $\mathcal{D}$ does not have to be treated as a variance-covariance matrix. Consequently, it does not have to be positive-definite, as long as the matrix $\mathcal{V}_i$ is positive-definite. The matrix $\mathcal{D}$ does need to be symmetric, though, to assure that the matrix $\mathcal{V}_i$ is symmetric. It follows that, while every LMM of the form, specified in (13.1)–(13.4), implies a marginal model, defined by (13.26), not every model of the form (13.26) can be interpreted as resulting from an LMM. Thus, the two models are *not* equivalent.

From the above it follows that LMs with fixed effects and correlated residual errors, presented in Chap. 10, are less restrictive than LMMs. Thus, in this respect, the former are more flexible than the latter. On the other hand, in general, LMs with fixed effects and correlated residual errors do not allow making inference about the variability that may be related to different levels of the data hierarchy.

It is worth noting that the effects of the covariates, included in the design matrix $\boldsymbol{X}_i$, are quantified by the *same* parameters $\boldsymbol{\beta}$ in both the conditional (13.12) and unconditional (13.24) mean. Thus, although the parameters are defined in the context of the subject-specific model (13.1), they can also be interpreted as quantifying effects at the population level. This possibility of a dual interpretation of fixed-effects $\boldsymbol{\beta}$ is a unique feature of the classical LMM, given by (13.1)–(13.4). It does not hold, for instance, for GLMMs, not described in this book.

The fact that the classical LMM implies the marginal model (13.26) is also important from a practical point of view. This is because it allows the construction of effective estimation approaches for the LMM. This topic is discussed in the next section.

13.5.2 Maximum-Likelihood Estimation

In general, the ML estimation involves constructing the likelihood function based on appropriate probability distribution function for the *observed data*. The unconditional distribution of $\mathbf{b}_i$ and the conditional distribution of $\mathbf{y}_i$ given $\mathbf{b}_i$, which were defined in Sect. 13.4 for the classical LMM, are not suitable for constructing the likelihood function, because the random effects $\mathbf{b}_i$ are not observed. For a similar reason, the joint distribution of $\mathbf{y}_i$ and $\mathbf{b}_i$ cannot be used.

Instead, estimation of LMMs is based on the marginal distribution of $\mathbf{y}_i$.

In fact, it coincides with the distribution given in (13.26). For this reason, the estimation of parameters of the classical LMM can be accomplished by using the ML or REML estimation for the implied marginal model, along the lines similar to those described in Sect. 10.4.2.

In particular, the ML estimation is based on the marginal log-likelihood resulting from (13.26). Following (10.25), the log-likelihood can be expressed as follows:

$$\ell_{\text{Full}}(\boldsymbol{\beta},\sigma^2,\boldsymbol{\theta}) \equiv -\frac{N}{2}\log(\sigma^2) - \frac{1}{2}\sum_{i=1}^{N}\log[\det(\boldsymbol{V}_i)] - \frac{1}{2\sigma^2}\sum_{i=1}^{N}(\mathbf{y}_i - \boldsymbol{X}_i\boldsymbol{\beta})'\boldsymbol{V}_i^{-1}(\mathbf{y}_i - \boldsymbol{X}_i\boldsymbol{\beta}), \tag{13.27}$$

where $\boldsymbol{V}_i$, defined in (13.25), depends on $\boldsymbol{\theta}$.

Estimates of $\boldsymbol{\beta}$, σ^2, and $\boldsymbol{\theta}$ are usually obtained using a log-profile-likelihood for $\boldsymbol{\theta}$ (see Sect. 10.4.2). The log-profile-likelihood results from plugging into (13.27) the estimators of $\boldsymbol{\beta}$ and σ^2, given by

$$\widehat{\boldsymbol{\beta}}(\boldsymbol{\theta}) \equiv \left(\sum_{i=1}^{N}\boldsymbol{X}_i'\boldsymbol{V}_i^{-1}\boldsymbol{X}_i\right)^{-1}\sum_{i=1}^{N}\boldsymbol{X}_i'\boldsymbol{V}_i^{-1}\mathbf{y}_i, \tag{13.28}$$

$$\widehat{\sigma}^2_{\text{ML}}(\boldsymbol{\theta}) \equiv \sum_{i=1}^{N}\mathbf{r}_i'\boldsymbol{V}_i^{-1}\mathbf{r}_i/n, \tag{13.29}$$

where $\mathbf{r}_i \equiv \mathbf{r}_i(\boldsymbol{\theta}) = \mathbf{y}_i - \boldsymbol{X}_i\widehat{\boldsymbol{\beta}}(\boldsymbol{\theta})$. Note that the expressions correspond to (10.26) and (10.27), presented in Sect. 10.4.2. By maximizing the log-profile-likelihood function over $\boldsymbol{\theta}$, we obtain estimators of these parameters. Plugging $\widehat{\boldsymbol{\theta}}$ into (13.28) and (13.29) yields the corresponding estimators of $\boldsymbol{\beta}$ and σ^2, respectively.

As has been mentioned in Sects. 4.4.2, 7.4.2, and 10.4.2, the ML estimates of the variance-covariance parameters are biased. For this reason, the parameters are better estimated using the REML estimation. Toward this end, the log-restricted-likelihood function, corresponding to (10.30), is considered. From this function, the parameter σ^2 is profiled out by replacing it by the following estimator, corresponding to (10.31):

$$\widehat{\sigma}^2_{\text{REML}}(\boldsymbol{\theta}) \equiv \sum_{i=1}^{N} \mathbf{r}_i' \boldsymbol{V}_i^{-1} \mathbf{r}_i/(n-p), \tag{13.30}$$

with $\mathbf{r}_i$ defined as in (13.29). This leads to a log-profile-restricted-likelihood function, which only depends on $\boldsymbol{\theta}$:

$$\begin{aligned}\ell^*_{\text{REML}}(\boldsymbol{\theta}) \equiv & -\frac{n-p}{2}\log\left(\sum_{i=1}^{N} \mathbf{r}_i'\mathbf{r}_i\right) - \frac{1}{2}\sum_{i=1}^{N}\log[\det(\boldsymbol{V}_i)] \\ & -\frac{1}{2}\log\left[\det\left(\sum_{i=1}^{N} \boldsymbol{X}_i'\boldsymbol{V}_i^{-1}\boldsymbol{X}_i\right)\right]. \end{aligned}\tag{13.31}$$

Maximization of (13.31) yields an estimator of $\boldsymbol{\theta}$, which is then plugged into (13.28) and (13.30) to provide estimators of $\boldsymbol{\beta}$ and σ^2, respectively.

For the mean-variance model, i.e., when the conditional variance of random errors is defined with the use of a variance function (13.15) that does depend on μ_{ij} (Sect. 7.3.1), the estimates of the parameters $\boldsymbol{\beta}$, σ^2, and $\boldsymbol{\theta}$ can be obtained using GLS approaches similar to those described in Sects. 7.8.1.1 and 10.4.2. We discuss these approaches in Sect. 13.8.

13.5.3 *Penalized Least Squares*

In this section, we outline a slightly different approach to the estimation of parameters $\boldsymbol{\beta}$, σ^2, $\boldsymbol{\theta}$ for the classical LMM, defined by (13.1)–(13.4). Essentially, the approach is based on the log-profile-restricted-likelihood for θ, as defined in Sect. 13.5.2. However, the numerical algorithm based on sparse matrices allows for a numerically efficient implementation of this *penalized least squares* (PnLS) approach. In our presentation we follow Bates (2012) who describes in detail a more general version of this algorithm, namely, *penalized weighted least squares* (PWLS) used in the context of GLMMs and NLMMs and implemented in the package **lme4.0**. For the sake of future reference and simplicity, we briefly summarize the methodology. We consider a single-level LMM, specified for all data (Sect. 13.2.2). Moreover, we assume the conditional independence model and homogeneous residual-error variance, i.e., $\boldsymbol{R} \equiv \boldsymbol{I}_n$.

In the PnLS estimation approach, the starting point is the density of the joint distribution of $\mathbf{y}$ and random effects $\mathbf{b}$ introduced in general terms in Sect. 13.4.3. The logarithm of the density of the joint distribution of $\mathbf{y}$ and random effects $\mathbf{b}$ is given by

$$\begin{aligned}h_{\text{Joint}}(\mathbf{y},\mathbf{b};\boldsymbol{\beta},\sigma^2,\boldsymbol{\theta}) \equiv & -\frac{n+Nq}{2}\log(\sigma^2) - \frac{1}{2}\log[\det(\mathbf{D})] \\ & -\frac{(\mathbf{y}-\boldsymbol{X}\boldsymbol{\beta}-\mathbf{Z}\mathbf{b})'(\mathbf{y}-\boldsymbol{X}\boldsymbol{\beta}-\mathbf{Z}\mathbf{b})+\mathbf{b}'\mathbf{D}^{-1}\mathbf{b}}{2\sigma^2}, \end{aligned}\tag{13.32}$$

where $\boldsymbol{X}$ and $\boldsymbol{Z}$ were defined in (13.7), while $\mathbf{D}$ was specified in (13.9). Note that, given the assumption that $\boldsymbol{R} \equiv \boldsymbol{I}_n$, in (13.32) for the remainder of the section, we have $\boldsymbol{\theta} \equiv \boldsymbol{\theta}_D$.

Upon applying the following form of the Cholesky representation:

$$\mathbf{D} = \boldsymbol{TSST}', \tag{13.33}$$

where $\boldsymbol{T}$ is a lower-triangular matrix with all diagonal elements equal to 1 and $\boldsymbol{S}$ is a diagonal matrix with nonnegative diagonal elements, we can express $\mathbf{b}$ as follows:

$$\mathbf{b} = \boldsymbol{TS}\mathbf{u}, \quad \text{with} \quad \mathbf{u} \sim \mathcal{N}_{Nq}(\mathbf{0}, \sigma^2 \mathbf{I}_{Nq}).$$

By allowing for zero elements on the diagonal matrix $\boldsymbol{S}$ used in (13.33), we consider a general case with a potentially singular (positive semi-definite) matrix $\mathbf{D}$.

It follows that, conditionally on $\mathbf{u}$, $\mathbf{y}$ is normally distributed with

$$\begin{aligned} \mathrm{E}(\mathbf{y} \mid \mathbf{u}) &= \boldsymbol{X\beta} + \boldsymbol{ZTS}\mathbf{u} \equiv \boldsymbol{X\beta} + \boldsymbol{A}'\mathbf{u}, \\ \mathrm{Var}(\mathbf{y} \mid \mathbf{u}) &= \sigma^2 \boldsymbol{I}_n, \end{aligned} \tag{13.34}$$

while the marginal mean and variance of $\mathbf{y}$ can be expressed as

$$\begin{aligned} \mathrm{E}(\mathbf{y}) &= \boldsymbol{X\beta}, \\ \mathrm{Var}(\mathbf{y}) &= \sigma^2(\boldsymbol{A}'\boldsymbol{A} + \boldsymbol{I}_n). \end{aligned} \tag{13.35}$$

Using the representation introduced above, (13.32) can be written as follows:

$$\begin{aligned} h_{\text{PnLS}}(\mathbf{y}, \mathbf{b}; \boldsymbol{\beta}, \sigma^2, \boldsymbol{\theta}) &\equiv -\frac{n+Nq}{2}\log(\sigma^2) \\ &\quad - \frac{(\mathbf{y} - \boldsymbol{X\beta} - \boldsymbol{A}'\mathbf{u})'(\mathbf{y} - \boldsymbol{X\beta} - \boldsymbol{A}'\mathbf{u}) + \mathbf{u}'\mathbf{u}}{2\sigma^2} \\ &\equiv -\frac{n+Nq}{2}\log(\sigma^2) - \frac{d(\boldsymbol{\beta}, \boldsymbol{\theta})}{2\sigma^2}. \end{aligned} \tag{13.36}$$

Note that term $d(\boldsymbol{\beta}, \boldsymbol{\theta})$ in (13.36) resembles a penalized sum of squares. In fact, it can be seen as a residual sum of squares in a linear regression model

$$\mathrm{E}\begin{pmatrix} \mathbf{y} \\ \mathbf{0} \end{pmatrix} = \begin{bmatrix} \boldsymbol{A}' & \boldsymbol{X} \\ \mathbf{I}_{Nq} & \mathbf{0} \end{bmatrix} \begin{pmatrix} \mathbf{u} \\ \boldsymbol{\beta} \end{pmatrix} \equiv \boldsymbol{X}^* \begin{pmatrix} \mathbf{u} \\ \boldsymbol{\beta} \end{pmatrix}.$$

The solution $(\widetilde{\mathbf{u}}', \widetilde{\boldsymbol{\beta}}')'$ for the linear regression problem satisfies

$$(\boldsymbol{X}^*)'\boldsymbol{X}^* \begin{pmatrix} \widetilde{\mathbf{u}} \\ \widetilde{\boldsymbol{\beta}} \end{pmatrix} = (\boldsymbol{X}^*)' \begin{pmatrix} \mathbf{y} \\ \mathbf{0} \end{pmatrix},$$

which can be explicitly written as

$$\begin{bmatrix} \boldsymbol{A}\boldsymbol{A}' + \mathbf{I}_{Nq} & \boldsymbol{A}\boldsymbol{X} \\ \boldsymbol{X}'\boldsymbol{A}' & \boldsymbol{X}'\boldsymbol{X} \end{bmatrix} \begin{pmatrix} \widetilde{\mathbf{u}} \\ \widetilde{\boldsymbol{\beta}} \end{pmatrix} = \begin{pmatrix} \boldsymbol{A}\mathbf{y} \\ \boldsymbol{X}'\mathbf{y} \end{pmatrix}. \tag{13.37}$$

It is worth noting that (13.37) corresponds to a general form of the LMM equations considered by Henderson (1984), which allow for a singular estimate of $\mathbf{D}$.

To reduce the storage space requirements and numerical complexity, it is advantageous to introduce a sparse lower-triangular Cholesky decomposition matrix

$$\boldsymbol{L} = \begin{bmatrix} \boldsymbol{L}_Z & \mathbf{0} \\ \boldsymbol{L}_{ZX} & \boldsymbol{L}_X \end{bmatrix},$$

which satisfies

$$\boldsymbol{L}\boldsymbol{L}' = \boldsymbol{P}(\boldsymbol{X}^*)'\boldsymbol{X}^*\boldsymbol{P}', \tag{13.38}$$

where the orthogonal matrix $\boldsymbol{P}$ is a "fill-reducing" permutation matrix, determined from the pattern of nonzero elements in $\boldsymbol{Z}$. The matrix reduces the number of nonzero elements in $\boldsymbol{L}$ and hence has a large impact on the storage space required for $\boldsymbol{L}$. It is important to stress that, although this has not been explicitly indicated in (13.38), $\boldsymbol{L}$ depends on $\boldsymbol{\theta}$.

If we assume that the matrix $\boldsymbol{P}$ is of a block-diagonal form

$$\boldsymbol{P} = \begin{bmatrix} \boldsymbol{P}_Z & \mathbf{0} \\ \mathbf{0} & \boldsymbol{P}_X \end{bmatrix},$$

then we get

$$\begin{bmatrix} \boldsymbol{P}_Z'\boldsymbol{L}_Z & \mathbf{0} \\ \boldsymbol{P}_X'\boldsymbol{L}_{ZX} & \boldsymbol{P}_X'\boldsymbol{L}_X \end{bmatrix} \begin{bmatrix} \boldsymbol{L}_Z'\boldsymbol{P}_Z & \boldsymbol{L}_{ZX}'\boldsymbol{P}_X \\ \mathbf{0} & \boldsymbol{L}_X'\boldsymbol{P}_X \end{bmatrix} = \begin{bmatrix} \boldsymbol{A}\boldsymbol{A}' + \mathbf{I}_{Nq} & \boldsymbol{A}\boldsymbol{X} \\ \boldsymbol{X}'\boldsymbol{A}' & \boldsymbol{X}'\boldsymbol{X} \end{bmatrix}. \tag{13.39}$$

Consequently, we can rewrite (13.36) as follows:

$$\begin{aligned} h_{\text{PnLS}}(\mathbf{y}, \mathbf{b}; \boldsymbol{\beta}, \sigma^2, \boldsymbol{\theta}) &= -\frac{n + Nq}{2}\log(\sigma^2) - \frac{\widetilde{d}(\boldsymbol{\theta})}{2\sigma^2} \\ &\quad - \frac{1}{2\sigma^2}\begin{pmatrix} \boldsymbol{P}_Z(\mathbf{u} - \widetilde{\mathbf{u}}) \\ \boldsymbol{P}_X(\boldsymbol{\beta} - \widetilde{\boldsymbol{\beta}}) \end{pmatrix}' \boldsymbol{L}\boldsymbol{L}' \begin{pmatrix} \boldsymbol{P}_Z(\mathbf{u} - \widetilde{\mathbf{u}}) \\ \boldsymbol{P}_X(\boldsymbol{\beta} - \widetilde{\boldsymbol{\beta}}) \end{pmatrix}, \end{aligned} \tag{13.40}$$

where $\widetilde{d}(\boldsymbol{\theta})$ is the value of penalized sum of squares $d(\boldsymbol{\beta}, \boldsymbol{\theta})$, defined in (13.36), computed at solution $(\widetilde{\mathbf{u}}', \widetilde{\boldsymbol{\beta}}')'$ of system of (13.37). Thus, $\widetilde{d}(\boldsymbol{\theta})$ is the minimum value of penalized sum of squares, assuming $\boldsymbol{\theta}$ is known.

The marginal log-likelihood, corresponding to (13.40), is given by

$$\ell_{\rm ML}(\boldsymbol{\beta},\sigma^2,\boldsymbol{\theta}) \equiv -\frac{n}{2}\log(\sigma^2) - \frac{1}{2}\log\{[\det(\boldsymbol{L}_Z)]^2]\} - \frac{\widetilde{d}(\boldsymbol{\theta})}{2\sigma^2}$$
$$-\frac{1}{2\sigma^2}\left[\boldsymbol{L}_X'\boldsymbol{P}_X(\boldsymbol{\beta}-\widetilde{\boldsymbol{\beta}})\right]'\boldsymbol{L}_X'\boldsymbol{P}_X(\boldsymbol{\beta}-\widetilde{\boldsymbol{\beta}}). \tag{13.41}$$

Essentially, it is a re-parameterized form of (13.27).

Given $\boldsymbol{\theta}$, the resulting estimator of $\boldsymbol{\beta}$ is $\widetilde{\boldsymbol{\beta}}$, defined in (13.37), while for σ^2 the estimator is given by

$$\widetilde{\sigma}^2_{\rm ML} \equiv \frac{\widetilde{d}(\boldsymbol{\theta})}{n}. \tag{13.42}$$

By plugging $\widetilde{\boldsymbol{\beta}}$ and $\widetilde{\sigma}^2$ into (13.41), we obtain the log-profile-likelihood for $\boldsymbol{\theta}$:

$$\ell^*_{\rm ML}(\boldsymbol{\theta}) \equiv -\frac{1}{2}\log\{[\det(\boldsymbol{L}_Z)]^2\} - \frac{n}{2}\log[\widetilde{d}(\boldsymbol{\theta})]. \tag{13.43}$$

The log-profile-likelihood is a re-parameterized version of the function obtained from plugging the estimators (13.28) and (13.29) into (13.27) (see Sect. 13.5.2).

Maximization of (13.43) over $\boldsymbol{\theta}$ yields the ML estimator of the parameter vector. The estimator is then used to obtain the ML estimators of σ^2 and $\boldsymbol{\beta}$ from (13.42) and (13.37), respectively. The estimators correspond to those given in (13.28) and (13.29).

The REML estimator of $\boldsymbol{\theta}$ is obtained by maximizing the log-profile-restricted-likelihood:

$$\ell^*_{\rm REML}(\boldsymbol{\theta}) \equiv -\frac{1}{2}\log\{[\det(\boldsymbol{L}_Z)\det(\boldsymbol{L}_X)]^2\} - \frac{n-p}{2}\log[\widetilde{d}(\boldsymbol{\theta})]. \tag{13.44}$$

The function is, essentially, a re-parameterized form of (13.31). The resulting estimator is then used to obtain the estimator of σ^2, which corresponds to the one given in (13.30):

$$\widetilde{\sigma}^2_{\rm REML} \equiv \frac{\widetilde{d}(\boldsymbol{\theta})}{n-p}. \tag{13.45}$$

An estimate of $\boldsymbol{\beta}$ is computed from (13.37).

As mentioned at the beginning of this section, the PnLS approach, described above, has been implemented in the package **lme4.0**, a developmental branch version of **lme4**. In the latter package, the implementation has been modified in several ways (Bates et al. 2012). First, the decomposition (13.33) of the matrix $\mathbf{D}$ has been replaced by the classical Cholesky decomposition $\mathbf{D} = \mathbf{Q}\mathbf{Q}'$. Additionally, the lower-triangular matrix on the right-hand side of (13.39) has been assumed to take the following form:

$$\begin{bmatrix} P_Z' T_Z & \mathbf{0} \\ T_{ZX}' & T_X' \end{bmatrix},$$

where matrices T_Z (lower-triangular), P_Z (permutation), T_X, and T_{ZX} (upper-triangular) are defined by the following relationships:

$$T_Z T_Z' \equiv P_Z(AA' + \mathbf{I}_{Nq})P_Z',$$
$$T_Z T_{ZX} \equiv P_Z A X,$$
$$T_{ZX}' T_{ZX} \equiv X'X - T_X' T_X,$$

with $A \equiv \mathbf{Q}'Z'$. By using the resulting decomposition, we obtain formulae for the log-profile-likelihood and log-profile-restricted-likelihood similar to (13.43) and (13.44), respectively, but with $\det(L_Z)$ and $\det(L_X)$ replaced, respectively, by $\det(T_Z)$ and $\det(T_X)$. Moreover, Equation (13.37), defining the PnLS estimates $\widetilde{\mathbf{u}}$ and $\widetilde{\boldsymbol{\beta}}$, can now be equivalently expressed as

$$T_X \widetilde{\boldsymbol{\beta}} = \mathbf{c}_\beta,$$
$$T_Z' P_Z \widetilde{\mathbf{u}} = \mathbf{c}_u - T_{ZX} \widetilde{\boldsymbol{\beta}},$$

where the vectors $\mathbf{c}_\beta$ and $\mathbf{c}_u$ are defined by

$$T_Z \mathbf{c}_u = P_Z A \mathbf{y},$$
$$T_X' \mathbf{c}_\beta = X'\mathbf{y} - T_{ZX}\mathbf{c}_u.$$

13.5.4 Constrained Versus Unconstrained Parameterization of the Variance-Covariance Matrix

Solving of maximization problems, necessary to obtain the estimators described in Sects. 13.5.2 and 13.5.3, is difficult from a numerical point of view. This is because the solution should lead to symmetric and positive-definite variance-covariance matrices $\mathcal{D}$ and $\mathcal{R}_i$, defined in (13.3). A possible solution is to parameterize the matrices in such a way that the optimization problem becomes unconstrained.

Toward this end, for matrix $\mathcal{R}_i$, one can use representation (13.4) and parameterize R_i by the parameterizations described in Sect. 10.4.3. For matrix $\mathcal{D}$, several solutions are possible (Pinheiro and Bates 1996).

For instance, we could consider parameterizing $\mathcal{D}$ in terms of variances and correlations. By using the log-transformation for the variances and Fisher's z-transform, defined in (10.33), for correlations, we would obtain a set of unconstrained parameters. This parameterization would reflect the individual constraints, i.e., that variances need to be positive and correlation coefficients are constrained to lie within the $[0,1]$ interval. However, in general, it would not reflect the *joint* restriction, i.e., that the set of back-transformed parameters has to define a positive-definite matrix. Thus, while this parameterization could be used for positive-definite matrices of some particular structure like, e.g., the compound-symmetry structure, i.e., with equal diagonal elements and equal off-diagonal elements (see Sect. 11.4.1), it is not suitable for the numerical optimization purposes in general. However, it is useful for the construction of confidence intervals for the elements of matrix $\mathcal{D}$, as it will be explained in Sect. 13.7.3.

An alternative, which addresses the issue, uses (13.4) and considers the representation of $\boldsymbol{D}$ in terms of the elements of its Cholesky decomposition, i.e., in terms of the elements of the upper-triangular matrix $\boldsymbol{U}$, where $\boldsymbol{D} = \boldsymbol{U}'\boldsymbol{U}$. The main advantage of this approach is that it is computationally simple and stable. However, one of its disadvantages is that the resulting parameterization is not unique. This problem is removed by requiring that the diagonal elements of $\boldsymbol{U}$ are positive. In that case, an unconstrained parameterization of $\boldsymbol{D}$ is obtained using the logarithms of the diagonal elements of $\boldsymbol{U}$ together with the off-diagonal elements of $\boldsymbol{U}$. Pinheiro and Bates (1996) call this parameterization a "log-Cholesky parameterization". Another disadvantage of this approach, however, is that there is no straightforward relationship between the elements of $\boldsymbol{D}$ and $\boldsymbol{U}$, except for the fact that $\mid \boldsymbol{U}_{1,1} \mid = \sqrt{\boldsymbol{D}_{1,1}}$. This latter relationship does allow deriving confidence intervals for the diagonal elements of $\boldsymbol{D}$, i.e., variances, but not for the off-diagonal elements, i.e., covariances.

Another approach to obtaining an unconstrained parameterization of $\boldsymbol{D}$ is to use the *matrix logarithm* (Pinheiro and Bates (1996, 2000), pp. 78–79). In particular, $\boldsymbol{D}$ is expressed using its *singular value decomposition* (SVD):

$$\boldsymbol{D} = \boldsymbol{Q}\boldsymbol{T}\boldsymbol{Q}', \tag{13.46}$$

where $\boldsymbol{T}$ is a diagonal matrix with all diagonal elements positive, and $\boldsymbol{Q}$ is an orthogonal matrix. Let us denote by $\log(\boldsymbol{T})$ a diagonal matrix with the diagonal elements equal to the logarithms of diagonal elements of $\boldsymbol{T}$. Next, define

$$\boldsymbol{D}^* \equiv \boldsymbol{Q}\log(\boldsymbol{T})\boldsymbol{Q}'. \tag{13.47}$$

The $\boldsymbol{D}^*$ matrix is logarithm of $\boldsymbol{D}$, i.e., $\boldsymbol{D} = \exp(\boldsymbol{D}^*)$, where

$$\exp(\boldsymbol{D}^*) \equiv \sum_{k=0}^{\infty} \frac{(\boldsymbol{D}^*)^k}{k!}. \tag{13.48}$$

The relationship $\boldsymbol{D} = \exp(\boldsymbol{D}^*)$ allows expressing the parameters $\boldsymbol{\theta}_D$ (Sect. 13.4), which define the matrix, as a function of the elements of the upper triangle of the matrix $\boldsymbol{D}^*$. The latter elements form a set of unconstrained parameters that can be used for numerical optimization purposes. However, there is no straightforward relation between the elements of $\boldsymbol{D}$ and $\boldsymbol{D}^*$. Thus, the matrix-logarithm parameterization is not suitable for the construction of confidence intervals for the elements of $\boldsymbol{D}$.

In some situations, it may not be possible to find a solution of the optimization problem that would lead to a positive-definite $\boldsymbol{D}$. This may happen if, e.g., the assumed form of the LMM, defined by (13.1)–(13.4), is not correct. In this case, a possible alternative is to consider the implied marginal model (13.26). As was mentioned in Sect. 13.5.1, in the marginal model, the important constraint is that the marginal variance-covariance matrix $\mathcal{V}_i$, given in (13.25), is positive-definite; the positive-definiteness of $\boldsymbol{D}$ is not a necessary condition. Thus, in principle, one could consider fitting the LMM with the only constraint that $\mathcal{V}_i$ is positive-definite. The resulting solution, if feasible, may lead, however, to a non-positive-definite $\boldsymbol{D}$, which would violate the interpretation of the model as a hierarchical one (Sect. 13.5.1), but would lead to a valid marginal model nevertheless. Such an option is not routinely available in functions used for fitting LMMs in R.

13.5.5 Uncertainty in Parameter Estimation

Similar to the case of the LM for correlated data (Sect. 10.4.4), the variance-covariance matrix of $\widehat{\boldsymbol{\beta}}$ is estimated by

$$\widehat{\mathrm{Var}}(\widehat{\boldsymbol{\beta}}) \equiv \widehat{\sigma}^2 \left(\sum_{i=1}^{N} \boldsymbol{X}_i' \widehat{\boldsymbol{V}}_i^{-1} \boldsymbol{X}_i \right)^{-1}, \tag{13.49}$$

where $\widehat{\sigma}^2$ and $\widehat{\boldsymbol{V}}_i$ are estimated by one of the methods described in Sect. 13.5.2. Note that, in the computation of $\widehat{\mathrm{Var}}(\widehat{\boldsymbol{\beta}})$, given in (13.49), the extra variability resulting from the use of the estimate $\widehat{\boldsymbol{V}}_i$ is not accounted for. For this reason, the computed variance underestimates the true variability of $\widehat{\boldsymbol{\beta}}$.

The variance-covariance matrix of $\widehat{\sigma}^2$ and $\widehat{\boldsymbol{\theta}}$ can be estimated in various ways. A possible solution, implemented in the `lme()` function of the **nlme** package in R, is to use the inverse of the negative Hessian of the log-restricted-likelihood (see (10.30) and Sect. 13.5.2), evaluated at the estimated values of σ^2 and $\boldsymbol{\theta}$. An alternative is to consider the inverse of the negative Hessian of the log-profile-likelihood, which results from replacing $\boldsymbol{\beta}$ in (13.27) with the estimator given by (13.28). Obviously, the validity of these computations depends on the validity of the likelihood functions, i.e., on the correct specification of the model.

It is worth mentioning that the variance-covariance matrix of $\widehat{\boldsymbol{\beta}}$ can be expressed in a form similar to (10.45), with $\mathcal{R}_i$ replaced by the model-based marginal variance-covariance $\mathcal{V}_i$, given in (13.25) (see, e.g., Equation (6.2) in Verbeke and Molenberghs (2000)). Consequently, similarly to the situation described Sect. 10.7 for LMs for correlated data, if $\mathcal{V}_i \not\equiv \mathrm{Var}(\mathbf{y}_i)$, i.e., if the LMM is not correctly specified, (13.49) will result in a biased (underestimated) assessment of the variability of $\widehat{\boldsymbol{\beta}}$.

13.5.6 Alternative Estimation Approaches

Although the ML- and REML-based approaches, described in Sect. 13.5.2, are the most popular estimation methods for LMMs, other approaches are also possible. Among them one discerns, for instance, a Bayesian approach, a noniterative minimum variance quadratic unbiased estimation (MIVQUE), and the EM-algorithm. Neither the Bayesian approach nor MIVQUE is implemented in the most popular packages used to fit LMMs in R. The EM-algorithm is used in the function `lme()` from the package **nlme** only to refine the initial values of the parameters $\boldsymbol{\theta}_D$ in the first iterations of the optimization routine. For these reasons, we will not provide a more detailed description of these approaches here. The interested reader is referred to the monographs by Davidian and Giltinan (1995), Gelman et al. (1995), or Verbeke and Molenberghs (2000).

13.6 Model Diagnostics

In analogy with other types of LMs (see Sects. 4.5, 7.5, and 10.5), after fitting an LMM, and before making any inferences based on it, it is important to check whether the model assumptions are fulfilled. The two main distributional assumptions pertain to the normality of the random effects $\mathbf{b}_i$ and of the residual errors $\boldsymbol{\varepsilon}_i$. Evaluation of the influence of individual observations on the model fit (Sect. 4.5.3) may also be of importance. These topics are discussed in this section. Note that, as in Sect. 13.5, we focus on the classical LMM, defined by (13.1)–(13.4), in which the matrix $\boldsymbol{R}_i$ is specified with the use of a variance function, which does not depend on the mean values μ_{ij} (Table 7.2).

13.6.1 Normality of Random Effects

In the LMM defined by (13.1)–(13.4), it is assumed that the random effects $\mathbf{b}_i$ are normally distributed with the mean zero and the variance-covariance matrix $\sigma^2\boldsymbol{D}$. To check the assumption, some "estimates" of the random effects $\mathbf{b}_i$ are needed. Toward this end, usually the conditional expectations of the random effects, given the observed responses of $\mathbf{y}_i$, are used:

$$\widehat{\mathbf{b}}_i \equiv \widehat{\boldsymbol{D}}\boldsymbol{Z}_i'\widehat{\boldsymbol{V}}_i^{-1}(\mathbf{y}_i^{(obs)} - \boldsymbol{X}_i\widehat{\boldsymbol{\beta}}). \tag{13.50}$$

The conditional expectations are often called *empirical Bayes* (EB) estimates, because they are obtained by using the estimated values of the fixed parameters $\boldsymbol{\beta}$ and variance-covariance parameters $\boldsymbol{\theta}$ in (13.20). Note that, strictly speaking, the random effects $\mathbf{b}_i$ are not parameters, so that rather than estimating their values, we are predicting them. Following this convention, the conditional expectations (13.50) might be called "predictors." In fact, they are often referred to as *best linear unbiased predictors* (BLUPs) or *empirical BLUPs* (EBLUPs). This term follows from the fact that it can be shown that the conditional expectations are BLUPs of $\mathbf{b}_i$ in the sense that they are unbiased and have minimum variance among all unbiased estimators, which are linear combinations of $\mathbf{y}_i$ (see, e.g., Verbeke and Molenberghs (2000, Sect. 7.4)). In what follows, we will be referring to the random-effects predictors, given by (13.50), as EBLUPs.

Similarly to (13.23), shrinkage of EBLUPs can be illustrated by noting that the following inequality,

$$\text{var}(\boldsymbol{\lambda}'\widehat{\mathbf{b}}_i) \leq \boldsymbol{\lambda}'\widehat{\mathcal{D}}\boldsymbol{\lambda}, \tag{13.51}$$

is true for any linear transformation $\boldsymbol{\lambda}$. We refer to this inequality in Fig. 17.1 and Panel R19.7.

It appears that using histograms or Q-Q plots of the predicted random errors for the purpose of checking their normality is of limited value. That is because the observed distribution of $\widehat{\mathbf{b}}_i$ does not necessarily reflect the true distribution of $\mathbf{b}_i$ (Verbeke and Molenberghs 2000, Sec.7.8.1). However, the plots of the conditional modes can be used to detect, e.g., outlying values that might warrant further inspection. Also, if the histogram is, e.g., bimodal, it may indicate that a covariate has been omitted from the $\boldsymbol{Z}_i$ matrix.

In practice, checking the normality assumption for $\mathbf{b}_i$ should be based on the comparison of the results obtained for a LMM with and without assuming the normality (Verbeke and Molenberghs 2000, Sec.7.8.4). This requires software for fitting LMMs with relaxed distributional assumptions about the random effects. Such an approach will not be presented in our book.

It is worth noting, however, that if the inferential goal focuses on the marginal model (13.26), and especially on the fixed effects $\boldsymbol{\beta}$, valid inference can be obtained even if the random effects do not follow a normal distribution (Verbeke and Molenberghs 2000, Sec.7.8.4).

13.6.2 Residual Diagnostics

The main tools for checking the assumption of the normality of residual errors $\boldsymbol{\varepsilon}_i$ are based on residuals. Note that, given the structure of the classical LMM, defined in (13.1)–(13.4), various types of raw residuals can be defined.

One set is the *conditional residuals*, which follow from the conditional mean representation (13.12), and are defined as

$$\widehat{\boldsymbol{\varepsilon}}_{(c)i} \equiv \mathbf{y}_i - \boldsymbol{X}_i\widehat{\boldsymbol{\beta}} - \boldsymbol{Z}_i\widehat{\mathbf{b}}_i, \tag{13.52}$$

where the formula for $\widehat{\mathbf{b}}_i$ is given in (13.50).

Another set is the *marginal residuals*, resulting from the marginal mean representation, given by (13.24). The marginal residuals are defined as

$$\widehat{\boldsymbol{\varepsilon}}_{(m)i} \equiv \mathbf{y}_i - \boldsymbol{X}_i \widehat{\boldsymbol{\beta}}. \tag{13.53}$$

The raw residuals are useful to check heterogeneity of the conditional or marginal variance. They are less recommended, however, for checking normality assumptions and/or detecting outlying observations. This is because, usually, raw residuals will be correlated and their variances will differ. Therefore, studentized and Pearson residuals are more often used (see Sects. 4.5.1 and 7.5). However, as in the case of the LM for correlated data (see Sect. 10.5), even the scaled residuals are not appropriate for, e.g., checking the normality of the residual errors. This is because the model (13.1)–(13.4) allows for a correlation between the errors. An approximate solution is to consider the transformation of the raw conditional or marginal residuals, which were defined in (13.52) and (13.53), respectively, based on the Cholesky decomposition of the (estimate of) residual variance-covariance matrix $\sigma^2 \boldsymbol{R}_i$ or the marginal variance-covariance matrix $\sigma^2 \boldsymbol{V}_i$, respectively (see Sects. 4.5.1 and 10.5). That is, to define

$$\widehat{\boldsymbol{\varepsilon}}^*_{(c)i} \equiv (\widehat{\sigma} \widehat{\boldsymbol{U}}'_{(c)i})^{-1} \widehat{\boldsymbol{\varepsilon}}_{(c)i}, \tag{13.54}$$

or

$$\widehat{\boldsymbol{\varepsilon}}^*_{(m)i} \equiv (\widehat{\sigma} \widehat{\boldsymbol{U}}'_{(m)i})^{-1} \widehat{\boldsymbol{\varepsilon}}_{(m)i}, \tag{13.55}$$

where the upper-triangular matrices $\widehat{\boldsymbol{U}}_{(c)i}$ and $\widehat{\boldsymbol{U}}_{(m)i}$ are defined by $\widehat{\boldsymbol{U}}'_{(c)i} \widehat{\boldsymbol{U}}_{(c)i} = \widehat{\boldsymbol{R}}_i$ and $\widehat{\boldsymbol{U}}'_{(m)i} \widehat{\boldsymbol{U}}_{(m)i} = \widehat{\boldsymbol{V}}_i$, respectively. Then, $\widehat{\boldsymbol{\varepsilon}}^*_{(c)i}$ (Pinheiro and Bates 2000, pp. 239) and $\widehat{\boldsymbol{\varepsilon}}^*_{(m)i}$ (Schabenberger 2004) should be approximately normally distributed with mean zero and variance-covariance matrix equal to an identity matrix. Thus, e.g., the normal Q-Q plot of the residuals should show approximately a straight line. Also, the scatterplot of the residuals against the estimated marginal mean values can be used to detect patterns suggesting a possible problem in the specification of the mean structure of the data or to check for outliers. Note that in R, in the **nlme** package, the transformed conditional residuals (13.54) are available.

Santos Nobre and da Motta Singer (2007) argue that the marginal residuals are *pure*, in the sense that the residuals are a function of only the marginal errors $\boldsymbol{\varepsilon}_{(m)i} \equiv \mathbf{y}_i - \boldsymbol{X}_i \boldsymbol{\beta}$, which they are supposed to estimate. On the other hand, the conditional residuals, which estimate the residual errors $\boldsymbol{\varepsilon}_i$, are *confounded* with the random effects $\mathbf{b}_i$, because the residuals are a function of $\mathbf{b}_i$ *and* $\boldsymbol{\varepsilon}_i$. For this reason, they suggest that the conditional residuals may not be suitable for checking, e.g., the normality of $\boldsymbol{\varepsilon}_i$. In particular, Santos Nobre and da Motta Singer (2007) recommend to use the plots of marginal residuals against covariates to check the linearity assumption for the covariates. On the other hand, the plots of the conditional residuals against the estimated conditional means $\widehat{\boldsymbol{\mu}}_i$ can be used to detect outlying observations or heteroscedasticity of the residual errors.

13.6.3 *Influence Diagnostics*

The basic tool to investigate the influence of a given observation on the estimates of $\boldsymbol{\beta}$, $\boldsymbol{\theta}$, and σ^2 is the likelihood displacement. It was introduced in the context of the classical LM in Sect. 4.5.3. Recall that the likelihood displacement, LD_i, as in (4.27), is defined as the change between the maximum log-likelihood computed when using all data and when excluding the i-th observation. For the LMM, given by (13.1)–(13.4), the likelihood-displacement definition (4.27) is modified by specifying $\widehat{\Theta} \equiv (\widehat{\boldsymbol{\beta}}', \widehat{\boldsymbol{\theta}}', \widehat{\sigma}^2)'$ and using the log-likelihood (13.27).

13.7 Inference and Model Selection

The inference for the classical LMM, specified by (13.1)–(13.4), focuses on the fixed-effect parameters $\boldsymbol{\beta}$ and/or the variance-covariance parameters $\boldsymbol{\theta}$. For these models, as described in Sect. 13.5.2, the estimation of the parameters uses the methods based on the marginal log-likelihood (13.27). Consequently, the inferential tools are very similar to those used for the LMs for correlated data. Thus, in what follows, we will frequently refer to the material contained in Sect. 10.6.

In Sect. 13.7.1, we describe statistical significance tests for the fixed effects, while in Sect. 13.7.2 we discuss the tests for variance-covariance parameters. Section 13.7.3 briefly discusses the construction of confidence intervals for the parameters of the model (13.1)–(13.4).

13.7.1 *Testing Hypotheses About the Fixed Effects*

Hypotheses about the parameters $\boldsymbol{\beta}$ are tested using the same methods that are applied for LMs for correlated data (see Sect. 10.6). In particular, linear hypotheses may be tested using the F-test, given by (4.36). The issue related to the computation of the degrees of freedom for the approximation of the distribution of the F-statistic by a central F distribution applies here as well. In the package **nlme**, this issue is ignored and the functions, available for fitting LMMs, and the null distribution of the F-statistic is crudely approximated with $\mathrm{rank}(\boldsymbol{L})$ numerator and $n-p$ denominator degrees of freedom. In the package **lme4.0**, the issue is addressed using a Bayesian approach (see, e.g., Davidian and Giltinan (1995), Sect. 3.2.3) and by applying the Markov chain Monte Carlo (MCMC) technique to sample from the posterior distribution of the parameters (Baayen et al. 2008).

An alternative is to use ML-based LR tests (Sect. 7.6.1). Pinheiro and Bates (2000, Sect. 2.4.2) argue that the LR tests for hypotheses about $\boldsymbol{\beta}$ can be "anti-conservative", i.e., yield p-values smaller than those resulting from the postulated χ^2 distribution. For this reason they suggest to *condition* on the estimates of variance-covariance parameters $\boldsymbol{\theta}$ and use the F-tests. Based on the example they provided, it appears, though, that the problem is more pronounced in the case when several fixed effects are tested at once and sample size is relatively small. Therefore, for some models considered in Chap. 18, fitted to a large sample size data, we in fact used the LR test to test the selected hypotheses about fixed effects.

Note that, instead of using a χ^2 distribution for an LR test, one could use an empirical distribution of the test statistic, obtained by fitting the alternative and null models to multiple datasets simulated under the null model (Pinheiro and Bates 2000, Sect. 2.4.1).

Finally, if the hypothesis about the parameters $\boldsymbol{\beta}$ cannot be expressed in a way that it would lead to alternative and null models, we can apply information criteria, like AIC or BIC (Sect. 4.7.2), to select the model that seems to best fit the data. Of course, strictly speaking, this is not a formal statistical testing approach. In this respect, it is also worth mentioning that the use of the log-restricted-likelihood-based criteria for LMMs with different mean structures is generally not advocated (see, e.g., Verbeke and Molenberghs (2000, Sect. 6.4)). For such cases, the use of the ML-based criteria is recommended. However, Gurka (2006) provides empirical arguments that this may not be necessarily a general recommendation. Thus, the issue is still debatable.

13.7.2 Testing Hypotheses About the Variance-Covariance Parameters

Similarly to the case of testing hypotheses about the parameters $\boldsymbol{\beta}$ (Sect. 13.7.1), inference about $\boldsymbol{\theta}$ uses the methods, which are applied for LMs for correlated data (Sect. 10.6). In particular, LR tests (Sect. 4.6.1) and information criteria (Sect. 4.7.2) are used for this purpose. The comments related to the need of the use of the REML-based tests apply to the LMMs as well. However, for the latter models, several additional issues need to be mentioned.

One issue concerns the distribution of the LR tests for testing null hypotheses about parameters $\boldsymbol{\theta}_D$, related to the matrix $\boldsymbol{D}$. The distribution depends on the type of the null hypothesis. In this respect, two cases can be considered. The first one pertains to the situation when the values of the variance-covariance parameters, compatible with the null hypothesis, do not lie on the boundary of the parameter space. This is, e.g., the case when we test a hypothesis that a correlation coefficient is equal to 0. In this case, the null distribution of the LR test is a χ^2 distribution with the number of degrees of freedom equal to the difference in the number of variance-covariance parameters between the null and alternative models.

The second case pertains to the situation when the values of the variance-covariance parameters, compatible with the null hypothesis, do lie on the boundary of the parameter space. In such situations, the null distribution of the LR test is not a χ^2 distribution. In certain cases (see, e.g., Verbeke and Molenberghs 2000, Sect. 6.3.4), it is possible to show that the null distribution is a mixture of several χ^2 distributions. As an example, consider the case of the model (13.1)–(13.4) with only the group-level random effects, i.e., random intercepts. In this case, the vectors $\mathbf{b}_i \equiv b_i$ and $\boldsymbol{\theta}_D \equiv \theta_D$ are unidimensional (scalars), and $\boldsymbol{D} \equiv \theta_D$. Now, let us consider the null hypothesis, which specifies that no group-level random effects are needed. We can express the null hypothesis as $H_0 : \theta_D = 0$. The alternative is that a random effect is required, expressed as $H_A : \theta_D > 0$. Clearly, H_0 specifies a value of the variance parameter θ_D on the boundary of the parameter space, as variance cannot be negative. In this case, the results developed by Self and Liang (1987), Stram and Lee (1994), and Liang and Self (1996), suggested that the null distribution of the LR test statistic is a 50:50 mixture of χ_0^2 and χ_1^2 distributions. However, Crainiceanu and Ruppert (2004) show that the mixture is actually a conservative approximation to the finite-sample distribution of the LR test, which they derive.

It is important to note that the issue of testing a hypothesis on the boundary of the parameter space applies only when a fully hierarchical view of the classical LMM, specified by (13.1), (13.2), (13.10), and (13.4), is taken. When a purely marginal view, corresponding to (13.26), is adopted, the issue does not apply. Indeed, in the latter case, as it was argued in Sect. 13.5.4, $\boldsymbol{D}$ does not have to be positive-definite, as long as $\mathcal{V}_i$, given in (13.25), is positive definite. This means that, in our example, values of $\theta_d \leq 0$ are possible for the alternative hypothesis as well. More details on this issue can be found in, e.g., Verbeke and Molenberghs (2003) and Molenberghs and Verbeke (2007).

If the null distribution of the LR test cannot be derived analytically, a potential solution is to use an empirical distribution obtained by fitting the alternative and null model to multiple datasets, with the dependent variable simulated under the null model.

Another issue is related to the approach based on the information criteria. The approach is used when the hypothesis about $\boldsymbol{\theta}$ cannot be expressed in the way that it would lead to alternative and null models. In this case, we can apply information criteria, like AIC or BIC (Sect. 4.7.2), to select the model that seems to best fit the data. Strictly speaking, this is not a formal statistical testing approach. Also, recent work (Gurka 2006) suggests that none of the information criteria is optimal to select LMMs, and that more work is still needed to understand the role that information criteria play in the selection of LMMs.

Obviously, irrespectively of the approach selected, before conducting any statistical significance tests, the fit of the chosen final model should be formally checked using the residual diagnostic methods described in Sect. 13.6.

13.7.3 Confidence Intervals for Parameters

Confidence intervals for the individual components of the parameter vector $\boldsymbol{\beta}$ can be constructed based on the t-distribution, used as an approximate distribution for the t-test statistic (see Sects. 4.6.2, 10.6, and 13.7.1). On the other hand, confidence intervals for the parameters $\boldsymbol{\theta}_R$, related to the matrix $\boldsymbol{R}_i$, and for σ can be obtained in the same way as it was described for the case of LMs for correlated data (Sect. 10.6).

Confidence intervals for parameters describing the structure of the matrix $\mathcal{D}$ can be obtained by considering a representation of the matrix in terms of variances (or standard deviations) and correlations. As explained in Sect. 13.5.4, application of the logarithmic transformation to the variances and Fisher's z-transform to the correlations yields a set of unconstrained parameters. After fitting an LMM, confidence intervals for the transformed parameters can be constructed using the normal approximation to the distribution of the ML or REML estimators (Sect. 10.6). The confidence intervals can then be back-transformed (see (7.33) in Sect. 7.6.2 and (10.41) in Sect. 10.6) to yield the corresponding intervals for variances (or standard deviations) and correlations.

13.8 Mean-Variance Models

In this section, we discuss the estimation approaches and inferential issues related to the use of the extended, mean-variance LMM.

As mentioned in Sect. 13.2, to define the model, we specify the conditional variance of residual errors with the help of a mean-dependent variance function, defined in (13.15), i.e., a function from the <δ, μ>- or <μ>-group of variance functions (Sect. 7.3.1). The use of a mean-dependent variance function implies that, in the hierarchical model, defined by (13.1), (13.2), (13.10), and (13.4), residual errors and random effects for the same group are no longer independent. This violates the assumption of the classical LMM (13.1)–(13.4) and raises theoretical and computational issues. In this section, we describe the issues and possible solutions.

In particular, in Sect. 13.8.1, we focus on the single-level mean-variance LMM. Section 13.8.2 briefly describes the formulation of multilevel mean-variance LMMs. In Sect. 13.8.3, issues related to the inference, model diagnostics, and other aspects of the use of the mean-variance LMMs are summarized.

13.8.1 Single-Level Mean-Variance Linear Mixed-Effects Models

In Sect. 13.4, we pointed out that the marginal distribution for the classical LMM was a normal distribution of the form given in (13.26). Thus, the estimation of the model could be based on the use of the likelihood functions derived from the normal distribution (Sect. 13.5.2).

However, for the mean-variance LMM, defined by (13.1), (13.2), (13.10), (13.4), and (13.15), the marginal distribution does not have a closed form expression. In fact, it is not obvious that it is a normal distribution. Thus, the estimation approaches, described in Sect. 13.5.2, cannot be applied.

To estimate mean-variance LMMs, we might use the fact that, following (13.10) and (13.16),

$$\text{Var}(\boldsymbol{\varepsilon}_i) = \text{E}\left[\sigma^2 \boldsymbol{R}_i(\boldsymbol{\mu}_i, \boldsymbol{\theta}_R; \mathbf{v}_i)\right], \tag{13.56}$$

where the expected value is taken with respect to the distribution of the random effects $\mathbf{b}_i$, indicated in (13.10). Thus, the unconditional variance (13.56) does not depend on the random effects, which could, in principle, simplify the estimation of the model. However, analytical computation of the unconditional variance is, generally, not feasible, because the variance function (13.15) is usually nonlinear in $\mathbf{b}_i$.

A computationally feasible alternative is to estimate (13.56) by plugging-in suitable predictors of $\mathbf{b}_i$. Then, mean-variance LMMs can be estimated by algorithms similar to those presented in Sects. 7.8.1 and 10.7.

In particular, consider an LMM, defined by (13.1), (13.2), (13.10), (13.4), and (13.15), with a variance function from the $<\delta, \mu>$-group (see Table 7.3). Then, the following algorithm, similar to the one described in Sects. 7.8.1 and 10.7, can be used:

1. Assume an initial value $\widehat{\boldsymbol{\beta}}^{(0)}$ of $\boldsymbol{\beta}$, $\widehat{\boldsymbol{\theta}}^{(0)}$ of $\boldsymbol{\theta}$, $\widehat{\mathbf{b}}_i^{(0)}$ of $\mathbf{b}_i$, and set the iteration counter $k = 0$.
2. Increase k by 1.
3. *Use* $\widehat{\boldsymbol{\beta}}^{(k-1)}$ *to compute* $\widehat{\boldsymbol{\mu}}_i^{(k)}$ *and (re)define the matrix function* $\boldsymbol{V}_i^{(k)}(\boldsymbol{\theta})$.
   Calculate $\widehat{\boldsymbol{\mu}}_i^{(k)} \equiv \mathbf{x}_i\widehat{\boldsymbol{\beta}}^{(k-1)} + \mathbf{z}_i\widehat{\mathbf{b}}_i^{(k-1)}$ (see Sect. 13.6.1).
   (Re)define the variance function $\lambda^{(k)}(\boldsymbol{\delta}; \widehat{\mu}_{ij}^{(k)}) \equiv \lambda(\widehat{\mu}_{ij}^{(k)}, \boldsymbol{\delta})$.
   (Re)define diagonal elements of matrix function $\boldsymbol{\Lambda}^{(k)}(\boldsymbol{\delta}; \widehat{\boldsymbol{\mu}}_i^{(k)})$.
   (Re)define the matrix function $\boldsymbol{R}_i^{(k)}(\boldsymbol{\theta}_R) \equiv \boldsymbol{\Lambda}^{(k)}(\boldsymbol{\delta}; \widehat{\boldsymbol{\mu}}_i^{(k)})\mathbf{C}(\boldsymbol{\varrho})\boldsymbol{\Lambda}^{(k)}(\boldsymbol{\delta}; \widehat{\boldsymbol{\mu}}_i^{(k)})$, and $\boldsymbol{V}_i^{(k)}(\boldsymbol{\theta}) \equiv \boldsymbol{Z}_i\boldsymbol{D}(\boldsymbol{\theta}_D)\boldsymbol{Z}_i' + \boldsymbol{R}_i^{(k)}(\boldsymbol{\theta}_R)$.
4. *Keep* $\widehat{\boldsymbol{\beta}}^{(k-1)}$ *fixed and compute* $\widehat{\boldsymbol{\theta}}^{(k)}$.
   While keeping the value $\widehat{\boldsymbol{\beta}}^{(k-1)}$ of $\boldsymbol{\beta}$ fixed and using the matrix functions $\boldsymbol{V}_i^{(k)}(\boldsymbol{\theta})$, compute an estimate $\widehat{\boldsymbol{\theta}}^{(k)}$ of $\boldsymbol{\theta}$ by maximizing an appropriate log-profile-likelihood.
5. *Keep* $\widehat{\boldsymbol{\theta}}^{(k)}$ *fixed. Compute* $\boldsymbol{\beta}^{(k)}$ *and* $\widehat{\mathbf{b}}_i^{(k)}$.
   Based on the formula in step 3, compute the matrices $\boldsymbol{V}_i^{(k)}(\widehat{\boldsymbol{\theta}}^{(k)})$ and use them to obtain estimates $\widehat{\boldsymbol{\beta}}^{(k)}$ of $\boldsymbol{\beta}$ from (13.28) and $\widehat{\mathbf{b}}_i^{(k)}$ of $\mathbf{b}_i$ from (13.50).
6. Iterate between steps 2 and 5 until convergence or until a predetermined number of iterations k.
7. Compute the estimate of σ^2 from (13.29) using the estimates of $\boldsymbol{\theta}$ and $\boldsymbol{\beta}$.

An ML-based version of the algorithm is obtained by using, in step 4, the fixed value $\widehat{\boldsymbol{\beta}}^{(k-1)}$ of $\boldsymbol{\beta}$ and the redefined form of the matrices $\boldsymbol{V}_i^{(k)}(\boldsymbol{\theta})$ to express σ^2 as in (13.29). The resulting expression for σ^2 is then plugged in (13.27), and the estimate $\widehat{\boldsymbol{\theta}}^{(k)}$ of $\boldsymbol{\theta}$ is computed by maximizing the so-obtained log-profile-likelihood.

A REML-based version of the algorithm results from using, in step 4, the log-profile-restricted-likelihood (13.31) to compute the estimate $\widehat{\boldsymbol{\theta}}^{(k)}$ of $\boldsymbol{\theta}$ (Sects. 7.8.1 and 10.7). Then, in step 7, σ^2 is computed from (13.30).

Note that the algorithm involves two iterative loops: the "external" one, related to the computation of the values $\widehat{\boldsymbol{\beta}}^{(k)}$ of $\boldsymbol{\beta}$ and $\widehat{\boldsymbol{\mu}}_i^{(k)}$ of $\widehat{\boldsymbol{\mu}}_i$, and the "internal" one, related to the computation (in step 4) of the value $\widehat{\boldsymbol{\theta}}^{(k)}$ of $\boldsymbol{\theta}$. It is also worth noting that, in step 3, the redefined variance function $\lambda^{(k)}(\cdot)$ does not depend on μ_{ij}, and therefore it belongs to the <δ>-group. This allows for the use of the marginal-model-based log-(profile)-likelihood functions, like (13.31), in step 4.

LMMs, defined by (13.1), (13.2), (13.4)–(13.10), and (13.15), with a variance function that depends only on μ_{ij}, i.e., which belongs to the <μ>-group (see Table 7.4), can be estimated by using an IRLS procedure similar to the one described in Sect. 7.8.1, with obvious modifications.

Algorithms equivalent to the PL-GLS and IRLS procedures, described above, can be formulated for the PnLS estimation technique (Sect. 13.5.3), resulting in a *penalized, iteratively re-weighted, least squares* (PnIRLS) approach (Bates 2012).

13.8.2 Multilevel Hierarchies

To extend the mean-variance formulation to, e.g., two-level LMMs, which were defined in (13.5), the definition of the variance function, given in (13.15), needs to be modified. In particular, we can assume that

$$\text{Var}(\varepsilon_{ijk} \mid \mathbf{b}_i, \mathbf{b}_{ij}) = \sigma^2 \lambda^2(\mu_{ijk}, \boldsymbol{\delta}; \mathbf{v}_{ijk}), \tag{13.57}$$

where

$$\text{E}(y_{ijk} \mid \mathbf{b}_i, \mathbf{b}_{ij}) \equiv \mu_{ijk} = \mathbf{x}'_{ijk}\boldsymbol{\beta} + \mathbf{z}'_{1,ijk}\mathbf{b}_i + \mathbf{z}'_{2,ijk}\mathbf{b}_{ij}, \tag{13.58}$$

where $\mathbf{x}_{ijk}$, $\mathbf{z}_{1,ijk}$, and $\mathbf{z}_{2,ijk}$ are column vectors containing the values of the x-, z_1-, and z_2-covariates for the k-th observation from the j-th subgroup of the i-th group.

The estimation algorithms, described in Sect. 13.8.1 for a single-level LMM, can be adapted to the two-level model case. Needless to say, they become more involved numerically.

13.8.3 Inference

For the mean-variance LMMs, estimates of $\boldsymbol{\beta}$ are asymptotically approximately normally distributed with a variance-covariance matrix, which can be estimated as in (13.49). Consequently, tests for linear hypotheses and CIs for the elements of $\boldsymbol{\beta}$ can be obtained along the lines described in Sect. 13.7.1. Note, however, that the use of LR tests is problematic given the fact that the algorithms described in Sect. 13.8.1 are not likelihood-based. It should also be born in mind that, especially for small sample sizes, standard errors computed from (13.49) may be too small, as the precision of estimation of $\boldsymbol{\beta}$ is influenced by the precision of the estimation of $\boldsymbol{\theta}$ (see Sects. 7.8.2 and 10.7).

Inference on the parameters $\boldsymbol{\theta}$ and σ^2 is complicated by problems similar to those described in Sects. 7.8.2 and 10.7. Thus, we do not discuss it further here.

13.9 Chapter Summary

In this chapter, we reviewed the essential concepts and methods underlying the formulation of an LMM for hierarchical data. In our presentation, we were focusing on the theoretical constructions, which are linked to the implementation of LMMs in R. Readers interested in a more detailed account of the theory of LMMs are referred to the monographs mentioned in the introduction to this chapter.

In Sect. 13.2, we described the formulation of the classical LMM, while in Sect. 13.3 we discussed the formulation of the extended model. Details of the formulation of both types of LMMs were discussed in Sect. 13.4. In the formulation, the concepts of variance function and correlation structure, developed in Chaps. 7 and 10, respectively, were used. An additional, novel element was the introduction of the random effects in the mean structure of the model. The use of the random effects allows to directly address the hierarchical structure of the data.

Sections 13.5, 13.6, and 13.7 were devoted to, respectively, the estimation approaches, diagnostic tools, and inferential methods used for the classical LMM. This type of LMMs is most commonly used in practice. In Sect. 13.8, we described the estimation and inferential techniques used for the extended LMM defined using a mean-dependent variance function.

To the extent possible, we used in our presentation the concepts and theory introduced in Chaps. 4, 7, and 10. Especially relevant was the material from Chap. 10, because, as mentioned in Sect. 13.5, the estimation of classical LMM is based primarily on the marginal likelihood or restricted likelihood functions, which are special cases of the likelihood functions presented in Sect. 10.4.2. We note that, if the missing at random (MAR) assumption about missing data mechanism is tenable, the ML estimation for the classical linear mixed-effect models yields valid estimates. Thorough discussion of this important topic can be found in Verbeke and Molenberghs (2000).

As compared to LMs for correlated data, described in Part III of the book, LMMs address directly the hierarchy present in grouped data. They allow drawing conclusions about the partition of the total variability of observations between the different levels of the hierarchy. This additional insight can be considered as an advantage of LMMs. On the other hand, as mentioned in Sect. 13.5.1, LMs for correlated errors are more flexible than LMMs. Thus, the choice between them depends on the goals of a particular analysis.

In the next two chapters, we describe the tools available for fitting LMMs in R.

Chapter 14
Fitting Linear Mixed-Effects Models: The `lme()` Function

14.1 Introduction

In Chap. 13, we summarized the main theoretical concepts underlying the construction of LMMs. Compared to the LMs introduced in Chaps. 4, 7, and 10, LMMs allow taking the hierarchical structure of data into account in the analysis. This is achieved by introducing, in addition to the mean (fixed-effects) structure, a random-effects structure.

There are several packages in R, which contain tools for fitting LMMs, like, e.g., **nlme**, **lme4.0**, or **MCMCglmm**. In the current chapter, we describe the use of the popular and well-established package **nlme**. The primary tool to fit LMMs in this package is the function `lme()`. In the next chapter, we will describe the use of the package **lme4.0**. Note that both packages allow to fit GLMMs and NLMMs, but these models are outside of the scope of this book.

The chapter is organized as follows. In Sect. 14.2, we describe objects of class *pdMat*, which represent positive-definite matrices. In particular, the class is used to represent variance–covariance matrices of random effects. In Sect. 14.3, we describe the class *reStruct*, used to represent the random-effects structure of an LMM. The random part of an LMM is represented using the *lmeStruct* class described in Sect. 14.4. All the aforementioned classes are related to the function `lme()`, which is the key function in the **nlme** package to fit LMMs. The use of the function is reviewed in Sect. 14.5. On the other hand, in Sect. 14.6, we summarize the methods, which allow extracting information from model-fit objects of class *lme*. Section 14.7 is devoted to the implementation of inferential tools for LMMs. A summary of the chapter is provided in Sect. 14.8.

As the basic example, we use the classical, single-level LMM, defined by (13.1)–(13.4). However, to illustrate the features of the syntax, we refer, in a few instances, to the two-level LMM, specified in (13.5).

A. Gałecki and T. Burzykowski, *Linear Mixed-Effects Models Using R: A Step-by-Step Approach*, Springer Texts in Statistics, DOI 10.1007/978-1-4614-3900-4_14,

14.2 Representation of a Positive-Definite Matrix: The *pdMat* Class

As compared to the LMs, introduced in Chaps. 4, 7, and 10, an important, new component of LMMs is the random-effects structure. By the random-effects structure, we mean the levels of the model hierarchy, the $\mathbf{Z}_i$ design matrices for the random effects $\mathbf{b}_i$, and the parameterized form of the matrix $\boldsymbol{D}$, as defined in Equations (13.1)–(13.4) for the classical, single-level LMM. In this section, we provide details of the implementation of various forms of positive-definite matrices available in the package **nlme**. The forms are used for specifying the matrix $\boldsymbol{D}$.

In particular, in Sect. 14.2.1, we describe the classes of such matrices and the corresponding constructor functions, while in Sect. 14.2.2, we discuss the methods to extract information from the objects constructed with the help of the functions.

14.2.1 Constructor Functions for the pdMat Class

Positive-definite matrices are represented in the package **nlme** by objects inheriting from the *pdMat* class. By issuing the `?pdClasses` command at the command prompt we obtain a list of standard classes of *pdMat* structures. We itemize them below for the reader's reference:

pdIdent	a multiple of identity
pdDiag	a diagonal matrix
pdCompSymm	compound symmetry
pdLogChol	a general positive-definite matrix using the log-Cholesky parameterization
pdSymm	a general positive-definite matrix with a parameterization based on SVD
pdNatural	a general-positive-definite matrix with the "natural" parameterization, i.e., in terms of standard deviations and correlations
pdBlocked	a blocked-diagonal matrix, with blocks defined by structures/classes defined above

The classes listed above are ordered roughly according to the increasing order of complexity of the represented matrix structures. The main difference between the *pdLogChol*, *pdSymm*, and *pdNatural* classes, which all represent a general variance–covariance matrix, lies in the used parameterization (Sect. 13.5.4) and will be illustrated in Sect. 14.2.2.

The constructor function, used to create or to modify objects that inherit from a particular class, is named after the corresponding class. For example, the `pdDiag()` function creates/modifies an object of class *pdDiag*. Note that the created object inherits also from the *pdMat* class. The *pdMat* constructors are primarily used in the specification of the random-effects structure of an LMM, with the help of the `random` argument of the model-fitting function `lme()`.

In the next section, we describe the arguments of the constructor functions.

14.2.1.1 Arguments of the Constructor Functions

The arguments of the *pdMat* constructor functions are `value`, `form`, `nam`, and `data`. We will focus on the `value` and `form` arguments. Arguments `data` and `nam` are merely used to assign names to the rows and columns of a positive-definite matrix. As such, they are less important, so we do not describe them. A description of all the arguments for a specific *pdMat* class is obtained by issuing a command like, e.g., `?pdSymm`.

The argument `form` is simply an optional one-sided formula. When used together with the `data` argument, the formula is evaluated and, subsequently, the appropriate names are assigned to the rows/columns of the matrix represented by the object. By default, the value of the `form` argument is `NULL`. If the `value` argument contains a one-sided formula, the argument `form` is ignored.

Although `value` is the main argument of a *pdMat* constructor function, we describe it as the last, because it can be used to specify the components of *pdMat* objects, defined by the arguments described above. The main role of this argument is to assign coefficients to *pdMat* objects by supplying a positive-definite matrix or a numeric vector. Other possible values of the `value` argument include: a *pdMat*-class object, a one-sided linear formula, or a vector of character strings. By default, its value is `numeric(0)`, which results in an uninitialized object.

The code in Panel R14.1 presents examples of application of constructor functions `pdCompSymm()` and `pdSymm()` to create objects of class *pdCompSymm* and *pdSymm*, respectively, which inherit from the *pdMat* class.

In Panel R14.1a, the `pdCompSymm()` function applies the argument `value` in the form of a one-sided formula `~agex`. Thus, it does not assign any numeric values. Consequently, the resulting object `pdCS0` of class *pdCompSymm*, representing a compound-symmetry matrix with constant diagonal and off-diagonal elements, is uninitialized.

In Panel R14.1b, the object `mtxUN` is a positive-definite, 2×2 matrix, while `dt1` is an auxiliary data frame with a single numeric variable `agex` with four observations. The `pdSymm()` function uses the matrix `mtxUN` as the `value` argument. Additionally, it specifies the one-sided formula `~agex` as the `form` argument and uses the data frame `dt1` to evaluate the variable `agex`. The resulting object `pdSm` of class *pdSymm*, which represents a general positive-definite matrix, is initialized.

To explain the names of the rows and columns of the matrix, contained in the object `pdSm`, we note that the formula `~agex`, used in Panel R14.1b, assumes the presence of an intercept and is equivalent to the formula `~1 + agex`. It follows that object `pdSm` can be interpreted as the variance–covariance matrix of a vector of random effects generated by the formula. Given that the variable `agex` is numeric (continuous), the formula implies the use of random intercepts and of random slopes (for `agex`). Hence, the use of the names `(Intercept)` and `agex` for the rows and columns of the matrix, contained in the object `pdSm`.

In Panel R14.1c, we illustrate how to construct an initialized object of class *pdCompSymm*. The object `mtxCS` is a positive-definite, 3×3 compound-symmetry

R14.1 *R syntax*: Creating objects inheriting from the *pdMat* class

(a) *Uninitialized object of class pdCompSymm*

```
> library(nlme)
> (pdCS0     <- pdCompSymm(~agex))
  Uninitialized positive definite matrix structure of class pdCompSymm
> isInitialized(pdCS0)                           # Not initialized
  [1] FALSE
```

(b) *Initialized object of class pdSymm*

```
> mtxUN      <- matrix(c(4, 1, 1, 9), nrow = 2)             # pdSymm matrix
> dt1        <- data.frame(agex = c(15, 45, 71, 82))        # Numeric age
> (pdSm      <- pdSymm(mtxUN, ~agex, data = dt1))
  Positive definite matrix structure of class pdSymm representing
              (Intercept) agex
  (Intercept)           4    1
  agex                  1    9
> isInitialized(pdSm)                                       # Initialized
  [1] TRUE
```

(c) *Initialized object of class pdCompSymm*

```
> mtxCS      <- matrix(4 * diag(3) + 1, nrow = 3)           # CompSymm matrix
> dt2        <- data.frame(agef=c("Y", "M", "O", "O")) # Factor age
> (pdCSf     <- pdCompSymm(mtxCS, ~-1 + agef, data = dt2))
  Positive definite matrix structure of class pdCompSymm representing
         agefM agefO agefY
  agefM      5     1     1
  agefO      1     5     1
  agefY      1     1     5
```

matrix, with all diagonal elements equal to 5 and all off-diagonal elements equal to 1. The auxiliary object `dt2` is a data frame with a single-variable `agef`, which is a factor with three levels and four observations. The `pdCompSymm()` constructor function uses the object `mtxCS` as the `value` argument. Additionally, it specifies the one-sided formula `~-1+agef` in the argument `form` and uses the data frame `dt2` to evaluate the variable `agef`. The resulting object `pdCsf` of class *pdCompSymm*, which represents a compound-symmetry matrix, is initialized. Note that its row and column names are defined by the levels of the factor `agef`. This is because the formula, used in the `form` argument, does not use an intercept and includes the factor `agef` (Sect. 5.2.1). Thus, the formula implies the use of three random effects, associated with the levels of the factor. Hence, the use of the factor-level names for the rows and columns of pdCSf. Note that the names are obtained by referring to the data frame dt2, which was provided in the argument data. Had the formula been changed to, e.g., ~agef, it would have implied the use of an intercept, and the name of the first row and column of the object pdSm would have been changed to (Intercept). The names of the remaining two rows and columns would remain unchanged.

In many cases, e.g., when specifying an LMM, it is sufficient to work with uninitialized *pdMat* objects, such as the one defined in Panel R14.1a. Initialized objects, i.e., objects with defined, known numerical values, such as those shown in Panels R14.1b and R14.1c, can be useful if initial values for coefficients of a positive-definite matrix need to be specified for a model-fitting routine.

14.2.2 Inspecting and Modifying Objects of Class pdMat

A list of methods available for probing and modifying objects of class *pdMat* is obtained by issuing the command methods(class="pdMat"). In Panel R14.2, we present examples of the use of selected methods, which can be applied to extract information from such objects.

More specifically, in Panel R14.2a, we present methods for an object of class *pdSymm*. As an example, we use the object pdSm defined in Panel R14.1. The summary() function displays the standard deviations and correlations associated with the positive-definite matrix represented by the object. Specific attributes of the object can be displayed using appropriate functions like, e.g., formula(), Names(), or Dim(). The function logDet() prints out the value of the logarithm of the determinant of the Cholesky factor of the positive-definite matrix represented by the *pdSymm*-class object.

In Panel R14.2b, we show the results of the application of selected methods to the object pdCSf of class *pdCompSymm*. Essentially, all the methods used in Panel R14.2a could be applied to the object as well.

Panel R14.3 is devoted entirely to the use of the coef() method for extracting coefficients of an initialized *pdMat*-class object. The method returns a vector with coefficients associated with the object. It allows for an optional logical argument unconstrained. If unconstrained=FALSE, a vector of constrained coefficients is returned (Sect. 13.5.4). Depending on the class of the object, the vector may contain upper-triangular elements of the positive-definite matrix represented by a *pdSymm*-class object or the standard deviation and correlation coefficient corresponding to a compound-symmetry matrix from a *pdCompSymm*-class object. If unconstrained=TRUE, the coefficients are returned in unconstrained form, suitable for the optimization purposes. By default, unconstrained=TRUE.

In Panel R14.3a, we extract coefficients from the object pdSm. This is an object of class *pdSymm*, i.e., it represents a positive-definite matrix of a general form. First, by specifying the argument unconstrained=FALSE, we obtain the upper-triangular elements of the matrix. Subsequently, using the default value of

R14.2 *R syntax*: Probing objects inheriting from the *pdMat* class. Objects pdSm and pdCSf were created in Panel R14.1

(a) *Extracting information from the object pdSm of class pdSymm*

```
> summary(pdSm)                    # Summary
  Formula: ~agex
  Structure: General positive-definite
             StdDev Corr
 (Intercept) 2      (Intr)
 agex        3      0.167
> formula(pdSm)                    # Formula
  ~agex
> Names(pdSm)                      # Row/col names
  [1] "(Intercept)" "agex"
> (Dmtx <- as.matrix(pdSm))        # D matrix
              (Intercept) agex
 (Intercept)            4    1
 agex                   1    9
> Dim(pdSm)                        # Dimensions of D
  [1] 2 2
> logDet(pdSm)                     # log|D^(1/2)|
  [1] 1.7777
> # VarCorr(pdSm)                  # Variances, correlation coefficients
> # corMatrix(pdSm)                # Corr(D)
```

(b) *Extracting information from the object pdCSf of class pdCompSymm*

```
> Names(pdCSf)                      # Row/col names
  [1] "agefM" "agefO" "agefY"
> as.matrix(pdCSf)                  # D matrix
        agefM agefO agefY
 agefM      5     1     1
 agefO      1     5     1
 agefY      1     1     5
```

the argument, we obtain the unconstrained coefficients, which result from applying the matrix-logarithm transformation (Sect. 13.5.4). We will explain the computation of the unconstrained coefficients in Panel R14.4.

Panel R14.3b illustrates the method of obtaining coefficients from the object pdCSf. The object is of class *pdCompSymm* and represents a compound-symmetry matrix. First, by setting the argument unconstrained=FALSE, we obtain the standard deviation and the correlation coefficient, which define the compound-symmetry

R14.3 *R syntax*: Extracting coefficients from an object inheriting from a *pdMat* class. Objects pdSm and pdCSf were created in Panel R14.1

(a) *Extracting coefficients from the object pdSm of class pdSymm*

```
> coef(pdSm, unconstrained = FALSE)    # Constrained coefficients
        var((Intercept)) cov(agex,(Intercept))
                       4                     1
               var(agex)
                       9
> coef(pdSm)                           # Unconstrained coefficients
  [1] 0.68424 0.08184 1.09344
```

(b) *Extracting coefficients from the object pdCSf of class pdCompSymm*

```
> coef(pdCSf, unconstrained = FALSE)  # Constrained coefficients
  std. dev     corr.
    2.2361    0.2000
> coef(pdCSf)                          # Unconstrained coefficients
  [1]  0.80472 -0.13353
> log(5)/2                             # First coefficient verified
  [1] 0.80472
> rho <- 0.2                           # ϱ
> nc  <- 3                             # No. of columns
> aux <- (rho + 1/(nc - 1))/(1 - rho) # Modified Fisher's z: (10.35)
> log(aux)                             # Second coefficient verified
  [1] -0.13353
```

structure. The use of the default value of the argument, TRUE, results in two coefficients. The first one is the logarithm of the standard deviation. The second one is the modified Fisher's z-transform (10.35) of the correlation coefficient. The computations of the values of the two unconstrained coefficients are verified at the end of Panel R14.3b.

In Panel R14.4, we illustrate different parameterizations of a general positive-definite matrix, represented by different *pdMat* classes. The parameterizations were described in Sect. 13.5.4.

First, in Panel R14.4a, we show explicitly the link between the unconstrained coefficients of an object of class *pdSymm* and the logarithm of a positive-definite matrix (Sect. 13.5.4). Toward this end, we create the object pdSm0 of class *pdSymm* from the matrix mtxUN and we list the unconstrained coefficients by applying the coef() method to the object. Next, by applying the function pdMatrix(), we obtain the positive-definite matrix, represented by the object, and store it in the object Dmtx. With the help of the chol() function, we compute the Cholesky

R14.4 *R syntax*: Various unconstrained parameterizations of a general positive-definite (variance–covariance) matrix. The matrix mtxUN was created in Panel R14.1

(a) *The matrix-logarithm parameterization – pdSymm class*

```
> pdSm0 <- pdSymm(mtxUN)
> coef(pdSm0)                        # Unconstrained θ_D
  [1] 0.68424 0.08184 1.09344
> Dmtx <- pdMatrix(pdSm0)            # Matrix D
> CholD <- chol(Dmtx)                # Cholesky factor U of D: D=U'U
> vd <- svd(CholD, nu=0)             # SVD of U: (13.46)
> vd$v %*% (log(vd$d) * t(vd$v))     # (13.47)
          [,1]    [,2]
  [1,] 0.68424 0.08184
  [2,] 0.08184 1.09344
```

(b) *The log-Cholesky parameterization – pdLogChol class*

```
> pdLCh <- pdLogChol(mtxUN)
> coef(pdLCh)                        # Unconstrained coefficients θ_D
  [1] 0.69315 1.08453 0.50000
> LChD  <- CholD                     # U
> diag(LChD) <- log(diag(LChD))      # diag(U) log-transformed
> LChD
          [,1]   [,2]
  [1,] 0.69315 0.5000
  [2,] 0.00000 1.0845
```

(c) *The "natural" parameterization – pdNatural class*

```
> pdNat <- pdNatural(mtxUN)
> coef(pdNat)                        # Unconstrained θ_D
  [1] 0.69315 1.09861 0.33647
> log(sqrt(diag(Dmtx)))              # log(SDs)
  [1] 0.69315 1.09861
> corD  <- cov2cor(Dmtx)             # Corr(D)
> rho   <- corD[upper.tri(corD)]     # ϱ_ij (for i<j)
> log((1+rho)/(1-rho))               # Fisher's z: (10.33)
  [1] 0.33647
```

decomposition of the matrix Dmtx and store the resulting Cholesky factor in the object CholD. Then, we apply the function svd() to compute SVD of CholD. The components of the decomposition are stored in the object vd. By extracting the components vd$v and vd$d, we compute the logarithm of the matrix CholD.

The upper-triangular elements of the resulting matrix-logarithm correspond to the unconstrained coefficients obtained by applying the `coef()` method to the object `pdSm0`. It is worth mentioning that the matrix logarithm of the matrix `Dmtx` can be obtained by simply doubling the elements of the matrix logarithm of the matrix `CholD`.

In Panel R14.4b, we show the *pdLogChol* representation of the matrix `mtxUN`. Toward this end, we apply the `pdLogChol()` constructor function to `mtxUN` and display the resulting unconstrained coefficients using the `coef()` method. The representation is based on the Cholesky decomposition of the matrix, obtained with the requirement that the diagonal elements of the resulting Cholesky factor are positive (Sect. 13.5.4). The coefficients are obtained from the elements of the Cholesky factor matrix, but with the diagonal elements replaced by their logarithms. To illustrate the computation of the coefficients explicitly, we reuse the matrix `CholD`, created in Panel R14.4a. Then, we replace the diagonal elements of the resulting matrix, `LChD`, by their logarithms. The upper-triangular elements of the so-obtained matrix correspond to the unconstrained coefficients of the *pdLogChol* representation.

Finally, in Panel R14.4c, we present the *pdNatural* representation of the matrix `mtxUN`. Toward this end, we apply the `pdNatural()` constructor function. The representation is based on the use of standard deviations and correlation coefficients, which correspond to `mtxUN`. The coefficients are obtained by log-transforming the standard deviations and by applying Fisher's z-transform to the correlation coefficients (Sect. 13.5.4). The transformations are shown explicitly in this subpanel. In the process, the function `cov2cor()` is used to compute the correlation matrix, corresponding to `mtxUN` (Sect. 12.3.2), while the function `upper.tri()` is applied to define the correlation coefficients as the upper-triangular elements of the computed correlation matrix.

Following the discussion, presented in Sect. 13.5.4, and the description given above, it is clear that the *pdLogChol*- and *pdSymm*-class representations are suitable for the numerical optimization purposes. On the other hand, the representation used in the *pdNatural* class does not guarantee that the represented matrix is positive definite. Thus, it should not be used in numerical optimization. However, it is suitable for the construction of the confidence intervals for the elements of the matrix, as explained in Sect. 13.7.3.

14.3 Random-Effects Structure Representation: The *reStruct* class

As mentioned in Sect. 14.2, the random-effects structure of an LMM includes the information about the levels of the model hierarchy, the $\mathbf{Z}_i$ design matrices, and the parameterized form of the matrix (or matrices) $\boldsymbol{D}$.

In the package **nlme**, the structure is represented by specialized list-objects of class *reStruct*. Every component of the list is in itself an object of class *pdMat*, corresponding to an appropriate level of model hierarchy.

14.3.1 Constructor Function for the reStruct Class

The function `reStruct()` is a constructor function for an object of class *reStruct*. The arguments of the function include `object`, `pdClass`, `REML`, `data`, `x`, `sigma`, `reEstimates`, and `verbose`. Description of these arguments can be obtained by issuing the command `?reStruct`.

The argument `object` is the most important one. We will describe its use in more detail, because the syntax is very similar to that of the `random` argument of the `lme()` function, which is the key function to fit LMMs in the package **nlme**. The syntax of the `lme()` function will be described in Sect. 14.5.

The essential role of the `object` argument is to pass the information necessary for the specification of the random-effects structure. In particular, the argument is used to provide the information about the model hierarchy and about the formulae associated with the `pdMat` objects, which are later used to create the design matrices $\boldsymbol{Z}_i$. In addition, the argument can be used to specify the information about the structure of the matrix (or matrices) $\boldsymbol{D}$, including the values of their elements.

In Table 14.1, we provide examples of four forms of the syntax that can be used for the argument `object` of the `reStruct()` constructor function. To maintain generality of the presentation, the examples are given for a hypothetical, two-level LMM, as defined in (13.5). We assume that the two levels of grouping are defined by grouping factors `g1` and `g2`. The variables `z1` and `z2`, together with random intercepts, are used as random-effects covariates at the grouping levels defined by `g1` and `g2`, respectively.

All forms of the syntax, shown in Table 14.1, allow a direct specification of the hierarchical structure of the model using grouping factors, such as `g1` and `g2` in our example. However, they differ in the flexibility of specifying other components of the random-effects structure. To illustrate the differences, we consider the use of the variables `z1` and `z2` to introduce random effects associated with covariates.

In Table 14.2, we point to the limitations of the different forms of the syntax, which were presented in Table 14.1. In part (a) of the table, we present an example of syntax for a single-level LMM, with grouping defined by the factor `g1` and a single random-effects covariate `z1`. In part (b) of the table, we show the four forms of the syntax for the same setting as in Table 14.1, i.e., for a two-level LMM.

The syntax (a) is the most flexible. It essentially allows incorporating the information about all components of the random-effects structure, which are supported by the `lme()` function. In particular, for a two-level LMM (see (13.5)), it allows specifying different structures of the $\boldsymbol{D}$ matrices at different levels of the model hierarchy. In the example presented in Table 14.1, the different matrix structures are represented by objects of classes *pdSymm* and *pdDiag*. That is, the matrix $\boldsymbol{D}_1$ is assumed to have a general form, while the matrix $\boldsymbol{D}_2$ is assumed to be diagonal.

Table 14.1 *R syntax*: Syntax[a] for the argument `object` of the `reStruct()` constructor function

Syntax form	Description
(a)	List with named components of class *pdMat* and with grouping factors[b,c] used as names of the components, e.g., `list(g1 = pdSymm(~z1), g2 = pdDiag(~z2))`
(b)	Unnamed list of one-sided formulae with `|` operator, e.g., `list(~z1 | g1, ~z2 | g2)`
(c)	Named list of one-sided formulae without `|` operator, with grouping factors used as names of the components, e.g., `list(g1 = ~z1, g2 = ~z2)`
(d)	One-sided formula with `|` operator, e.g., `~z1	g1/g2`

[a]The examples of the syntax are given for a hypothetical two-level model (13.5)
[b]Variables `z1` and `z2` are used as the random-effects covariates.
[c]Variables `g1` and `g2` are considered grouping factors

Table 14.2 *R syntax*: Limitations of the different forms of the syntax for the `object` argument of the `reStruct()` function

Form	Syntax of the argument	Limitation
(a) A single-level LMM. Grouping factor: `g1`. *Z*-covariate: `z1`		
Form	Syntax of the argument	Limitation
(a)	`list(g1 = pdSymm(~z1))`	Most flexible
(b)	`list(~z1 | g1)`	No structure for $\boldsymbol{D}$; *pdLogChol* class by default[a]
(c)	`list(g1 = ~z1)`	*Same as above*
(d)	`~z1	g1`	*Same as above*
(b) A two-level LMM. Grouping factors: `g1`, `g2`. *Z*-covariates: `z1`, `z2`		
Form	Syntax of the argument	Limitation
(a)	`list(g1 = pdSymm(~z1), g2 = pdDiag(~z2))`	Most flexible
(b)	`list(~z1 | g1, ~z2 | g2)`	The same $\boldsymbol{D}$ structure (*pdLogChol* class) used for both grouping factors[a]
(c)	`list(g1 = ~z1, g2 = ~z2)`	*Same as above*
(d)	`~z1	g1/g2`	*Same as above* Additionally, the same *Z*-covariate(s) for both levels.

[a] The default value of the second argument, `pdClass = "pdLogChol"`, is assumed

The remaining forms of the syntax, (b)–(d), are notationally simpler, but also less flexible, as compared to (a). One complication is that the structure of the matrix (matrices) $\boldsymbol{D}$ has to be determined from the value of another argument of the function `reStruct()`, namely, `pdClass`. By default, the argument specifies the *pdLogChol* class, which results in a general positive-definite matrix. To change this default choice, the argument `pdClass` needs to be specified explicitly, and the call to the `reStruct()` function has to assume the form `reStruct(`*`object,pdClass`*`)`.

For LMMs for data with two or more levels of grouping, an additional limitation of the forms (b)–(d) of the syntax is that the structures of matrices $\boldsymbol{D}$ at different levels of grouping are forced to be *the same*.

A specific limitation of the syntax (d) for multilevel LMMs is that it also requires that the random-effects covariates are assumed to be the same at different grouping levels. For some models, this limitation is irrelevant; however, this is the case for, e.g., LMMs with random intercepts only.

It is worth mentioning that, regardless of the form of the syntax used, the order of specifying the grouping factors is important. More specifically, even if the grouping factors are coded as crossed with each other, they are effectively treated as nested, with the nesting order corresponding to the order, in which the factors are specified in the syntax. In particular, the grouping factors specified later in the syntax are nested within the factors specified earlier. For example, according to the syntax (a) in Table 14.1, the factor g2 would be treated as nested within the factor g1.

14.3.2 Inspecting and Modifying Objects of Class reStruct

In Panel R14.5, we demonstrate how to create and extract information from objects of class *reStruct*. We use the syntax form (a) (Table 14.1) to create the *reStruct*-class object reSt. The object is constructed for a hypothetical two-level LMM, as defined in (13.5). We assume that the two levels of grouping are defined by the grouping factors g1 and g2. The structures of the variance–covariance matrices $\boldsymbol{D}_1$ and $\boldsymbol{D}_2$ of random effects at the two levels of grouping are defined by the objects pdSm of class *pdSymm* and pdCSf of class *pdCompSymm*, respectively.

Using the function isInitialized(), we verify whether the object reSt is initialized. Given that both pdSm and pdCSf were initialized objects that inherited from the *pdMat* class (see Panel R14.1), the resulting *reStruct*-class object is also initialized. By applying the function names(), we get the names of the components of the list, contained in reSt, i.e., the names of factors g1 and g2. The function formula() extracts the formula from each of the components. The displayed formulae correspond to those used in the definition of the objects pdSymm and pdDCf in Panel R14.1.

The function getGroupsFormula() provides information about the grouping of the data, used in the definition of the *reStruct*-class object. It refers to the conditioning expression, i.e., the expression used after the | operator in the formula(e) defining the object (see the syntax forms shown in Table 14.2). In our example, the structure is defined by the factors g1 and g2, with levels of g2 nested within the levels of g1. Note that the function getGroupsFormula() allows two optional arguments, asList and sep. Information about the use of these arguments can be obtained by issuing the command ?getGroupsFormula.

In Panel R14.5, we also apply the function Names() to the object reSt. The function returns the names of rows/columns for the matrices, represented by the *pdMat*-class objects, which define the *reStruct*-class object (see also Panel R14.2).

R14.5 *R syntax*: Creating an object of class *reStruct*, representing a two-level LMM for data with two levels of grouping, and extracting information from the object. Auxiliary objects pdSm and pdCSf, which inherit from the *pdMat* class, were created in Panel R14.1

```
> reSt <- reStruct(list(g1=pdSm,     # D1
+                       g2=pdCSf)) # D2
> isInitialized(reSt)
  [1] TRUE
> names(reSt)                        # Note: order g1, g2 reversed
  [1] "g2" "g1"
> formula(reSt)                      # Formulae for pdMat components
  $g2
  ~-1 + agef

  $g1
  ~agex
> getGroupsFormula(reSt)             # Model hierarchy
  ~g1/g2
  <environment: 0x0000000003d6efd8>
> Names(reSt)                        # Row/col names for pdMat components
  $g2
  [1] "agefM" "agefO" "agefY"

  $g1
  [1] "(Intercept)" "agex"
```

In Panel R14.6, we show the methods of extracting information about the matrices corresponding to the *pdMat*-class objects, which define a *reStruct*-class object. As an example, we use the object reSt, which was created in Panel R14.5. The function as.matrix() used in Panel R14.6a displays the positive-definite matrices, corresponding to the two variance–covariance matrices of random effects at the two levels of grouping. The displayed matrices are, obviously, equivalent to those stored in the objects pdSm and pdCSf, which were used to define the object reSt (see Panel R14.1). By applying the function coef(), we list the unconstrained coefficients corresponding to the matrices. They correspond to the values displayed in Panel R14.3.

The individual *pdMat*-class objects, defining the *reStruct*-class object, can be obtained by extracting the appropriate components of the list, which is contained in the latter object. One possible way to achieve that goal is illustrated in Panel R14.6b. Additionally, using the all.equal() function, we confirm that the object, extracted as the g2 component of reSt, is equivalent to the *pdMat*-class object pdCsf.

R14.6 *R syntax*: Extracting information about *pdMat*-class objects directly from an object of class *reStruct*, representing a two-level LMM for data with two-levels of grouping. The object `reSt`, which inherits from the *reStruct* class, was created in Panel R14.5

(a) *Listing information about positive-definite matrices from a reStruct object*

```
> as.matrix(reSt)                         # D1, D2
  $g1
              (Intercept) agex
  (Intercept)           4    1
  agex                  1    9

  $g2
        agefM agefO agefY
  agefM     5     1     1
  agefO     1     5     1
  agefY     1     1     5
> coef(reSt)                              # Unconstrained coeff. for D2, D1
       g21      g22      g11      g12      g13
   0.80472 -0.13353  0.68424  0.08184  1.09344
```

(b) *Extracting individual pdMat-class components from a reStruct object*

```
> reSt[["g1"]]                            # See pdSm in Panel R14.1b
  Positive definite matrix structure of class pdSymm representing
              (Intercept) agex
  (Intercept)           4    1
  agex                  1    9
> g2.pdMat  <- reSt[["g2"]]               # See pdCSf in Panel R14.1c
> all.equal(pdCSf, g2.pdMat)              # g2.pdMat and pdCSf are equal
  [1] TRUE
```

Panel R14.7 demonstrates how to evaluate an object of class *reStruct* in the context of a dataset. Toward this end, we use data frames `dt1` and `dt2`, which were created in Panel R14.1, together with the object `reSt`, which was created in Panel R14.5.

In Panel R14.7, we first apply the default method of the generic `model.matrix()` function (Sect. 5.3.2) to formulae extracted from the *pdMat*-class objects `pdSm` and `pdCSf`. The formulae are evaluated using the data stored in data frames `dt1` and `dt2`, respectively. The created random-effects design matrices, $\boldsymbol{Z}_1$ and $\boldsymbol{Z}_2$, are stored in the objects `Zmtx1` and `Zmtx2`, respectively, and displayed with the help of the matrix-printing function `prmatrix()`.

Next, we create the random-effects design matrix $\boldsymbol{Z}$ corresponding to the object `reSt`. Toward this end, we first create the data frame `dtz` by merging the data frames `dt1` and `dt2`. Then, we apply the function `model.matrix()` with

R14.7 *R syntax*: Creation of the design matrix **Z** by evaluating an object of class *reStruct* for (hypothetical) data containing random-effects covariates. Objects dt1, dt2, pdSm, and pdCSf were created in Panel R14.1. The object reSt was created in Panel R14.5

```
> Zmtx1 <- model.matrix(formula(pdSm), dt1)
> prmatrix(Zmtx1)                      # Design matrix Z1 for pdSm
  (Intercept) agex
1           1   15
2           1   45
3           1   71
4           1   82
> Zmtx2 <- model.matrix(formula(pdCSf),dt2)
> prmatrix(Zmtx2)                      # Design matrix Z2 for pdCSf
  agefM agefO agefY
1     0     0     1
2     1     0     0
3     0     1     0
4     0     1     0
> dtz  <- data.frame(dt1,dt2)          # Data frame to evaluate reSt
> Zmtx <- model.matrix(reSt, dtz) # Design matrix Z for reSt
> prmatrix(Zmtx)                       # Matrix Z w/out attributes
  g2.agefM g2.agefO g2.agefY g1.(Intercept) g1.agex
1        0        0        1              1      15
2        1        0        0              1      45
3        0        1        0              1      71
4        0        1        0              1      82
```

arguments object=reSt and data=dtz. Note that, because the object reSt is of class *reStruct*, the generic function model.matrix() does *not* dispatch its default method, but the model.matrix.reStruct() method from the **nlme** package. As a result, the random-effects design matrices for the objects pdSm and pdCSf, which define the object reSt, are created and merged. The outcome is stored in the matrix-object Zmtx, which is displayed with the use of the function prmatrix(). Note that, in Zmtx, the three first columns come from the design matrix corresponding to the object pdSm, which was used to define the variance–covariance matrix of random effects present at the level of grouping corresponding to the factor g2. When defining the object reSt, the factor was specified as the second one, after the factor g1 (see Panel R14.5).

It is worth noting that, as compared to the default method of the function model.matrix(), the model.matrix.reStruct() method also allows for an optional argument contrast. The argument can be used to provide a named list of the contrasts, which should be used to decode the factors present in the definition of the *reStruct*-class object. Unless the argument is explicitly used, the default contrast specification is applied (see Sect. 5.3.2).

Table 14.3 *R syntax*: Extracting results from a hypothetical object reSt of class *reStruct*

Random- effects structure component to be extracted	Syntax
Summary	`summary(reSt)`
The *reStruct* formula	`formula(reSt)`
Groups formula	`getGroupsFormula(reSt)`
Constrained coefficients	`coef(reSt, unconstrained=FALSE)`
Unconstrained coefficients	`coef(reSt)`
List of $\boldsymbol{D}$ matrices	`as.matrix(reSt)` `pdMatrix(reSt)`
Log-determinants of $\boldsymbol{D}^{1/2}$ matrices	`logDet(reSt)`

For the reader's convenience, in Table 14.3, we summarize the methods used to extract the information about the components of an *reStruct*-class object.

14.4 The Random Part of the Model Representation: The *lmeStruct* Class

The *lmeStruct* class is an auxiliary class, which allows us to compactly store the information about the random part of an LMM, including the random effects structure, correlation structure, and variance function. Objects of this class are created using the lmeStruct() function with three arguments: reStruct, corStruct, and varStruct. The arguments are given as objects of class *reStruct*, *corStruct*, and *varFunc*, respectively. The classes were described in Sects. 14.3, 11.2, and 8.2, respectively.

The argument reStruct is mandatory, while corStruct and varStruct are optional, with the default value equal to NULL.

The function lmeStruct() returns a list determining the model components. The list contains at least one component, namely, reStruct.

When specifying an LMM with the help of the lme() function, the use of an *lmeStruct*-class object is not needed. Such an object is nevertheless created very early during the execution of the lme()-function call. The importance of the *lmeStruct* class will become more apparent in Sect. 14.6, where we demonstrate how to extract results from an object containing a fit of an LMM.

In Panel R14.8, we demonstrate how to create and extract information from an object of class *lmeStruct*.

First, we create an object of class *reStruct*. Toward this end, we use the reStruct() constructor function (Sect. 14.3.1). The created object, reSt, is the same as the one constructed in Panel R14.5. It defines the random-effects structure of a two-level LMM, with grouping specified by factors g1 and g2 (Sect. 14.3.2). The variance–covariance matrices of random effects at the two levels of grouping are defined by the objects pdSm of class *pdSymm* (a general positive-definite matrix) and pdCSf of class *pdCompSymm* (a compound-symmetry matrix), respectively.

R14.8 *R syntax*: Creating and probing objects of class *lmeStruct*. Objects pdSm and pdCSf, which inherit from the *pdMat* class, were created in Panel R14.1

```
> reSt  <- reStruct(list(g1=pdSm, g2=pdCSf))          # reStruct class
> corSt <- corExp(c(0.3,0.1), form=~tx, nugget=TRUE) # corStruct class
> vF    <- varExp(0.8, form=~agex)                   # varFunc class
> (lmeSt<- lmeStruct(reStruct=reSt, corStruct=corSt, # lmeStruct class
+           varStruct = vF))                         # ... created.
 reStruct  parameters:
      g21      g22      g11      g12      g13
  0.80472 -0.13353  0.68424  0.08184  1.09344
 corStruct  parameters:
  range nugget
    0.3    0.1
 varStruct  parameters:
 expon
   0.8
> coefs <- coef(lmeSt,unconstrained=FALSE)# Constrained coefficients...
> (as.matrix(coefs))                      # ... printed more compactly
                                   [,1]
 reStruct.g2.std. dev           2.23607
 reStruct.g2.corr.              0.20000
 reStruct.g1.var((Intercept))   4.00000
 reStruct.g1.cov(agex,(Intercept)) 1.00000
 reStruct.g1.var(agex)          9.00000
 corStruct.range                1.34986
 corStruct.nugget               0.52498
 varStruct.expon                0.80000
```

In the next step, we create an object of class *corStruct* (Sect. 11.2). As an example, we consider the *corExp* class, which represents the exponential correlation structure (Sect. 11.4.3). The object corSt corresponds to an exponential structure with the range parameter $\varrho = 0.3$ and the nugget equal to 0.1 (see Sects. 10.3.2 and 11.2.1 and Panel R11.1).

Finally, we specify the object vF of class *varFunc* (Sect. 8.2). As an example, we consider the *varExp* class, which represents a variance structure defined by an exponential function of the covariate agex (see Table 7.2 in Sect. 7.3.1).

Using the objects reSt, corSt, and vF as the arguments reStruct, corStruct, and varStruct, respectively, of the lmeStruct() function, we create the object lmeSt of class *lmeStruct*. With the help of the coef() function, combined with the use of the unconstrained=FALSE argument, we display the coefficients, defining the various components of the *lmeStruct*-class object, in the constrained form (see Sects. 8.3, 11.3.1, and 14.2.2).

14.5 Using the Function `lme()` to Specify and Fit Linear Mixed-Effects Models

The generic `lme()` function is the most frequently used function to fit LMMs in R. It allows to specify and fit models described in Sect. 13.2 with nested random effects and correlated and/or heteroscedastic within-group residual errors. In general, to define the LMM in full, we need at least to specify the mean structure and the random-effects structure, including the grouping factors defining model hierarchy. In addition, the correlation structure, variance function, and model frame need to be defined.

In Table 14.4, we summarize selected arguments used by the function `lme()`, together with a reference to the section describing the appropriate syntax and an indication of the implied LMM components. In what follows, we briefly summarize the use of the arguments.

The principal argument `fixed` is primarily used to define the mean structure of an LMM. The argument can accept objects of classes *formula*, *groupedData*, or *lmList*. Depending on the class of the object, the corresponding method of the `lme()` function, i.e., `lme.formula()`, `lme.groupedData()`, or `lme.lmList()`, is used.

The most common choice for the `fixed` argument is a two-sided formula (Sect. 5.2).

The argument can be specified using an object of class *groupedData*. This way allows providing the information about the mean structure and about the model hierarchy defined by (nested) grouping factors. An important limitation of this form of specification of the `fixed` argument is that it allows only for mean structures with one (primary) covariate.

The argument can also be specified by providing an *lmList*-class object, i.e., a list of *lm*-class linear-model-fit objects (Sect. 5.5) for all levels of a grouping factor. This method is rarely used and we do not present it here.

Table 14.4 *R syntax*: Selected arguments of the function `lme()` used to specify a linear mixed-effects model defined in Sect. 13.2

	Argument		
Name	Class	Syntax	Component(s) created/defined
`fixed`	*formula*	Sect. 5.2	Mean structure
	groupedData	Sect. 2.6	Mean structure; grouping factors
	lmList	–	–
`random`	*reStruct*[a]	Sect. 14.3	Random-effects structure; grouping factors (optionally)
`correlation`	*corStruct*[a]	Sect. 11.2	Correlation structure
`weights`	*varFunc*[a]	Sect. 8.2	Variance function
`data`	*data.frame*	Sect. 5.4	Data
	groupedData	Sect. 2.6	Data; grouping factors
`method`		Sect. 5.4	Estimation method

[a]Other choices of the class for the corresponding argument are possible but not listed.

The `random` argument is the primary argument used to define the random-effects structure. In the argument, the syntax forms (a)–(d), shown in Table 14.1, can be used. They allow specifying all aspects of the random-effects structure, *including* the model hierarchy defined by grouping factors.

The specification of the `random` argument can be simplified by omitting the reference to grouping factors in the forms (a)–(c) of syntax from Table 14.1. For example, in the syntax (a), the names of the list components `g1` and `g2` can be omitted. That is, a list with *unnamed* components, i.e., `list(pdSymm(~z1), pdDiag(~z2))`, can be used. The simplified syntax has disadvantages. For instance, it does not include the information about the grouping factors defining the model hierarchy. The information needs to be supplemented using a *groupedData*-class object in the `fixed` or `data` argument.

If the `random` argument is not specified, then, by default, it is assumed that the design matrices for the fixed and random effects are equal ($\boldsymbol{X}_i \equiv \boldsymbol{Z}_i$) and that the variance–covariance matrix for the random effects is defined by an object of class *pdSymm*. That is, a general variance–covariance matrix is assumed. Also in this case, the information about the model hierarchy needs to be supplemented using a *groupedData*-class object in the `fixed` or `data` argument.

The syntax of the arguments `weights` and `correlation` is the same as for the corresponding arguments of the `gls()` function (Sects. 8.4 and 11.5). The two arguments allow to specify the residual variance–covariance matrix $\mathbf{R}_i$, as defined in (13.4), using the decomposition given by (10.8). The default values for the `weights` and `correlation` arguments imply independent and homoscedastic conditional residual errors.

The `data` argument is used to provide the raw data and, optionally, the information about the data hierarchy. Similarly to other model-fitting functions, additional arguments `subset` and `na.action` can be used together with `data` to define the model frame (Sect. 5.3.1). For the definition of these arguments, we refer to Sect. 5.3.1.

The default value of the `method` argument is `method="REML"`. That is, the model parameters are estimated using the restricted likelihood (Sect. 13.5.2). An alternative is `method="ML"`. It is worth mentioning that the initial values for the $\boldsymbol{\theta}_D$ parameters are refined using an EM-based algorithm (Sect. 13.5.6).

14.6 Extracting Information from a Model-Fit Object of Class *lme*

In Table 14.5, we present several methods to extract information from a model-fit object of class *lme*. We assume that an object `lme.fit` is available, which contains the results of fitting a single-level LMM, defined in (13.1)–(13.4).

Table 14.5 *R syntax*: Extracting results from a hypothetical object lme.fit of class *lme*, which represents a single-level linear mixed-effects model fitted using the lme() function

Model-fit component to be extracted	Function: lme() Package: **nlme** Object: lme.fit Class: *lme*
Summary	`(summ <- summary(lme.fit))`
Estimation method	`lme.fit$method`
$\widehat{\boldsymbol{\beta}}$	`fixef(lme.fit)`
$\widehat{\boldsymbol{\beta}}$, se($\widehat{\boldsymbol{\beta}}$), *t*-test	`summ$tTable`
$\widehat{\text{Var}}(\widehat{\boldsymbol{\beta}})$	`vcov(lme.fit)`
95% CI for $\boldsymbol{\beta}$	`intervals(lme.fit, which="fixed")`
$\widehat{\sigma}$	`summ$sigma`
95% CI for $\boldsymbol{\theta}, \sigma$	`intervals(lme.fit, which="var-cov")`
95% CI for $\boldsymbol{\theta}_D$	`intervals(lme.fit, which="var-cov")$reStruct`
$\widehat{\mathbf{b}}_i$	`ranef(lme.fit)`
$\widehat{\boldsymbol{\beta}}$ + "coupled" $\widehat{\mathbf{b}}_i$	`coef(lme.fit)`
	`coef(summ)`
$\widehat{\mathcal{D}}$	`getVarCov(lme.fit)`
$\widehat{\mathcal{D}}$ and $\widehat{\sigma}$	`VarCorr(lme.fit)`
$\widehat{\mathcal{R}}_i$	`getVarCov(lme.fit, type="conditional")`
$\widehat{\mathcal{V}}_i$	`getVarCov(lme.fit, type="marginal")`
ML value	`logLik(lme.fit, REML = FALSE)`
REML value	`logLik(lme.fit, REML = TRUE)`
AIC	`AIC(lme.fit)`
BIC	`BIC(lme.fit)`
Fitted values:	
- conditional, (13.12)	`fitted(lme.fit)`
- marginal, (13.24)	`fitted(lme.fit, level=0)`
Raw residuals:	
- conditional, (13.52)	`resid(lme.fit, type="response")`
- marginal, (13.53)	`resid(lme.fit, type="response", level=0)`
Normalized residuals:	
- conditional, (13.54)	`resid(lme.fit, type="normalized")`
- marginal, (13.55)	–
Pearson residuals, Sect. 7.5.1	`resid(lme.fit, type="pearson")`
Predicted values:	
- conditional	`predict(lme.fit, newdata)`
- marginal	`predict(lme.fit, newdata, level= 0)`

Using the generic function summary() allows us to obtain general information about the fitted form of the model, including the information about the estimated values of the fixed effects, the fitted random-effects structure, and the estimated residual variance–covariance matrix.

If information about only a specific aspect of the fitted model is needed, it can be obtained by extracting a specific component of the model-fit object or by applying a special function to the object. For instance, the estimation method, used to fit the model, is displayed by extracting the `lme.fit$method` component. Estimates of the fixed effects $\boldsymbol{\beta}$ are displayed using the function `fixef()`, while the estimated variance–covariance of the estimates (Sect. 13.5.5) is obtained using the function `vcov()`.

Confidence intervals for the model parameters (see Sect. 13.7.3) are obtained by applying the generic function `intervals()`. Intervals for a specific subgroup of the parameters are selected using the argument `which`. For instance, `which="fixed"` provides CIs for the fixed effects, while `which="var-cov"` yields the intervals for all variance–covariance parameters. By default, `which="all"`, i.e., CIs for all model parameters are provided. The confidence level can be chosen with the help of the `level` argument. By default, `level=0.95`. The result of the application of the function `intervals()` to a model-fit object of class *lme* is a list with named components. Each of the components is a data frame, with rows corresponding to the parameters of the model, and columns representing the estimated values and the confidence limits for the parameters. The possible components are the following: `fixed` (fixed effects), `reStruct` (parameters of the variance-covariance matrices of the random effects), `corStruct` (residual correlation-structure parameters), `varFunc` (residual variance-function parameters), and `sigma` (scale parameter). In Table 14.5, we present how to display CIs only for the parameters of the variance–covariance matrices of the random effects by extracting the `reStruct` component of the object resulting from the `intervals()` function call.

By applying the function `ranef()` to a model-fit object of class *lme*, the estimated random effects are displayed. By default, the effects at all levels of grouping are displayed. The levels can be selected with the help of the `level` argument. Information about other arguments, available for the function `ranef()`, can be obtained by issuing the command `?ranef`.

The function `coef()`, applied to an *lme*-class model-fit object, displays the estimated coefficients for a particular (or all) levels of grouping. The coefficients are obtained by summing the fixed effects and, if appropriate, the "coupled" random effects (Sect. 13.2.1). The levels can be selected with the help of the `level` argument. Information about other arguments of the function `coef()` can be obtained by issuing command `?coef.lme`.

Estimates of the variance–covariance matrices of the random effects and residual errors, as well as the marginal variance–covariance matrix, are obtained using the function `getVarCov()`. The argument `type` allows choosing the matrix to be displayed. In particular, `type="random.effects"` (the default) prints out the estimates of the variance–covariance matrices of random effects, `type="conditional"` prints out the estimate of the residual variance–covariance matrix, and `type="marginal"` provides the estimate of the marginal variance-covariance matrix. With the help of the `individuals` argument, it is possible to select the group(s) of observations, for which the function `getVarCov()` should display the (residual or marginal) variance–covariance matrices.

An alternative method to extract the variance–covariance matrix of the random effects is to use the function `VarCorr()`. When applied to an *lme*-class model-fit object, the function extracts the estimated variances, standard deviations, and correlations of the random effects. Additionally, it provides the estimates of σ^2 and σ. The function uses three arguments: `x`, `sigma`, and `rdig`. The first one specifies the model-fit object; the second one is an optional numeric value that indicates a multiplier for the standard deviations and assumes the value of 1 as default; and the last one is an optional integer value, which indicates the number of digits (by default, 3) that are to be used to represent the correlation estimates.

Fitted values, residuals, and predicted values are obtained by applying the functions `fitted()`, `resid()`, and `predict()`, respectively. All the functions allow for an optional argument `level`, in the form of a vector of integers, which indicates the level(s) of grouping, for which the values are to be extracted. The levels increase from 0, i.e., the population level, to the highest level of grouping, i.e., the level corresponding to the grouping factor, which is nested within all the other factors. Thus, `level=0` yields the estimates of the marginal mean values or marginal residuals, while for nonzero levels, the conditional mean values or conditional residuals are provided. In particular, the conditional mean values at a particular level, k say, are obtained by adding the marginal mean values and the predictors of the random effects at the grouping levels lower or equal to k. The conditional residuals at level k are obtained by subtracting the conditional mean values at that level from the dependent-variable vector. By default, `level` specifies the highest level of grouping.

An important argument of the function `resid()` is `type`. It indicates the type of residuals to be computed (Sect. 13.6.2). The possible choices are `type="response"` (raw residuals), `type="pearson"` (Pearson residuals), and `type="normalized"` (normalized residuals). It is worth mentioning that the Pearson/normalized residuals are standardized/transformed based on the elements of the residual variance–covariance matrix $\widehat{\mathcal{R}}_i$, and not on the marginal variance–covariance matrix $\widehat{\mathcal{V}}_i$. Hence, the use of arguments `type="pearson"` or `type="normalized"` in combination with a nondefault value of `level` argument is not meaningful. This remark applies, for example, to marginal residuals obtained using `level=0`.

The function `predict()` allows for an optional argument `newdata`. It indicates a data frame for which the predictions are to be calculated. The data frame should contain all variables that were used to specify the fixed effects and the random effects of the fitted LMM, as well as the grouping factors. If the argument is missing, the function will employ data used to fit the model. Consequently, it returns the fitted values corresponding to the level specified in the `level` argument.

In Table 14.6, we present methods of extracting the details about the `lme()`-function call which was used to create a model-fit object of class *lme*. The methods are similar to those presented in Tables 5.5, 8.2, and 11.1 for a *gls*-class model-fit object. Note that the function `model.matrix()` (Sect. 5.3.2) provides the design matrix for the mean structure, i.e., the matrix $\boldsymbol{X}_i$. Extracting the design matrix for the random effects, $\boldsymbol{Z}_i$, from an *lme*-class model-fit object is difficult. As it requires extra programming, we do not present the required code.

Table 14.6 *R syntax*: Extracting components of the `lme()`-function call from a hypothetical object `lme.fit`, which represents a fitted single-level linear mixed-effect model

R call component	Syntax
R call	`(cl <- getCall(lme.fit))`
Formula for the mean structure	`(form <- formula(lme.fit))`
Argument `random`	`cl$random`
Argument `correlation`	`cl$correlation`
Argument `weights`	`cl$weights`
Data name	`(df.name <- cl$data)`
Data frame	`(df <- eval(df.name))`
Model frame	`(mf <- model.frame(form, df))`
Design matrix	`model.matrix(form, mf)`

Table 14.7 *R syntax*: Extracting information about the random components from a hypothetical object `lme.fit`, which represents a fitted single-level linear mixed-effect model

Auxiliary objects to be extracted (R class)	Syntax
Random part of the model (*lmeStruct*) *see* Section 14.4	`lmeSt <- lme.fit$modelStruct`
Random-effects structure (*reStruct*) *see* Section 14.3.2	`reSt <- lmeSt$reStruct`
Variance-function structure (*varFunc*) *see* Table 8.2b	`vF <- lmeSt$varStruct`
Correlation structure (*corStruct*) *see* Table 11.1b	`cSt <- lmeSt$corStruct`

In Table 14.7, we summarize methods to extract information about the random components of a fitted LMM. As mentioned in Sect. 14.4, the random part of the model, which includes the random effects structure, the residual correlation structure, and the residual variance function, is represented by an object of class *lmeStruct*. The object can be accessed by referring to the `modelStruct` component of the *lme*-class model-fit object. The random effects structure, as described in Sect. 14.3, is represented by an object of class *reStruct*, which is stored as the `reStruct` component of the *lmeStruct*-class object. If a correlation structure and/or a variance function were used in defining the residual variance–covariance matrix of the LMM, they are represented by objects of classes *corStruct* and *varFunc*, respectively, which are stored as components `corStruct` and `varStruct`, respectively, of the *lmeStruct*-class object.

14.7 Tests of Hypotheses About the Model Parameters

As was the case for LMs for independent, heteroscedastic observations (Sect. 8.5) or fixed-effects LMs for correlated data (Sect. 11.6), results of the F-tests for linear hypotheses about the fixed effects (Sect. 13.7.1), based on a fitted LMM, are accessed by applying the `anova()` method to the model-fit object of class *lme*. By default, the sequential-approach tests are obtained (Sect. 4.7.1). To obtain the marginal-approach tests, the argument `type="marginal"` should be used. F-tests for individual/multiple terms can be obtained by applying the argument `Terms` in the form of an integer vector or a character vector that specifies the terms in the model that should be jointly tested. If a character vector is used, it should contain the names of the terms used in the model formula. If an integer vector is used, its elements should correspond to the order in which terms are included in the model formula. Additional arguments that can be used in the `anova()` method for *lme*-class model-fit objects include `test`, `adjustSigma`, `L`, and `verbose`. The information about the use of these arguments can be obtained by issuing the command `?anova.lme`.

The t-tests for individual coefficients are provided by applying, e.g., the `summary()` method to the model-fit object. Note that, in this case, the marginal-approach tests are obtained.

As was discussed in Sect. 13.7.1, neither the p values for the F-tests, nor the ones for the t-tests, adjust for the fact that the null distribution of the test statistics is only approximated by F- or t-distributions, respectively. Thus, the degrees of freedom for the tests are computed as in a balanced, multilevel ANOVA design (Schluchter and Elashoff 1990; Pinheiro and Bates 2000). In particular, assuming G levels of grouping, the number of denominator degrees of freedom ddf_g for the tests of fixed effects at level g ($g = 1, \ldots, G+1$) is equal to

$$ddf_g = N_g - (N_{g-1} + p_g), \tag{14.1}$$

where N_g is the number of groups at the g-th grouping level and p_g is the number of fixed-effects coefficients estimated at that level. The latter is the number of coefficients related to the variables whose values change across the values of the grouping factor(s) at the grouping level g, but do not change across the values of the grouping factor(s) at the level $g-1$. Note that the intercept, if present in the model, is treated as being estimated at the level $g = 0$, but its denominator degrees of freedom are calculated from the level $G+1$, i.e., at the level of observations. An example of the calculation of the denominator degrees of freedom is presented in Sect. 16.7.1.

When the function `anova()` is applied to two or more objects of class *lme*, it provides LR-test statistics, calculated based on pairs of the LMMs represented by the consecutive objects. If the models are nested, have the same structure of random effects and of residual variance–covariance matrix, and are fitted using ML, the results of the LR tests, reported by the function, provide valid tests for hypotheses about the fixed effects (Sect. 7.6.1). On the other hand, if the nested models are fitted using REML and have the same mean structure, but different random structures, the reported LR tests are valid tests of hypotheses about the parameters defining random-effects structure.

In Sect. 13.7.1 it was mentioned that, instead of using a χ^2 distribution for an LR test of a hypothesis about fixed effects, one could use an empirical distribution of the test statistic, obtained by fitting the alternative and null models to multiple datasets simulated under the null model. A similar comment was made in Sect. 13.7.2 for the LR test of hypotheses about random effects, when the values of the variance–covariance parameters, compatible with the null hypothesis, lie on the boundary of the parameter space. To address this issue, the function `simulate()` from the package **nlme** can be used. It computes the ML- and/or REML-based log-likelihood values for multiple datasets simulated from the null and alternative LMMs. This allows the calculation of the empirical distribution of the LR-test statistic.

The function admits the following arguments: `object`, `m2`, `nsim`, `method`, `seed`, `niterEM`, and `useGen`. The first four are the most important ones and we describe them in more detail below. A description of all of the arguments can be obtained by issuing the command `?simulate`.

The argument `object` defines the null model. The argument can be provided either as an object of class *lme*, which represents a fitted LMM, or as a named list with components `fixed`, `data`, and `random`, which should define a valid call of the function `lme()` to fit the LMM (Sect. 14.5). The argument `m2` defines the alternative model and can be specified in a similar way as the `object` argument. If it is specified as a list, only those components that change between the null and alternative models need to be specified. The argument `nsim` is a positive integer which indicates the number of simulations to be performed. By default, `nsim=1`. Thus, although the arguments is optional, in practice, it should be always specified.

Finally, the argument `method`, which is an optional character array, allows choosing the form of the likelihood on which the LR-test statistic is to be based. By default, `method=c("REML", "ML")`, i.e., both ML- and REML-based LR-test statistics are used.

The function returns an object of class *simulate.lme*, which is a named list with two components: `null` and `alt`. Each of them has components `ML` and/or `REML`, which are matrices. The matrices contain, in particular, the column `logLik` which provides the ML- or REML-based log-likelihood value for each of the `nsim` simulations. Additional attributes of the *simulate.lme*-class object include, among others, `seed` and `df`. The former gives the random seed used in the random number generator, while the latter gives the difference in the number of parameters between the null and alternative models.

One way to present the result of the `simulate()`-function call is to plot the empirical and nominal p values. The former are obtained from the empirical distribution of the LR-test statistic values corresponding to the simulated values of the ML- or REML-based log-likelihood, while the latter are computed from applying a χ^2 distribution or a mixture of χ^2 distributions to the simulated values of the LR-test statistic. The plot can be obtained by a call like `plot(`*`object`*`,`*`df`*`)`, where *`object`* is a *simulate.lme*-class object, while *`df`* is a vector of integers, which defines the degrees of freedom of a χ^2 distribution to be used to compute the nominal p values. If the vector contains more than one integer, multiple plots of nominal *versus* empirical p values are created by computing the nominal p values from the χ^2 distribution with the number of degrees of freedom equal to each of the integers. Additionally, a plot for an equal-weight mixture of the χ^2 distributions is created.

An example of the use of the function `simulate()` is provided in Sect. 16.6.2.

For the case of using the REML-based LR test for testing a hypothesis about the random-effects structure, a possible alternative is the function `exactRLRT()` from the package **RLRsim**. The function simulates values of the REML-based LR-test statistic for testing the null hypothesis that the variance of a random effect is 0 in an LMM with a known correlation structure of the tested random effect and independent and identically distributed random errors. The simulations are based on the finite-sample distribution of the test statistic (Sect. 13.7.2) which was derived by Crainiceanu and Ruppert (2004). The performance of the simulations was studied by Scheipl (2010).

The main arguments of the function `exactRLRT()` are `m`, `m0`, and `mA`. The first one is a model-fit object of class *lme* or *lmer*. For LMMs with a single variance component (random effect), it provides the fitted model under the alternative hypothesis. For models with multiple variance components, it should provide the model containing only the random effect to be tested. Arguments `mA` and `m0` apply only to models with multiple variance components. The former specifies the model fitted under the alternative hypothesis, while the latter gives the model fitted under the null hypothesis. Additional arguments include `nsim`, which is used to specify the number of values of the test statistic to be simulated. By default, `nsim=10000`. The list of all arguments of the function `exactRLRT()` can be obtained by issuing the command `?exactRLRT` (after attaching the package **RLRsim**). An example of the use of the function `exactRLRT()` is provided in Sect. 16.6.1.

It is worth noting that the functions `simulate()` and `exactRLRT()` have important limitations. For instance, they both only apply to conditional independence LMMs (Sect. 13.4). Additionally, the function `exactRLRT()` allows only for independent random effects; `simulate()` can accommodate correlated random effects, i.e., LMMs with nondiagonal variance–covariance matrices of random effects $\boldsymbol{D}$.

Finally, it is worth mentioning that the function `anova()`, when applied to two or more objects of class *lme*, also provides the information criteria (Sect. 4.7.2) that can be used to choose the best-fitting models from a set of nonnested models with different mean and/or variance–covariance structures. The AIC and BIC can also be obtained using the functions `AIC()` and `BIC()`, respectively (see Table 14.5).

14.8 Chapter Summary

In this chapter, we presented the tools available for fitting LMMs in the R package **nlme**. In particular, we focused on the function `lme()`.

The use of the function involves the concepts of model formula, grouped data, variance function, and correlation structure. The concepts were introduced in the previous chapters in the context of simpler LMs to facilitate their description and explanation. A new, important component was the random-effects structure.

We described several tools related to it, including the *pdMat* class for representing positive-definite matrices (Sect. 14.2) and the *reStruct* class for representing the random-effects structure (Sect. 14.3). In Sect. 14.4, we explained the representation of the random part of an LMM in the form of objects of class *lmeStruct*. The objects are created when the function `lme()` is used to fit an LMM (Sect. 14.5). In Sect. 14.6, we described how to extract information about the estimated model from an *lme*-class model-fit object. Finally, in Sect. 14.7, we briefly reviewed the tools available for inference based on a fitted LMM.

The use of the R tools presented in this chapter will be illustrated in Chap. 16, where the application of LMMs to the analysis of the ARMD case study will be described.

It is worth noting that, as was mentioned in Sect. 14.3.1, when a *reStruct*-class object is created, the grouping factors are effectively treated as nested. This means that the function `lme()` is not particularly suitable to fit LMMs with, e.g., crossed random effects. Such models can be easily fitted by applying the function `lmer()` from the package **lme4.0**, which we present in the next chapter.

Chapter 15
Fitting Linear Mixed-Effects Models: The `lmer()` Function

15.1 Introduction

In Chap. 14, we introduced the `lme()` function from the **nlme** package. The function is a popular and well-established tool to fit LMMs. It is especially suitable for fitting LMMs to data with hierarchies defined by nested grouping factors.

In the current chapter, we present the function `lmer()` from the package **lme4.0**. The function is especially suitable to fit LMMs with crossed random effects. It can fit LMMs to data with hierarchies defined by nested grouping factors, too.

There are several important differences between the two functions. An important, technical difference is that `lme()` has been programmed in the S3 system (Sect. 1.2), while `lmer()` has been implemented in the S4 system. As a consequence, the methods of extracting results for an LMM fitted by applying the `lmer()` function are different than the methods used for the *lme*-class model-fit objects. Another important difference is that `lmer()` employs computations based on sparse matrices implemented in **Matrix** package. Consequently, it can be used for large-scale computational problems, requiring high speed of calculations and efficient storage of the data.

In what follows, we describe in more detail the use of the function `lmer()` to fit LMMs. Note that, because of space restrictions, we cannot extensively discuss all the technical issues related to the use of the S4 system features in the implementation of `lmer()`. Thus, we limit ourselves to providing only the necessary information, which can be used by the reader as a guideline for further reading and study.

The chapter is organized as follows. In Sect. 15.2, we briefly introduce LMMs with crossed random effects. Section 15.3 presents the syntax used by `lmer()` to specify LMMs. The structure of the model-fit objects and the methods of extracting results from the objects are described in Sect. 15.4. Inferential tools available for the LMMs that were fitted with the use of the `lmer()` function are discussed in

A. Gałecki and T. Burzykowski, *Linear Mixed-Effects Models Using R: A Step-by-Step Approach*, Springer Texts in Statistics, DOI 10.1007/978-1-4614-3900-4_15,

Sect. 15.5. In Sect. 15.6, we describe some details related to the implementation of the computational algorithms used by `lmer()` in the **lme4.0** package. A summary of the chapter is provided in Sect. 15.7.

15.2 Specification of Models with Crossed and Nested Random Effects

In this section, we describe the specification of LMMs with crossed random effects and compare it to the specification of LMMs with nested random effects. The material presented in this section will help us to understand the syntax used by the `lmer()` function.

It is worth noting that, strictly speaking, we should use the terms "LMMs with random effects associated with levels of crossed/nested grouping factors" rather than "LMMs with crossed/nested random effects." However, the latter terms are commonly used in the LMM literature. For this reason, we will adopt them.

To explain the difference between the nested and crossed random effects, it is worthwhile to introduce a concrete example. Thus, we will consider hypothetical experiments aimed at assessing the precision of machines which cut shapes from steel plates. In the experiments, each machine, from a randomly selected group of N machines, cuts $n > 1$ times the same shape out from each of P steel plates. Let us denote by y_{ijs} the precision measurement obtained for the s-th shape ($s = 1,\ldots,n$) from the j-th plate ($j = 1,\ldots,P$) for the i-th machine ($i = 1,\ldots,N$). Thus, for each machine, we obtain in total $P \cdot n$ measurements.

We may be interested in assessing the influence of the effect of machine and plate on the measurements. Given that the machines and plates are selected at random, we should treat their effects as random. In what follows, we will first consider a hypothetical experiment involving the effects of plates that are nested within machines. Then we will consider an experiment, in which the effects of plates and machines are crossed.

15.2.1 A Hypothetical Experiment with the Effects of Plates Nested Within Machines

Let us imagine that the experiment is run so that each of the P series of n shapes for each machine is obtained from a *different* plate. We can then propose the following model for the measurements generated in the experiment:

$$y_{ijs} = \mu + b_{1,i} + b_{2,ij} + \varepsilon_{ijs}, \tag{15.1}$$

where $b_{1,i} \sim \mathcal{N}(0, d_M)$ is the random effect corresponding to machine i, $b_{2,ij} \sim \mathcal{N}(0, d_P)$ is the random effect corresponding to plate j *specific* to machine i

and independent of $b_{1,i}$, and $\varepsilon_{ijs} \sim \mathcal{N}(0, \sigma^2)$ is the residual (measurement) error, independent of both $b_{1,i}$ and $b_{2,ij}$. Note that the plate effects are specific to, i.e., *nested* within, each machine. As a result, the model (15.1) includes $N \cdot P$ plate effects. To indicate the nesting, we use the index ij in the symbolic representation of the random plate effect $b_{2,ij}$ in (15.1).

The resulting marginal variances and covariances are as follows:

$$\begin{aligned} \mathrm{Var}(y_{ijs}) &= d_M + d_P + \sigma^2, \\ \mathrm{Cov}(y_{ijs}, y_{ijs'}) &= d_M + d_P, \\ \mathrm{Cov}(y_{ijs}, y_{ij's}) = \mathrm{Cov}(y_{ijs}, y_{ij's'}) &= d_M, \\ \mathrm{Cov}(y_{ijs}, y_{i'js}) = \mathrm{Cov}(y_{ijs}, y_{i'js'}) &= 0, \qquad (15.2) \\ \mathrm{Cov}(y_{ijs}, y_{i'j's}) = \mathrm{Cov}(y_{ijs}, y_{i'j's'}) &= 0. \qquad (15.3) \end{aligned}$$

Note that from (15.2) and (15.3) it follows that the measurements for different machines, indexed by i, are independent. Thus, model (15.1) can be written in the form, presented in (13.5), upon putting $\boldsymbol{\beta} = \mu$, $\boldsymbol{X}_{ij} = \boldsymbol{Z}_{1,ij} \equiv \boldsymbol{Z}_{1,*} = \mathbf{1}_n$ (a column vector of n 1s), $\boldsymbol{Z}_{2,ij} \equiv \boldsymbol{Z}_{2,*} = \boldsymbol{I}_n$ ($n \times n$ identity matrix), and assuming that $\mathbf{b}_i \sim \mathcal{N}_1(0, d_M)$, $\mathbf{b}_{ij} \sim \mathcal{N}_m(0, d_P \boldsymbol{I}_n)$, and $\boldsymbol{\varepsilon}_{ij} \sim \mathcal{N}_n(0, \sigma^2 \boldsymbol{I}_n)$.

Moreover, model (15.1) can also be expressed in the form specified in (13.6)–(13.9) with block-diagonal matrices $\boldsymbol{Z}$, $\mathbf{D}$, and $\boldsymbol{R}$, where the blocks of the matrices are defined, respectively, by $\boldsymbol{Z}_i = (\mathbf{1}_m \otimes \boldsymbol{Z}_{1,*}, \boldsymbol{I}_m \otimes \boldsymbol{Z}_{2,*}) = (\mathbf{1}_{n \cdot m}, \boldsymbol{I}_m \otimes \mathbf{1}_n)$, $\boldsymbol{D} = \mathrm{diag}(d_M/\sigma^2, \{d_P/\sigma^2\}\boldsymbol{I}_m)$, and $\boldsymbol{R}_i = \boldsymbol{I}_{n \cdot m}/\sigma^2$. Note that the resulting matrix $\mathbf{D} = \sigma^2 \boldsymbol{I}_N \otimes \boldsymbol{D}$ is of dimension $\{N \cdot (m+1)\} \times \{N \cdot (m+1)\}$.

15.2.2 A Hypothetical Experiment with the Effects of Plates Crossed with the Effects of Machines

Let us now assume that the experiment is run with only P steel plates, so that the first n shapes for each machine are obtained from (the same) plate 1, the second n shapes from (the same) plate 2, and so on. We can then propose the following model for the measurements generated in the experiment:

$$y_{ijs} = \mu + b_{1,i} + b_{2,j} + \varepsilon_{ijs}, \qquad (15.4)$$

where $b_{1,i} \sim N(0, d_M)$ is the random effect corresponding to machine i, $b_{2,j} \sim \mathcal{N}(0, d_P)$ is the random effect corresponding to plate j and independent of $b_{1,i}$, and $\varepsilon_{ijs} \sim \mathcal{N}(0, \sigma^2)$ is the residual (measurement) error independent of both $b_{1,i}$ and $b_{2,j}$. Note that, as compared to (15.1), the plate effects are no longer specific to machines, but each plate effect b_j remains the same for all machine effects b_i. We can say that the effects are *crossed*. As a result, model (15.4) includes only m plate effects. To indicate the crossing of the random effects, we use the index j in the symbolic representation of the random plate effect $b_{2,j}$ in (15.4).

The marginal variances and covariances, corresponding to the model (15.4), are as follows:

$$
\begin{aligned}
\text{Var}(y_{ijs}) &= d_M + d_P + \sigma^2, \\
\text{Cov}(y_{ijs}, y_{ijs'}) &= d_M + d_P, \\
\text{Cov}(y_{ijs}, y_{ij's}) = \text{Cov}(y_{ijs}, y_{ij's'}) &= d_M, \\
\text{Cov}(y_{ijs}, y_{i'js}) = \text{Cov}(y_{ijs}, y_{i'js'}) &= d_P, \qquad (15.5) \\
\text{Cov}(y_{ijs}, y_{i'j's}) = \text{Cov}(y_{ijs}, y_{i'j's'}) &= 0. \qquad (15.6)
\end{aligned}
$$

In contrast to (15.2) and (15.3), the equation (15.5) implies that the measurements for different machines are correlated. Hence, the model (15.4) cannot be written in the form, presented in (13.5). It can be expressed in the form specified in (13.6)–(13.9), but with a non-block-diagonal matrix $\mathbf{Z}$, given by $\mathbf{Z} = (\boldsymbol{I}_N \otimes \mathbf{1}_{n \cdot m}, \mathbf{1}_N \otimes \{\boldsymbol{I}_m \otimes \mathbf{1}_n\})$. Note that the resulting matrix $\mathsf{D} = \text{diag}(d_M \boldsymbol{I}_N, d_P \boldsymbol{I}_m)$ is block diagonal, but of dimension $(N+m) \times (N+m)$ rather than $\{N \cdot (m+1)\} \times \{N \cdot (m+1)\}$, as was the case for the model (15.1).

The hypothetical examples presented above illustrate that the group-level formulation of an LMM, as described in Sect. 13.2.1, can only be used in the case of LMMs with nested random effects. It is generally not appropriate when crossed random effects are included in the model. On the other hand, the whole-data formulation, presented in Sect. 13.2.2, can be used for LMMs with nested and/or crossed random effects. From the example it is clear, however, that we need to pay attention to the notation, as it should allow indicating which effects are nested and which are crossed.

15.2.3 General Case

We continue with the notational considerations in a more general case.

Toward this end, we will follow the observation that, for the identification of nested random effects, the use of compound indices is needed, as was the case for the model (15.1). More specifically, for that model, we may use the following notation for the model formulation for all data:

$$
\mathbf{y} = \boldsymbol{X}\boldsymbol{\beta} + \mathbf{Z}_1 \mathbf{b}_1 + \mathbf{Z}_{12} \mathbf{b}_{12} + \boldsymbol{\varepsilon}, \qquad (15.7)
$$

where $\mathbf{b}_1$ and $\mathbf{b}_{12}$ are vectors of random machine effects and plate effects, respectively. By using the single subscript in $\mathbf{b}_1$, we indicate that the random effects contained in the vector $\mathbf{b}_1$ are related to the levels of the first grouping factor, while by using the double subscript in $\mathbf{b}_{12}$, we indicate that the random effects contained in the vector $\mathbf{b}_{12}$ are related to the levels of the second grouping factor and are nested within the levels of the first factor. Note that $\mathbf{b}_1$ and $\mathbf{b}_{12}$ are constructed by "stacking", for all machines, the vectors of random machine and plate effects, respectively, in a way similar to how the vector $\mathbf{b}$ was constructed in Sect. 13.2.2. Also, the block-diagonal matrices $\boldsymbol{Z}_1$ and $\boldsymbol{Z}_{12}$ are constructed in a way similar to the matrix $\boldsymbol{Z}$ in (13.6).

On the other hand, for the model (15.4), we use the following notation for the formulation of the model for all data:

$$\mathbf{y} = \boldsymbol{X}\boldsymbol{\beta} + \boldsymbol{Z}_1\mathbf{b}_1 + \boldsymbol{Z}_2\mathbf{b}_2 + \boldsymbol{\varepsilon}, \tag{15.8}$$

where we indicate that the random machine effects, contained in the vector $\mathbf{b}_1$, are related to the levels of the first grouping factor, while the random plate effects, contained in the vector $\mathbf{b}_2$, are related to the levels of the second grouping factor and are crossed with the random effects $\mathbf{b}_1$.

Following this convention, the following hypothetical model equation:

$$\mathbf{y} = \boldsymbol{X}\boldsymbol{\beta} + \boldsymbol{Z}_1\mathbf{b}_1 + \boldsymbol{Z}_{12}\mathbf{b}_{12} + \boldsymbol{Z}_{13}\mathbf{b}_{13} + \boldsymbol{\varepsilon}, \tag{15.9}$$

would be interpreted as implying an LMM with three sets of random effects: the first set, represented by the vector $\mathbf{b}_1$, related to the levels of the grouping factor indexed by the first index; the second set, represented by the vector $\mathbf{b}_{12}$, related to the levels of the grouping factor indexed by the second index, with the effects nested within the levels of the first factor; and the third set, represented by the vector $\mathbf{b}_{13}$, related to the levels of the grouping factor indexed by the third index, with the effects nested within the levels of the first factor. Note that the random effects contained in the vectors $\mathbf{b}_{12}$ and $\mathbf{b}_{13}$ are crossed within the levels of the first grouping factor.

Common assumptions for the random-effects structure for models of the types presented in (15.7)–(15.9) are:

- Random effects associated with different grouping factors are independent
- Random effects associated with different levels of a grouping factor are independent
- The vector of random effects associated with a given level of a grouping factor has (typically) a general positive semidefinite variance-covariance matrix, which is common to all the levels of the grouping factor.

Obviously, each of the model specifications, defined in (15.7)–(15.9), can be expressed in the general form, given in (13.7), upon appropriate concatenation of the $\boldsymbol{Z}$ matrices and stacking of the random-effects vectors. Note that the $\boldsymbol{Z}$ matrix in (13.7) would no longer be block diagonal for models (15.8) and (15.9). However, the disadvantage of the general form is that the information about crossing and/or nesting of random effects, readily accessible from the notation used in the specifications (15.7)–(15.9), is hidden inside the structure of the $\boldsymbol{Z}$ matrix.

As we have already mentioned, LMMs with crossed random effects are an important class of LMMs, which can be fitted with the help of the function `lmer()` from the package **lme4.0**. In the next sections, we describe the specific features of the function.

15.3 Using the Function lmer() to Specify and Fit Linear Mixed-Effects Models

The use of the lmer() function to fit an LMM usually involves a call like lmer(*formula*, *data*), where *formula* specifies the model we want to fit, and *data* indicates the data frame containing the data. In this respect, the use of the function lmer() is more similar to the use of the function lm() (Sect. 5.4) than lme() (Sect. 14.5). However, as it will be explained later, the syntax of the formula argument used in the lmer() function is very much different from the one used in the lm() function.

In lmer(), the model is specified by the formula argument. As is the case for most model-fitting functions in R, this is the first argument. The second argument is data. It is advisable that the data frame, indicated in the argument, is not a *groupedData*-class object. This is in contrast to the function lme() (Sect. 14.5).

Another important argument, relevant for the estimation of LMMs, is REML. It is logical and is used to specify whether the ML- or REML-based estimation is to be applied (Sect. 13.5.3). The default value is REML=TRUE, indicating the use of the REML estimation.

Additional useful arguments are subset, na.action, weight, offset, and contrasts. Their meaning and use is the same as for the lm() function (Sect. 5.4). In particular, subset and na.action are used to specify rows of the data that are used to build the model frame, while weight is used to assign *known* weights to observations. The meaning of the offset argument was explained in Sect. 5.4.

A full list of the arguments of the lmer() function is obtained by issuing, after attaching the package **lme4.0**, the command ?lmer from R's command prompt.

The form of the formula, used in the lmer() function, differs from the one used to specify LMMs in the lme() function and from the one described in Sect. 5.2. Thus, we will discuss it now.

15.3.1 The lmer() *Formula*

The primary goal of an lmer() formula is to express *both* the fixed- and random-effects structures of an LMM using *one* expression. The syntax is of the form:

R expression ~ *termX.1* +···+ *termX.k* + (*termZ.1*) +···+.(*termZ.l*)

Similar to the classical two-sided formula, described in Sect. 5.2, the lmer() formula consists of two expressions separated by the ~ (tilde) symbol. As in the classical formula, the expression on the left, which is typically the name of a variable, is evaluated as the continuous response.

Similar to the classical formula, the right-hand side of the formula consists of one or more terms, separated by the + symbols. The main difference is that there are two types of terms, namely, X- and Z-terms. The X-terms, *termX.i*, are used to specify the fixed-effects part of the model. They have the same syntax as in the classical formula (Sect. 5.2).

In contrast, the Z-terms, *(termZ.j)*, are enclosed in parentheses and are used to specify the part of the model involving random effects. Every Z-term is associated with a grouping factor. The term's syntax is of the form *(Zf | gf)*, where *Zf* and *gf* are two expressions enclosed in parentheses and separated by the vertical bar | symbol, which can be read as "given" or "by". The expression *Zf* on the left of the | symbol is a linear-model expression used to specify the model matrices for random effects. The expression's syntax is essentially the same as for right-hand side of the classical formula. The expression *gf*, to the right of the | symbol, is evaluated to a (grouping) factor.

The structure of the `lmer()` formula clearly corresponds to the notation introduced in Sect. 15.2. Every Z-term in the formula corresponds to a **Zb** term used in a specification similar to the one used in (15.7)–(15.9). It generates the **Z** matrices for a model specified as in (15.7)–(15.9) or contributes a set of columns to the **Z** matrix in the general formulation (13.7).

In Table 15.1, we present examples of syntax of the Z-terms, *(termZ.j)*, used in the `lmer()` formulae to specify random-effects structures in a single-level LMM. For the reader's reference, we also provide the corresponding syntax for the `random` argument of the `lme()` function (Sect. 14.5).

The first example of syntax in Table 15.1, labeled as (1a), shows the `lmer()` formula for a simple single-level LMM with random intercepts. The Z-term `(1|g1)` indicates the use of random intercepts by specifying `1` to the left of the | symbol. The random intercepts correspond to the levels of the grouping factor `g1`, as indicated by specifying `g1` to the right of the | symbol. In this respect, the syntax is similar to the one used for specifying random effects in the `random` argument of the `lme()` function (Sect. 14.3.1).

The syntax (1b) illustrates a specification of a single-level LMM with random intercepts and slopes. Note that the formula of the Z-term `(z1|g1)` implicitly assumes an inclusion of an intercept. Thus, the Z-term defines, in fact, two random effects. As mentioned earlier (Sect. 14.2), it is also implicitly assumed that the 2×2 variance-covariance matrix of the two random effects is of a general form, i.e., allowing random intercepts and slopes to have different variances and to be correlated.

The syntax (1c), on the other hand, illustrates how to obtain a diagonal variance-covariance matrix for random intercepts and slopes in a single-level LMM. This is achieved by specifying separate Z-terms for the intercepts and the slopes, but using the same grouping factor. Moreover, in the Z-term `(0+z1|g1)`, we indicate that we do not want to include an intercept. As a result, the Z-term includes in the model equation only the random slopes, which are assumed to be independent of the random intercepts, specified in the other Z-term.

Table 15.1 Basic examples of the Z-terms syntax used in the lmer() formulae for single-level linear mixed-effects models and the corresponding syntax for the random argument of the lme() function

Syntax Z-term(s) in the lmer() formula[a]	The random argument of the lme() function	Comment
(1a) (1|g1)	~1|g1	Random intercepts only 1×1 $\boldsymbol{D}$ matrix
(1b) (z1|g1)	~z1|g1	Random intercepts and slopes General 2×2 $\boldsymbol{D}$ matrix
(1c) (1|g1) + (0 + z1|g1)	pdDiag(~z1|g1)	Random intercepts and slopes Diagonal 2×2 $\boldsymbol{D}$ matrix

[a]The variable z1 is a random-effects covariate and g1 is a grouping factor

The syntax shown in Table 15.1 can be easily modified to accommodate more than two random effects. Note that in the table we demonstrate how to specify a general and diagonal variance-covariance matrix of random effects. The current version of the **lme4.0** package does not allow for other structures. In particular, it does not allow for a compound-symmetry structure of the matrix $\boldsymbol{D}$.

In Table 15.2, we present additional examples of Z-terms used in the lmer() formula for models with nested and/or crossed random effects.

For the sake of simplicity of presentation, we illustrate the syntax for models with random intercepts only. Needless to say, the presented examples, combined with the syntax given in Table 15.1, can be generalized for other models, including those containing different sets of random effects.

First, in examples (2a)–(2c), we present the syntax for specifying a two-level LMM with nested random intercepts. The implied random-effects structure corresponds to the one specified in the models defined by (15.1) or (15.7). As mentioned in Sect. 2.6, factors representing nested design can be coded in two different ways. Consequently, the model in question can be specified using different syntax.

The simpler syntax (2a) can be used only if the grouping factors involved in defining nested hierarchy, say g1 and g12, are explicitly coded as nested. In contrast, the syntaxes (2b) and (2c) do not impose any requirements regarding the coding of the factors. Instead, in (2b), the expression g1:g2 in the second Z-term is evaluated to an auxiliary factor factor(g1:g2), which, by definition, is nested within g1. The version (2c) simply expands to (2b).

Note that the form of the random argument of the lme() function, corresponding to (2a)–(2c), is relatively simple and uniform.

To specify an LMM with two crossed random effects, syntax (3) is used. The random-effects structure of the model corresponds to the one implied by (15.4) or, more generally, by (15.8). Obviously, this syntax requires that the factors g1 and g2 are coded as crossed. Although there is no direct way to specify this model in the syntax of an lme()-function call, in Sect. 19.4 we will demonstrate how to overcome this limitation.

Examples (4a) and (4b) in Table 15.2 present the syntax for a two-level LMM including both nested and crossed random effects. The random-effects structure

Table 15.2 Additional examples of the Z-terms syntax for linear mixed-effects models with nested and/or crossed effects. The corresponding syntax for the `random` argument of the `lme()` function is also given. We also assume that grouping factors `g1` and `g2` are coded as crossed

<table>
<tr><th>Random effects in the model</th><th>Syntax label</th><th>Z-terms in the lmer() formula</th><th>The random argument of the lme() function</th></tr>
<tr><td>Nested (15.7)</td><td>(2a)</td><td>(1|g1) + (1|g12)[a]</td><td>~1|g1/g12</td></tr>
<tr><td></td><td>(2b)</td><td>(1|g1) + (1|g1:g2)</td><td>~1|g1/g2</td></tr>
<tr><td></td><td>(2c)</td><td>(1|g1/g2)</td><td>~1|g1/g2</td></tr>
<tr><td>Crossed (15.8)</td><td>(3)</td><td>(1|g1) + (1|g2)</td><td>Not definable[b]</td></tr>
<tr><td>Crossed/ nested (15.9)</td><td>(4a)</td><td>(1|g1) +
(1|g12) +
(1|g13)[a]</td><td>list(g1 =
pdBlocked(list(
pdIdent(~1),
pdIdent(~1|g12),
pdIdent(~1|g13)
)))</td></tr>
<tr><td></td><td>(4b)</td><td>(1|g1) +
(1|g1:g2) +
(1|g1:g3)</td><td>list(g1 =
pdBlocked(list(
pdIdent(~1),
pdIdent(~1|g2),
pdIdent(~1|g3)
)))</td></tr>
</table>

[a]Factors `g12` and `g13` are explicitly coded as nested within `g1`
[b]In Sect. 19.4, we demonstrate how to circumvent this limitation

corresponds to the one implied by (15.9). Similar to the syntax (2a) for the two-level LMM with nested random effects, (4a) requires that the factors `g12` and `g13` are coded as nested within `g1`. On the other hand, the syntax (4b) does not impose any requirements regarding the coding of the factors. Instead, using the Z-term `(1|g1)` together with `(1|g1:g2)` and `(1|g1:g3)`, allows us to explicitly specify that the random effects for the grouping factors defined in the second and third Z-term are nested in the factor `g1`. Note that the random-effects structure of the model can, actually, be specified using the `random` argument of the `lme()` function, but with a rather complex syntax.

As illustrated in Table 15.2, for LMMs involving nested random effects, different syntax of the Z-terms in the `lmer()`-function model formula can be used. In general, it is recommended to choose the simpler syntax when writing the R code. This would imply choosing the syntax (2a) or (4a). However, if, for any reason, the factor `g12`, used in (2a) and (4a), is not coded as nested, the model formula will be mistakenly interpreted by the function `lmer()` as specifying an LMM with crossed effects. Hence, for reasons of transparency and unambiguity, one might actually prefer the syntax (2b) or (4b) because, in that case, the nesting is explicit in the formula, and not hidden in the structure of the data.

Finally, it is worth reiterating that all LMMs, which can be specified using the `lmer()`-model formula, are conditional independence models with homoscedastic residual errors (Sect. 13.4). This is an important limitation, as compared to the set of LMMs available for fitting with the use of the function `lme()`. It is possible that in the future versions of the package **lme4** this limitation may be removed.

15.4 Extracting Information from a Model-Fit Object of Class *mer*

Results of fitting an LMM using the `lmer()` function from the **lme4.0** package are stored in an object of class *mer*. The class represents fits of mixed-effects models, including linear, generalized linear, or nonlinear mixed-effects models. Information about the structure of the objects of class *mer* can be obtained by issuing the command `help("mer-class")`. An important part of the structure is the *slots*, which contain results for the specific components of the fitted model. The list of names of the slots for *mer*-class objects can be obtained by applying the command `slotNames("mer")`. Note that this structure results from the fact that the *mer*-class objects are part of the S4 system (Sect 1.2).

In Table 15.3, we present several methods to extract information from a hypothetical model-fit object `mer.fit` of class *mer*.

It can be observed that many methods, listed in Table 15.3, correspond to those used for extracting information from a model-fit object of class *lme* (see Table 14.5). Thus, a description of their use can be found in Sect. 14.6.

Several differences between Tables 14.5 and 15.3 can be noted, though. For instance, a method for the `intervals()` generic function is not available for the *mer*-class objects. Thus, to obtain confidence intervals for the parameters of interest, additional programming is necessary. In contrast, a `sigma()` extractor function is available for *mer*-class objects with no corresponding method for *lme*-class objects. Other method functions available for both classes of objects, e.g., `fitted()` and `residuals()`, do not necessarily use the same set of arguments. Several other components of the *mer*-class objects that are not accessible by using the standard method functions can be extracted by applying the `getME()` function. The function uses two arguments. The first is the name of the *mer*-class model-fit object, while the second one is a character string specifying the name of the component to be extracted. For example, the estimates of the $\boldsymbol{\theta}_D$ parameters are obtained by applying the command `getME(mer.fit,"theta")`.

In Table 15.3, we also list the methods that can be used to extract the details of the model formulation. For instance, to obtain the `lmer()`-function call, used to fit an LMM, we can use the `getCall()` function to extract the `"call"` slot of the *mer*-class object. The formula used to specify the design matrices for mean- and random-effects structures of the model is obtained by applying the generic function `formula()` to the *mer*-class object.

There are several useful plot functions, applicable to *mer*-class objects. For instance, the function `plot()`, applied to the result of the application of the function

Table 15.3 Extracting results from a hypothetical object `mer.fit` of class *mer*, which represents a linear mixed-effects model, fitted using the `lmer()` function

Model fit component to be extracted	Function: `lmer()` Package: **lme4.0** Object: `mer.fit` Class: *mer*
Summary	`(summ <- summary(mer.fit))`
Print	`show(mer.fit)`
Estimation method	`isREML(mer.fit)`
$\widehat{\boldsymbol{\beta}}$	`fixef(mer.fit)`
$\widehat{\boldsymbol{\beta}}$, se($\widehat{\boldsymbol{\beta}}$), t-test	`coef(summ)`
$\widehat{\text{Var}}(\widehat{\boldsymbol{\beta}})$	`vcov(mer.fit)`
$\widehat{\boldsymbol{\theta}}_D$	`getME(mer.fit, "theta")`
$\widehat{\sigma}$	`sigma(mer.fit)`
$\widehat{\mathbf{b}}_i$	`ranef(mer.fit)`
$\widehat{\boldsymbol{\beta}}$ + coupled $\widehat{\mathbf{b}}_i$	`coef(mer.fit)`
$\widehat{\mathcal{D}}$ and $\widehat{\sigma}$	`VarCorr(mer.fit)`
ML value	`logLik(mer.fit, REML = FALSE)`
REML value	`logLik(mer.fit, REML = TRUE)`
Deviance	`deviance(mer.fit)`
Fitted values:	
- subject-specific	`fitted(mer.fit)`
Raw residuals:	
- subject-specific	`residuals(mer.fit)`
Predicted values:	*Not implemented*
R call	`(cl <- getCall(mer.fit))`
`lmer()` formula	`formula(mer.fit)`
Data name	`(df.name <- cl$data)`
Data frame	`eval(df.name)`
Design matrix	`model.matrix(mer.fit)`

of `ranef()` to a *mer*-class object produces a normal Q-Q plot (or scatterplot) of the predicted random effects for the grouping factors. Similarly, the use of `plot()` for the result of the application of `coef()` to a *mer*-class object produces a normal Q-Q plot (or scatterplot) of the predicted random coefficients, i.e., of the sums of the estimated random effects and "coupled" random-effects predictors. Examples of these plots are given in Figs. 18.9, 19.1, and 19.2.

Another useful graphical function is `dotplot()`. When used in combination with, e.g., the function `ranef()`, it produces a dotplot, i.e., the plot of the (ordered) predicted random effects (EBLUPs) for each level of a grouping factor. The EBLUPs are used as x-coordinates, while the corresponding levels of the grouping factor are shown on the y-axis. The plot allows checking whether there are levels of a grouping factor with extremely large or small predicted random effects. Examples of these plots are given in Fig. 19.3.

15.5 Tests of Hypotheses About the Model Parameters

One of the important differences in using functions lme() and lmer() for fitting LMMs is that the latter does not automatically provide any p values for the statistical-significance tests based on the fitted model.

Thus, to compute the asymptotic p values for the F- or LR tests for the fixed effects (Sect. 13.7.1), manipulation of the results, extracted from a *mer*-class model-fit object, is needed. This applies also to the LR tests for the variance-covariance parameters (Sect. 13.7.2).

As an alternative, empirical p values, based on simulations, may be used. In fact, one can argue that, given the problems with the approximation of the asymptotic null distribution of the tests statistics for the fixed effects (Sects. 7.6.1 and 13.7.1) and for the variance-covariance parameters (Sect. 13.7.2), the use of the simulation-based estimates of p values might actually be more appropriate. A disadvantage is that the approach is numerically more involved.

To compute the empirical p values, the simulate() function (Sect. 14.7) can be applied to a *mer*-class model-fit object. Note that the simulate.mer() method, suitable for *mer* class objects, is different, in many respects, from the simulate.lme() method, available for the *lme*-class objects. Most importantly, simulate.mer() simulates values of the dependent variable for a fitted model. In contrast, simulate.lme() simulates log-likelihood values for the null and alternative models (Sect. 14.7).

Efficient use of the simulate.mer() method is enhanced by combining it with the application of the function refit(). The latter is a generic function which fits a model to a new response vector. For a model fitted using the lmer() function, it is much faster to refit the model with the help of the function refit() than to do it using the model formula. Application of refit() usually involves a call of the form refit(*object*, *newresp*). The first argument, *object*, is a model-fit object returned by the function lmer(). The second argument, *newresp*, is, typically, a numerical vector containing new values of the dependent variable to which the model is to be refitted. Note that the model specification, including the formula, covariates, and their values, etc., stays the same as in the *object*. Only the values of the dependent variable change.

The use of simulate.mer() typically involves the following syntax/sequence of commands:

```
> simY    <- simulate(mer.fit, nsim)
> simInfo <- apply(simY, 2,
+       function(y){
+       simFit <- refit(mer.fit, y)
+       summ   <- summary(simFit)
        ... extract and return info from summ object
+  })
```

First, the object simY is created by applying the function simulate() to the *mer*-class model-fit object mer.fit. The object simY is a matrix with rows corresponding to the rows of the data frame used to fit the model and nsim columns, which contain the simulated values of the dependent variable. In other words, the number of rows in simY is determined by the number of rows in the model frame, and the number of columns corresponds to the number of simulations nsim.

Second, the object simInfo is created using the apply() function to refit the model, represented by the mer-class object, to each of the columns of the simY matrix. This is achieved by applying repeatedly the function refit() and storing the results. Note that refitting of the model is time-consuming. Therefore, it is recommended that all the required information about the refitted models is extracted for further processing and stored in the auxiliary object simInfo. For instance, the values of the fixed-effects test statistics should be stored. Based on these values, the null distribution of the test statistics can be approximated.

An illustration of the application of the simulate() function is given, e.g., in Sect. 16.7.1.

Empirical p values for tests of hypotheses about variance-covariance parameters can also be obtained using the function exactRLRT() from the package **RLRsim** (Sect. 14.7).

15.6 Illustration of Computations

In this section, we illustrate several aspects of the implementation of the PnLS estimation approach (Sect. 13.5.3) in the function lmer(). In Panel R15.1, we simulate data for an LMM with crossed random effects. Toward this end, we first use the function gl() to create two factors, i and j, each of length 6. The former has n1=2 levels, and the latter has n2=3 levels. The factors play the role of indices of the levels of two grouping factors with crossed random effects. The random effects are stored in the numerical vectors b1x and b2x. The former contains n1=2 values, generated according to the standard normal distribution. The latter contains n2=3 values, generated according to the mean-zero normal distribution with standard deviation 2.

Next, we create the data frame dt0, with factors i and j as the only variables. Then, we construct the data frame dtc by adding the following variables to the data frame dt0: eps, b1, b2, y, g2, and g1. The variable eps contains six random values generated from the normal distribution with mean zero and standard deviation 0.2. The variables b1 and b2 are created from the random-effects variables b1x and b2x, respectively, by replicating the appropriate random values according to the levels of the factors i and j, respectively. The variable y is a dependent variable, resulting from the following LMM with crossed random effects:

$$y = 10 + b_1 + b_2 + \varepsilon.$$

Finally, the variables g1 and g2 are factors corresponding to i and j, respectively, with the levels labeled by letters.

R15.1 *R syntax*: Simulating data for a linear mixed-effects model with two crossed random effects

```
> n1 <- 2                    # Number of levels for the factor g1
> n2 <- 3                    # Number of levels for the factor g2
> i <- gl(n1, n2)            # i
> j <- gl(n2, 1, n1*n2)      # j
> b1x <- rnorm(n1, 0, 1)     # b_i
> b2x <- rnorm(n2, 0, 2)     # b_j
> dt0 <- data.frame(i, j)
> (dtc <-
+   within(dt0,
+          {                 # g1 and g2 are crossed
+           eps <- rnorm(nrow(dt0), 0, 0.2)
+           b1 <- b1x[i]
+           b2 <- b2x[j]
+           y <- 10 + b1 + b2 + eps
+           g2 <- factor(j, labels = letters[1:n2])
+           g1 <- factor(LETTERS[i])
+           }))
  i j g1 g2       y        b2      b1       eps
1 1 1  A  a 10.1444 -0.055214 0.13972  0.059933
2 1 2  A  b 12.2306  2.012640 0.13972  0.078235
3 1 3  A  c  9.3669 -0.601585 0.13972 -0.171216
4 2 1  B  a 11.2304 -0.055214 1.25311  0.032463
5 2 2  B  b 13.1089  2.012640 1.25311 -0.156847
6 2 3  B  c 10.8498 -0.601585 1.25311  0.198232
```

R15.2 *R syntax*: Constructing the random-effects design matrices and the matrix $AA' + I$ for further processing

```
> Zg1 <- model.matrix(~ 0 + g1, data = dtc)  # Z_1 for g1
> Zg2 <- model.matrix(~ 0 + g2, data = dtc)  # Z_2 for g2
> Z0 <- cbind(Zg1, Zg2)                       # Z for g1 and g2
> A0 <- t(Z0)                                 # A = Z'
> A0c <- tcrossprod(A0)                       # AA'
> Dg <- diag(nrow(A0))
> (A0q <- A0c + Dg)                           # AA' + I
    g1A g1B g2a g2b g2c
g1A   4   0   1   1   1
g1B   0   4   1   1   1
g2a   1   1   3   0   0
g2b   1   1   0   3   0
g2c   1   1   0   0   3
```

R15.3 *R syntax*: The number of nonzero elements in the Cholesky factor without and with permutation (S3 system). Objects (matrices) A0 and A0q were created in Panel R15.2

(a) *Cholesky factor without permutation*

```
> L0 <- t(chol(A0q))                # L such that LL' = AA' + I
> sum(L0 != 0.0)                    # Count of nonzero elements
  [1] 14
> max(abs(L0 %*% t(L0)- A0q))       # Verify LL' = AA' + I
  [1] 4.4409e-16
```

(b) *Permutation achieved in two different ways*

```
> pvec <- c(3, 4, 5, 1, 2)          # Permutation vector
> A1    <- A0[pvec, ]               # Rows permuted in A
> A1c   <- tcrossprod(A1)
> (A1q <- A1c + Dg)                 # AA' + I (permuted)
       g2a g2b g2c g1A g1B
  g2a    3   0   0   1   1
  g2b    0   3   0   1   1
  g2c    0   0   3   1   1
  g1A    1   1   1   4   0
  g1B    1   1   1   0   4
> A1q. <- A0q[pvec, pvec]           # Cols and rows permuted in AA'
> identical(A1q, A1q.)
  [1] TRUE
> L1    <-  t(chol(A1q.))           # LL'= AA' + I (permuted)
> sum(L1 != 0.0)                    # Count of nonzero elements
  [1] 12
```

In Panel R15.2, we illustrate the creation of the random-effects design matrices corresponding to the factors `g1` and `g2`. Toward this end, we use the data frame `dtc`, created in Panel R15.1, and the function `model.matrix()` (Sect. 5.3.2). The matrices `Zg1` and `Zg2` are then combined column-wise to obtain the matrix `Z0`, which becomes the random-effects design matrix corresponding to the formulation of the LMM for all data, as in (13.7). Using the matrix `Z0` allows us to create the matrix `A0q`, which, upon scaling by the scale parameter, provides the marginal variance-covariance matrix for all data, as defined in (13.35).

In Panel R15.3, we present in more detail the use of the sparse Cholesky decomposition to reduce storage requirements and numerical complexity related to fitting LMMs (Sect. 13.5.3). In particular, we use S3-system functions to illustrate the issue.

First, in Panel R15.3a, we compute an ordinary Cholesky decomposition of the matrix `A0q` using the function `chol()`. The number of nonzero elements in the resulting lower-triangular Cholesky-factor matrix is equal to 14.

Then, in Panel R15.3b, we illustrate how the number of nonzero elements in the Cholesky-factor matrix can be reduced using a permutation. In fact, we present two methods to use the permutation toward this end.

To present the first method, we create the matrix A1 by permuting the rows of the matrix A0. Then, we compute the permuted counterpart of the matrix A0q.

To present the second method, we create the matrix A1q. by directly permuting the rows and columns of the matrix A0q.

Note that resulting matrices A1q and A1q. are identical. Thus, their lower-triangular Cholesky-factor matrices are identical, too. As can be seen from the last two lines of the code, presented in Panel R15.3b, the Cholesky-factor matrices contain only 12 nonzero elements, as compared to 14 nonzero elements for the Cholesky-factor matrix of A0q.

In Panel R15.4, we present the use of the sparse Cholesky decomposition by applying S4 functions and classes defined in the package **Matrix**.

First, in Panel R15.4a, we investigate the number of nonzero elements in the "ordinary" Cholesky-factor matrix. Toward this end, we coerce the matrix A0 to become an object of class *dgCMatrix*. This is a class of sparse numeric matrices for which nonzero elements in the columns are sorted into increasing row order. The *dgCMatrix* class is the "standard" class for sparse numeric matrices in the **Matrix** package (Bates and Maechler 2012). More information about the class can be obtained by issuing the command help("dgCMatrix-class").

By using the sparse-matrix representation of A0, we compute, with the help of the function tcrossprod(), the matrix A0c, which is the cross product of A0 and its transpose. Then, we apply the function Cholesky() to compute L0, the Cholesky-factor matrix of A0c. By default, the function uses a fill-reducing permutation and applies it to the rows and columns of the sparse matrix, specified as the first argument of the function. To prevent the use of the permutation, we change the value of the perm argument to FALSE. Note that we specify Imult=1. This means that we want a decomposition for the sum of the matrix A0c and of an identity matrix. By default, the numerical argument Imult is equal to 0, implying the decomposition of only the sparse matrix, specified as the first argument of the function. Finally, by putting LDL=FALSE, we use the $\boldsymbol{LL}'$ form of the Cholesky decomposition. By default, LDL=TRUE and the decomposition of the form $\boldsymbol{LDL}'$ is used, where $\boldsymbol{L}$ is a unit lower-triangular matrix, as in (13.33). More information about the function Cholesky() can be obtained by issuing the command ?Cholesky.

With the help of the function nnzero(), we verify that the number of nonzero elements in the Cholesky-factor matrix L0 is equal to 14, as it was the case shown in Panel R15.3a. Note that we apply the function to the object L0. which is a sparse-matrix representation of the matrix L0. The function nnzero() returns the number of nonzero values of a numeric-like R object. More information about the function can be obtained by issuing the command ?nnzero.

Note that the matrix A0q, equal to the sum of the cross-product matrix A0c and an identity matrix, shown in Panel R15.4a, is a permuted version of the matrix A1q, shown in Panel R15.3b. In the last line of the code of Panel R15.4a, we show that the matrix L0 is indeed the Cholesky-factor matrix of A0q.

R15.4 *R syntax*: The number of nonzero elements in a Cholesky factor without and with permutation (using the **Matrix** package in the S4 system). The object (matrix) `A0` was created in Panel R15.2

(a) *Cholesky factor without permutation*

```
> library(Matrix)
> A0 <- as(A0, "dgCMatrix")                    # A0 matrix coerced to sparse
> A0c <- tcrossprod(A0)                        # AA'
> L0 <- Cholesky(A0c, perm = FALSE, Imult = 1, LDL = FALSE)
> nnzero(L0. <- as(L0, "sparseMatrix"))        # Coerced to verify
  [1] 14
> Dg <- Diagonal(nrow(A0))
> (A0q   <- A0c + Dg)
  5 x 5 sparse matrix of class "dsCMatrix"
      g1A g1B g2a g2b g2c
  g1A   4   .   1   1   1
  g1B   .   4   1   1   1
  g2a   1   1   3   .   .
  g2b   1   1   .   3   .
  g2c   1   1   .   .   3
> max(abs(L0. %*% t(L0.) - A0q))               # LL' = AA' + I
  [1] 4.4409e-16
```

(b) *Permutation of rows and columns of the matrix* $\boldsymbol{AA}'$. *The same as in R15.3b*

```
> pvec <-  c(3, 4, 5, 1, 2)                    # Permutation vector
> P1   <-  as(pvec, "pMatrix")                 # Permutation matrix
> A1c  <-  P1 %*% A0c %*% t(P1)
> L1   <-  Cholesky(A1c, perm = FALSE, Imult = 1, LDL = FALSE)
> nnzero(as(L1, "sparseMatrix"))
  [1] 12
```

(c) *Suboptimal permutation obtained using the* `Cholesky()` function

```
> L2 <-  Cholesky(A0c, perm=TRUE, Imult =1, LDL = FALSE)
> nnzero(as(L2, "sparseMatrix"))
  [1] 13
> slot(L2,"perm") + 1L                         # Permutation
  [1] 5 1 3 4 2
> detach(package:Matrix)
```

In Panel R15.4b, we illustrate how the number of nonzero elements in the Cholesky-factor matrix can be reduced using a permutation of the rows and columns of the cross-product matrix. Toward this end, we first create a permutation matrix

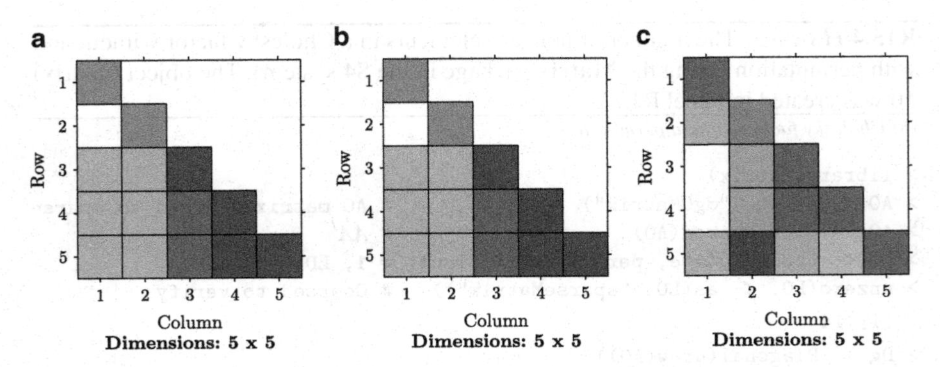

Fig. 15.1 Illustration of the number of nonzero elements in a Cholesky factor for three different permutations of the rows and columns of the matrix $\boldsymbol{AA}' + \boldsymbol{I}$. (**a**) No permutation (**b**) (34512) (**c**) (51342)

P1, corresponding to the required permutation. The matrix is obtained by coercing a permutation vector pvec into an object of class *pMatrix*. The class represents permutation matrices, stored as one-dimensional permutation vectors. We then permute the rows and columns of the matrix A0c using the matrix P1 and store the resulting matrix as the object A1c. Finally, we compute L1, the Cholesky-factor matrix of A1c, and check that it contains only 12 nonzero elements.

In Panel R15.4c, we illustrate an alternative way of using permutations to reduce the number of nonzero elements in a Cholesky-factor matrix. Toward this end, when applying the function Cholesky() to the matrix A0c, we set the argument perm equal to TRUE. This implies that a fill-reducing permutation is computed and applied to the rows and columns of A0c. Note that the resulting Cholesky-factor matrix L2 contains 13 nonzero elements. This indicates that the permutation, chosen automatically by the function, is not optimal. We extract the permutation vector from the slot perm of the object L2. The resulting vector is different from the vector pvec, which was defined in Panel R15.4b.

In Fig. 15.1, we present the structures of the Cholesky-factor matrices, which were computed in Panels R15.3 and R15.4. Figure 15.1a shows the structure of the Cholesky-factor matrix obtained for the matrix A0q, defined in Panel R15.2, without applying any permutation. Thus, the structure corresponds to the matrix L0, computed in Panels R15.3a and R15.4a. It contains 14 nonzero elements, with their magnitude indicated by different shades of gray. Figure 15.1b shows the structure of the Cholesky-factor matrix, obtained by applying the permutation 34512. It corresponds to the matrix L1, computed in Panels R15.3b and R15.4b. The structure contains 12 nonzero elements. Finally, Fig. 15.1c shows the structure of the Cholesky-factor matrix, obtained by applying the permutation 51342. It corresponds to the matrix L2, computed in Panel R15.4c, and contains 13 nonzero elements.

As it can be seen from Panels R15.3 and R15.4 and Fig. 15.1, identification of the optimal "fill-reducing" permutation matrix that can be used to reduce the required storage volume and numerical complexity is not that obvious. In our example, involving a small amount of data, the gain in storage obtained by using the "fill-reducing" permutation matrix appears to be fairly modest. However, the gain is much more apparent for large data.

R15.5 A linear mixed-effects model with crossed random intercepts, fitted to the simulated data. The data frame `dtc` was created in Panel R15.1

```
> require(lme4.0)
> fmc <- lmer(y ~ 1 + (1|g1) + (1|g2), data = dtc)
> summary(fmc)
  Linear mixed model fit by REML
  Formula: y ~ 1 + (1 | g1) + (1 | g2)
     Data: dtc
    AIC  BIC logLik deviance REMLdev
   21.1 20.3  -6.56     14.8    13.1
  Random effects:
   Groups   Name        Variance Std.Dev.
   g2       (Intercept) 1.7807   1.334
   g1       (Intercept) 0.6444   0.803
   Residual             0.0472   0.217
  Number of obs: 6, groups: g2, 3; g1, 2

  Fixed effects:
              Estimate Std. Error t value
  (Intercept)   11.155      0.961    11.6
> gf <- getME(fmc, "flist")       # Grouping factors
> xtabs(~g1 + g2, gf)             # g1 and g2 fully crossed
     g2
  g1  a b c
    A 1 1 1
    B 1 1 1
> (Zt <- getME(fmc, "Zt"))        # Z'
  5 x 6 sparse matrix of class "dgCMatrix"

  a 1 . . 1 . .
  b . 1 . . 1 .
  c . . 1 . . 1
  A 1 1 1 . . .
  B . . . 1 1 1
```

In Panel R15.5, we fit an LMM with a fixed overall intercept and crossed random intercepts for the grouping factors `g1` and `g2` (see Panel R15.1) to the simulated data from the data frame `dtc`. We store the results in the model-fit object `fmc` of class *mer*. A summary of the results is displayed by applying the generic function `summary()` to the model-fit object (Sect. 15.4).

By extracting the `flist` slot of the object `fmc` with the help of the `getME()` function (Sect. 15.4), we obtain a list of values of the grouping factors and store it in the data frame `gf`. The latter object is, in turn, used in the function `xtabs()` to obtain a contingency table for the factors `g1` and `g2`. The displayed table contains one observation in each cell, indicating that the levels of the two factors (and the corresponding random intercepts) are completely crossed.

By referring, with the help of the `getME()` function, to the `Zt` slot of the object `fmc`, we extract and display the transpose of the random-effects design matrix. Note that the matrix is stored as an object of class *dgCMatrix* (see the explanation of the code in Panel R15.4). The sparse structure of the matrix is clearly seen.

Note that, although the syntax included in Panel R15.5 can also be executed in the **lme4** package, we require the use of **lme4.0**. In this way, the model-fit object `fmc` of class *mer* is created, as required by the code used in in the next two panels, i.e., R15.6 and R15.7.

Panel R15.6 presents an approach to extract matrices involved in the decompositions (13.33) and (13.38) from an object of class `mer`. A list of the $\boldsymbol{S}$ and $\boldsymbol{T}$ factors involved in the decomposition (13.33) of the variance-covariance matrix for each random-effects term in the model formula is stored in the `ST` slot of the `mer`-class object. The unit lower-triangular matrix, $\boldsymbol{T}$, and the diagonal matrix, $\boldsymbol{S}$, for each term are stored as a single matrix with diagonal elements from $\boldsymbol{S}$ and off-diagonal elements from $\boldsymbol{T}$.

In Panel R15.6, we first use the function `expand()` to obtain the list of terms in the expansion of the `ST` slot of the object `fmc`. We store the expansion in the object `STs`. Note that, apart from the *mer*-class object to expand, the function `expand()` admits one extra argument, `sparse`. By default, `sparse=TRUE`, in which case the elements of the list are the numeric scalar `sigma`, which contains the REML or ML estimate of the scale parameter, and three sparse matrices: `P`, the permutation matrix involved in the decomposition (13.38); `S`, the diagonal scale matrix involved in (13.33); and `T`, the unit lower-triangular matrix involved in (13.33). These components are listed in Panel R15.6 by applying the generic function `summary()` to the object `STs`. When `sparse=FALSE`, each element of the list, produced by the `expand()` function, is the expansion of the corresponding element of the `ST` slot into a list of `S`, the diagonal matrix, and `T`, the (dense) unit lower-triangular matrix.

After displaying the contents of the object `STs`, in Panel R15.6, we display the components `P`, `S`, and `T` of `STs`. Note that the permutation matrix `P` implies the permutation 31245.

In Panel R15.7, we further illustrate the details of the implementation of the PnLS estimation approach, described in Sect. 13.5.3.

Toward this end, we first create the product of the matrices $\boldsymbol{S}$ and $\boldsymbol{T}$. Then, by referring to the component `sigma` of the object `STs`, i.e., of the expanded `ST` slot of the model-fit object `fmc`, we extract the estimate of the scale parameter σ. The extracted value corresponds to the one displayed in Panel R15.5. With the help of the function `tcrossprod()`, we compute the cross-product $\boldsymbol{TSST}'$ and multiply it by σ^2. As a result, and in accordance with the equations (13.9) and (13.33), we

R15.6 *R syntax*: Extracting information about the matrices involved in decompositions (13.33) and (13.38) from a model-fit object of class *mer*. The model-fit object `fmc` was created in Panel R15.5

```
> STs <- expand(fmc)          # Expand the ST slot
> summary(STs)
        Length Class     Mode
  sigma  1     -none-    numeric
  P     25     pMatrix   S4
  T     25     dtCMatrix S4
  S     25     ddiMatrix S4
> (P <- STs$P)                # Permutation matrix P
  5 x 5 sparse matrix of class "pMatrix"

  [1,] . . | . .
  [2,] | . . . .
  [3,] . | . . .
  [4,] . . . | .
  [5,] . . . . |
> S <- STs$S                  # Diagonal scale-matrix S
> summary(S)
  5 x 5 sparse matrix of class "ddiMatrix", with 5 entries
    i j      x
  1 1 1 6.1439
  2 2 2 6.1439
  3 3 3 6.1439
  4 4 4 3.6959
  5 5 5 3.6959
> T <- STs$T                  # Unit lower-triangular matrix T
> summary(T)                  # All off-diagonal elements equal to 0
  5 x 5 sparse matrix of class "dtCMatrix", with 0 entries
  [1] i j x
  <0 rows> (or 0-length row.names)
```

obtain the variance-covariance matrix $\mathbf{D}$ of the random intercepts included in the LMM fitted in Panel R15.5. Note that the resulting matrix is of dimension 5×5, as the model included five intercepts: three corresponding to the levels of the grouping factor `g1` and two for the levels of the grouping factor g2. The diagonal elements of the matrix correspond to the estimates of variances of the random intercepts, displayed in Panel R15.5.

By using the function `getME()`, we extract from the slot `A` of the model-fit object `fmc` the information about the estimated form of the matrix $\boldsymbol{A}$, defined in (13.34). As shown in Panel R15.7, the transpose of the matrix is, indeed, equal to $\boldsymbol{ZTS}$.

R15.7 *R syntax*: Extracting information about matrices and transformations involved in the implementation of the PnLS approach from a model-fit object of class *mer*. The model-fit object fmc was created in Panel R15.5, while the objects P, S, T, and STs were created in Panel R15.6

```
> TS <- T %*% S
> (sig <- STs$sigma)                        # σ
  [1] 0.2172
> sig * sig * tcrossprod(TS)                # D=σ²TSST': (13.9), (13.33)
  5 x 5 sparse matrix of class "dsCMatrix"

  [1,] 1.7807 .      .      .      .
  [2,] .      1.7807 .      .      .
  [3,] .      .      1.7807 .      .
  [4,] .      .      .      0.6444 .
  [5,] .      .      .      .      0.6444
> A  <- getME(fmc, "A")
> ZTS <- t(Zt) %*% TS                       # ZTS
> max(abs(t(A) - ZTS ))                     # A'=ZTS: (13.34)
  [1] 0
> Ac <- tcrossprod(A)                       # AA'
> AcI <- Ac + diag(nrow(A))                 # AA'+I
> Ls <- slot(fmc, "L")                      # L_Z: (13.38)
> PP <- P %*% AcI %*% t(P)                  # P(AA'+I)P'
> L <- as(Ls, "sparseMatrix")
> max(abs(tcrossprod(L) - PP))              # L_Z L_Z'=P(AA'+I)P': (13.38)
  [1] 0.0024641
> detach(package:lme4.0)
```

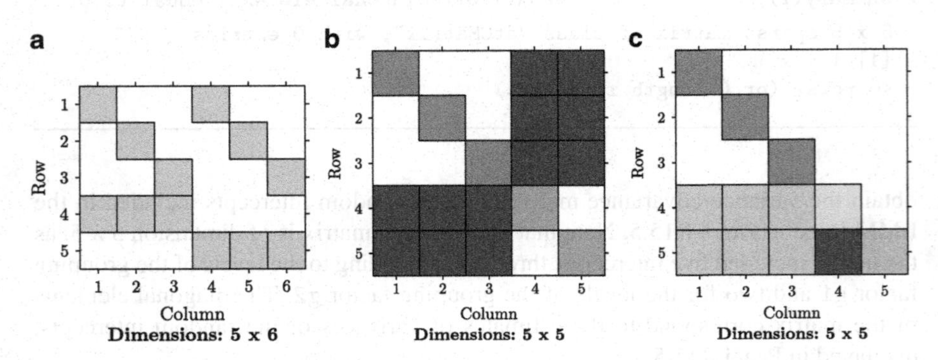

Fig. 15.2 Patterns of nonzero elements in various matrices involved in the PnLS estimation and extracted from a model-fit object of class *mer*. **(a)** $\boldsymbol{Z}'$ **(b)** $\boldsymbol{AA}'$ **(c)** $\boldsymbol{L}_Z$

Finally, by applying the function `getME()`, we refer to the slot `L` of the model-fit object `fmc` and extract the information about the estimated form of the matrix $\boldsymbol{L}_Z$, defined in (13.38). By using the permutation matrix $\boldsymbol{P}$, we compute the matrix $\boldsymbol{P}(\boldsymbol{A}\boldsymbol{A}'+\boldsymbol{I})\boldsymbol{P}'$ and show that it is equal to the product $\boldsymbol{L}_Z\boldsymbol{L}_Z'$, as follows from (13.38).

In Figure 15.2, we present the structures of the matrices $\boldsymbol{Z}'$, $\boldsymbol{A}\boldsymbol{A}'$, and $\boldsymbol{L}_Z$, represented, respectively, by the object `Zt` in Panel R15.5 and `Ac` and `L` in Panel R15.7. The figure clearly shows the sparse nature of the matrices.

15.7 Chapter Summary

In this chapter, we reviewed the main features of the function `lmer()` from the package **lme4.0**. The function allows fitting LMMs. It can be seen as an alternative to the function `lme()` from the package **nlme**.

As compared to `lme()`, the function `lmer()` differs in several important aspects. First of all, it is especially suitable for fitting LMMs with crossed random effects (Sect. 15.2). It can also fit LMMs with nested random effects. However, in the latter case, the range of the available variance-covariance matrices for the random effects is restricted to diagonal or general matrices. Hence, the range is more limited than it is the case of the `lme()` function. Also, `lmer()` can only fit LMMs with independent residual errors.

The ability to deal with LMMs with crossed random effects is reflected in the syntax used by `lmer()` for model specification (Sect. 15.3). Namely, the mean- and random-effects structures are defined in a single formula; the interpretation of random-structure terms depends on whether involved grouping factors are coded as crossed or nested (Sect. 15.3.1).

An important feature of the function `lmer()` is that it has been programmed in the S4 system. This implies differences, as compared to the function `lme()`, in the structure of the model-fit objects and in the methods of extracting results from the objects (Sect. 15.4). It is worth noting here that the p-values for tests, based on an LMM fitted using the function `lmer()`, are not provided by default and need to be evaluated empirically (Sect. 15.5).

It is also worth mentioning that `lmer()` uses efficient computational algorithms which are based on sparse-matrix representations, as described in Sect. 13.5.3. Some details on the implementation of the elements of the algorithm were presented in Sect. 15.6.

The use of the function `lmer()` will be illustrated in the next chapters, where the application of LMMs to the analysis of the case studies will be described.

Finally, it is worth mentioning that, as was indicated in Sect. 1.2, the package **lme4.0** is an early version of the package **lme4**. It is our understanding that the former will not undergo any major changes and the latter will continue to be developed. The current version of the **lme4** package uses a slightly different implementation of the PnLS estimation (see Sect. 13.5.3). Consequently, fitted models are represented by objects of the *lmerMod* class with a different internal structure as compared to the *mer* class used in **lme4.0** (Sect. 15.4). As a result, e.g., some slots, available in the *mer*-class objects created by the function `lmer()` from the **lme4.0** package, are not available in the objects created by the function from the **lme4** package. To alleviate this problem, an auxiliary function `getME()` has been developed in both **lme4** and **lme4.0** packages. The function allows extracting selected components from a model-fit object using a unified syntax across the two packages. Hence, the syntax pertaining to the `lmer()` function, presented in this chapter, to a large extent can be used both in **lme4.0** and **lme4**.

Chapter 16
ARMD Trial: Modeling Visual Acuity

16.1 Introduction

In Chap. 12, we presented an analysis of the age-related macular degeneration (ARMD) data using LM with fixed effects for correlated data. The analysis took into account the correlation between visual acuity measurements obtained for the same patient. To apply the models, we used the function `gls()` from the R package **nlme**. Note that the models can be seen as population-averaged (marginal) models.

An alternative approach to the analysis of the ARMD data, which allows taking into account the correlation between the measurements, is to use linear mixed-effects models (LMMs). In this approach, the hierarchical structure of the data is directly addressed, with random effects that describe the contribution of the variability at different levels of the hierarchy to the total variability of the observations.

In this chapter, we fit several LMMs to the ARMD data. We primarily use the function `lme()` from the package **nlme**. For illustration purposes, several models are refitted by applying the function `lmer()` from the package **lme4.0**.

In particular, in Sect. 16.2, we consider a random-intercept model with homoscedastic residual errors, while in Sect. 16.3, we present a random-intercept model with heteroscedastic errors, with residual variance specified as a power function of time. A series of models with random intercepts and random slopes for time and with heteroscedastic residual errors is described in Sects. 16.4 and 16.5. In Sect. 16.6, we look at the issue of testing hypotheses about the random effects. In Sect. 16.7, we repeat the analysis for selected models using the function `lmer()` from the package **lme4.0**. A summary of the analyses is provided in Sect. 16.8.

16.2 A Model with Random Intercepts and Homogeneous Residual Variance

We start with a simple model, which we label **M16.1**. It contains subject-specific random intercepts and homoscedastic residual errors. Consequently, observations for the same individual, which share the random intercept, are correlated with

A. Gałecki and T. Burzykowski, *Linear Mixed-Effects Models Using R: A Step-by-Step Approach*, Springer Texts in Statistics, DOI 10.1007/978-1-4614-3900-4_16,

a constant (positive) correlation coefficient. Marginally, this corresponds to a compound-symmetry correlation structure with a correlation parameter greater than zero. A compound-symmetry marginal model was fitted to the ARMD data as model **M12.1** in Sect. 12.3. As was mentioned in that section, the compound-symmetry structure is too simple to describe the variance-covariance structure of the visual acuity measurements. Thus, we present model **M16.1** mainly to illustrate the fundamental steps in specifying and fitting an LMM.

16.2.1 Model Specification

Model **M16.1** is specified as follows:

$$\begin{aligned}\text{VISUAL}_{it} = {} & \beta_0 + \beta_1 \times \text{VISUAL0}_i + \beta_2 \times \text{TIME}_{it} + \beta_3 \times \text{TREAT}_i \\ & + \beta_4 \times \text{TREAT}_i \times \text{TIME}_{it} \\ & + b_{0i} + \varepsilon_{it}. \end{aligned} \tag{16.1}$$

The term VISUAL_{it} in (16.1) denotes the value of visual acuity measured for patient i ($i = 1, \ldots, 240$) at time t ($t = 1, 2, 3, 4$, corresponding to values of 4, 12, 24, and 52 weeks, respectively). In the fixed-effects part of the model, given by the first two lines of (16.1), VISUAL0_i is the value of visual acuity measured at baseline; TIME_{it} is the time of the measurement, corresponding to t; TREAT_i is the treatment indicator, equal 1 for the active group and 0 otherwise; and $\text{TREAT}_i \times \text{TIME}_{it}$ is their interaction. The parameter β_0 is an overall intercept, β_1 describes the change in the mean visual acuity due to a unit increase in visual acuity at baseline, β_2 describes the change due to a one week change in time, β_3 gives an overall treatment effect, and β_4 describes the additional change due to a one week change in time for patients treated with the active treatment. Note that we use a linear time effect, following the findings based on the final marginal model **M12.3**. However, contrary to these findings, we include the interaction between time and treatment in (16.1), to "enrich" the fixed part of the mean structure. Also, as it will become clear shortly, we simplify the variance-covariance structure.

In the random-effects part of the model, given by the last line of (16.1), b_{0i} is a patient-specific random intercept, assumed to be normally distributed with mean 0 and variance d_{11}, while ε_{it} is a residual random error assumed to be normally distributed with mean 0 and variance σ^2. Note that, formally speaking, the random intercept b_{0i} is a subject-specific *deviation* from the fixed intercept β_0. It is, however, customary to call b_{0i} a subject-specific random intercept, despite the fact that, actually, β_0 *and* b_{0i} are "coupled" (Sect. 13.2.1) and they *both* contribute to the subject-specific intercept. Typically, this convention does not lead to any misunderstanding.

The fixed part of model **M16.1** assumes that the average profile is linear in time, with different intercepts and slopes for the placebo and active treatment groups. The subject-specific profiles are assumed to also be linear in time, with subject-specific (random) intercepts that shift the individual profiles from the average linear trend.

In matrix notation, the model for the subject i with a complete set of four visual acuity measurements is expressed as follows:

$$\begin{pmatrix} \text{VISUAL}_{i1} \\ \text{VISUAL}_{i2} \\ \text{VISUAL}_{i3} \\ \text{VISUAL}_{i4} \end{pmatrix} = \left(\begin{array}{c|c|c|c|c} 1 & \text{VISUAL0}_i & 4 & \text{TREAT}_i & 4 \cdot \text{TREAT}_i \\ 1 & \text{VISUAL0}_i & 12 & \text{TREAT}_i & 12 \cdot \text{TREAT}_i \\ 1 & \text{VISUAL0}_i & 24 & \text{TREAT}_i & 24 \cdot \text{TREAT}_i \\ 1 & \text{VISUAL0}_i & 52 & \text{TREAT}_i & 52 \cdot \text{TREAT}_i \end{array}\right) \begin{pmatrix} \beta_0 \\ \beta_1 \\ \beta_2 \\ \beta_3 \\ \beta_4 \end{pmatrix}$$

$$+ \begin{pmatrix} 1 \\ 1 \\ 1 \\ 1 \end{pmatrix} \times b_{0i} + \begin{pmatrix} \varepsilon_{i1} \\ \varepsilon_{i2} \\ \varepsilon_{i3} \\ \varepsilon_{i4} \end{pmatrix}. \tag{16.2}$$

Solid vertical lines in (16.2) are used to separate the columns in the subject-specific design matrix $\boldsymbol{X}_i$ for the subject i.

Note that (16.2) can easily be written in the form of (13.1)–(13.3), upon defining

$$\mathbf{y}_i \equiv \begin{pmatrix} \text{VISUAL}_{i1} \\ \text{VISUAL}_{i2} \\ \text{VISUAL}_{i3} \\ \text{VISUAL}_{i4} \end{pmatrix}, \tag{16.3}$$

$$\boldsymbol{X}_i \equiv \left(\begin{array}{c|c|c|c|c} 1 & \text{VISUAL0}_i & 4 & \text{TREAT}_i & 4 \cdot \text{TREAT}_i \\ 1 & \text{VISUAL0}_i & 12 & \text{TREAT}_i & 12 \cdot \text{TREAT}_i \\ 1 & \text{VISUAL0}_i & 24 & \text{TREAT}_i & 24 \cdot \text{TREAT}_i \\ 1 & \text{VISUAL0}_i & 52 & \text{TREAT}_i & 52 \cdot \text{TREAT}_i \end{array}\right), \; \boldsymbol{Z}_i \equiv \begin{pmatrix} 1 \\ 1 \\ 1 \\ 1 \end{pmatrix}, \tag{16.4}$$

$$\boldsymbol{\varepsilon}_i \equiv \begin{pmatrix} \varepsilon_{i1} \\ \varepsilon_{i2} \\ \varepsilon_{i3} \\ \varepsilon_{i4} \end{pmatrix}, \quad \boldsymbol{\beta} \equiv \begin{pmatrix} \beta_0 \\ \beta_1 \\ \beta_2 \\ \beta_3 \\ \beta_4 \end{pmatrix}, \quad \text{and} \quad \mathbf{b}_i \equiv b_{0i}, \tag{16.5}$$

with

$$\mathcal{D} \equiv d_{11}, \quad \text{and} \quad \mathcal{R}_i \equiv \sigma^2 \boldsymbol{I}_4, \tag{16.6}$$

where $\boldsymbol{I}_4$ is the 4×4 identity matrix.

The random part of model **M16.1**, specified by (16.6), leads, according to (13.25), to the following marginal variance-covariance matrix for the subject i with four observations:

$$\mathcal{V}_i \equiv \boldsymbol{Z}_i \mathcal{D} \boldsymbol{Z}_i' + \sigma^2 \boldsymbol{I}_4 = \begin{pmatrix} 1 \\ 1 \\ 1 \\ 1 \end{pmatrix} d_{11} \left(1\ 1\ 1\ 1 \right) + \begin{pmatrix} \sigma^2 & 0 & 0 & 0 \\ 0 & \sigma^2 & 0 & 0 \\ 0 & 0 & \sigma^2 & 0 \\ 0 & 0 & 0 & \sigma^2 \end{pmatrix}$$

$$= \begin{pmatrix} \sigma^2 + d_{11} & d_{11} & d_{11} & d_{11} \\ d_{11} & \sigma^2 + d_{11} & d_{11} & d_{11} \\ d_{11} & d_{11} & \sigma^2 + d_{11} & d_{11} \\ d_{11} & d_{11} & d_{11} & \sigma^2 + d_{11} \end{pmatrix}. \tag{16.7}$$

Consequently, the implied marginal variance-covariance structure is that of compound symmetry with a common correlation equal to $\varrho = d_{11}/(\sigma^2 + d_{11})$. Note that, because the variance component d_{11} is constrained to be nonnegative, ϱ is also forced to be nonnegative.

16.2.2 R Syntax and Results

In Panel R16.1, we use the function `lme()` to fit model **M16.1**, specified by (16.1)–(16.6).

The formula `lm2.form`, used in Panel R16.1, defines the fixed part of the model, as specified in (16.1), including an interaction between `time` and treatment. The factor `treat.f` is parameterized with "Placebo" as the reference level. The argument `random=~1|subject` specifies random subject-specific intercepts. By default, `lme()` assumes independent residual errors with a constant variance, σ^2. Also, because there is no `method` argument in the `lme()` function call, the default REML estimation is used. To change it to the ML estimation, we should add the `method="ML"` argument to the function call.

In addition to the model specification, in Panel R16.1, we also display the results of the fit of model **M16.1**. Note that the model formula is explicitly displayed in the printout. This was achieved by applying, before printing the results, the function `update()` to the object `fm16.1` to evaluate the formula `lm2.form` with the help of the function `eval()`. To simplify the code, this step is *not* shown in Panel R16.1.

Additionally, we print out the estimated fixed-effects table using the `printCoefmat()` function. The argument `has.Pvalue=TRUE` specifies that the last column of the table contains p-values which should be printed (`P.values= TRUE`). A description of all of the arguments of the function `printCoefmat()` can be obtained by issuing the command `?printCoefmat`.

R16.1 *ARMD Trial*: Model **M16.1** fitted using the function `lme()`

```
> lm2.form <-                                    # (16.1)
+    formula(visual ~ visual0 + time + treat.f + treat.f:time)
> (fm16.1 <-                                     # M16.1
+     lme(lm2.form,
+     random = ~1|subject, data = armd))         # b0i:(16.5)
  Linear mixed-effects model fit by REML
    Data: armd
    Log-restricted-likelihood: -3289
    Fixed: visual ~ visual0 + time + treat.f + time:treat.f
         (Intercept)              visual0                time
            9.288078             0.826440           -0.212216
      treat.fActive time:treat.fActive
           -2.422000            -0.049591

  Random effects:
   Formula: ~1 | subject
          (Intercept) Residual
  StdDev:      8.9782   8.6275

  Number of Observations: 867
  Number of Groups: 234
> printCoefmat(summary(fm16.1)$tTable, # Print fixed-effects, etc.
+         has.Pvalue = TRUE, P.values = TRUE) # ... with p-values
                       Value Std.Error       DF t-value p-value
  (Intercept)         9.2881    2.6819 631.0000    3.46 0.00057
  visual0             0.8264    0.0447 231.0000   18.50 < 2e-16
  time               -0.2122    0.0229 631.0000   -9.26 < 2e-16
  treat.fActive      -2.4220    1.5000 231.0000   -1.61 0.10774
  time:treat.fActive -0.0496    0.0336 631.0000   -1.48 0.14002
```

Results presented in Panel R16.1 indicate that the standard deviation $\sqrt{d_{11}}$ of the random intercepts, as specified in (16.6), is estimated to be equal to 8.98, while the residual standard deviation, σ, is estimated to be equal to 8.63. Note that, as was mentioned in Sect. 14.7, the p-values, corresponding to the t-test statistics for the fixed-effects coefficients, are for the marginal-approach tests. A summary of the REML-based estimates for model **M16.1** is also given in Table 16.1.

In Panel R16.2, we demonstrate how to extract information about the grouping of data or, equivalently, about the data hierarchy implied by the fitted model. By using the function `getGroupsFormula()` (see Panel R14.5), we obtain the conditioning expression used in the specification of the `random` argument. It indicates a single level of grouping, defined by the levels of the factor `subject`. By applying the function `getGroups()` to the model-fit object, we extract the grouping factor and store it in the object `grpF`. With the help of the function `str()`, we display the structure of the object. The printout implies that the grouping factor had 234 levels (subjects). Moreover, we can conclude that, e.g., for the first subject we had

Table 16.1 *ARMD Trial*: REML-based parameter estimates[a] for models **M16.1** and **M16.2** with subject-specific random intercepts

	Parameter	fm16.1	fm16.2
Model label		**M16.1**	**M16.2**
Log-REML value		−3288.99	−3260.56
Fixed effects			
Intercept	β_0	9.29(2.68)	7.07(2.30)
Visual acuity at t=0	β_1	0.83(0.04)	0.87(0.04)
Time (in weeks)	β_2	−0.21(0.02)	−0.21(0.03)
Trt(Actv *vs.* Plcb)	β_3	−2.42(1.50)	−2.31(1.24)
Tm × Treat(Actv)	β_4	−0.05(0.03)	−0.05(0.04)
reStruct(subject)			
SD(b_{i0})	$\sqrt{d_{11}}$	8.98(7.99,10.09)	7.71(6.83,8.69)
Variance function			
Power (TIME$^\delta$)	δ		0.31(0.23,0.39)
Scale	σ	8.63(8.16,9.12)	3.61(2.87,4.54)

[a] Approximate SE for fixed effects and 95% CI for covariance parameters are included in parentheses

R16.2 *ARMD Trial*: Data grouping/hierarchy implied by model **M16.1**. The model-fit object `fm16.1` was created in Panel R16.1

```
> getGroupsFormula(fm16.1)                # Grouping formula
  ~subject
  <environment: 0x000000001a310670>
> str(grpF <- getGroups(fm16.1))          # Grouping factor
  Factor w/ 234 levels "1","2","3","4",..: 1 1 2 2 2 2 3 3 3 4...
  - attr(*, "label")= chr "subject"
> grpF[1:17]
  [1] 1 1 2 2 2 2 3 3 3 4 4 4 4 6 6 6 6
  234 Levels: 1 2 3 4 6 7 8 9 10 11 12 13 14 15 16 17 18 19... 240
> levels(grpF)[1:5]
  [1] "1" "2" "3" "4" "6"
> range(xtabs(~grpF))                    # Min, Max no. of observations
  [1] 1 4
```

two observations, for the second subject we had four observations, etc. Similar information is obtained by listing a subset of elements of the grouping factor. The minimum and maximum number of observations across all subjects are obtained by applying the function `range()` to the result of a cross tabulation of the levels of the factor `grpF`, provided by the function `xtabs()`.

To get more insight into the estimated variance-covariance structure of model **M16.1**, we use the `getVarCov()` and `VarCorr()` functions, as shown in Panel R16.3.

R16.3 *ARMD Trial*: The estimated variance-covariance matrices for random effects ($\mathcal{D}$) and residual errors ($\mathcal{R}_i$) for model **M16.1**. The model-fit object `fm16.1` was created in Panel R16.1

(a) *The $\mathcal{D}$-matrix estimate*

```
> getVarCov(fm16.1, individual = "2")     # d̂11:(16.6)
  Random effects variance covariance matrix
               (Intercept)
  (Intercept)       80.608
    Standard Deviations: 8.9782
> VarCorr(fm16.1)                          # d̂11, σ̂2
  subject = pdLogChol(1)
              Variance StdDev
  (Intercept) 80.608   8.9782
  Residual    74.434   8.6275
```

(b) *The $\mathcal{R}_i$-matrix estimate*

```
> getVarCov(fm16.1,
+           type = "conditional",          # R̂i:(16.6)
+           individual = "2")
  subject 2
  Conditional variance covariance matrix
         1      2      3      4
  1 74.434  0.000  0.000  0.000
  2  0.000 74.434  0.000  0.000
  3  0.000  0.000 74.434  0.000
  4  0.000  0.000  0.000 74.434
    Standard Deviations: 8.6275 8.6275 8.6275 8.6275
```

The `getVarCov()`-function call, used in Panel R16.3a, does not include the `type` argument (see Sect. 14.6 and Table 14.5). This means that the default value of the argument, i.e., `type="random.effect"`, is employed. As a result, the function provides the estimated variance-covariance matrix $\mathcal{D}$ of the random effects. In the case of model **M16.1**, it gives the estimated variance and standard deviation of the subject-specific random intercepts. The argument `individual="2"`, used in the `getVarCov()`-function call, requests the random effects variance-covariance matrix for the second individual, i.e., `subject==2`, in the analyzed dataset. In fact, in our case, the subject number is not of importance, as the variance-covariance structure of random effects is assumed to be the same for all individuals.

In Panel R16.3a, we also illustrate how to extract estimates of the $\mathcal{D}$ matrix elements using the function `VarCorr()` (see Sect. 14.6 and Table 14.5).

In Panel R16.3b, we specify the `type="conditional"` and `individual="2"` arguments in a call to the `getVarCov()` function. As a result, we obtain the estimated variance-covariance matrix $\mathcal{R}_i$ of the residual random errors for the second subject. As noted previously, this subject has all four post-randomization visual acuity measurements, so a 4×4 matrix is reported. Because model **M16.1** assumes independent residual errors with the same variance at all measurement times, a diagonal matrix $\widehat{\mathcal{R}}_i = \widehat{\sigma}^2 \boldsymbol{I}_4 = 74.434 \times \boldsymbol{I}_4$ is displayed, as specified in (16.6).

Finally, in Panel R16.4, we obtain the estimated marginal variance-covariance matrix, defined in (16.7), by applying the function `getVarCov()` with the `type="marginal"` argument. The result, for `individual="2"`, is stored in the object `fm16.1cov` and displayed. The marginal variance is estimated by the sum of the estimated residual variance $\widehat{\sigma}^2 = 74.434$ and the variance of the random intercepts $\widehat{d}_{11} = 80.608$. The latter variance component becomes the covariance, as seen from (16.7).

The resulting marginal correlation matrix is obtained by applying the `cov2cor()` function (see Panel R14.4) to the first component of the list-object `fm16.1cov`, which contains the estimated marginal variance-covariance matrix. As noted earlier, the estimated marginal correlation matrix implies a constant, positive correlation coefficient equal to 0.52 for any two visual acuity measurements obtained for the same patient at different timepoints.

16.3 A Model with Random Intercepts and the varPower$(\cdot)$ Residual Variance-Function

As noted in the exploratory analysis (Sect. 3.2) and, e.g., in Chap. 12, the variability of visual acuity measurements increases in time. Therefore, we consider a model with variance of random errors expressed as a power function of the TIME covariate.

16.3.1 Model Specification

To specify the new model, labeled **M16.2**, we use the same fixed-effects part as in model **M16.1**. However, we modify the variance-covariance structure of residual random errors, specified in (16.6). More specifically, following the results obtained in Chaps. 9 and 12, we consider the use of the varPower$(\cdot)$ variance function, introduced in Sect. 7.3.1. Thus, we assume that

$$\mathcal{R}_i = \sigma^2 \begin{pmatrix} (\text{TIME}_{i1})^{2\delta} & 0 & 0 & 0 \\ 0 & (\text{TIME}_{i2})^{2\delta} & 0 & 0 \\ 0 & 0 & (\text{TIME}_{i3})^{2\delta} & 0 \\ 0 & 0 & 0 & (\text{TIME}_{i4})^{2\delta} \end{pmatrix}. \tag{16.8}$$

R16.4 *ARMD Trial*: The estimated marginal variance-covariance matrix and the corresponding correlation matrix for model **M16.1**. The model-fit object `fm16.1` was created in Panel R16.1

```
> (fm16.1cov <-
+     getVarCov(fm16.1,
+               type = "marginal",             # V̂i:(16.7)
+               individual = "2"))
  subject 2
  Marginal variance covariance matrix
          1       2       3       4
  1 155.040  80.608  80.608  80.608
  2  80.608 155.040  80.608  80.608
  3  80.608  80.608 155.040  80.608
  4  80.608  80.608  80.608 155.040
    Standard Deviations: 12.452 12.452 12.452 12.452
> (cov2cor(fm16.1cov[[1]]))                     # Corr(V̂i)
          1       2       3       4
  1 1.00000 0.51991 0.51991 0.51991
  2 0.51991 1.00000 0.51991 0.51991
  3 0.51991 0.51991 1.00000 0.51991
  4 0.51991 0.51991 0.51991 1.00000
```

Note that $\mathcal{R}_i$, defined in (16.8), can be decomposed as $\mathcal{R}_i = \sigma^2 \boldsymbol{\Lambda}_i \boldsymbol{C}_i \boldsymbol{\Lambda}_i$ using $\boldsymbol{\Lambda}_i$, given in (12.3), and by setting $\boldsymbol{C}_i = \boldsymbol{I}_4$. It should be stressed here that the parameter σ^2, used in (16.8), can only be interpreted as a (unknown) scale parameter. This is in contrast to (16.6), where it could be interpreted as the variance of residual errors.

The matrix $\mathcal{R}_i$, given in (16.8), is diagonal with unequal elements defined by the varPower(·) function. Consequently, as compared to model **M16.1**, the marginal variance-covariance and correlation matrices of model **M16.2** have different structures. In particular, the marginal variance-covariance matrix becomes equal to

$$\mathcal{V}_i = \begin{pmatrix} \sigma_1^2 + d_{11} & d_{11} & d_{11} & d_{11} \\ d_{11} & \sigma_2^2 + d_{11} & d_{11} & d_{11} \\ d_{11} & d_{11} & \sigma_3^2 + d_{11} & d_{11} \\ d_{11} & d_{11} & d_{11} & \sigma_4^2 + d_{11} \end{pmatrix}, \tag{16.9}$$

where

$$\sigma_t^2 = \sigma^2 (\text{TIME}_{it})^{2\delta}.$$

It is worth observing that, because the variance changes with time, the marginal correlation coefficients between observations made at different times are no longer equal.

R16.5 *ARMD Trial*: Model **M16.2** fitted using the function `lme()`. The model-fit object `fm16.1` was created in Panel R16.1

```
> (fm16.2 <-                                          # M16.2 ← M16.1
+    update(fm16.1,
+           weights = varPower(form = ~ time),        # (9.4)
+           data = armd))
  Linear mixed-effects model fit by REML
    Data: armd
    Log-restricted-likelihood: -3260.6
    Fixed: visual ~ visual0 + time + treat.f + time:treat.f
         (Intercept)             visual0                time
            7.066881            0.866544           -0.212627
       treat.fActive time:treat.fActive
           -2.305034           -0.050888

  Random effects:
   Formula: ~1 | subject
          (Intercept) Residual
  StdDev:      7.7056   3.6067

  Variance function:
   Structure: Power of variance covariate
   Formula: ~time
   Parameter estimates:
    power
  0.31441
  Number of Observations: 867
  Number of Groups: 234
```

16.3.2 R Syntax and Results

In Panel R16.5, we fit model **M16.2**. More specifically, we update the object `fm16.1`, representing the fitted model **M16.1**, using the `weights = varPower(form = ~time)` argument in a call to the `update()` function. Note the use of the `varPower()` variance-function constructor (see Sect. 8.2) in the `weights` argument (see Sect. 14.5). Results of fitting model **M16.2** using REML are stored in the object `fm16.2`. Panel R16.5 presents a summary of the estimates of the model parameters. More detailed results are shown in Table 16.1.

The results, presented in Panel R16.5, indicate that the scale parameter σ is estimated to be equal to 3.607. The power coefficient δ of the `varPower()` variance function, as specified in (9.4), is estimated to be equal to 0.314. The estimate of the standard deviation of the random intercepts equals 7.706.

Panel R16.6 presents the estimates of the variance-covariance matrices associated with model **M16.2**. To obtain the estimate of the $\mathcal{D}$ matrix, we apply the

R16.6 *ARMD Trial*: The estimated $\mathcal{D}$, $\mathcal{R}_i$, and $\mathcal{V}_i$ matrices for model **M16.2**. The model-fit object `fm16.2` was created in Panel R16.5

```
> VarCorr(fm16.2)                                # d̂11: (16.6), σ̂^2
  subject = pdLogChol(1)
              Variance StdDev
  (Intercept) 59.376   7.7056
  Residual    13.008   3.6067
> getVarCov(fm16.2,                              # R̂i: (16.8)
+           type = "conditional",
+           individual = "2")
  subject 2
  Conditional variance covariance matrix
         1      2      3      4
  1 31.103  0.000  0.000   0.00
  2  0.000 62.062  0.000   0.00
  3  0.000  0.000 95.966   0.00
  4  0.000  0.000  0.000 156.05
    Standard Deviations: 5.577 7.8779 9.7962 12.492
> (fm16.2cov <-                                  # V̂i: (16.9)
+    getVarCov(fm16.2,
+              type = "marginal",
+              individual = "2"))
  subject 2
  Marginal variance covariance matrix
         1       2       3       4
  1 90.479  59.376  59.376  59.376
  2 59.376 121.440  59.376  59.376
  3 59.376  59.376 155.340  59.376
  4 59.376  59.376  59.376 215.430
    Standard Deviations: 9.512 11.02 12.464 14.677
> cov2cor(fm16.2cov[[1]])                        # Corr(V̂i)
          1       2       3       4
  1 1.00000 0.56645 0.50083 0.42529
  2 0.56645 1.00000 0.43230 0.36710
  3 0.50083 0.43230 1.00000 0.32457
  4 0.42529 0.36710 0.32457 1.00000
```

`VarCorr()` function. The estimated variance of random intercepts is equal to 59.376. Note that it is smaller than the value of 80.608, obtained for model **M16.1** (see Panel R16.3). This is expected, because, by allowing for heteroscedastic residual random errors, a larger part of the total variability is explained by the residual variances. The estimated variance-covariance matrix of the residual errors $\mathcal{R}_i$ is obtained using the `getVarCov()` function with the `type="conditional"`

Table 16.2 *ARMD Trial*: REML-based estimates[a] for linear mixed-effects models[b] with random intercepts and time slopes

	Parameter	fm16.3	fm16.4	fm16.5
Model label		**M16.3**	**M16.4**	**M16.5**
Log-REML value		−3215.30	−3215.90	−3214.47
Fixed effects				
Intercept	β_0	4.74(2.26)	5.26(2.27)	5.44(2.26)
Visual acuity at t=0	β_1	0.91(0.04)	0.90(0.04)	0.90(0.04)
Time (in weeks)	β_2	−0.22(0.03)	−0.22(0.03)	−0.24(0.02)
Trt(Actv *vs.* Plcb)	β_3	−2.26(1.15)	−2.28(1.17)	−2.66(1.13)
Tm × Treat(Actv)	β_4	−0.06(0.05)	−0.06(0.05)	
reStruct(`subject`)				
SD(b_{i0})	$\sqrt{d_{11}}$	6.98(5.99,8.13)	7.23(6.33,8.26)	7.24(6.33,8.27)
SD(b_{i1})	$\sqrt{d_{22}}$	0.27(0.23,0.32)	0.28(0.24,0.33)	0.28(0.24,0.33)
cor((Intercept),time)	ϱ_{12}	0.14(−0.13,0.38)		
Variance function				
Power (TIME$^\delta$)	δ	0.11(0.02,0.20)	0.11(0.01,0.21)	0.11(0.02,0.21)
Scale	σ_1	5.12(4.00,6.56)	5.03(3.90,6.49)	5.04(3.92,6.48)

[a]Approximate SE for fixed effects and 95% CI for covariance parameters are included in parentheses
[b]The variance function `varPower()` of the `time` covariate was used in all three models

argument. It corresponds to the matrix specified in (16.8). Thus, for instance, the first diagonal element of the $\widehat{\mathcal{R}}_i$ matrix is equal to $\hat{\sigma}^2 \cdot 4^{2\hat{\delta}} = 3.6067^2 \cdot 4^{2\cdot 0.3144} = 31.103$.

The estimated marginal variance-covariance matrix, shown in Panel R16.6, corresponds to the matrix $\mathcal{V}_i$, given in (16.9). It is obtained by applying the `getVarCov()` function with the `type="marginal"` argument to the `fm16.2` model-fit object. The corresponding estimated marginal correlation matrix indicates a decreasing correlation between visual acuity measurements made at more distant timepoints. This agrees with the conclusion drawn for the final marginal model **M12.3**, defined by (12.3), (12.6), and (12.9), for which results are displayed in Table 12.2 and Panel R12.12. Note, however, that the direct comparison of the marginal variance-covariance matrices for models **M12.3** and **M16.2** is not appropriate. This is because the marginal variance-covariance matrix of model **M16.2**, displayed in Panel R16.6, is much more structured than that of model **M12.3**, printed in Panel R12.12. On the other hand, they both allow for marginal correlation coefficients, which depend on the time "distances", or "positions", of visual-acuity measurements.

To summarize the results of analyses presented in the current and the previous section, Table 16.1 displays REML-based parameter estimates for models **M16.1** and **M16.2**.

R16.7 *ARMD Trial*: Residual plots for model **M16.2**. The model-fit object `fm16.2` was created in Panel R16.5

(a) *Default residual plot of conditional Pearson residuals*

```
> plot(fm16.2)                          # Fig. 16.1
```

(b) *Plots (and boxplots) of Pearson residuals per time and treatment*

```
> plot(fm16.2,                          # Figure not shown
+      resid(., type = "pearson") ~ time | treat.f,
+      id = 0.05)
> bwplot(resid(fm16.2, type = "p") ~ time.f | treat.f, # Fig. 16.2
+        panel = panel.bwxplot2,        # User-defined panel (not shown)
+        data = armd)
```

(c) *Normal Q-Q plots of Pearson residuals and predicted random effects*

```
> qqnorm(fm16.2, ~resid(.) | time.f)   # Fig. 16.3
> qqnorm(fm16.2, ~ranef(.))            # Fig. 16.4
```

16.3.3 Diagnostic Plots

At this point, we might want to take a look at the goodness of fit of model **M16.2**. The fitted model is represented by the object `fm16.2`. The syntax for several residual plots is given in Panel R16.7.

The default residual plot for the object is obtained using the `plot()` command in Panel R16.7a and presented in Fig. 16.1. The plot displays the conditional Pearson residuals (Sect. 13.6.2) *versus* fitted values. As such, the plot is not very informative, because it pools all the residuals together, despite the fact that residuals obtained from the same individual are potentially correlated. However, it can serve for detecting, e.g., outliers. In Fig. 16.1, a group of such residuals can be seen in at the bottom and the top of the central part of the scatterplot.

A modified plot of the residuals for each timepoint and treatment group might be more helpful. Toward this end, we use the form of the `plot()`-function call shown in Panel R16.7b. Note that, in the plot formula, we apply the `type="pearson"` argument in the `resid()` function, which indicates the use of the Pearson residuals. Moreover, in the formula, we use the term `~time|treat` to obtain plots per treatment group over time in separate panels. Additionally, by applying the argument `id=0.05` to the `plot()` statement, we label the residuals larger, in absolute value, than the 97.5th percentile of the standard normal distribution by the number of the corresponding observation from the `armd` data frame.

Note that we do not present the resulting plot. Instead, in Fig. 16.2, we present its enhanced version, with box-and-whiskers plots superimposed over a stripplot of the

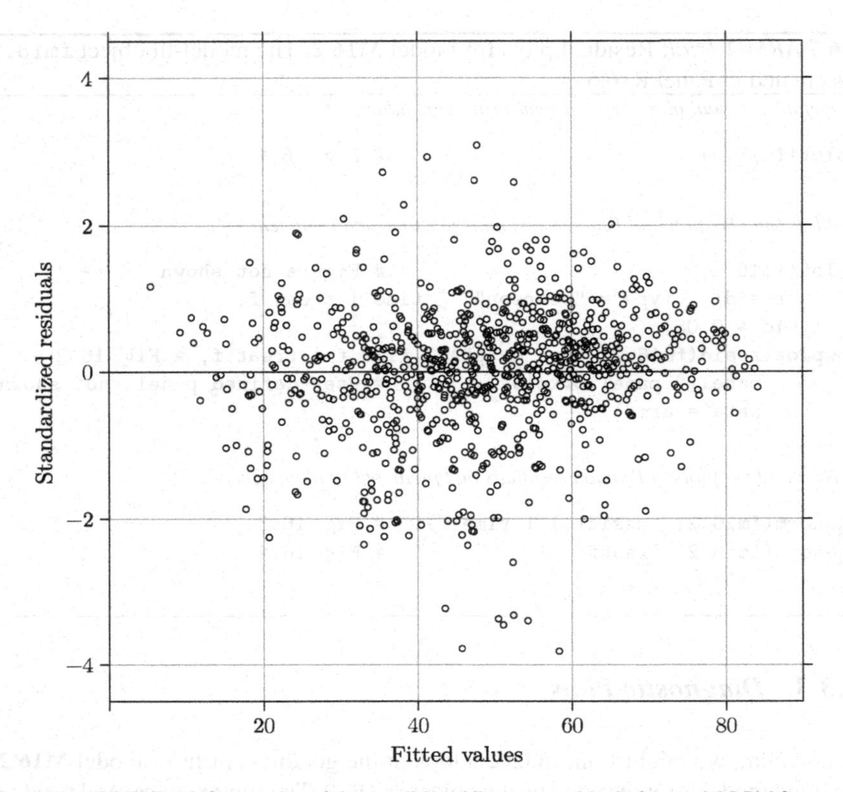

Fig. 16.1 *ARMD Trial*: Scatterplot of the conditional Pearson residuals for model **M16.2**

residuals for each timepoint and treatment group. Toward this end, in Panel R16.7b, we use the function `bwplot()` from the package **lattice** (Sect. 3.2.2). In the first argument of `bwplot()`, we use a formula requesting a plot of the Pearson residuals *versus* the levels of the `time.f` factor, separately for the levels of the `treat.f` factor. The residuals are extracted from the model-fit object `fm16.2` by applying the `resid()` function (Sect. 5.5). The key component of the `bwplot()`-function call is an auxiliary panel-function `panel.bwxplot2`. Due to the complexity of the R code used to create the panel function, we do not present it; however, the code is available in the package **nlmeU** containing the supplementary materials for the book.

Figure 16.2 allows for an evaluation of the distribution of the conditional Pearson residuals for each timepoint and treatment group. Despite standardization, the variability of the residuals seems to vary. The plot reveals also a number of outliers, i.e., residuals larger, in absolute value, than the 97.5th percentile of the standard normal distribution (they have been labeled in the plot by the corresponding observation number). However, given the large number of observations, one might

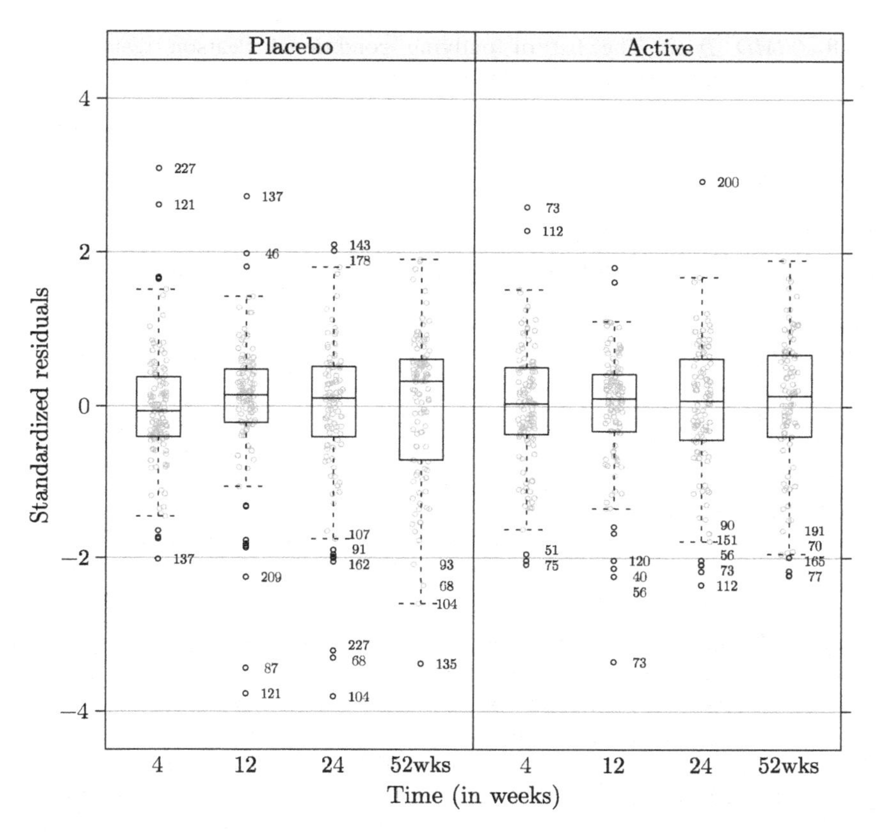

Fig. 16.2 *ARMD Trial*: Stripplots (and box-and-whiskers plots) of the conditional Pearson residuals for each timepoint and treatment group for model **M16.2**

expect a group of outlying values. It is worth noting that the outliers are present in all treatment groups and at all timepoints.

Panel R16.8 lists the subjects for whom outlying residuals were labeled in Fig. 16.2. Toward this end, the conditional Pearson residuals are extracted from the model-fit object `fm16.2` and stored in the vector `resid.p`. Indices for the residuals larger, in absolute value, than the 97.5th percentile of the standard normal distribution are stored in the logical vector `idx`. The data frame `outliers.idx` contains selected variables from the `armd` dataset together with the residuals and the logical index vector. The data frame `outliers` is a subset of `outliers.idx` and contains observations for which the value of the variable `idx`, given as the second argument of the function `subset()`, is equal to 1. There are 38 such observations, for which the value of the `subject` number is printed out. Note that, for several subjects, there is more than one outlying residual, because there is more than one visual acuity measurement possible per subject.

R16.8 *ARMD Trial*: The list of outlying conditional Pearson residuals for model **M16.2**. The model-fit object `fm16.2` was created in Panel R16.5

```
> id <- 0.05                                      # Argument for qnorm()
> outliers.idx <-
+   within(armd,
+          {
+          resid.p <- resid(fm16.2, type = "pearson") # Pearson resids.
+          idx <- abs(resid.p) > -qnorm(id/2)         # Indicator vector
+          })
> outliers  <- subset(outliers.idx, idx)         # Data with outliers
> nrow(outliers)                                  # Number of outliers
  [1] 38
> outliers$subject                                # IDs of outliers
   [1] 40  46  51  56  56  68  68  70  73  73  73  75  77  87  90
  [16] 91  93  104 104 107 112 112 120 121 121 135 137 137 143 151
  [31] 162 165 178 191 200 209 227 227
  234 Levels: 1 2 3 4 6 7 8 9 10 11 12 13 14 15 16 17 18 19... 240
```

Figure 16.3 shows the normal Q-Q plot of the conditional Pearson residuals per timepoint. The plot was obtained using the first `qqnorm()`-function call shown in Panel R16.7c. The patterns do show some deviations from a linear trend.

We can also look at the normal Q-Q plot of the predicted random effects (random intercepts). The effects are estimated by EBLUPs (Sect. 13.6.1). They can be extracted from the `fm16.2` model-fit object using the function `ranef()` (see Sect. 14.6 and Table 14.5), as shown in the second `qqnorm()`-function call shown in Panel R16.7c. The resulting Q-Q plot is shown in Fig. 16.4 and is slightly curvilinear. This could be taken as an indication of nonnormality of the random effects. However, as mentioned in Sect. 13.6.1, such a plot may not necessarily reflect the true distribution of the random effects. Hence, it should be interpreted with caution.

An important diagnostic plot is presented in Fig. 16.5. It shows the observed and predicted values of the visual acuity measurements for selected patients. Panel R16.9 demonstrates how to generate the object containing the data necessary for constructing the figure using the `augPred()` function.

The function `augPred()` allows obtaining predicted values for the object specified as the first argument. The object can be of class *lmList* (14.5), *gls* (11.6), and *lme* (14.6). If the object has a grouping structure, the predicted values are obtained for each group. Conveniently, the function adds the original observations to the returned object, which is a data frame with four columns containing the values of the primary covariate, the groups, the predicted or observed values, and the indicator of the type of the value from the third column.

The optional arguments of the function `augPred()` include: `primary`, `level`, `length.out`, `minimum`, and `maximum`. The argument `primary` is a one-sided formula indicating the covariate at which values the predicted values should be

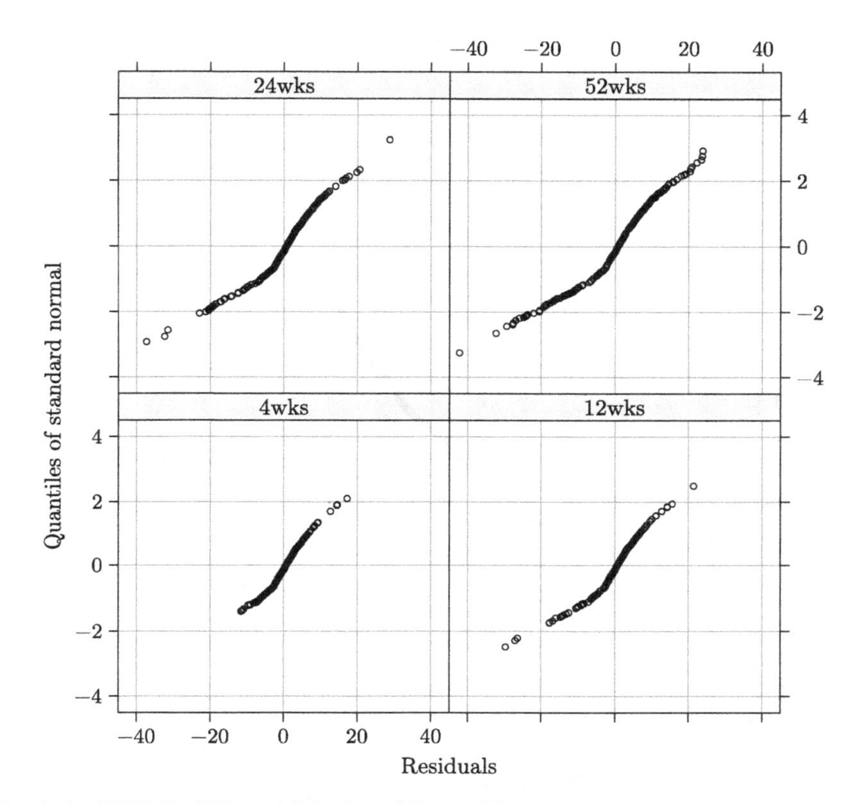

Fig. 16.3 *ARMD Trial*: Normal Q-Q plots of the conditional Pearson residuals for each timepoint for model **M16.2**

computed. In the call presented in Panel R16.9, we indicate that the predicted values should be computed at the values of the variable `time`.

The arguments `minimum` and `maximum` allow for providing the lower and upper limit, respectively, for the values of the primary covariate at which the predicted values are to be computed. By default, the arguments become equal to, respectively, the minimum and maximum of the values of the covariate. In the call presented in Panel R16.9, we use the default values of the arguments, i.e., the minimum and maximum values of the `time` variable, which are equal to, respectively, 4 and 52 weeks.

The argument `level` of the function `augPred()` is an integer vector specifying the grouping levels for which the predicted values are to be computed. Its interpretation is the same as for the function `predict()` (Sect. 14.6). In the `augPred`-function call shown in Panel R16.9, we use `level=0:1`, which amounts to specifying that the predicted values should be computed at the level 0, i.e., the population level, and at the level 1, i.e., the subject level.

Finally, the `length.out` argument is an integer indicating the number of values of the primary covariate at which the predictions should be evaluated. By default,

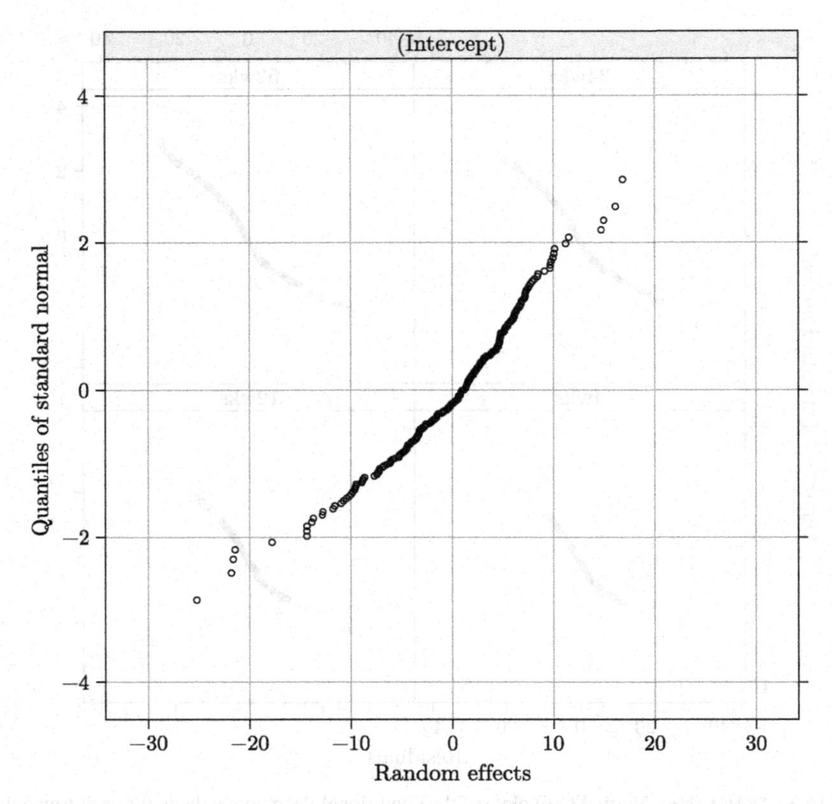

Fig. 16.4 *ARMD Trial*: The normal Q-Q plot of the predicted random intercepts for model **M16.2**

R16.9 *ARMD Trial*: Predicted visual acuity values for model **M16.2**. The model-fit object `fm16.2` was created in Panel R16.5

```
> aug.Pred <-                    # augPred for M16.2
+    augPred(fm16.2,
+           primary = ~time,      # Primary covariate
+           level = 0:1,          # Marginal(0) and subj.-spec.(1)
+           length.out = 2)
> plot(aug.Pred, layout = c(4, 4, 1), # Fig. 16.5
+      key = list(lines = list(lty = c(1,2)),
+                 text = list(c("Marginal", "Subject-specific")),
+                 columns = 2))
```

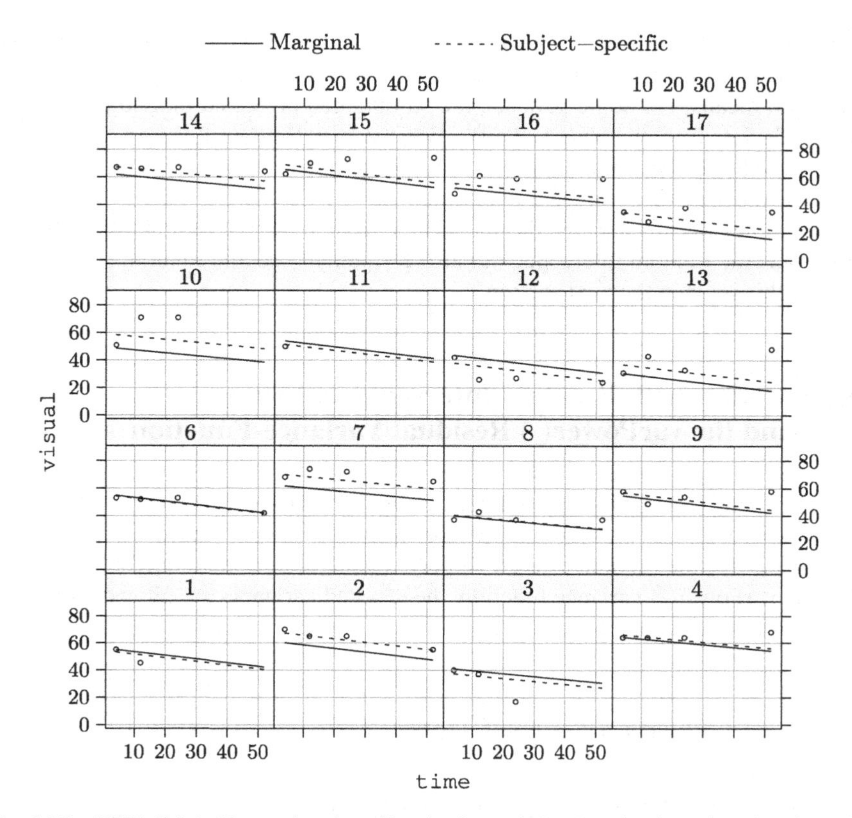

Fig. 16.5 *ARMD Trial*: Observed and predicted values of visual acuity for selected patients for model **M16.2**

it assumes the value 51. In Panel R16.9, we set `length.out=2`, i.e., the predicted values are obtained at two values `time`, i.e., at the minimum (4 weeks) and the maximum (52 weeks). The two predicted values are sufficient to describe the (population- and subject-specific) linear trend in (continuous) time, implied by the fitted form of model **M16.2**.

By applying, in Panel R16.9, the `plot()` function to the object `aug.Pred` with the `level=0:1` argument, a plot of the population-level and within-subject predicted values is obtained. The argument `layout=c(4,4,1)` requests one page of plots, arranged in four rows with four plots each. Each plot corresponds to a single subject; thus, the predictions for the first 16 subjects are plotted. Finally, the `key` argument allows specifying the legend, which is placed at the top of the graph. We refer the reader to the R help system for the `xyplot()` function from the **lattice** package for a detailed description of other available arguments.

The resulting plot is shown in Fig. 16.5. The predicted population means, shown in the plot, decrease linearly in time. This is consistent with the trend observed in Fig. 3.2. According to the assumed structure of the model, the population means are shifted for individual patients by subject-specific random intercepts.

Note that, as a result, the slopes of the individual profiles are the same for all subjects. Consequently, all subject-specific lines are parallel to lines representing the population means. For some patients, the so-obtained predicted individual profiles strongly deviate from the observed ones. For instance, for the subjects 4 and 15, the predicted individual patterns suggest a decrease of visual acuity over time, while the observed values actually increase over time.

A possible way to improve the individual predictions is to allow not only for patient-specific random intercepts, but also for patient-specific random slopes. We focus on this issue in the next section.

16.4 Models with Random Intercepts and Slopes and the varPower$(\cdot)$ Residual Variance-Function

In this section, we consider a model with two subject-specific random effects: a random intercept and a random slope for time. We use two variance-covariance structures $\mathcal{D}$ for the random effects, namely, a general one and a diagonal one. By using the varPower$(\cdot)$ variance function, the residual variances are allowed to differ between different timepoints.

16.4.1 Model with a General Matrix $\mathcal{D}$

To specify model **M16.3** with a general variance-covariance matrix $\mathcal{D}$, we modify the model equation (16.1) as follows:

$$\begin{aligned}\text{VISUAL}_{it} &= \beta_0 + \beta_1 \times \text{VISUAL0}_i + \beta_2 \times \text{TIME}_{it} + \beta_3 \times \text{TREAT}_i \\ &\quad + \beta_4 \times \text{TREAT}_i \times \text{TIME}_{it} \\ &\quad + b_{0i} + b_{2i} \times \text{TIME}_{it} + \varepsilon_{it}. \qquad (16.10)\end{aligned}$$

The equation (16.10) can be written in the form of (13.1)–(13.3), upon defining $\mathbf{y}_i$, $\boldsymbol{X}_i$, $\boldsymbol{\varepsilon}_i$, and $\boldsymbol{\beta}$ as in (16.3)–(16.5), but with

$$\mathbf{Z}_i = \begin{pmatrix} 1 & 4 \\ 1 & 12 \\ 1 & 24 \\ 1 & 52 \end{pmatrix}, \quad \mathbf{b}_i = \begin{pmatrix} b_{0i} \\ b_{2i} \end{pmatrix}, \qquad (16.11)$$

and with the variance-covariance structure of the random effects given by

$$\mathbf{b}_i \sim \mathcal{N}(\mathbf{0}, \mathcal{D}) \quad \text{and} \quad \boldsymbol{\varepsilon}_i \sim \mathcal{N}(\mathbf{0}, \mathcal{R}_i), \qquad (16.12)$$

where

$$\mathcal{D} = \begin{pmatrix} d_{11} & d_{12} \\ d_{21} & d_{22} \end{pmatrix} \tag{16.13}$$

and $\mathcal{R}_i$ is given by (16.8).

Note that the assumed form of $\mathcal{D}$ implies that the random intercepts and slopes are correlated. For instance, a positive correlation between b_{0i} and b_{2i} means that, for individuals with a higher initial value of visual acuity, the post-randomization measurements will increase more rapidly or decrease more slowly than for patients with a lower initial value.

It is worth reflecting on the marginal variance-covariance structure implied by model **M16.3**. According to this model, the marginal covariance between visual acuity measurements for the subject i at times t_1 and t_2 $(t_1, t_2 = 1, 2, 3, 4)$ can be written as follows:

$$\begin{aligned} \mathrm{Cov}(y_{it_1}, y_{it_2}) &= \begin{pmatrix} 1 & \mathrm{TIME}_{it_1} \end{pmatrix} \mathcal{D} \begin{pmatrix} 1 \\ \mathrm{TIME}_{it_2} \end{pmatrix} + \mathrm{I}(t_1 = t_2)\sigma^2(\mathrm{TIME}_{it_1})^{2\delta} \\ &= d_{11} + d_{12}(\mathrm{TIME}_{it_1} + \mathrm{TIME}_{it_2}) + d_{22}\mathrm{TIME}_{it_1}\mathrm{TIME}_{it_2} \\ &\quad + \mathrm{I}(t_1 = t_2)\sigma^2(\mathrm{TIME}_{it})^{2\delta}, \end{aligned} \tag{16.14}$$

where $\mathrm{I}(A)$ is the indicator function for condition A. Hence, the marginal variance of visual acuity measurements for the subject i at the time t can be expressed as

$$\mathrm{Var}(y_{it}) = d_{11} + 2d_{12}\mathrm{TIME}_{it} + d_{22}\mathrm{TIME}_{it}^2 + \sigma^2(\mathrm{TIME}_{it})^{2\delta}. \tag{16.15}$$

Thus, the variance becomes a power function, including a quadratic component, of the measurement time.

In Panel R16.10, we fit model **M16.3**, defined by (16.10)–(16.13), by updating the object `fm16.2`. Specifically, we use the syntax `random = ~ 1 + time | subject` to specify the random-effects structure (Sect. 14.3.1). By applying this particular formula in the `random` argument, we imply that, for each level of the `subject` grouping variable, a random intercept and a random slope for time are to be considered, with a (default) general variance-covariance matrix $\mathcal{D}$ represented by an object of class *pdLogChol*.

The basic results of fitting model **M16.3** are displayed in Panel R16.10. More details are shown in Table 16.2. In Panel R16.10, we also present the 95% CIs for the variance-function and correlation-structure parameters. They are computed using the methods described in Sect. 13.7.3. The results show a low estimated value of the correlation coefficient for the random effects b_{0i} and b_{2i}, equal to 0.138. The confidence interval for the correlation coefficient suggests that, in fact, the two random effects can be uncorrelated. Therefore, in the next section, we consider a simplified form of the $\mathcal{D}$ matrix.

R16.10 *ARMD Trial*: The estimated $\widehat{\mathcal{D}}$ matrix and confidence intervals for the $\boldsymbol{\theta}_D$ parameters for model **M16.3**. The model-fit object `fm16.2` was created in Panel R16.5

```
> fm16.3 <-                                         # M16.3 ← M16.2
+     update(fm16.2,
+            random = ~1 + time | subject,
+            data = armd)
> getVarCov(fm16.3, individual = "2")          # D̂: (16.16)
 Random effects variance covariance matrix
             (Intercept)     time
 (Intercept)    48.70500 0.26266
 time            0.26266 0.07412
   Standard Deviations: 6.9789 0.27225
> intervals(fm16.3, which = "var-cov")    # 95% CI for θD, δ: (16.8), σ
 Approximate 95% confidence intervals

  Random Effects:
   Level: subject
                              lower     est.    upper
 sd((Intercept))            5.99019 6.97891 8.13082
 sd(time)                   0.23009 0.27225 0.32213
 cor((Intercept),time) -0.12564 0.13824 0.38386

  Variance function:
          lower     est.  upper
 power 0.015191 0.10744 0.1997
 attr(,"label")
 [1] "Variance function:"

  Within-group standard error:
  lower    est.  upper
 3.9993 5.1222 6.5604
```

16.4.2 Model with a Diagonal Matrix $\mathcal{D}$

In this section, we consider model **M16.4**, which, similarly to model **M16.3**, is defined by (16.10), but for which we specify that

$$\mathcal{D} = \begin{pmatrix} d_{11} & 0 \\ 0 & d_{22} \end{pmatrix}. \tag{16.16}$$

Thus, we assume that random intercepts b_{0i} and random slopes b_{1i} have different variances and are uncorrelated.

R16.11 *ARMD Trial*: Confidence intervals for the parameters of model **M16.4**. The model-fit object fm16.3 was created in Panel R16.10

```
> fm16.4 <-                                # M16.4 ← M16.3
+     update(fm16.3,
+            random = list(subject = pdDiag(~time)), # Diagonal D
+            data = armd)
> intervals(fm16.4)                        # 95% CI for β, θ_D, δ, σ
Approximate 95% confidence intervals

 Fixed effects:
                       lower       est.     upper
(Intercept)          0.81277  5.262213  9.711655
visual0              0.82464  0.899900  0.975157
time                -0.27954 -0.215031 -0.150524
treat.fActive       -4.58882 -2.278756  0.031308
time:treat.fActive -0.15055 -0.056451  0.037646
attr(,"label")
[1] "Fixed effects:"

 Random Effects:
  Level: subject
                   lower    est.   upper
sd((Intercept)) 6.33067 7.23195 8.26153
sd(time)        0.24108 0.28096 0.32744

 Variance function:
         lower    est.   upper
power 0.014823 0.11108 0.20733
attr(,"label")
[1] "Variance function:"

 Within-group standard error:
 lower   est.  upper
3.8979 5.0312 6.4939
```

To fit model **M16.4**, we use the constructor-function pdDiag(). The function creates an object of class *pdDiag*, representing a diagonal positive-definite matrix (Sect. 14.2.1). Thus, in Panel R16.11, we update the object fm16.3, which represents model **M16.3**, using the argument random=pdDiag(~time). By specifying the argument, we imply a diagonal form of the variance-covariance matrix $\mathcal{D}$ of the random intercepts and slopes (Sect. 14.3.1).

Panel R16.11 presents the 95% CIs for all the parameters of model **M16.4**. They suggest that the mean structure could be simplified by removing the time:treat.f interaction. More detailed results for the model are provided in Table 16.2.

R16.12 *ARMD Trial*: Testing a null hypothesis about the $\boldsymbol{\theta}_D$ parameters for model **M16.4**. The model-fit object `fm16.3` was created in Panel R16.10

```
> anova(fm16.4, fm16.3)                  # H0: d12=0 (M16.4 ⊂ M16.3)
       Model df    AIC    BIC  logLik   Test L.Ratio p-value
fm16.4     1  9 6449.8 6492.6 -3215.9
fm16.3     2 10 6450.6 6498.2 -3215.3 1 vs 2   1.194  0.2745
```

In Panel R16.12, we use the REML-based LR test (Sect. 13.7.2) to verify the null hypothesis that in the matrix $\mathcal{D}$, defined in (16.13), the element $d_{12} = 0$. Toward this end, we apply the `anova()` function to the objects `fm16.4` and `fm16.3`, which represent the fitted models **M16.4** (null) and **M16.3** (alternative), respectively. We note that both models have the same mean structure so that the use of the REML-based LR test is justified. In addition to information criteria and REML values for both models, the results of the LR test, which is based on models **M16.3** and **M16.4**, are displayed. Given that the null hypothesis specifies a value inside the parameter space, the asymptotic χ^2 distribution with one degree of freedom can be used to assess the outcome of the test (Sect. 13.7.2). The result is not statistically significant at the 5% significance level. It indicates that, by assuming a simpler, diagonal structure of the matrix $\mathcal{D}$, we do not worsen the fit of the model. This conclusion is in agreement with the computed values of AIC: the value of 6,450.6 for model **M16.3** is slightly larger than the value of 6,449.8 for model **M16.4**, which indicates a slightly better fit of the latter model.

Note that, according to model **M16.4** and (16.15), the marginal variance of visual acuity for the subject i at time t can be written as

$$\text{Var}(y_{it}) = d_{11} + d_{22}\text{TIME}_{it}^2 + \sigma^2(\text{TIME}_{it})^{2\delta}.$$

Consequently, given that $\hat{\delta} = 0.11$, the implied marginal variance function is predominantly a quadratic function over time. As d_{11}, d_{22}, and σ^2 are necessarily positive, the function increases with time, which is in agreement with the observation made in the exploratory analysis (see, e.g., Panel R3.6 in Sect. 3.2).

Figure 16.6 presents the conditional Pearson residuals for model **M16.4**. As compared to the similar plot for model **M16.2** (see Fig. 16.1), it shows fewer residuals with an absolute value larger than the 97.5th percentile of the standard normal distribution.

Figure 16.7 presents the normal Q-Q plot of the conditional Pearson residuals per timepoint for model **M16.4**. The plot looks comparable to the corresponding plot for model **M16.2** shown in Fig. 16.3.

Note that Figs. 16.6 and 16.7 were constructed using the syntax similar to the one presented in Panels R16.7b and R16.7c, respectively. Thus, we do not present the details of the syntax for the two figures.

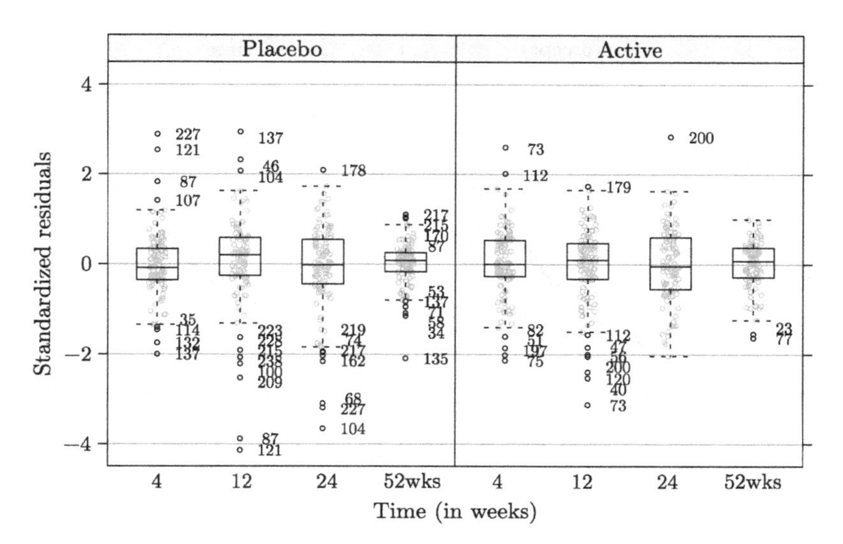

Fig. 16.6 *ARMD Trial*: Stripplots (and box-and-whiskers plots) of the conditional Pearson residuals for each timepoint and treatment group for model **M16.4**

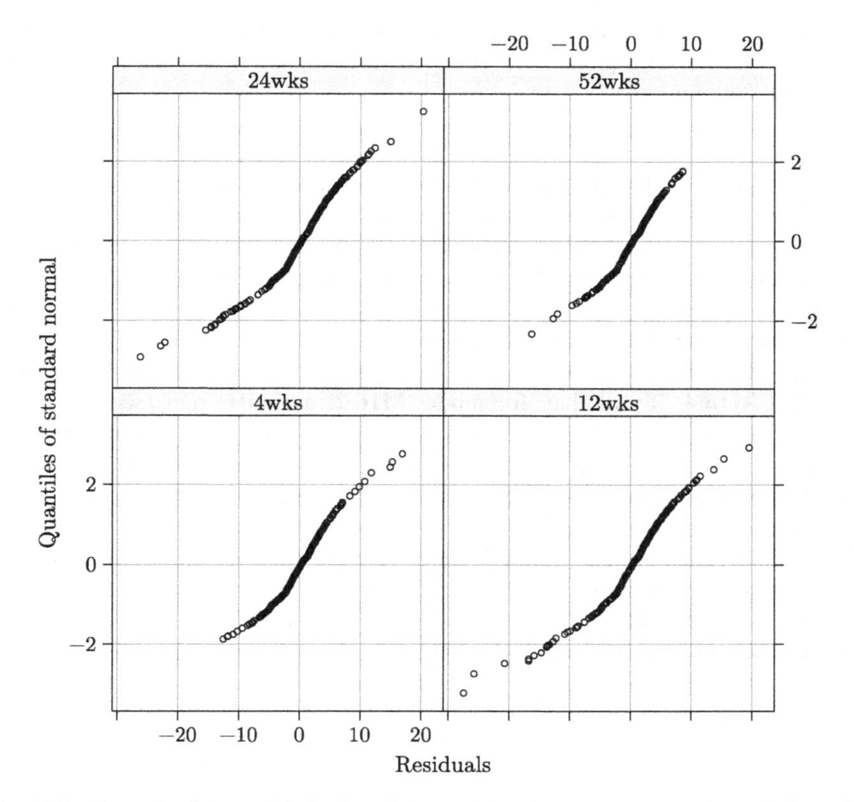

Fig. 16.7 *ARMD Trial*: Normal Q-Q plots of the conditional Pearson residuals for each timepoint for model **M16.4**

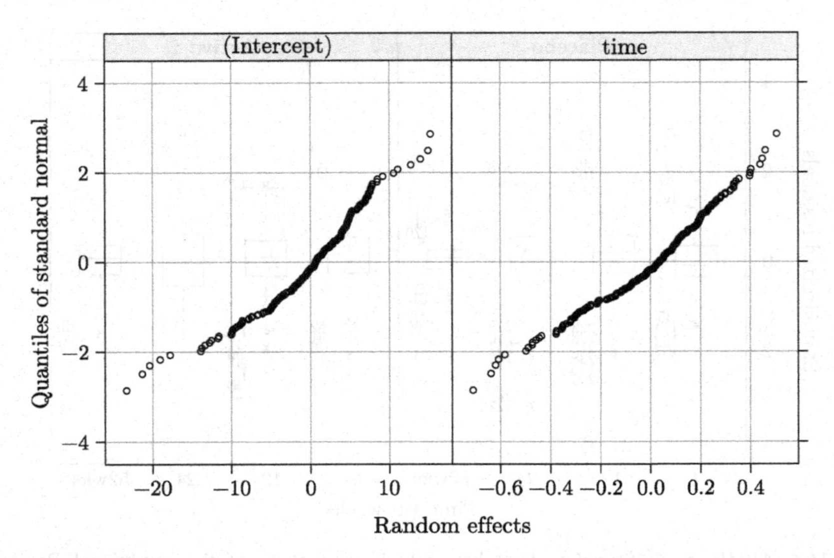

Fig. 16.8 *ARMD Trial*: Normal Q-Q plots of the predicted random effects for model **M16.4**

Figure 16.8 presents the normal Q-Q plots of the predicted random effects for model **M16.4**. The plots were obtained by using the following form of the `qqnorm()`-function call:

```
> qqnorm(fm16.4, ~ranef(.))          # Fig. 16.8
```

As a result, two plots are produced: one for the random intercepts and one for the random slopes. The latter is slightly closer to a straight line than the former. It is worth noting that the plot for the random intercepts resembles the one obtained for model **M16.2** (see Fig. 16.4). Recall that, as mentioned in Sect. 13.6.1, we should interpret the graphs with caution, because they may not necessarily reflect the correct distribution of the random effects.

Finally, Fig. 16.9 presents the predicted marginal and subject-specific values for model **M16.4**. Recall that, for model **M16.2**, a similar plot (see Fig. 16.5) showed a decreasing slope of the individual profiles, the same for all subjects. As a consequence, for some patients, e.g., no. 4 and 15, the so-obtained predicted individual profiles strongly deviated from the observed ones. This is not the case of the profiles shown in Fig. 16.9, for which the slopes vary. As a result, the predicted individual profiles follow more closely the observed values and capture, e.g., increasing trends in time. This illustrates that model **M16.4** offers a better fit to the data than model **M16.2**.

Given the satisfactory fit of model **M16.4**, in the next section, we focus on the inference about the mean structure of the model.

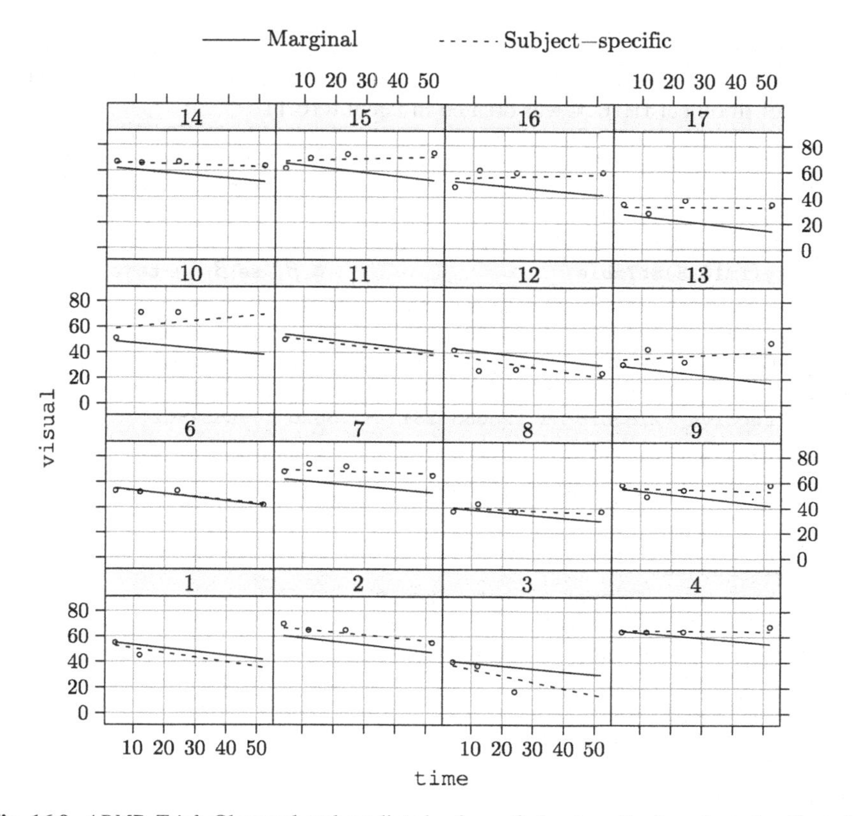

Fig. 16.9 *ARMD Trial*: Observed and predicted values of visual acuity for selected patients for model **M16.4**

16.4.3 Model with a Diagonal Matrix $\mathcal{D}$ and a Constant Treatment Effect

As mentioned in Sect. 16.4.2, the mean structure of model **M16.4** could be simplified by removing the $\text{TREAT}_i \times \text{TIME}_{it}$ interaction (see Panel R16.11). Toward this end, we specify model **M16.5** by modifying (16.10) as follows:

$$\begin{aligned}\text{VISUAL}_{it} &= \beta_0 + \beta_1 \times \text{VISUAL0}_i + \beta_2 \times \text{TIME}_{it} + \beta_3 \times \text{TREAT}_i \\ &\quad + b_{0i} + b_{2i} \times \text{TIME}_{it} + \varepsilon_{it}. \end{aligned} \tag{16.17}$$

As compared to (16.10), (16.17) does not contain the $\beta_4 \times \text{TREAT}_i \times \text{TIME}_{it}$ interaction term in the fixed-effects part. Note that we keep all other elements of the model specification as for model **M16.4**. In particular, the variance-covariance matrix $\mathcal{D}$ is given by (16.16).

To fit model **M16.5**, we modify the LM formula and update the object `fm16.4` using the new formula object. The syntax is presented in Panel R16.13. The panel

R16.13 *ARMD Trial*: Fixed-effects estimates, their approximate standard errors, and 95% confidence intervals for the variance-covariance parameters of model **M16.5**. The model-fit object `fm16.4` was created in Panel R16.11

```
> lm3.form <- formula(visual ~ visual0 + time + treat.f) # (12.9)
> fm16.5  <-                                     # M16.5 ← M16.4
+    update(fm16.4,
+           lm3.form, data = armd)
> summary(fm16.5)$tTable                         # β̂, se(β̂), t-test
                   Value Std.Error  DF  t-value    p-value
 (Intercept)     5.44156  2.261866 632   2.4058 1.6424e-02
 visual0         0.89983  0.038215 231  23.5464 2.5503e-63
 time           -0.24156  0.023917 632 -10.0997 2.4641e-22
 treat.fActive -2.65528  1.128683 231  -2.3525 1.9485e-02
> intervals(fm16.5, which = "var-cov")           # 95% CI for θ_D, δ, σ
 Approximate 95% confidence intervals

  Random Effects:
   Level: subject
                     lower    est.  upper
 sd((Intercept)) 6.33448 7.23570 8.2651
 sd(time)        0.24121 0.28102 0.3274

  Variance function:
          lower    est.   upper
 power 0.015687 0.11052 0.20535
 attr(,"label")
 [1] "Variance function:"

  Within-group standard error:
  lower   est.  upper
 3.9177 5.0391 6.4815
```

also shows the results of the t-tests for the fixed effects (Sect. 13.7.1). Note that these are the marginal-approach tests (Sect. 5.6). Thus, the effect of each covariate is tested under the assumption that all other covariates are included in the model as well. The result of the test for the `treat.f` factor is statistically significant at the 5% significance level. It suggests a time-independent, negative average effect of the active treatment. This finding is in agreement with the results of the exploratory analysis (Sect. 3.2) and of the previous analysis using an LM with fixed effects for correlated data (Chap. 12). Note that the point estimates of the fixed effects, shown in Panel R16.13, are close to the corresponding estimates obtained for the final model **M12.3** for correlated data (see Table 12.2).

Panel R16.13 also presents the 95% CIs for all the variance-covariance parameters of model **M16.5**. The point estimates and intervals are very close to

R16.14 *ARMD Trial*: The estimates of matrices $\mathcal{D}$, $\mathcal{R}_i$, and $\mathcal{V}_i$ for model **M16.5**. The model-fit object fm16.5 was created in Panel R16.13

```
> VarCorr(fm16.5)                                  # D^: (16.16), σ^
  subject = pdDiag(time)
              Variance  StdDev
  (Intercept) 52.355293 7.23570
  time         0.078974 0.28102
  Residual    25.392868 5.03913
> getVarCov(fm16.5,                                # R^_i: (16.8)
+           type = "conditional", individual = "2")
  subject 2
  Conditional variance covariance matrix
         1      2      3      4
  1 34.498  0.00  0.000  0.000
  2  0.000 43.98  0.000  0.000
  3  0.000  0.00 51.262  0.000
  4  0.000  0.00  0.000 60.816
    Standard Deviations: 5.8735 6.6317 7.1597 7.7984
> (fm16.5cov <-                                    # V^_i: (16.9)
+    getVarCov(fm16.5,
+              type = "marginal",
+              individual = "2"))
  subject 2
  Marginal variance covariance matrix
         1       2       3       4
  1 88.117  56.146  59.937  68.782
  2 56.146 107.710  75.100 101.640
  3 59.937  75.100 149.110 150.920
  4 68.782 101.640 150.920 326.720
    Standard Deviations: 9.387 10.378 12.211 18.075
> cov2cor(fm16.5cov[[1]])                          # Corr(V^_i)
          1       2       3       4
  1 1.00000 0.57633 0.52290 0.40538
  2 0.57633 1.00000 0.59261 0.54180
  3 0.52290 0.59261 1.00000 0.68375
  4 0.40538 0.54180 0.68375 1.00000
```

those displayed in Panel R16.11 for model **M16.4**. This is not surprising, given that the two models differ only slightly with respect to their mean structure.

Another summary of estimates of the parameters of model **M16.5** is given in Table 16.2, which also contains estimated parameters of models **M16.3** and **M16.4**.

Panel R16.14 displays the estimated forms of matrices $\mathcal{D}$, $\mathcal{R}_i$, and $\mathcal{V}_i$ for model **M16.5**. The estimated marginal variance-covariance matrix $\widehat{\mathcal{V}}_i$ indicates an increasing trend of variances of visual acuity measurements over time, while the corresponding correlation matrix suggests a decreasing correlation between the measurements obtained at more distant timepoints. These findings are in agreement with the results of the exploratory analysis (Sect. 3.2) and with the results obtained for model **M12.3** for correlated data (Table 12.2). Note, however, that a direct comparison of the estimated marginal matrices to their counterparts obtained for model **M12.3** is not appropriate, because the matrices for model **M16.5** are much more structured than those of model **M12.3** (see a similar comment in Sect. 16.3.2).

16.5 An Alternative Residual Variance Function: varIdent(·)

The LMMs, presented in Sects. 16.3 and 16.4, were specified with the use of the varPower(·) variance function (see the definition of the matrix $\boldsymbol{\Lambda}_i$ in (16.8)). This may be an overly constrained function, because it assumes that the variances of the visual acuity measurements change as a power function of the measurement time. The choice was motivated by the results obtained in Chaps. 9 and 12, where models, defined with the use of the varPower(·) variance function, fitted the ARMD data better than models with unconstrained variances, specified with the use of the varIdent(·) function (see, e.g., Sect. 12.5.2). However, it is possible that, in the framework of LMMs, a more general variance function might allow obtaining a better fit than the power function.

To verify this hypothesis, we will use the LR test constructed based on models **M16.3** and **M16.6**. Both of the models have the same fixed- and random effects structure, given by (16.10). They differ with respect to $\mathcal{R}_i$ matrix specification. Specifically, in the former model, the matrix $\mathcal{R}_i$ is defined using power function. In contrast, in the latter model the matrix $\mathcal{R}_i$ is defined as follows:

$$\mathcal{R}_i = \sigma_1^2 \begin{pmatrix} 1 & 0 & 0 & 0 \\ 0 & \frac{\sigma_2^2}{\sigma_1^2} & 0 & 0 \\ 0 & 0 & \frac{\sigma_3^2}{\sigma_1^2} & 0 \\ 0 & 0 & 0 & \frac{\sigma_4^2}{\sigma_1^2} \end{pmatrix} \equiv \sigma^2 \begin{pmatrix} \delta_1^2 & 0 & 0 & 0 \\ 0 & \delta_2^2 & 0 & 0 \\ 0 & 0 & \delta_3^2 & 0 \\ 0 & 0 & 0 & \delta_4^2 \end{pmatrix}, \tag{16.18}$$

where $\delta_t \equiv \sigma_t/\sigma_1$ $(t = 1, \ldots, 4)$ is the ratio of SD of the visual acuity measurements at occasion t relative to SD of the measurements at the first occasion, and where $\sigma^2 \equiv \sigma_1^2$. This parameterization corresponds to a *varIdent*-class variance function (Sect. 7.3.1) and is specified in such a way that it allows identifying the variance-function parameters δ_t (Sect. 7.3.2).

To fit model **M16.6**, we update the object `fm16.3` using an appropriate form of the `varIdent()` constructor function in the `weights` argument of the `lme()` function. The suitable syntax and results of fitting of the model are displayed in Panel R16.15a. Additional results are provided in Table 16.3. Panel R16.15b also includes the result of the LR test obtained with the use of the `anova()` function,

Table 16.3 *ARMD Trial*: REML-based estimates[a] for linear mixed-effects models with random intercepts and slopes

	Parameter	fm16.6	fm16.7
Model label		**M16.6**	**M16.7**
Log-REML value		−3204.05[b]	−3218.57
Fixed effects			
Intercept	β_0	5.10(2.18)	5.35(2.33)
Visual acuity at t=0	β_1	0.90(0.04)	0.90(0.04)
Time (in weeks)	β_2	−0.21(0.03)	−0.22(0.03)
Trt(Actv *vs.* Plcb)	β_3	−2.18(1.12)	−2.31(1.21)
Tm × Treat(Actv)	β_4	−0.06(0.05)	−0.06(0.05)
reStruct(`subject`)			
SD(b_{i0})	$\sqrt{d_{11}}$		7.35(6.41,8.43)
SD(b_{i1})	$\sqrt{d_{22}}$		0.28(0.24,0.33)
Scale	σ_1		6.68(6.25,7.14)

[a]Approximate SE for fixed effects and 95% CI for covariance parameters are included in parentheses
[b]Likelihood optimization did not converge

which is based on the likelihoods of models **M16.6** and **M16.3**. Note that the latter (null) model is nested in the former. The outcome of the test is statistically significant at the 5% significance level and suggests that the use of the more general varIdent(·) variance function to define matrix $\mathcal{R}_i$, as in (16.18), gives a better fit than the use of the varPower(·) function.

We need to be careful before accepting this conclusion, though. A closer inspection of the results displayed in Panel R16.15 reveals that the estimated value of parameter δ_4 is extremely small and substantially differs from the estimated values of δ_2 and δ_3. This is surprising, because all previous analyses indicated that the variance of the last visual acuity measurement (at week 52) was the largest.

A signal of the problems with the estimation of model **M16.6** can be also obtained by, e.g., attempting to compute confidence intervals for the variance-covariance parameters. In particular, issuing the command

```
> intervals(fm16.6, which = "var-cov")
```

results in an error message indicating problems with estimating the variance-covariance matrix for the estimates of the parameters.

Finally, the problem with convergence of the estimation algorithm for model **M16.6** is also clearly reflected in the normal Q-Q plot of the conditional Pearson residuals, shown in Fig. 16.10 and obtained by issuing the command

```
> qqnorm(fm16.6, ~resid(.)|time.f)          # Fig. 16.10
```

Note that the residuals for week 52 are all equal to 0.

To investigate the source of the problem, we present, in Fig. 16.11, plots of the cross-sections of the restricted-likelihood surface for δ_2, δ_3, δ_4, and σ. For brevity, we do not show the R code used to create the figure. Each plot is obtained by

R16.15 *ARMD Trial*: Fitting model **M16.6** and testing its variance function using a REML-based likelihood-ratio test. The model-fit object `fm16.3` was created in Panel R16.10

(a) *Fitting of model* **M16.6**

```
> (fm16.6 <-                                              # M16.6 ← M16.3
+     update(fm16.3, weights = varIdent(form = ~1 | time.f)))
  Linear mixed-effects model fit by REML
    Data: armd
    Log-restricted-likelihood: -3204
    Fixed: visual ~ visual0 + time + treat.f + time:treat.f
         (Intercept)             visual0                 time
             5.10354             0.90120             -0.21041
      treat.fActive time:treat.fActive
            -2.18434            -0.05931

  Random effects:
   Formula: ~1 + time | subject
   Structure: General positive-definite, Log-Cholesky parametrization
              StdDev  Corr
  (Intercept) 7.34621 (Intr)
  time        0.31104 -0.132
  Residual    4.62311

  Variance function:
   Structure: Different standard deviations per stratum
   Formula: ~1 | time.f
   Parameter estimates:
        4wks       12wks       24wks       52wks
  1.00000000 1.62525293 1.74357631 0.00051508
  Number of Observations: 867
  Number of Groups: 234
```

(b) *REML-based LR test for the variance function*

```
> anova(fm16.3, fm16.6)      # varPower (M16.3) ⊂ varIdent (M16.6)
         Model df    AIC    BIC  logLik   Test L.Ratio p-value
  fm16.3     1 10 6450.6 6498.2 -3215.3
  fm16.6     2 12 6432.1 6489.2 -3204.0 1 vs 2  22.499  <.0001
```

fixing the other parameters at the reported REML estimates. The panel for δ_4, the ratio of the residual SD of the visual acuity measurements at 52 weeks relative to week 4, shows an approximately flat horizontal line close to zero. More precisely, the line shows that the difference between the log-restricted-likelihood for the values of δ_4 within the interval presented in the plot and the reported log-REML value

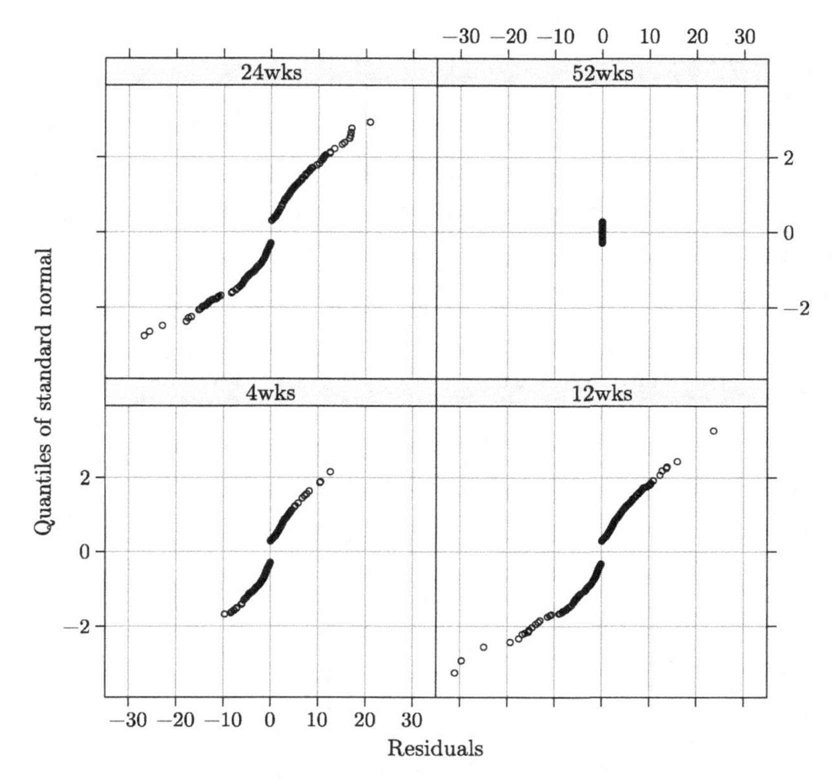

Fig. 16.10 *ARMD Trial*: The normal Q-Q plot of the conditional Pearson residuals for model **M16.6**. Panel for 52 weeks indicates the problem with model fit

of -3204.05 ranges between 2.4×10^{-7} and -4.0×10^{-7}. This indicates that, if we assume model **M16.6**, the data contain very little information about this particular parameter, because the log-restricted-likelihood function surface is virtually flat in the corresponding direction of the parameter space. Moreover, the plot for δ_4, unlike the other plots shown in Fig. 16.11, does not suggest any maximum of the likelihood function within the presented interval of δ_4 values. This means that the REML estimate, reported by the `lme()` function in Panel R16.14a, is not an optimum value.

Given the close similarity of the structure of models **M16.6** and **M16.3**, a question is: Why were there no apparent problems with fitting of the latter model? Although the models are similar, they differ with respect to the form of the marginal variance-covariance structure of visual acuity measurements. The form of the covariance of the measurements obtained for the subject i at different times, implied by model **M16.6**, is the same as the one resulting from model **M16.3** and is given by (16.14). However, the variance for a measurement obtained at the time t, implied by model **M16.6**, is

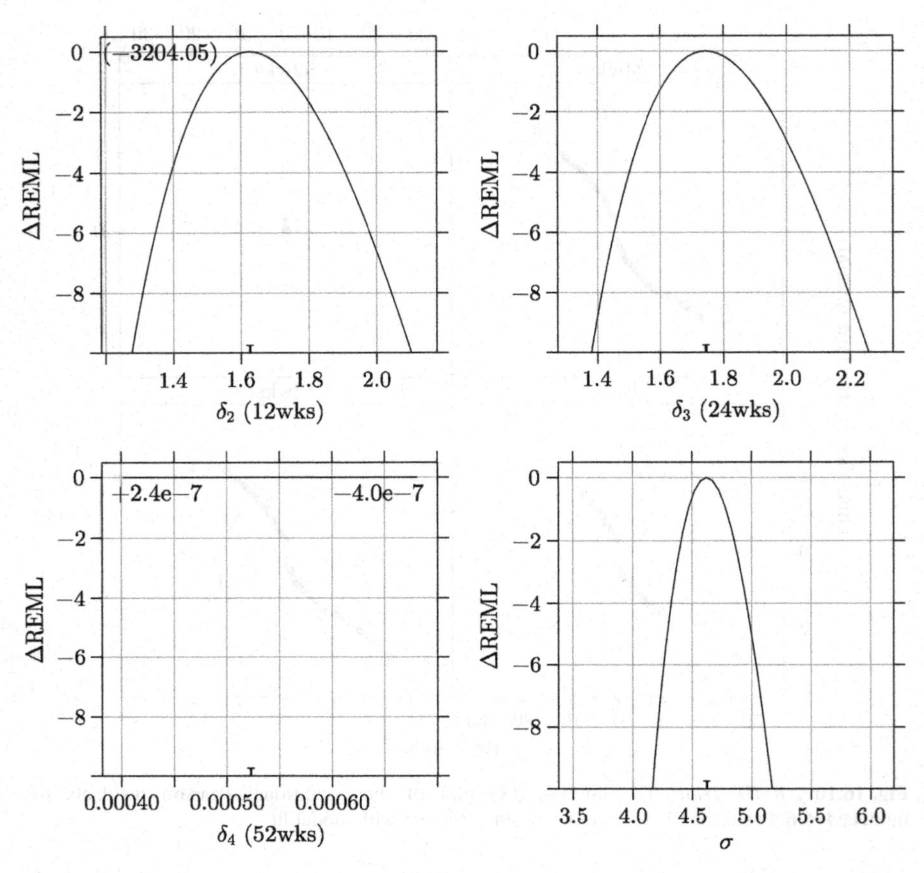

Fig. 16.11 *ARMD Trial*: Differences (ΔREML) between the values of the log-restricted-likelihood for model **M16.6** at the reported REML estimate and at different values of parameters δ_2, δ_3, δ_4, and σ. The value of 0 at the vertical axis corresponds to the reported log-REML value of −3204.05. Flat line in panel for δ_4 indicates problem with model fit

$$\text{Var}(y_{it}) = d_{11} + 2d_{12}\text{TIME}_{it} + d_{22}\text{TIME}_{it}^2 + \sigma^2\delta_t^2. \tag{16.19}$$

Equations (16.14) and (16.19) define the ten unique elements of the marginal variance-covariance matrix $\mathcal{V}_i$ for model **M16.6** as linear functions of seven parameters: d_{11}, d_{12}, d_{22}, σ^2, δ_2, δ_3, and δ_4. Given that the number of parameters is close to the number of equations, collinearity among the parameters may result, with consequences in the form of convergence problems of the estimation algorithm.

On the other hand, the right-hand side of (16.15) for model **M16.3** has fewer parameters and involves a power function of time, which is nonlinear in terms of the parameter δ. Hence, in this case, the collinearity is less likely to appear.

Thus, as compared to model **M16.6**, model **M16.3** imposes an additional restriction on the form of the marginal variance-covariance structure. The restriction limits the parameter space, in which it is possible to find an optimum solution, because the data become more informative.

It follows that, to use the Ident(·) function, some additional restrictions on the form of model **M16.6** would need to be introduced. However, we will not pursue this direction further but rather leave it as an exercise to the reader.

16.6 Testing Hypotheses About Random Effects

As mentioned in Sect. 13.7.2, formal tests of hypotheses about the variance-covariance structure can be performed using the LR test based on the restricted likelihood function. An important issue is the null distribution of the test statistics. In particular, when the values of the variance-covariance parameters, compatible with the null hypothesis, lie in the interior of the parameter space, the null distribution is a χ^2 distribution with the number of degrees of freedom equal to the difference in the number of (independent) variance-covariance parameters between the null and alternative models (Sect. 13.7.2). Examples of such tests were shown in Sects. 16.4.2 (Panel R16.11) and 16.5 (Panel R16.15).

However, when the values of the variance-covariance parameters, compatible with the null hypothesis, lie on the boundary of the parameter space, the exact form of the null distribution is difficult to obtain. As it was mentioned in Sect. 13.7.2, in certain cases (see, e.g., Verbeke and Molenberghs 2000, Sect. 6.3.4), the distribution is given by a mixture of several χ^2 distributions. Note that this result has been obtained by assuming that the residual errors are independent and homoscedastic. In other cases, the only practical alternative is to simulate the null distribution. In R this can be done using the `simulate()` function from the **nlme** package or using the function `exactRLRT()` from the package **RLRsim**.

The use of the functions was briefly reviewed in Sect. 14.7. It was noted there that both functions only allow for independent, homoscedastic residual errors. Moreover, the function `exactRLRT()` accommodates only independent random effects, while the function `simulate()` is not defined for model-fit objects of class *gls*.

These limitations preclude us from testing, e.g., whether inclusion of random intercepts is improving the fit of model **M16.2**, as compared to a model without random effects, but with residual variances expressed with the help of the Power(·) variance function. For the same reason, we cannot test the statistical significance of extending model **M16.2** by the inclusion of random slopes, which leads to model **M16.3**.

In these cases, the plausibility of the modifications of the random effects structure needs to be assessed using, e.g., residual diagnostics and/or by applying the information criteria (Sect. 13.7.2). Panel R16.16 presents the values of the AIC for models **M16.1**–**M16.5**. It can be seen that, e.g., the AIC for model **M16.2**,

R16.16 *ARMD Trial*: The values of Akaike's Information Criterion for models **M16.1–M16.5**

```
> AIC(fm16.1, fm16.2,                          # M16.1, M16.2
+      fm16.3, fm16.4)                         # M16.3, M16.4
        df    AIC
  fm16.1  7 6592.0
  fm16.2  8 6537.1
  fm16.3 10 6450.6
  fm16.4  9 6449.8
> fm16.4ml <- update(fm16.4, method = "ML")
> fm16.5ml <- update(fm16.5, method = "ML")
> anova(fm16.4ml, fm16.5ml)                    # M16.4 ⊂ M16.5
          Model df    AIC    BIC  logLik   Test L.Ratio p-value
  fm16.4ml    1  9 6438.0 6480.9 -3210.0
  fm16.5ml    2  8 6437.4 6475.5 -3210.7 1 vs 2  1.3972  0.2372
```

i.e., 6,537.1, is much larger than the value of 6,450.6 for model **M16.3**. This points to a better fit of the latter model. Also, as suggested by, e.g., Fig. 16.9, the predicted values obtained for model **M16.3** follow more closely the observed ones, as compared to model **M16.2** (see Fig. 16.5).

Note that the lowest value of the AIC is obtained for model **M16.5**, suggesting that the model provides the best overall fit to the data. This reflects the choices we made with respect to the random-effects structure in the process of arriving at the model.

In the remainder of this section, we illustrate the use of the analytic results and of the R simulation functions for testing hypotheses about the random effects structure with parameter values at the boundary of the parameter space. Toward this end, we consider several models for the ARMD data which assume homoscedasticity of the residual errors.

16.6.1 Test for Random Intercepts

Let us first consider model **M16.1** containing random intercept. To test whether subject-specific random intercepts are needed, we might use a REML-based LR test based on the alternative model **M16.1** and a null model that assumes homoscedastic residual errors and no random effects.

In Panel R16.17, we conduct the REML-based LRT by referring the LR-test statistic to a null distribution obtained using a mixture of χ^2 distributions or a simulation technique.

In particular, for the first approach, presented in Panel R16.17a, we create the object `vis.gls1a`, which represents the fit of the null model. The model does not

R16.17 *ARMD Trial*: The REML-based likelihood-ratio test for no random intercepts in model **M16.1**. The formula-object `lm2.form` and the model-fit object `fm16.1` were created in Panel R16.1

(a) *Using* $0.5\chi_0^2 + 0.5\chi_1^2$ *as the null distribution*

```
> vis.gls1a   <-                                # Null model
+    gls(lm2.form, data = armd)
> (anova.res  <- anova(vis.gls1a, fm16.1))   # Null vs. M16.1
           Model df    AIC    BIC logLik   Test L.Ratio p-value
  vis.gls1a    1  6 6839.9 6868.5  -3414
  fm16.1       2  7 6592.0 6625.3  -3289 1 vs 2  249.97  <.0001
> (anova.res[["p-value"]][2])/2                # 0.5χ0^2 + 0.5χ1^2
  [1] 0
```

(b) *Using the function* `exactRLRT()` *to simulate the null distribution*

```
> library(RLRsim)
> exactRLRT(fm16.1)                            # M16.1 (alternative)
       simulated finite sample distribution of RLRT.  (p-value
       based on 10000 simulated values)

  data:
  RLRT = 249.97, p-value < 2.2e-16
```

---

include any random intercepts and is defined by the formula `lm2.form`. Thus, it has the same mean structure as the alternative model **M16.1**, which is represented by the object `fm16.1`. Then, we apply the `anova()` to calculate value of the REML-based LR test statistics.

Note that we are testing the null hypothesis that the variance of the random intercept is zero, which is on the boundary of the parameter space. Thus, the p-value reported by `anova()` is computed by referring the value of the LR-test statistic to the incorrect χ_1^2 null distribution. In this case, the appropriate asymptotic distribution is a 50%–50% mixture of the χ_0^2 and χ_1^2 distributions (Sect. 13.7.2). To obtain the correct p-value, we divided the χ_1^2-based p-value, extracted from the object `anova.res` containing the results of the `anova()`-function call, by 2. Clearly, in the current case, the adjusted p-value indicates that the result of the test is statistically significant. It allows us to reject the null hypothesis that the variance of the distribution of random intercepts is equal to 0.

An alternative, shown in Panel R16.17b, is to use the empirical null distribution of the LR test, obtained with the help of the function `exactRLRT()` from the package **RLRsim** (Sect. 14.7). In the panel, we show the result of application of the function to the object `fm16.1`. Because we test a random effect in model **M16.1**, which contains only a single random effect, we use the abbreviated form of the function call, with m as the only argument. The p-value of the REML-based LR test, estimated from 10,000 simulations (the default), clearly indicates that the result of the test is statistically significant. In this case, given the importance of including the random intercepts into the model, which are needed to adjust for the correlation between visual acuity measurements, there is not much difference with the p-value obtained using the asymptotic 50%–50% mixture of the χ_0^2 and χ_1^2 distributions.

To simulate the null distribution of the LRT, we could consider applying the `simulate()` function to objects `vis.gls1` (see Panel R6.3) and `fm16.1`. Unfortunately, the necessary `simulate.gls()` method is not developed for model-fit objects of class *gls*. In the next section, we will illustrate how to use the `simulate()` function to test for the need of random slope.

16.6.2 Test for Random Slopes

For illustrative purposes, we consider a model with uncorrelated subject-specific random intercepts and slopes and independent, homoscedastic residual errors. That is, we consider a model specified by (16.10)–(16.12), with $\mathcal{D}$ given by (16.16), and $\mathcal{R}_i = \sigma^2 \times \boldsymbol{I}_4$. We will refer to this newly defined model as **M16.7**. In this section, we will use the REML-based LR test to test whether random slopes are needed in model **M16.7**. The test involves comparison of two models, namely, **M16.1** (null) and **M16.7** (alternative).

In Panel R16.18, we introduce three approaches to perform the LR test for random slopes.

To begin, in Panel R16.18a, we fit model **M16.7**, which contains random slopes, by modifying model **M16.4**. More specifically, we assume a constant residual variance. The resulting model is stored in the model-fit object `fm16.7`. The results of fitting of the model are provided in Table 16.3.

In the first approach, shown in Panel R16.18b, we perform the REML-based LR test and explore the use of a 50%–50% mixture of the χ_1^2 and χ_2^2 distributions as the null distribution (see Verbeke and Molenberghs 2000, Sect. 6.3.4). To compute the corresponding p-value, we extract the LR-test statistic value from the object `an.res`, which contains the results of the `anova()`-function call, and we use it as an argument of the `pchisq()` function, which computes the upper tail probabilities of the χ^2 distributions with 1 and 2 degrees of freedom. Clearly, the adjusted p-value indicates that the result of the test is statistically significant. Thus, the test allows us to reject the null hypothesis that the variance of random slopes is equal to 0.

In Panels R16.18c and R16.18d, we consider simulating the null distribution of the REML-based LR-test statistic.

Toward this end, in Panel R16.18c, we use the `exactRLRT()` function. Note that, as it was mentioned earlier, the function allows only for independent random effects. This is the reason why we illustrate the use of the function for model **M16.7** with a diagonal matrix $\mathcal{D}$. Because we consider a model with two variance components, i.e., random intercepts and random slopes, we need to specify all three arguments

R16.18 *ARMD Trial*: The REML-based likelihood-ratio test for random slopes for model **M16.7**. The model-fit objects `fm16.1` and `fm16.4` were created in Panels R16.1 and R16.10, respectively

(a) *Fitting model* **M16.7**

```
> fm16.7 <-                                    # M16.7 ← M16.4
+    update(fm16.4, weights = NULL,            # Constant resid. variance
+           data = armd)
```

(b) *Using* $0.5\chi^2_1 + 0.5\chi^2_2$ *as the null distribution*

```
> (an.res <-                                   # M16.1 (null)
+      anova(fm16.1, fm16.7))                  # M16.7 (alternative)
         Model df    AIC    BIC  logLik   Test L.Ratio p-value
  fm16.1     1  7 6592.0 6625.3 -3289.0
  fm16.7     2  8 6453.1 6491.2 -3218.6 1 vs 2  140.83  <.0001
> (RLRT <- an.res[["L.Ratio"]][2])             # LR-test statistic
  [1] 140.83
> .5 * pchisq(RLRT, 1, lower.tail = FALSE) + # 0.5χ²₁+ 0.5χ²₂
+   .5 * pchisq(RLRT, 2, lower.tail = FALSE)
  [1] 1.3971e-31
```

(c) *Using the function* `exactRLRT()` *to simulate the null distribution*

```
> mAux <-                                      # Auxiliary model with ...
+  update(fm16.1, random = ~0 + time|subject,     # ... random slopes only.
+          data = armd)
> exactRLRT(m = mAux,                          # Auxiliary model
+          m0 = fm16.1,                        # M16.1 (null)
+          mA = fm16.7)                        # M16.7 (alternative)
     simulated finite sample distribution of RLRT.  (p-value
     based on 10000 simulated values)

  data:
  RLRT = 140.83, p-value < 2.2e-16
```

(d) *Using the function* `simulate()` *to simulate the null distribution*

```
> vis.lme2.sim <-                                  # M16.1 (null)
+    simulate(fm16.1, m2 = fm16.7, nsim = 10000) # M16.7 (alternative)
> plot(vis.lme2.sim, df = c(1, 2),                 # Fig. 16.12
+      abline = c(0,1, lty=2))
```

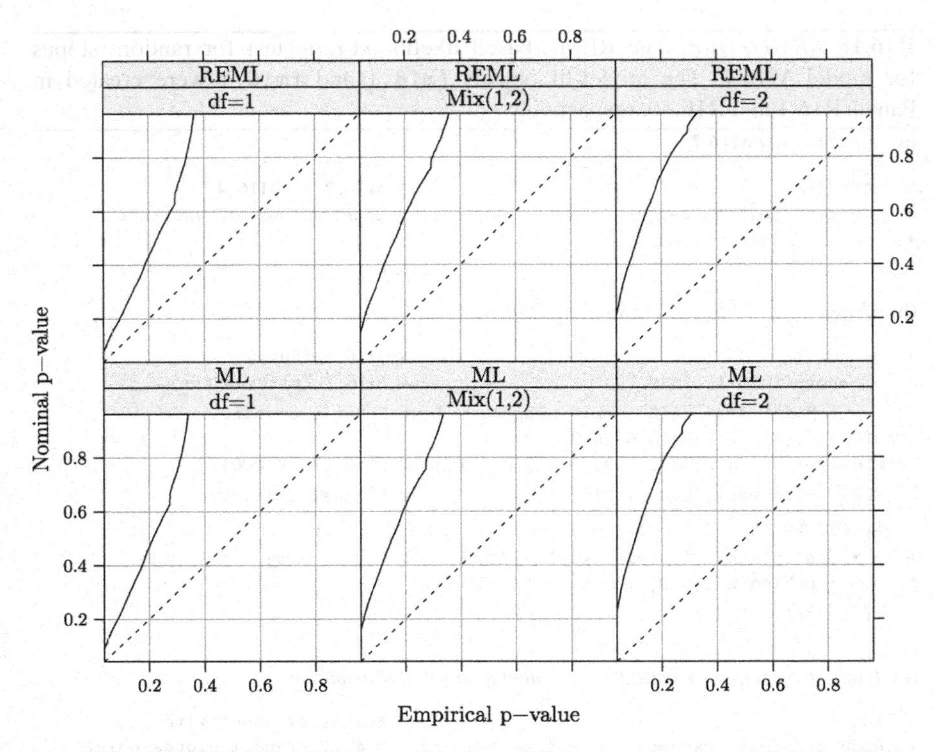

Fig. 16.12 *ARMD Trial*: Empirical and nominal p-values for testing the need of random slopes in model **M16.7**

m, m0, and mA of the function exactRLRT() (Sect. 14.7). The required form of the function call is shown in Panel R16.18b. The simulated p-value is essentially equal to 0, indicating that null hypothesis can be rejected.

Finally, in Panel R16.18d, the function simulate() is applied to obtain a plot of empirical and nominal p-values (Sect. 14.7). The former are generated by simulating the values of the REML-based LR-test statistic. The plot, in turn, can be used to choose the appropriate null distribution for the calculation of the p-value corresponding to the observed value of the test statistic.

More specifically, the function simulate() is applied to the objects fm16.1 and fm16.7, with the former specified as the null model and the latter indicated, with the help of the argument m2, as the alternative model. The number of the simulated test-statistic values is set, with the help of the nsim argument, at 10,000.

The plot() statement creates a plot of the empirical and nominal p-values of the LR-test statistic. The nominal p-values are computed using three distributions: χ^2_1, χ^2_2, and a 50%–50% mixture of χ^2_1 and χ^2_2. The required degrees of freedom are passed to the plot() function using the argument df in the form of a numeric vector (Sect. 14.7). To include in the plot, e.g., a 65–35% mixture, the argument weights=c(0.65,0.35) should explicitly be used.

The resulting plot is shown in Fig. 16.12. Note that two rows of three panels are displayed: one row for the REML and one for the ML estimation. As was mentioned in Sect. 14.7, by default, the function `simulate.lme()` uses both forms of the LR test.

The plot shows that the nominal p-values, obtained using χ_1^2, χ_2^2, or a 50%–50% mixture of χ_1^2 and χ_2^2 distributions, are larger than the corresponding simulated values. This implies that the use of any of those distributions would result in a conservative test.

16.7 Analysis Using the Function `lmer()`

In this section, we refit models **M16.1** and **M16.7**, presented in Sects. 16.2.1 and 16.6.2, respectively, using the function `lmer()` from the package **lme4.0**. The choice of the models is dictated by the fact that, at the time of writing of this book, the function allows only for independent, homoscedastic residual errors. Note that the two models do not adequately describe the ARMD data, as can be concluded from the results of the analyses obtained with the help of the `lme()` function. Thus, the results presented in the current section should be treated mainly as the illustration of the use of the `lmer()` function.

16.7.1 Basic Results

In Panel R16.19, we demonstrate how to fit model **M16.1** with the help of the function `lmer()`. The model included random intercepts and assumed that residual variance was constant. It was fitted using the `lme()` function in Panel R16.1, with the fit stored in the object `fm16.1`.

In Panel R16.19a, we present the `lmer()`-function syntax for fitting model **M16.1**. Note the direct specification of the random-effects structure in the `formula` argument (Sect. 15.3.1). Also, it is worth noting that the argument `data` is provided with a data frame, and not with a grouped data object. In fact, in contrast to the `lme()` function, the use of grouped data objects is neither needed nor recommended (Sect. 15.3). The model is fitted using REML, which is the default estimation method.

The results of the fitted model are printed using the generic `print()` function. It is worth noting that the values of the t-test statistics for the fixed effects are provided without any p-values (Sect. 15.5). Methods to calculate p-values will be presented later in this section.

The `corr=FALSE` argument, used in the `print()`-function call, excludes the estimated correlation matrix of the fixed effects from the printout. This is because the names of the fixed effects are long and the printout of the matrix would not be legible. Instead, in Panel R16.19b, the variance-covariance matrix of the fixed

R16.19 *ARMD Trial*: Model **M16.1** fitted using the function `lmer()`

(a) *Model fit and results*

```
> require(lme4.0)
> fm16.1mer  <-                                  # M16.1
+     lmer(visual ~ visual0 + time * treat.f + (1|subject),
+          data = armd)
> print(fm16.1mer, corr = FALSE)                 # Corr(β) not printed
  Linear mixed model fit by REML
  Formula: visual ~ visual0 + time * treat.f + (1 | subject)
     Data: armd
    AIC  BIC logLik deviance REMLdev
   6592 6625  -3289     6566    6578
  Random effects:
   Groups   Name        Variance Std.Dev.
   subject  (Intercept) 80.6     8.98
   Residual             74.4     8.63
  Number of obs: 867, groups: subject, 234

  Fixed effects:
                       Estimate Std. Error t value
  (Intercept)            9.2881     2.6817    3.46
  visual0                0.8264     0.0447   18.50
  time                  -0.2122     0.0229   -9.26
  treat.fActive         -2.4220     1.4999   -1.61
  time:treat.fActive    -0.0496     0.0336   -1.48
```

(b) *Correlation matrix for* $\widehat{\boldsymbol{\beta}}$

```
> vcovb <- vcov(fm16.1mer)                          # Var(β)
> corb <- cov2cor(vcovb)                            # Corr(β)
> nms <- abbreviate(names(fixef(fm16.1mer)), 5)
> rownames(corb) <- nms
> corb
  5 x 5 Matrix of class "dpoMatrix"
             [,1]       [,2]       [,3]      [,4]       [,5]
  (Int)  1.00000 -0.9200264 -0.1847104 -0.294970  0.1263825
  visl0 -0.92003  1.0000000 -0.0028807  0.022204  0.0017642
  time  -0.18471 -0.0028807  1.0000000  0.334926 -0.6832037
  trt.A -0.29497  0.0222042  0.3349262  1.000000 -0.4757390
  tm:.A  0.12638  0.0017642 -0.6832037 -0.475739  1.0000000
```

effects is extracted directly from the model-fit object using the function `vcov()`. Then the corresponding correlation matrix is obtained by applying the function `cov2cor()`. Before printing out the resulting object, `corb`, we abbreviate the names of the fixed effects with the use of the abbreviate() function. Note the use of the function fixef() for extracting the fixed effects from the model-fit object (Table 15.3). The abbreviated names are then used to label the rows of the correlation matrix corb. In this way, the printout of the matrix gains in transparency.

The results shown in Panel R16.19 correspond to those presented in Panel R16.1 and Table 16.1.

In Panel R16.20, we present the methods to compute various variance components, implied by model **M16.1**. In particular, in Panel R16.20a, we use the function VarCorr() (Table 15.3) to extract the estimated variance and standard deviation of random intercepts. Additionally, the estimate of the scale parameter σ is also displayed. Note that, given the fact that model **M16.1** is a conditional independence LMM with homoscedastic residual errors, σ can be interpreted as the residual standard deviation. All the estimates are similar to those reported in Panel R16.3.

Note that we also show how to extract and store the estimated value of the scale parameter from the model-fit object fm16.1mer, which is an object of class *mer*. Toward this end, we use the sigma() extractor function (Table 15.3) and store the result in the object sgma.

Computation of the marginal variance-covariance matrix is more complicated, because the function getVarCov(), which was used for this purpose for *lme*-class model-fit objects (see Table 14.5), does not work for *mer*-class objects. Consequently, we have to use a direct manipulation of the components of the model-fit object fm16.1mer. The calculations are presented in Panel R16.20b. In particular, we first extract the matrix $\boldsymbol{A}$, as defined in (13.34), by using the getME() function (Sect. 15.4). Then, we create a matrix identity of an appropriate dimension and compute the marginal variance-covariance matrix as in (13.35). The outcome is stored in the object V. Note that, as a result, we obtain a matrix for *all* observations, which is difficult to display. To obtain a legible printout, we have to select a few rows and columns of the matrix, corresponding to a particular level of the grouping factor (subject). Toward this end, with the help of the getME() function, we extract the flist slot, containing the grouping factor, of the model-fit object, and we display its structure with the help of the str()function. From the list of the levels of the grouping factor, i.e., subjects, we select the second subject, for which four post-randomization visual acuity measurements are available. The observations correspond to rows/columns 3–6 of the marginal variance-covariance matrix. The printout of these rows and columns of V corresponds to the printout shown in Panel R16.4; thus, we do not repeat it in Panel R16.20.

As was mentioned earlier (see Panel R16.19), the standard summary of the results, which are contained in an *mer*-class model-fit object, does not include *p*-values for the tests of individual fixed effects. Thus, in Panel R16.21, we present a method for the direct calculation of the *p*-values.

In Panel R16.21a, we begin with the computation of *p*-values of the *t*-test statistics, defined in (4.37) (see also Sect. 13.7.1). In particular, we calculate *p*-values corresponding to the marginal-approach tests (Sect. 4.7.1). Toward this end,

R16.20 *ARMD Trial*: Estimated variance components and the marginal varianc-covariance matrix for model **M16.1**. The model-fit object `fm16.1mer` was created in Panel R16.19

(a) *Variance-components estimates*

```
> VarCorr(fm16.1mer)                    # D^, Corr(D^), σ^
  $subject
                   (Intercept)
  (Intercept)          80.608
  attr(,"stddev")
  (Intercept)
        8.9782
  attr(,"correlation")
                   (Intercept)
  (Intercept)            1

  attr(,"sc")
  [1] 8.6275
> (sgma <- sigma(fm16.1mer))            # σ^
  [1] 8.6275
```

(b) *The marginal variance-covariance matrix* $\mathcal{V}$

```
> A <- getME(fm16.1mer, "A")            # A
> I.n <- Diagonal(ncol(A))              # I_N
> V <- sgma^2 * (I.n + crossprod(A)) # V  = σ^2(I_N+A'A)
> str(getME(fm16.1mer, "flist"))        # Grouping factor
  'data.frame':     867 obs. of   1 variable:
   $ subject: Factor w/ 234 levels "1","2","3","4",..: 1 1 2 2 2 2 3 ...
    - attr(*, "assign")= int 1
> # V[3:6, 3:6]                         # V_i not displayed (see R16.4)
```

we first extract the slot `coefs` from the object, which results from the application of the `summary()` function to the model-fit object `fm16.1mer`. The slot contains the matrix of estimates, standard errors, and t-test statistics for the fixed-effects coefficients. We store the content of the slot in the object `coefs`.

Then, we establish the number of degrees of freedom for the test statistics. Toward this end, we note that model **M16.1** is a single-level LMM. Thus, we have $G = 1$ grouping levels (Sect. 14.7).

First, we consider variables `visual0` and `treat.f`. Because their values change across subjects, but are constant for visual acuity measurements of the same subject, the fixed effects of these two variables are estimated at the subject level, i.e., at the grouping level $g = G = 1$, according to the notation used in (14.1). The effects of the two variables are described by $p_1 = 2$ coefficients. In total, there are $N_1 = 234$ subjects in the analyzed data frame (see Panel R16.20b). Thus, according to (14.1), the number of degrees of freedom for the t-tests for `visual0` and `treat.f` equals

R16.21 *ARMD Trial*: Calculation of "naïve" p-values for the tests for fixed effects for model **M16.1**. The model-fit object fm16.1mer was created in Panel R16.19

(a) *P-values for the marginal-approach t-tests*

```
> coefs <- coef(summary(fm16.1mer))                       # β̂, se(β̂), t-stat
> ddf <- c(631, 231, 631, 231, 631)                       # Denominator df
> pT <- 2 * (1 - pt(abs(coefs[, "t value"]), ddf))        # p-value
> tTable <- cbind(coefs, ddf, pT)
> printCoefmat(tTable, P.values = TRUE, has.Pvalue = TRUE)
                     Estimate Std. Error t value ddf      pT
  (Intercept)          9.2881     2.6817  3.4635 631 0.00057
  visual0              0.8264     0.0447 18.5035 231 < 2e-16
  time                -0.2122     0.0229 -9.2551 631 < 2e-16
  treat.fActive       -2.4220     1.4999 -1.6148 231 0.10772
  time:treat.fActive  -0.0496     0.0336 -1.4776 631 0.14002
```

(b) *P-values for the sequential-approach F-tests*

```
> (dtaov <- anova(fm16.1mer))
  Analysis of Variance Table
                 Df Sum Sq Mean Sq F value
  visual0         1  25578   25578  343.64
  time            1  14627   14627  196.51
  treat.f         1    516     516    6.94
  time:treat.f    1    163     163    2.18
> ddf1 <- ddf[-1]                                 # ddf for intercept omitted
> within(dtaov,
+        {
+         `Pr(>F)` <- pf(`F value`, Df, ddf1, lower.tail = FALSE)
+         denDf <- ddf1
+         })
  Analysis of Variance Table
                 Df Sum Sq Mean Sq F value denDf Pr(>F)
  visual0         1  25578   25578  343.64   231 <2e-16
  time            1  14627   14627  196.51   631 <2e-16
  treat.f         1    516     516    6.94   231  0.009
  time:treat.f    1    163     163    2.18   631  0.140
```

$ddf_1 = N_1 - (N_0 - p_1) = 234 - (1+2) = 231$, where $N_0 = 1$, because the model includes an intercept.

On the other hand, the values of variables time and time:treat.f change across visual acuity measurements for each subject. Thus, the fixed effects of these variables are estimated at the observation (visual acuity measurement) level, i.e., at the level $G+1 = 2$ (Sect. 14.7). The effects are expressed with the help of

$p_2 = 2$ coefficients. There are $N_2 = 867$ measurements in the analyzed dataset (see Panel R16.20b). Hence, following (14.1), the number of degrees of freedom for the t-tests for `time` and `time:treat.f` equals $ddf_2 = N_2 - (N_1 - p_2) = 867 - (234 + 2) = 631$. Note that the same number of degrees of freedom is used for the test for the intercept.

Based on the results of the computations, described above, we construct the numeric vector `ddf` with the number of degrees of freedom for the test statistics.

Finally, the p-values are obtained using a t-distribution with the number of degrees of freedom provided by the vector `ddf`. To enhance the legibility of display, we merge the objects `coefs`, `ddf`, and `pT`, and we print the resulting matrix with the use of the `printCoefmat()` utility function (see Panel R16.1). The p-values, presented in Panel R16.21a, correspond to those shown in Panel R16.1 for the `lme()` function.

In Panel R16.21b, we compute the p-values for the F-test statistics (4.36) (see also Sect. 13.7.1). Toward this end, we first apply the `anova()` function to the model-fit object `fm16.1mer` and store the result in the object `dtaov`. Note that, as seen from the printout of the contents of the object, `dtaov` does not include the p-values corresponding to the calculated values of the F-test statistics. It is also worth mentioning that the statistics correspond to the sequential-approach tests (Sect. 4.7.1). To compute the p-values, we need to establish the number of the denominator degrees of freedom for each of the test statistics. They were defined and stored in the vector `ddf` in Panel R16.21a. Note that we remove from the vector its first element, because the `anova()` table does not include the F-test for the intercept. Then, with the help of the `within()` generic function, we add two components to the `dtaov` object: `Pr(>F)`, with the p-values corresponding to the F-test statistics, and `denDf`, with the denominator degrees of freedom, as defined in the object `ddf1`.

It is worth mentioning that, as discussed in Sects. 13.7.1 and 14.7, the calculations, presented in Panel R16.21, ignore the fact that the null distribution of the t- and F-test statistics are, in fact, only approximated by the t- or F-distributions. Thus, the resulting p-values may be incorrect. An alternative is to estimate the p-values based on simulations. This is the approach which we discuss next.

16.7.2 Simulation-Based p-Values: The `simulate.mer()` *Method*

In Panel R16.22, we demonstrate how to simulate values of the dependent variable based on a fitted model **M16.1**.

Toward this end, in Panel R16.22a, we use the generic `simulate()` function. The function is applied to the model-fit object `fm16.1mer` of class *mer*. Consequently, the `simulate.mer()` method of the function is invoked. By specifying

R16.22 *ARMD Trial*: Simulations of the dependent variable based on the fitted form of model **M16.1** using the simulate.mer() method. The model-fit object fm16.lmer was created in Panel R16.19

(a) *Refitting the model to the simulated data*

```
> merObject <- fm16.lmer                          # M16.1 fit
> simD1 <- simulate(merObject, nsim = 1000)    # Simulated y from M16.1
> SimD1summ <- apply(simD1,
+       2,                                        # Over columns
+    function(y){
+      auxFit <- refit(merObject, y)              # Refit M16.1 with new y
+      summ <- summary(auxFit)                    # Summary
+      beta <- fixef(summ)                        # β̂
+      Sx <- getME(auxFit, "theta")               # S element
+      sgma <- sigma(auxFit)                      # σ̂
+      list(beta = beta, ST = Sx, sigma = sgma)
+             })
```

(b) *Matrices/vectors with estimates of* $\boldsymbol{\beta}$, $\sqrt{d_{11}/\sigma^2}$, *and* σ *for all simulations*

```
> betaE  <-                                       # Matrix with β̂
+    sapply(SimD1summ, FUN = function(x) x$beta)
> STe <- sapply(SimD1summ, FUN = function(x) x$ST)
> sigmaE <- sapply(SimD1summ, FUN = function(x) x$sigma)
```

---

the argument nsim=1000, we request 1,000 simulations. The result is stored in the matrix simD1. Next, we use the function apply() to iterate through the columns of the matrix simD1 and to use them as vectors of values of the dependent variable to refit the model. This step is time-consuming, as we are refitting the model a large number of times. Thus, in Panel R16.22a, we also demonstrate how to extract and store for further processing as much relevant information as possible with the help of the summary() function. Because objects of class *summary.mer* may use a lot of memory, we also show how to extract specific components of interest, like the estimates of $\boldsymbol{\beta}$, $\boldsymbol{D}$, σ. The selected estimates are stored in the list-object SimD1summ. The components of the list are named sim_1, sim_2, ..., sim_1000. Each of them is itself a named list with three components: beta, ST, and sigma. The components are vectors containing the estimates of, respectively, $\boldsymbol{\beta}$, $\sqrt{d_{11}/\sigma^2}$, and σ for model **M16.1** for a particular simulation.

It is worth noting that the code, presented in Panel R16.22a, uses specific features of model **M16.1**. In particular, the model contains only random intercepts. In this case, all the elements of the diagonal S matrix (see (13.33)) are exactly the same and are interpretable as $\sqrt{d_{11}/\sigma^2}$. Hence, extracting only the first element of the matrix is sufficient to obtain the information about the estimate of matrix $\boldsymbol{D}$ (see (13.9) in

Sect. 13.2.2). In Panel R16.22a, this is achieved by applying the function `getME()` to the `"theta"` component of the model-fit object `auxFit`.

The syntax, presented in Panel R16.22b, stores the simulation-based estimates of $\boldsymbol{\beta}$, $\sqrt{d_{11}/\sigma^2}$, and σ for further processing. To extract the estimates, we use the function `sapply()`, which applies the function, specified in the `FUN` argument, to each column of the matrix `SimD1summ` (see, e.g., Panel R3.10). The object `betaE` is a matrix with 1,000 columns containing the estimates of $\boldsymbol{\beta}$ for the simulations. The objects `STe` and `sigmaE` are numeric vectors with 1,000 elements each, containing the estimates of $\sqrt{d_{11}/\sigma^2}$ and σ, respectively, for the simulations.

In Panel R16.23, we compute the mean value, median, and 2.5th and 97.5th percentile of the estimates of the fixed-effects coefficients and ***ST*** variance-covariance parameters obtained from refitting model **M16.1** to the simulated data. Toward this end, to address fixed-effects coefficients, in Panel R16.23a we use the function `apply()`, which allows computation of the statistics for each row of the matrix `betaE`. We also use the function to calculate the empirical p-values for the coefficients. Toward this goal, for each coefficient, i.e., each row of the `betaE` matrix, we compute the proportion of estimated values larger than 0 and the corresponding two-sided p-value. Note that the so-obtained empirical p-value is restricted not to be smaller than `1/nsim`.

The printout of the summary statistics, presented in Panel R16.23a, provides information about the empirical distribution of the estimates for each fixed-effect coefficient. In particular, the last column displays the estimates of p-values. The empirical p-values are slightly larger (more conservative) than the values computed in Panel R16.21a.

Panel R16.23b presents the syntax to compute the mean value, median, and 2.5th and 97.5th percentiles of the simulation-based estimates of $\sqrt{d_{11}}$, the SD of random intercepts, and of σ, the SD of residual errors. The syntax is essentially similar to the one used in Panel R16.23a. Note that $\sqrt{d_{11}}$ is computed according to the representation (13.33). The means for $\sqrt{d_{11}}$ and σ are very close to the point estimates reported in Panel R16.19. The 2.5th and 97.5th percentiles can be used to assess the precision of the estimation of $\sqrt{d_{11}}$ and σ, in a similar way as the CIs provided by the function `intervals()` in the **nlme** package (Sect. 14.6). Note, however, that there is no counterpart of the `intervals()` function in the **lme4.0** package.

It might be of interest to present the distribution of the simulation-based estimates of the parameters of model **M16.1**. In Panel R16.24, we demonstrate the syntax that can be used to create plots of density functions corresponding to the empirical distribution functions for the fixed-effects coefficients and $\boldsymbol{ST} = \sqrt{d_{11}/\sigma^2}$. Toward this end, we begin with creating the matrix `parSimD1`, which contains, as columns, the simulation-based estimates of the parameters of interest. Next, we transpose it and save it in the data frame `parSimD1t`. Subsequently, we use the function `densityplot()` from the package **lattice** to create the plots. The package is automatically attached together with the **lme4.0**, so we do not need to load it separately. Note, however, that in the call to the function `densityplot()` we

R16.23 *ARMD Trial*: Simulation-based summary statistics of the distribution of the fixed effects and variance components for model **M16.1**. Objects betaE, STe, and sgmaE were created in Panel R16.22

(a) *Empirical means, quantiles, and p-values for fixed-effects coefficients*

```
> betaEm <- apply(betaE, 1, mean)                # Means (for each row)
> betaEq <-                                      # Quantiles
+     apply(betaE, 1, +
            FUN = function(x) quantile(x, c(0.5, 0.025, 0.975)))
> ptE <-                                         # p-values
+     apply(betaE, 1,
+           FUN = function(x){
+              prb <- mean(x > 0)
+              2 * pmax(0.5/ncol(betaE), pmin(prb, 1 - prb))
+                       })
> cbind(betaEm, t(betaEq), ptE)                  # Bind results columnwise
                         betaEm       50%     2.5%     97.5%   ptE
  (Intercept)          9.394788  9.492017  4.08912 14.660523 0.001
  visual0              0.825508  0.827323  0.74051  0.912949 0.001
  time                -0.213524 -0.213382 -0.26105 -0.168694 0.001
  treat.fActive       -2.452846 -2.424268 -5.26639  0.501710 0.110
  time:treat.fActive -0.048591 -0.049308 -0.11922  0.022705 0.144
```

(b) *Empirical means and quantiles for* $\sqrt{d_{11}}$ *and* σ

```
> d11E <- STe * sigmaE                           # d11=(d11/σ^2)^1/2 σ
> rndE <- rbind(d11E, sigmaE)                    # Matrix with two rows
> rndEm <- rowMeans(rndE)                        # Means (for each row)
> rndEq <- apply(rndE, 1,                        # Quantiles
+    FUN = function(x) quantile(x, c(0.5, 0.025, 0.975)))
> cbind(rndEm, t(rndEq))                         # Bind results
          rndEm    50%   2.5%   97.5%
  d11E   8.9627 8.9567 7.9279 10.0541
  sigmaE 8.6316 8.6265 8.1620  9.0935
```

apply the function melt() from the package reshape. Thus, we need to attach that package. The function is used to stack the variables, contained in the data frame parSimD1s, into one column, named (by default) value (see also the description of Panel R3.9). During the process, another variable, named (by default) variable, is created to identify the values of value, which correspond to the original variables from parSimD1s. The formula, provided as the first argument of the densityplot()-function call, requests the plot of a Gaussian kernel estimate (the default) of the density function for the empirical distribution of values of each variable. The resulting plots are presented in Fig. 16.13.

R16.24 *ARMD Trial*: Syntax to construct the density plots for the simulation-based estimates of the fixed-effects coefficients and variance-covariance parameters for model **M16.1**. Objects betaE, STe, and sigmaE were created in Panel R16.22

```
> names(sigmaE) <- names(STe) <- NULL          # For vectors
> parSimD1  <-                                  # Matrix
+     rbind(betaE, ST1 = STe, sigma = sigmaE)
> parSimD1t <-                                  # Transposed
+      data.frame(t(parSimD1), check.names=FALSE)
> parSimD1s <-                                  # Subset
+      subset(parSimD1t, select = -`(Intercept)`) # Intercept omitted
> require(reshape)                              # melt function needed
> densityplot(~value | variable,                # Fig.16.13
+              data = melt(parSimD1s),          # Molten data
+              scales = list(relation = "free"),
+              plot.points = FALSE)
> detach(package:reshape)
```

The density plots, presented in Fig. 16.13, are relatively symmetric. They suggest, for instance, that CIs, based on the normal-distribution approximation of the empirical distribution, might be adequate for construction of the interval estimates of the parameters.

16.7.3 Test for Random Intercepts

In Panel R16.25, we present different approaches to compute the *p*-value for the REML-based LR test for the need of including random intercepts in model **M16.1**.

As explained in Sect. 13.7.2, in this case, the null hypothesis specifies that the variance of the random effects is zero, which is a value on the boundary of the parameter space. Thus, asymptotically, the null distribution of the test is given by the 50%–50% mixture of the χ^2_0 and χ^2_1 distributions. In Panel R16.25a, we illustrate the computation of the *p*-value based on the mixture distribution. First, we fit the model corresponding to the null hypothesis, i.e., a classical, homoscedastic LM. We store the fit of the model in the object vis.lm2 of class *lm*. By applying the generic function logLik() we obtain the logarithm of the REML for the null and the alternative models and we compute the value of the LR-test statistic (4.29). Note that it corresponds to the value obtained with the help of the function anova() in Panel R16.17a. To calculate the *p*-value, we halve the *p*-value resulting from the χ^2_1 distribution. Obviously, it indicates a statistically significant result of the LR test.

In Panel R16.25b, we evaluate the *p*-value by simulating the finite-sample-size distribution of the LR-test statistic with the help of the function exactRLRT() from

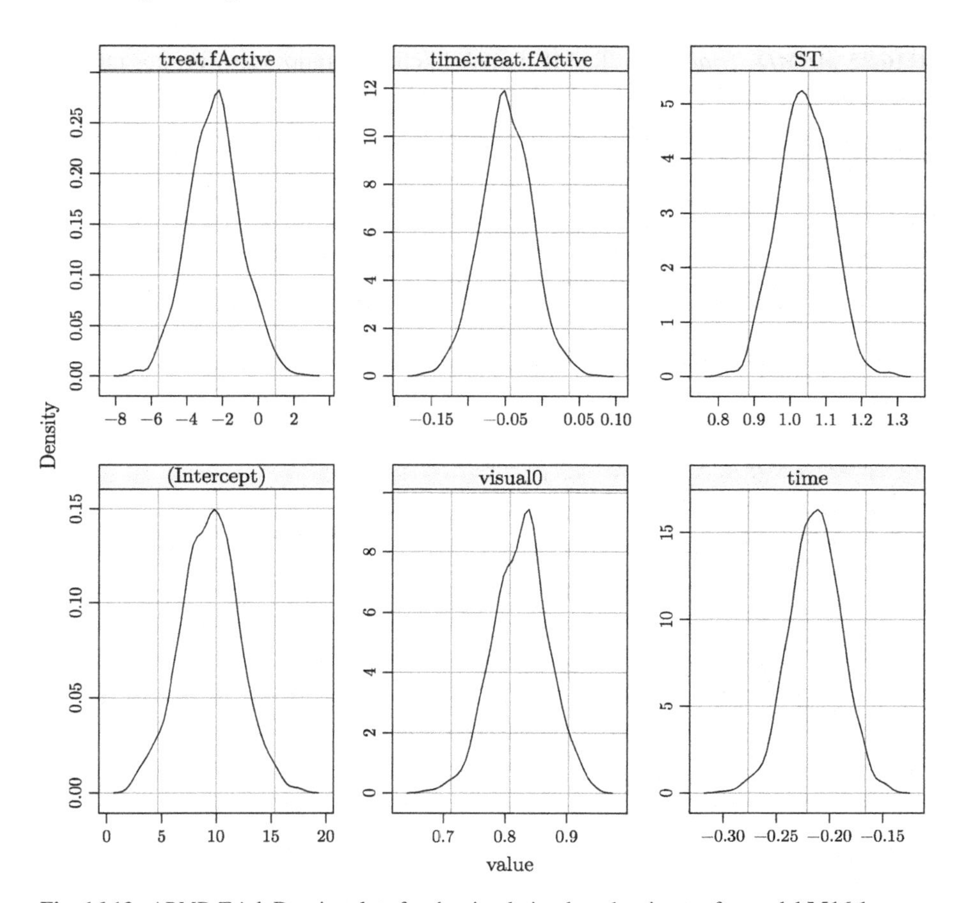

Fig. 16.13 *ARMD Trial*: Density plots for the simulation-based estimates for model **M16.1**

the package **RLRsim** (Sect. 14.7). The result is exactly the same as the one given in Panel R16.17b. Note, however, that the function is applied to the object `fm16.1mer`, which is an object of class *mer*.

Finally, we compute the empirical *p*-value by simulating a number of samples of the dependent variable from the null model, computing the value of the LR-test statistic for these samples, and using the so-obtained empirical distribution. This approach is presented in Panel R16.25c. To simulate `nsim=100` samples of the values of the dependent variable from the null model, we use the `simulate()` generic function. We store the samples as the columns of the data frame `lm2sim`. Then, with the help of the `apply()` function, we compute, for each column of the data frame, the value of the LR-test statistic. Toward this end, we create an auxiliary data frame `dfAux` by replacing the variable `visual` in the `armd` data frame with the simulated sample. We then fit the null model to the auxiliary data frame, and by applying the function `logLik()`, we extract the log-REML value from the model-fit

R16.25 *ARMD Trial*: The REML-based likelihood-ratio test for no random intercepts for model **M16.1**. The model-fit object fm16.1mer was created in Panel R16.19

(a) *Using* $0.5\chi^2_0 + 0.5\chi^2_1$ *as the null distribution*

```
> lm2.form  <- visual ~ visual0 + time + treat.f + treat.f:time
> vis.lm2 <- lm(lm2.form, data = armd)              # The null model
> (RLRTstat <-                                      # Compare to R16.17
+     -2 * as.numeric(logLik(vis.lm2, REML=TRUE)
+     - logLik(fm16.1mer)))      # log-REML for M16.1 (alternative)
  [1] 249.97
> 0.5 * pchisq(RLRTstat, 1, lower.tail = FALSE)  # p-value
  [1] 1.3211e-56
```

(b) *Using the function* exactRLRT() *to simulate the null distribution*

```
> require(RLRsim)
> exactRLRT(fm16.1mer)                              # M16.1 (alternative)
        simulated finite sample distribution of RLRT.  (p-value
        based on 10000 simulated values)

  data:
  RLRT = 249.97, p-value < 2.2e-16
```

(c) *Using the* simulate.mer() *method to obtain the empirical p-value*

```
> lm2sim <- simulate(vis.lm2, nsim = 100)# y simulated from the null model
> RLRTstatSim <- apply(lm2sim,
+   2,                                              # For each column
+   function(y){
+    dfAux  <- within(armd, visual <- y)            # Auxiliary data
+    lm0    <- lm(formula(vis.lm2), data = dfAux)# The null model
+    llik0  <- as.numeric(logLik(lm0, REML=TRUE))# log-REML, the null
+    llikA  <- as.numeric(logLik(refit(fm16.1mer, y)))
+    RLRTstat<- -2 * (llik0 - llikA)                # LR-test statistics
+              })
> mean(RLRTstat <= RLRTstatSim)                     # Empirical p-value
  [1] 0
```

object lm0. Next, we refit the alternative model **M16.1** to the sampled data with the help of the function refit() and we use the function logLik() to extract the log-REML. Finally, we compute the LR-test statistic, as defined in (4.29). To obtain the empirical p-value, we compute the proportion of the simulated values of the test statistic, which are larger than or equal to the observed value. The resulting p-value is equal to 0.

16.7.4 Test for Random Slopes

In this section, we revisit model **M16.7**, defined in Sect. 16.6.2. The model included random intercepts and random slopes for time and assumed that residual variance was constant. Additionally, the random intercepts and slopes were assumed to be independent.

In Panel R16.26, we fit model **M16.7** by employing the `lmer()` function and conduct the test for the interaction term. In particular, in Panel R16.26a, we note the use of two Z-terms, `(1|subject)` and `(0 + time|subject)`, in the `lmer()` formula (Sect. 15.3.1). By using the two terms, we essentially emulate a diagonal 2×2 matrix $\mathcal{D}$ (see Table 15.1). We also show selected results for the fitted model by extracting suitable components (see Table 15.3) of the model-fit object `fm16.2mer`.

Panel R16.26b demonstrates the construction of the LR test for the `treat.f:time` interaction. Toward this end, we first refit model **M16.7**, while omitting the interaction term from the model-defining formula. Then we apply the `anova()` generic function to obtain the result of the LR test. Note that the results, stored in model-fit objects `lmer2Dd` and `lmer3Dd`, were obtained using the default fitting method, i.e., REML. However, the LR test, reported by the `anova()` method, is constructed using the ML, because it pertains to a test of a hypothesis for a fixed effect. The computed p-value indicates that the result of the test is statistically not significant at the 5% significance level.

In Panel R16.27, we present different approaches to compute the p-value for the REML-based LR test for the need of including random slopes in model **M16.7**.

In this case, the null distribution of the test is, asymptotically, given by the 50%–50% mixture of the χ_1^2 and χ_2^2 distributions (Sect. 13.7.2). Panel R16.27a, we illustrate the computation of the p-value based on the mixture distribution. Toward this end, we use the function `logLik()` to extract the logarithm of REML for models **M16.1** and **M16.7**, which are represented by the model-fit objects `fm16.1mer` and `fm16.2mer`, respectively. Then we compute the value of the REML-based LR-test statistic (4.29). Note that it corresponds to the value obtained with the help of the function `anova()` in Panel R16.18a. To calculate the p-value, we sum the halves of the p-values resulting from the χ_1^2 and χ_2^2 distributions. The result indicates a statistically significant result of the LR-test.

In Panel R16.27b, we evaluate the p-value by simulating the finite-sample-size distribution of the REML-based LR-test statistic with the help of the function `exactRLRT()` from the package **RLRsim** (Sect. 14.7). Note that, in this case, we need to fit an auxiliary model, which includes random slopes as the only random effects (see the description of the function `exactRLRT()` in Sect. 14.7). The code and results are comparable to those given in Panel R16.18b. Note, however, that in Panel R16.27b the function `exactRLRT()` is applied to objects of class *mer*.

R16.26 *ARMD Trial*: Model **M16.7** fitted using the function `lmer()`

(a) *Fitting the model and extracting basic information*

```
> fm16.2mer  <-                                         # M16.7
+   lmer(visual ~ visual0 + time + treat.f + treat.f:time +
+          (1|subject) + (0 + time|subject),
+          data = armd)
> summ <- summary(fm16.2mer)
> coef(summ)                                            # t-Table
                     Estimate Std. Error t value
 (Intercept)          5.349030   2.332568  2.2932
 visual0              0.898460   0.039317 22.8519
 time                -0.215370   0.032266 -6.6749
 treat.fActive       -2.313752   1.209754 -1.9126
 time:treat.fActive -0.055059   0.047090 -1.1692
> unlist(VarCorr(fm16.2mer))                            # D. Short printout
   subject   subject
 54.071157  0.079359
> sigma(fm16.2mer)                                      # σ
 [1] 6.6834
```

(b) *Likelihood-ratio test for the* `treat.f:time` *interaction*

```
> fm16.2aux  <-                                         # Model M16.7 with ...
+    update(fm16.2mer, . ~ . - treat.f:time)            #... interaction omitted
> anova(fm16.2aux, fm16.2mer)
 Data: armd
 Models:
 fm16.2aux: visual ~ visual0 + time + treat.f +
 fm16.2aux:    (1 | subject) + (0 + time | subject)
 fm16.2mer: visual ~ visual0 + time + treat.f + treat.f:time +
 fm16.2mer:    (1 | subject) + (0 + time | subject)
           Df  AIC  BIC logLik Chisq Chi Df Pr(>Chisq)
 fm16.2aux  7 6441 6474  -3213
 fm16.2mer  8 6441 6480  -3213  1.38      1       0.24
```

16.8 Chapter Summary

In this chapter, we analyzed the ARMD data by applying LMMs. By using the models, the hierarchical structure of the data was directly addressed, which allowed taking into account the correlation between the visual acuity measurements obtained for the same individual.

Table 16.4 provides information about the models defined in this chapter.

The main tool that was used to fit the models in Sects. 16.2–16.6 was the function `lme()` from the package **nlme**. In Sect. 16.7, we refitted some of the models using the function `lmer()` from the package **lme4.0**. The latter function

Table 16.4 *ARMD Trial*: Summary of the models defined and fitted using the REML estimation, in Chap. 16

(**a**) Models fitted using the function `lme()` from the package **nlme**[a]

Model label	Section	Syntax	R object	Mean	Random effects (matrix $\mathcal{D}$)	Residual variance
M16.1	16.2	R16.1	`fm16.1`	(12.8)	Intercept (16.6)	Constant
M16.2	16.3	R16.5	`fm16.2`	(12.8)	Intercept (16.6)	*varPower*
M16.3	16.4.1	R16.10	`fm16.3`	(12.8)	Intercept, slope correlated (16.13)	*varPower*
M16.4	16.4.2	R16.11	`fm16.4`	(12.8)	Intercept, slope uncorrclated (16.16)	*varPower*
M16.5	16.4.3	R16.13	`fm16.5`	(12.9)	Intercept, slope uncorrelated (16.16)	*varPower*
M16.6	16.5	R16.15	`fm16.6`[b]	(12.8)	Intercept, slope correlated (16.13)	*varIdent*
M16.7	16.6.2	R16.18	`fm16.7`	(12.8)	Intercept, slope uncorrelated (16.16)	Constant

(**b**) Models fitted using the function `lmer()` from the package **lme4.0**

Model label	Section	Syntax	R object	Mean	Random effects (matrix $\mathcal{D}$)	Residual variance
M16.1	16.2	R16.19	`fm16.1mer`	(12.8)	Intercept (16.6)	Constant
M16.7	16.6.2	R16.26	`fm16.7mer`	(12.8)	Intercept, slope uncorrelated (16.16)	Constant

[a] The mean structure, defined in (12.8), is represented by the formula `visual ~ visual0 + time + treat.f + time:treat.f`

[b] Optimization of the likelihood for **M16.6** did not converge

R16.27 *ARMD Trial*: The REML-based likelihood-ratio test for no random slopes in model **M16.7**. Model-fit objects `fm16.1mer` and `fm16.2mer` were created in Panels R16.19 and R16.26, respectively

(a) *Using* $0.5\chi_1^2 + 0.5\chi_2^2$ *as the null distribution*

```
> RML0 <- logLik(fm16.1mer)            # log-REML, M16.1 (null)
> RMLa <- logLik(fm16.2mer)            # log-REML, M16.7 (alternative)
> (RLRTstat <- -2 * as.numeric(RML0 - RMLa))
  [1] 140.83
> .5 * pchisq(RLRTstat, 1, lower.tail = FALSE) + # p-value
+   .5 * pchisq(RLRTstat, 2, lower.tail = FALSE)
  [1] 1.3971e-31
```

(b) *Using the function* `exactRLRT()` *to simulate the null distribution*

```
> require(RLRsim)
> mAux  <- lmer(visual ~               # Auxiliary model with ...
+           visual0 + time + treat.f + treat.f:time +
+           (0 + time| subject),       # ... random slopes only.
+         data = armd)
> exactRLRT(m = mAux,                  # Auxiliary model
+         m0= fm16.1mer,               # M16.1 (null)
+         mA= fm16.2mer)               # M16.7 (alternative)
      simulated finite sample distribution of RLRT.  (p-value
      based on 10000 simulated values)

  data:
  RLRT = 140.83, p-value < 2.2e-16
```

is especially suited for, e.g., LMMs with crossed random effects, but it can only deal with conditional-independence models with homoscedastic residual errors. In this respect, it offers a more limited choice of models than `lme()`. For this reason, in our presentation, we primarily focused on the use of `lme()`.

In the process of arriving at the form of the final model **M16.5**, we fixed the mean structure as in (16.1) and built a series of models (see Table 16.4) with various random structures: model **M16.1** with random intercepts and homoscedastic residual variances (Sect. 16.2); model **M16.2** with random intercepts and residual variances described by a variance function defined as a power of the measurement time (Sect. 16.3); model **M16.3** with correlated random intercepts and random slopes and the power-of-time residual variances (Sect. 16.4.1); and model **M16.4** with independent random intercepts and random slopes and the power-of-time residual variances (Sect. 16.4.2). The last model gave a satisfactory fit to the data and allowed us to simplify the mean structure by adopting a constant treatment effect, as reflected in model **M16.5** in Sect. 16.4.3.

The presented approach was adopted mainly for illustrative purposes. In practice, we should start building the model using the most general fixed- and random-effects structures. Then, we might consider simplifying the random-effects structure while checking the fit of the simplified models using the LR test or information criteria (Sect. 13.7.2). When a more parsimonious structure with a satisfactory fit to the data has been found, we could consider in turn simplifying the mean structure. After arriving at a final model, we should check its fit by residual diagnostics (Sect. 13.6.2).

Thus, in the case of the visual acuity data, we might begin, for instance, from model **M16.3**, but with time included in the mean structure as a factor, and try to simplify the model by removing the random effects of time. We would most likely find that the simplification was worsening the fit of the model. Thus, we might settle for a model with random intercepts and time effects, and consider simplifying the mean structure by assuming, e.g., a continuous time effect and a constant treatment effect. This step would most likely lead us to model **M16.5** as the final model.

In Sect. 16.5, we additionally considered model **M16.6** with correlated random intercepts and random slopes and time-specific residual variances. As the model assumes a slightly more general residual-variance structure than model **M16.3**, it could offer a better fit. We discovered, however, that model **M16.6** could not be fitted to the data by the function `lme()`. From a practical point of view of using the function to fit LMMs, this example illustrates that the results of a model fit need always to be carefully checked for symptoms of nonconvergence. This is because the function may fail to report any apparent error messages that would indicate problems with convergence of the estimation algorithm.

In Sect. 16.6, we discussed the issue of testing hypotheses about the random-effects structure. This is a difficult issue, due to the problems with obtaining the null distribution of the LR-test statistic in situations when the null hypothesis involves values of parameters at the boundary of the parameter space. Exact analytical results are available only for a limited set of special cases. In practice, a simulation approach is often used. However, the R functions available for this purpose are also limited in their scope. For instance, they apply to models with homoscedastic residual errors. For this reason, their application to the models considered for the ARMD data, which specified the residual variances using the varPower($\cdot$) variance function, was not possible. In such a case, the choice of the random effects structure may need to be based on an informal comparison of the fit of the models based on residual diagnostics and/or the information criteria. To nevertheless illustrate the tools for testing hypotheses about the random-effects structure, we considered model **M16.7** with uncorrelated random intercepts and slopes and homoscedastic, independent residual errors.

As mentioned earlier, in Sect. 16.7, we refitted models **M16.1** and **M16.7** using the function `lmer()` from the package **lme4.0**. This allowed us to illustrate the differences in the use of the function, as compared to `lme()`. Important differences include, e.g., the form of the model-defining formula and the methods to extract components from a model-fit object. Also, `lmer()` does not report p-values, which means that the user needs to know additional tools that allow to evaluate results of significance tests. We have presented such tools in Sects. 16.7.2–16.7.4.

In the next chapter, we further illustrate the use of the function `lme()` for fitting LMMs by applying the models in the analysis of the PRT study data.

Chapter 17
PRT Trial: Modeling Muscle Fiber Specific-Force

17.1 Introduction

In Sect. 3.3, we presented an exploratory analysis of the measurements of muscle fiber isometric and specific force, collected in the PRT study. In this chapter, we use LMMs to analyze the data.

In particular, we first focus on data for the muscle fiber specific force. In Sect. 17.2, we consider type-1 fibers only and fit an LMM with two correlated, heteroscedastic, occasion-specific random effects for each individual and homoscedastic independent residual errors. We subsequently modify the model for residual variation using the power-of-the-mean variance function (Sect. 17.3). In the next step, we consider models for both fiber types. In Sects. 17.4 and 17.5, we construct conditional-independence LMMs with four correlated, heteroscedastic, fiber-type×occasion-specific random effects for each individual. In Sects. 17.6 and 17.7, the random-effects structure of the models is simplified by considering more parsimonious structures of variance covariance matrices of the random effects. Toward this end, we develop and use a new class of positive-definite matrices, the *pdKronecker* class. Finally, in Sect. 17.8, we construct the most comprehensive LMM, which takes into account the data for two dependent variables, i.e., the isometric and specific force, and for both fiber types. A summary of the chapter is presented in Sect. 17.9.

17.2 A Model with Occasion-Specific Random Intercepts for Type-1 Fibers

We begin with an analysis of a subset of the data pertaining to type-1 fibers only. Modeling these data should give us insight into, the variance-covariance structure of the pre- and post-training measurements for the particular type of fibers. The

A. Gałecki and T. Burzykowski, *Linear Mixed-Effects Models Using R: A Step-by-Step Approach*, Springer Texts in Statistics, DOI 10.1007/978-1-4614-3900-4_17,

information can be useful in constructing a more advanced model, which would take into account both fiber types.

17.2.1 Model Specification

Figures 3.5 and 3.6 indicate that there is a considerable variability between subjects with respect to the means and variances of the specific-force measurements for each of the four combinations of the fiber type and occasion levels. This suggests that an LMM with occasion-specific random intercepts for a subject might be reasonable to model the type-1 fiber data. Inclusion of the random intercepts should allow for adjusting for the possible correlation between the repeated measurements of the type-1 fibers for the same individual at the same occasion, i.e., pre- or post-training, as well as between the two different occasions.

The model should also take into account the factors used in the experimental design. In particular, it should include the effect of the intensity of training (intervention), which was the main effect of interest. Also, as the experiment was stratified for sex and age, these two factors should be included in the model as well. Possible effects of the occasion (pre- and post-training) should be taken into account. Finally, BMI of the subject can potentially influence the value of the fiber's specific force and should be adjusted for, too.

Taking into account the aforementioned considerations, we use model **M17.1**, defined by the following equation:

$$\begin{aligned} \text{SPEC.FO}_{itr} &= \beta_0 + \beta_1 \times \text{PRT}_i + \beta_2 \times \text{OCC}_{it} + \beta_3 \times \text{SEX}_i \\ &\quad + \beta_4 \times \text{AGE}_i + \beta_5 \times \text{BMI}_i \\ &\quad + \beta_{12} \times \text{PRT}_i \times \text{OCC}_{it} + \beta_{34} \times \text{SEX}_i \times \text{AGE}_i \\ &\quad + b_{0it} + \varepsilon_{itr} \\ &\equiv \mu_{it} + b_{0it} + \varepsilon_{itr}, \end{aligned} \tag{17.1}$$

where SPEC.FO_{itr} is the value of the specific force for the r-th ($r = 1, \ldots, n_{i1t}$) type-1 fiber sample measured for the subject i ($i = 1, \ldots, N$) at the occasion t ($t = 1, 2$ for pre- and post-training, respectively). In (17.1), PRT_i, OCC_{it}, SEX_i, and AGE_i are the values of the indicator variables for the i-th subject for, respectively, the "low-intensity" intervention (control) group, "post-intervention" measurement occasion t, females, and the older age group. BMI_i is the value of subject's BMI. The coefficient of interest is β_{12}, associated with the PRT × OCC interaction term. It quantifies the post- *versus* pre-intervention change in the specific force measurements for the low-intensity-training group, as compared to the high-intensity group. The SEX $\times$ AGE interaction term is included along with corresponding main effects to take into account stratification used in the study.

The residual random errors ε_{itr} are assumed to be independent and normally distributed with mean zero and variance σ^2. It follows that the variance-covariance matrix of the errors for the i-th subject is given by

$$\mathcal{R}_i = \sigma^2 \mathbf{I}_{n_{i1t}}. \tag{17.2}$$

In addition to the residual errors, the model equation specifies, for each subject, two occasion-specific, i.e., pre- and posttreatment, random intercepts: b_{0i1} and b_{0i2}, respectively. We assume that the vector $\mathbf{b}_i \equiv (b_{0i1}, b_{0i2})'$ is normally distributed with mean zero and variance-covariance matrix $\mathcal{D}$, that is,

$$\mathbf{b}_i \equiv \begin{pmatrix} b_{0i1} \\ b_{0i2} \end{pmatrix} \sim \mathcal{N}(\mathbf{0}, \mathcal{D}), \tag{17.3}$$

where

$$\mathcal{D} = \begin{pmatrix} d_{11} & d_{12} \\ d_{12} & d_{22} \end{pmatrix}.$$

17.2.1.1 The Marginal Interpretation

Model **M17.1** implies that the marginal expected value of SPEC.FO_{itr} is equal to μ_{it}, defined in (17.1).

Inclusion of b_{0i1} and b_{0i2} in the model allows for modeling of the correlation between the SPEC.FO measurements obtained for a particular individual. This can be seen from the resulting marginal variances and covariances:

$$\text{Var}(\text{SPEC.FO}_{i1r}) = d_{11} + \sigma^2, \quad \text{Var}(\text{SPEC.FO}_{i2r}) = d_{22} + \sigma^2,$$

$$\text{Cov}(\text{SPEC.FO}_{i1r}, \text{SPEC.FO}_{i2r}) = \text{Cov}(\text{SPEC.FO}_{i1r}, \text{SPEC.FO}_{i2r'}) = d_{12},$$

where $r \neq r'$. Thus, the marginal variance-covariance matrix $\mathcal{V}_i$, defined in (13.25), of the vector

$$\mathbf{y}_i = (\text{SPEC.FO}_{i11}, \ldots, \text{SPEC.FO}_{i1n_{i11}}, \text{SPEC.FO}_{i21}, \ldots, \text{SPEC.FO}_{i2n_{i12}})',$$

is of dimension $(n_{i11} + n_{i12}) \times (n_{i11} + n_{i12})$ and has the following structure:

$$
\begin{pmatrix}
d_{11}+\sigma^2 & d_{11} & \cdots & d_{11} & d_{12} & d_{12} & \cdots & d_{12} \\
d_{11} & d_{11}+\sigma^2 & \cdots & d_{11} & d_{12} & d_{12} & \cdots & d_{12} \\
\vdots & \vdots & \ddots & \vdots & \vdots & \vdots & \ddots & \vdots \\
d_{11} & d_{11} & \cdots & d_{11}+\sigma^2 & d_{12} & d_{12} & \cdots & d_{12} \\
 & & & & & & & \\
d_{12} & d_{12} & \cdots & d_{12} & d_{22}+\sigma^2 & d_{22} & \cdots & d_{22} \\
d_{12} & d_{12} & \cdots & d_{12} & d_{22} & d_{22}+\sigma^2 & \cdots & d_{22} \\
\vdots & \vdots & \ddots & \vdots & \vdots & \vdots & \ddots & \vdots \\
d_{12} & d_{12} & \cdots & d_{12} & d_{22} & d_{22} & \cdots & d_{22}+\sigma^2
\end{pmatrix}. \tag{17.4}
$$

It follows that any two pre-training type-1 fiber measurements for the same individual are positively correlated with the correlation coefficient equal to $d_{11}/(d_{11}+\sigma^2)$. Similarly, the correlation coefficient for any two posttreatment measurements is equal to $d_{22}/(d_{22}+\sigma^2)$. Thus, different strength of correlation between measurements taken at different occasions for the same individual is allowed by the model. Finally, the correlation coefficient for a pair of a pre- and post-training measurements is equal to $d_{12}/\sqrt{(d_{11}+\sigma^2)(d_{22}+\sigma^2)}$.

17.2.2 **R** *Syntax and Results*

In Panel R17.1, we fit model **M17.1** to the data for type-1 fibers and, based on the object representing model fit, we explore data grouping/hierarchy implied by the model.

Toward this end, in Panel R17.1a, we create the model formula, corresponding to the fixed-effects part of (17.1). Then, we create the data frame `prt1`, which contains the subset of the data frame `prt`, consisting of observations for type-1 fibers. Note that, in the `subset()`-function call, we use the argument `select=-fiber.f`, which removes the factor variable `fiber.f` from the subset. Finally, we fit model **M17.1** to the subset using the function `lme()` (Sect. 14.5). Note that, in the formula used in the argument `random`, we remove the intercept. As a result, we include in the model two random intercepts, corresponding to the levels of the factor `occ.f`. The two intercepts are defined at the levels of the single grouping-factor `id` (Sect. 14.3.1). The variance-covariance matrix of the random intercepts is assumed, by default, to have a general form. The results of fitting of the model are stored in the object `fm17.1`.

In Panel R17.1b, we extract information about the data hierarchy, implied by the syntax used in Panel R17.1a. By using the `getGroupsFormula()` function (Sect. 14.4), we verify that the grouping is defined by the levels of the factor `id`. With the help of the function `getGroups()`, we extract the grouping factor from the model-fit object, store it in the object `grpF`, and display the structure of `grpF` by applying the generic function `str()`. In particular, we learn that the factor has

R17.1 *PRT Trial*: Model **M17.1** fitted to the data for type-1 fibers using the function `lme()`

(a) *Fitting of the model*

```
> data(prt, package = "nlmeU")
> lme.spec.form1 <-
+    formula(spec.fo ~ (prt.f + occ.f)^2 + sex.f + age.f +
+              sex.f:age.f + bmi)
> prt1 <- subset(prt, fiber.f == "Type 1", select = -fiber.f)
> fm17.1 <-                                        # M17.1:(17.1)
+    lme(lme.spec.form1,
+        random = ~occ.f - 1|id,                   # D:(17.3)
+        data = prt1)
```

(b) *Data grouping/hierarchy implied by the model*

```
> getGroupsFormula(fm17.1)
  ~id
  <environment: 0x0000000019905c30>
> str(grpF <- getGroups(fm17.1))
   Factor w/ 63 levels "5","10","15",..: 1 1 1 1 1 1 1 1 1 1 ...
   - attr(*, "label")= chr "id"
> nF1 <- xtabs(~grpF)          # Number of type-1 fibers per subject
> range(nF1)                   # Min, max number of type-1 fibers
  [1]  6 36
> nF1[which.min(nF1)]          # Subject with the minimum number of fibers
  275
    6
> str(fm17.1$dims)             # Basic dimensions used in the fit
  List of 5
   $ N    : int 1299
   $ Q    : int 1
   $ qvec : num [1:3] 2 0 0
   $ ngrps: Named int [1:3] 63 1 1
    ..- attr(*, "names")= chr [1:3] "id" "X" "y"
   $ ncol : num [1:3] 2 8 1
```

63 levels, corresponding to the patients included in the study. With the help of the function `xtabs()`, we create a contingency table for the levels of the grouping factor and store it in the object `nF1`. By applying the function `range()`, we check that the minimum and maximum number of observations per patient are equal to 6 and 36, respectively. The minimum is obtained for the patient with `id` equal to 275. Finally, we show the structure of the list, which is contained in the `dims` component of the model-fit object. The list comprises five components:

R17.2 *PRT Trial*: Estimates of the fixed-effects coefficients for model **M17.1**. The model-fit object `fm17.1` was created in Panel R17.1

```
> fixed1 <- summary(fm17.1)$tTable                 # β̂, se(β̂), t-test
> nms <- rownames(fixed1)                          # β names
> nms[7:8] <- c("fLow:fPos", "fMale:fOld")         # Selected names shortened
> rownames(fixed1) <- nms                          # New names assigned
> printCoefmat(fixed1, digits = 3,                 # See also Table 17.1
+              has.Pvalue = TRUE, P.values = TRUE)
               Value Std.Error        DF t-value p-value
(Intercept)  127.724    15.416 1234.000    8.28   3e-16
prt.fLow       2.886     4.338   57.000    0.67   0.509
occ.fPos       4.703     2.667 1234.000    1.76   0.078
sex.fMale     -1.385     5.364   57.000   -0.26   0.797
age.fOld       8.984     5.155   57.000    1.74   0.087
bmi            0.491     0.578   57.000    0.85   0.399
fLow:fPos     -2.133     3.750 1234.000   -0.57   0.570
fMale:fOld   -12.680     7.553   57.000   -1.68   0.099
```

- `N` is the number of observations included in the data used for fitting the model
- `Q` is the number of levels of grouping
- `qvec` is a numeric vector, which provides the number of random effects at each level of grouping, from the innermost to the outermost level, where the last two values are equal to *zero* and correspond to the fixed effects and the response, respectively
- `ngrps` is a vector providing the number of groups at each grouping level, from the innermost to the outermost level, with the last two values equal to *one* and corresponding to the fixed effects and the response, respectively
- `ncol` is a numeric vector containing the number of columns in the model matrix for each level of grouping, from the innermost to outermost level, with the last two values equal to the fixed effects and to *one*

Thus, the printout in Panel R17.1b indicates that model **M17.1** was fitted to 1,299 observations, grouped according to a single factor named `id` with 63 levels, consistent with the results presented in Sect. 2.3. The model included two random effects at each level of the grouping factor. The model matrix included two columns for the random effects and eight columns for the fixed effects (including the intercept), as specified by the model equation (17.1).

In Panel R17.2, we present the estimates of the fixed-effects coefficients for model **M17.1**. To obtain a legible display, we first save the `tTable` component of the list resulting from applying the generic `summary()` function to the model-fit object `fm17.1` (Sect. 14.6). Then, we extract the names of the rows of the `tTable` component using the function `rownames()` and we use the same function to shorten the selected row names. Finally, we print out the fixed-effects table using the `print-Coefmat()` function. For the description of the use of the function, see the syntax in Panel **M16.1** and its explanation in Sect. 16.2.2. An explanation of the arguments of the function can be obtained by issuing the command `?printCoefmat`.

In the printout, presented in Panel R17.2, it is worth noting that there are two different numbers of degrees of freedom used for the fixed-effects coefficients. The t-tests of the coefficients for `prt.fLow`, `sex.fMale`, `age.fOld`, `bmi`, and `fMale:fOld`, are based on 57 degrees of freedom. These coefficients are estimated at the level $g = G = 1$ of grouping (see Sect. 14.7), i.e., at the individual level. At this level, there are 63 groups. Thus, according to (14.1), the number of degrees of freedom equals $63 - (1+5) = 57$, because there are five coefficients at this level of grouping, and the model includes an intercept (see also Sect. 16.7.1). On the other hand, the coefficients for `occ.fPos` and `fLow:fPos` are estimated at the level $G+1 = 2$ of grouping, i.e., at the observation level. Given that there are 1,299 observations in total, the number of degrees of freedom, according to (14.1), equals $1,299 - (63+2) = 1,234$. As explained in Sect. 14.7, this number is also assumed for the intercept, although the intercept is treated as being estimated at the level $g = 0$ of the data hierarchy. Note that, as was mentioned in Sect. 13.7.1, the computation of the degrees of freedom does not reflect the fact that the true distribution the test statistic is merely approximated by a central F-distribution (Sect. 7.6.1).

In Panel R17.3, we present estimates of the matrices $\mathcal{D}$ and $\mathcal{R}_i$ for model **M17.1**. They are extracted from the model-fit object `fm17.1` using the function `getVar-Cov()` (Sect. 14.6).

In Panel R17.3a, we present the estimate of the matrix $\mathcal{D}$. The matrix is of dimension 2×2, as defined in (17.3). Using the estimated values of the elements of the matrix, the correlation coefficient between the random intercepts corresponding to the pre- and post-training measurement occasions is estimated to be equal to $166.45/\sqrt{238.70 \cdot 201.48} = 0.759$. The result of the calculations is confirmed using the `VarCorr()` function.

In accordance with (17.2), the matrices $\mathcal{R}_i$ are diagonal with a constant element σ^2 on the diagonal. The dimension of the matrix $\mathcal{R}_i$ depends on the number of repeated observations for type-1 fibers for a particular individual. In Panel R17.3b, we extract the estimates of the matrix for the subjects with `id` equal to `"5"` and `"275"`. As we noted in Panel R17.1b, for the subject "275", there are only six measurements in total (five pre- and one post-training). Thus, we can actually print out the entire estimate of the matrix $\mathcal{R}_i$. However, this is not advisable for the subject "5", for whom 30 measurements were collected (12 pre- and 18 post-training). In this case, it is better if we print the first six elements of the diagonal of the matrix, which we can extract from the matrix with the help of the function `diag()`.

In fact, given the constant, diagonal structure of the matrices $\mathcal{R}_i$, we could avoid the use of the functions `getVarCov()` or `VarCorr()` altogether and simply compute the estimate of the residual variance σ^2, based on the component `sigma` of the object obtained by applying the function `summary()` to the model-fit object `fm17.1` (see Table 14.5 in Sect. 14.6). Note that, in model **M17.1**, σ can be interpreted as the SD of the residual random errors. The resulting estimate of σ^2 is equal to 505.59, which, obviously, corresponds to the values of the diagonal elements of estimated matrices $\mathcal{R}_i$ for subjects "5" and "275".

R17.3 *PRT Trial*: The estimated $\mathcal{D}$ and $\mathcal{R}_i$ matrices for model **M17.1**. The model-fit object `fm17.1` was created in Panel R17.1

(a) *The estimate of the matrix* $\mathcal{D}$

```
> getVarCov(fm17.1)                          # D^: (17.3)
Random effects variance covariance matrix
          occ.fPre occ.fPos
occ.fPre    238.68   166.42
occ.fPos    166.42   201.46
  Standard Deviations: 15.449 14.194
> VarCorr(fm17.1)
id = pdLogChol(occ.f - 1)
         Variance StdDev Corr
occ.fPre 238.68   15.449 occ.fPr
occ.fPos 201.46   14.194 0.759
Residual 505.59   22.485
```

(b) *The estimates of the matrix* $\mathcal{R}_i$ *for subjects "5" and "275"*

```
> Ri <-                                      # Ri is a list containing R^_i ...
+   getVarCov(fm17.1, c("5", "275"),# ... for subjects "5" and "275".
+             type = "conditional")
> Ri$"275"                                   # R^_i for the subject "275": (17.2)
       1      2      3      4      5      6
1 505.59   0.00   0.00   0.00   0.00   0.00
2   0.00 505.59   0.00   0.00   0.00   0.00
3   0.00   0.00 505.59   0.00   0.00   0.00
4   0.00   0.00   0.00 505.59   0.00   0.00
5   0.00   0.00   0.00   0.00 505.59   0.00
6   0.00   0.00   0.00   0.00   0.00 505.59
> Ri.5   <- Ri$"5"                           # R^_i for the subject "5" ...
> dim(Ri.5)                                  # ... with large dimensions ...
 [1] 30 30
> (Ri.5d <- diag(Ri.5)[1:6])                 # ... its first 6 diagonal elements.
     1      2      3      4      5      6
505.59 505.59 505.59 505.59 505.59 505.59
> sgma <- summary(fm17.1)$sigma # σ^
> sgma^2                        # σ^2
 [1] 505.59
```

In Panel R17.4, we extract information about the estimate of the marginal variance-covariance matrix $\mathcal{V}_i$ for model **M17.1** for the subject "5." Given that the matrix is of dimension 30×30, we need to construct an abbreviated printout of

R17.4 *PRT Trial*: The estimated marginal variance-covariance matrix $\mathcal{V}_i$ for model **M17.1**. The model-fit object `fm17.1` and the data frame `prt1` were created in Panel R17.1

(a) *Rows/cols names for the subject "5"*

```
> dt5 <-                               # Data with 30 observations
+    subset(prt1,
+           select = c(id, occ.f),     # ... and 2 variables
+           id == "5")                 # ... for the subject "5".
> auxF1 <- function(elv) {
+    idx <- 1:min(length(elv), 2)  # Up to two indices per vector
+    elv[idx]                          # ... returned.
+ }
> (i.u5 <-                             # Selected indices printed
+    unlist(
+       tapply(rownames(dt5),          # ... for the subject "5"
+              dt5$occ.f,              # ... by occ.f subgroups
+              FUN = auxF1)))
 Pre1 Pre2 Pos1 Pos2
  "1"  "2" "20" "22"
> dt.u5  <- dt5[i.u5, ]               # Raw data for selected indices
> (nms.u5 <-                          # Row names constructed
+    paste(i.u5, dt.u5$occ.f, sep = "."))
 [1] "1.Pre"  "2.Pre"  "20.Pos" "22.Pos"
```

(b) *The matrix $\mathcal{V}_i$ estimate for the subject "5"*

```
> Vi <-                                 # Vi is a list containing ...
+   getVarCov(fm17.1, "5",                # ... matrix V̂i for subject "5".
+             type = "marginal")
> Vi.5 <- Vi$"5"                        # Vi.5 is a V̂i matrix: (17.4)
> Vi.u5 <- Vi.5[i.u5, i.u5]             # A sub-matrix selected, ...
> rownames(Vi.u5) <- nms.u5               # ... row/column names changed,
> Vi.u5                                   # ... the sub-matrix printed.
             1      2     20     22
1.Pre   744.27 238.68 166.42 166.42
2.Pre   238.68 744.27 166.42 166.42
20.Pos  166.42 166.42 707.05 201.46
22.Pos  166.42 166.42 201.46 707.05
> cov2cor(Vi.u5)                          # Corr(V̂i)
               1       2      20      22
1.Pre   1.00000 0.32069 0.22941 0.22941
2.Pre   0.32069 1.00000 0.22941 0.22941
20.Pos  0.22941 0.22941 1.00000 0.28493
22.Pos  0.22941 0.22941 0.28493 1.00000
```

the elements of the matrix. Toward this end, in Panel R17.4a, we select the data for the variables `id` and `occ.f` for the subject with `id=="5"` and store them in the data frame `dt5`. Then we define an auxiliary function `auxF1()`, which, for a vector, selects at most two index values. Subsequently, with the help of the generic function `tapply()`, we apply `auxF1()` to the vectors which contain names of the rows of the data frame `dt5` for different values of the factor `occ.f`. The resulting list contains indices for two different `Pre` and `Pos` observations for the subject "5." With the help of the generic function `unlist()`, we simplify the list to a vector and store it in the object `i.u5`. We use the vector to select the corresponding rows from the data frame `dt5`. Finally, we use the generic function `paste()` to, first, convert the values of vectors `i.u5` and `dt5$occ.f` to characters and, second, to concatenate the character values using "." as a separator. The concatenated strings, stored in the vector `nms.u5`, will be used to label the rows/columns in abbreviated printouts.

In Panel R17.4b, we use the function `getVarCov()` (Sect. 14.6) to extract the estimate of the marginal variance-covariance matrix $\mathcal{V}_i$ for model **M17.1** for the subject "5." We store the estimate in the matrix-object `Vi.5`. The dimension of the matrix is large, 30×30, so we select from the matrix the rows and columns that correspond to the observations indexed by the values of the index vector `i.u5`, which was created in Panel R17.4a. The resulting submatrix is of dimension 4×4 and can easily be displayed. Before displaying it, however, we abbreviate the names of the rows of the sub-matrix using the generic function `rownames()` and the character-vector `nms.u5`. Finally, we print out the submatrix. The printout clearly shows the structure, indicated in (17.4). Using the results shown in Panel R17.3, we can check that the estimate of the variance of pre-training measurements is equal to $744.27 = 238.68 + 505.59$. The estimate of the correlation coefficient between two within-subject pre-training values is equal to $238.68/744.27 = 0.321$. For two post-training measurements, the estimated correlation coefficient is equal to $201.46/707.05 = 0.285$, while for a pair of a pre- and post-training observations, it is equal to $166.42/\sqrt{744.27 \cdot 707.05} = 0.229$. Thus, two pre-training or two post-training observations for the same individual exhibit about the same level of correlation, which is slightly higher than the correlation between a pair of pre-training and post-training observations. The result of the calculations is confirmed by using the `cov2cor(Vi.u5)` command.

Panel R17.5 presents the syntax to extract and plot EBLUPs (Sect. 13.6.1) of the random effects (intercepts) for model **M17.1**. The predictors are extracted from the model-fit object using the generic function `ranef()` (Sect. 14.6) and stored in the object `rnf`. By using the function `var()`, we obtain the estimates of the variances of $\widehat{b}_{0i1}$ and $\widehat{b}_{0i2}$. We note that the variances are smaller than the corresponding estimates in the $\widehat{\mathcal{D}}$ matrix, shown in Panel R17.3a. This is a well-known feature of the EBLUPs and illustrates the "shrinkage" phenomenon (see, e.g., (13.51) in the current volume and Sect. 7.5 in Verbeke and Molenberghs 2000).

By applying the `plot()` function to the object, we obtain the side-by-side plot, shown in Fig. 17.1a. The plot shows the random-effects predictors for the two levels of the `occ.f` factor. It indicates that the estimates are correlated: small (large) values

R17.5 *PRT Trial*: Empirical BLUPs for the random effects for model **M17.1**. The model-fit object `fm17.1` was created in Panel R17.1

```
> rnf <- ranef(fm17.1)              # b̂_i: (13.50)
> (vrnf <- var(rnf))                # var(b̂_i). Compare to D̂ in R17.3a.
           occ.fPre occ.fPos
  occ.fPre   184.31   141.70
  occ.fPos   141.70   150.59
> plot(rnf)                         # Side-by-side plot (Fig. 17.1a)
> library(ellipse)
> myPanel <- function(x,y, ...){
+   panel.grid(h = -1, v = -1)
+   panel.xyplot(x, y)
+   ex1 <-                          # Ellipse based on D̂: (17.3)
+      ellipse(getVarCov(fm17.1))
+   panel.xyplot(ex1[, 1], ex1[, 2], type = "l", lty = 1)
+   ex2 <- ellipse(vrnf)            # Ellipse based on var(b̂_i).
+   panel.xyplot(ex2[ ,1], ex2[, 2], type = "l", lty = 2)
+ }
> xyplot(rnf[, 2] ~ rnf[, 1],       # Scatterplot b̂_i1 vs. b̂_i0 (Fig. 17.1b)
+        xlab = "Pre-intervention",
+        ylab = "Post-intervention",
+        xlim = c(-40, 40), ylim = c(-40, 40),
+        panel = myPanel)
```

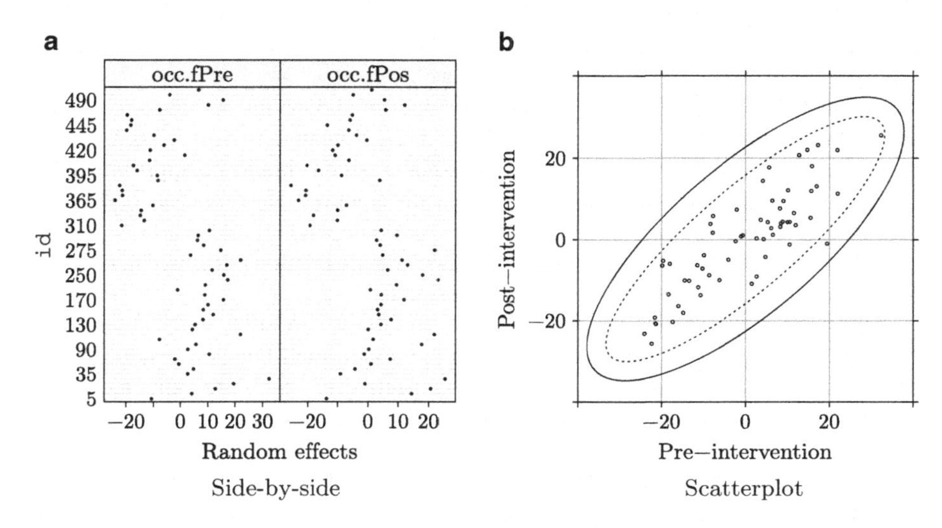

Fig. 17.1 Empirical BLUPs for the random effects for model **M17.1**

of the `Pre`-level random effect are associated with small (large) values of the `Pos`-level random effect. This is in agreement with the estimated value of the matrix

R17.6 *PRT Trial*: Plots of the conditional Pearson residuals for model **M17.1**. The model-fit object `fm17.1` and data frame `prt1` were created in Panel R17.1

```
> prt1r <-                                          # Auxiliary data
+     within(prt1,
+            {                                      # Pearson residuals
+              residP1 <- residuals(fm17.1, type = "p")
+              fitted1 <- fitted(fm17.1)
+            })
> range(prt1r$residP1)                            # Info for y-axis range
  [1] -2.8753  5.7108
> xyplot(residP1 ~ fitted1| occ.f,    # Resids  vs. fitted (Fig. 17.2a)
+        data = prt1r, ylim = c(-6, 6),
+        type = c("p", "smooth"),
+        grid = TRUE)
> qqnorm(prt1r$residP1); qqline(prt1r$residP1) # Q-Q plot (Fig. 17.3a)
```

$\mathcal{D}$ (see the discussion of the results presented in Panel R17.3). To illustrate the correlation, we plot the scatterplot of the estimated random intercepts using the function `xyplot()`. The resulting graph is shown in Fig. 17.1b and provides a visual interpretation of the correlation. We also enhanced the default scatterplot graph, with the help of the `ellipse()` function from the **ellipse** package. Specifically, we added two ellipses representing the 95% confidence regions corresponding to the 2×2 matrices $\widehat{\mathcal{D}}$ (solid line) and var$(\widehat{\boldsymbol{b}}_i)$ (dashed) line, respectively. The relationship between these two ellipses, with one being inner to the other, provides yet another, more comprehensive illustration of "shrinkage".

In Panel R17.6, we construct plots of the conditional Pearson residuals to evaluate the fit of model **M17.1** to the data (Sect. 13.6.2). Note that, in Sect. 13.6.2, it was argued in favor of using the normalized residuals, which should be approximately independent and follow the standard normal distribution. However, according to model **M17.1**, the conditional residual errors are independent and homoscedastic, so the use of the Cholesky-decomposition-based transformation presented in (13.54) creates residuals equivalent to the Pearson residuals.

To construct the plots, we first add the variables `residP1` and `fitted1` to the `prt1` data frame. The variables contain the conditional Pearson residuals and subject-specific fitted values (Sect. 13.6.2), respectively, which are extracted from the model-fit object `fm17.1` with the help of the functions `residuals()` and `fitted()`, respectively (Sect. 14.6). With the help of the `xyplot()` function, we plot a scatterplot of the conditional Pearson residuals *versus* the fitted values, separately for the `Pre` and `Pos` levels of the `occ.f` factor. Note that we use the `type` argument to add the loess-smoothed curve to the scatterplots. The resulting graph is shown in Fig. 17.2a. Especially for the post-training measurements, the scatterplot suggests a possibility of an increase of residual variance with an increasing mean value. We will attempt to address this issue in the next section.

We also construct a normal Q-Q plot of the Pearson residuals using the function `qqnorm()`. The plot is shown in Fig. 17.3a. A deviation from normality, especially for the right-hand tail, can be observed.

17.3 A Mean-Variance Model with Occasion-Specific Random Intercepts for Type-1 Fibers

In this section, we will modify model **M17.1** to address the issue of the increasing residual variance, suggested by the residual scatterplots shown in Fig. 17.2a.

Toward this end, we consider model **M17.2** which, similarly to model **M17.1**, is defined by (17.1), but which assumes that the residual variance is a power function of the conditional mean value (Sect. 13.4.2):

$$\mathrm{Var}(\varepsilon_{itr} \mid b_{0it}) = \sigma^2(\mu_{it} + b_{0it})^{2\delta}, \tag{17.5}$$

where μ_{it} was defined in (17.1). Thus, model **M17.2** is an example of a mean-variance model (Sect. 13.8). In what follows, we will fit model **M17.2** and compare it with model **M17.1**.

17.3.1 R *Syntax and Results*

The syntax used to fit model **M17.2** to the observations from the data frame `prt1` is presented in Panel R17.7. Note that we fit the model by updating the specification of model **M17.1**, represented by the object `fm17.1`, with an appropriate value of the

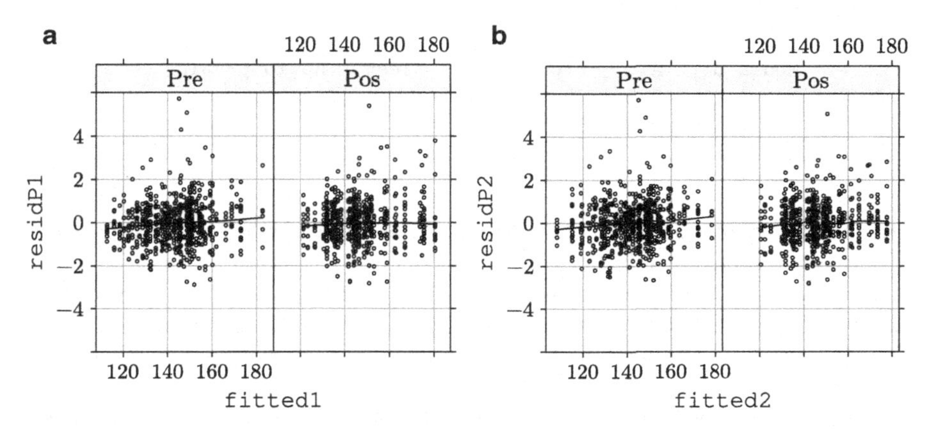

Fig. 17.2 Plots of the conditional Pearson residuals *versus* fitted values for models (**a**) **M17.1** and (**b**) **M17.2**

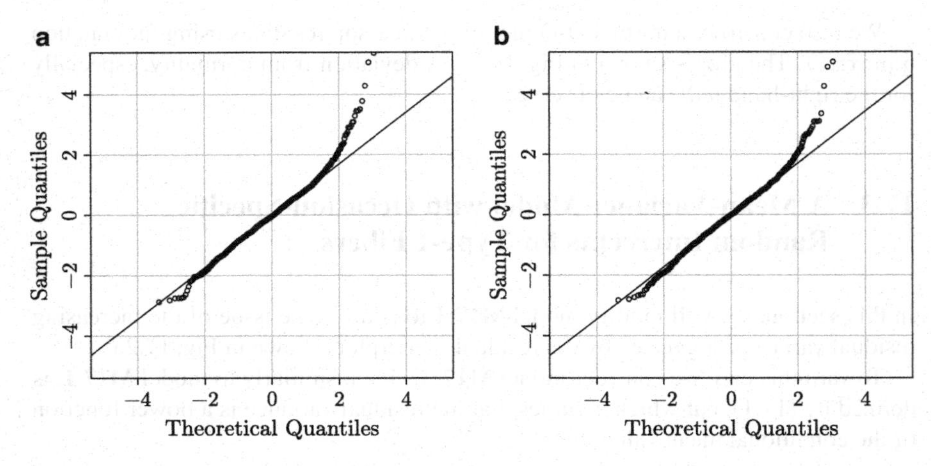

Fig. 17.3 Normal Q-Q plots for the conditional Pearson residuals for models (**a**) **M17.1** and (**b**) **M17.2**

R17.7 *PRT Trial*: Model **M17.2** fitted to the data for type-1 fibers using the function `lme()`. The model-fit object `fm17.1` and the data frame `prt1` were created in Panel R17.1

```
> fm17.2 <-                                    # M17.2 ← M17.1
+    update(fm17.1,
+           weights = varPower(form = ~fitted(.)),
+           data = prt1)
> intervals(fm17.2)$varStruct                  # 95% CI for δ, (17.5)
       lower  est.  upper
 power 1.1859 1.566 1.9461
 attr(,"label")
 [1] "Variance function:"
> anova(fm17.1, fm17.2)                        # H0: δ=0 (M17.1⊂M17.2)
        Model df   AIC   BIC  logLik   Test L.Ratio p-value
 fm17.1     1 12 11936 11998 -5956.2
 fm17.2     2 13 11895 11962 -5934.5 1 vs 2  43.548  <.0001
```

`weights` argument (see Table 14.4). The results are stored in the object `fm17.2` and details are presented in Table 17.1.

In Panel R17.7, by referring to the `varStruct` component of the object that results from the application of the `intervals()` function (Sect. 14.6) to the model-fit object `fm17.2`, we obtain the 95% CI for the parameter δ (Sect. 13.7.3). The CI indicates that $\delta > 1$, i.e., that the residual variance does increase with the mean value. Given that model **M17.1** is nested within **M17.2**, with the help of the `anova()` function (Sect. 14.7), we obtain the p-value of the REML-based LR test comparing

Table 17.1 *PRT Trial*: REML-based parameter estimates[a] for models **M17.1** and **M17.2** for type-1-fiber measurements with occasion-specific random intercepts

	Parameter	fm17.1	fm17.2
Model label		**M17.1**	**M17.2**
Log-REML value		−5,956.25	−5,934.47
Fixed effects:			
Intercept	β_0	127.72(15.42)	128.51(15.19)
PRT (low *vs.* high)	β_1	2.89(4.34)	3.02(4.35)
Occasion (post *vs.* pre)	β_2	4.70(2.67)	4.97(2.61)
Sex (M *vs.* F)	β_3	−1.38(5.36)	−1.51(5.27)
Age (old *vs.* yng)	β_4	8.98(5.15)	7.85(5.11)
BMI	β_5	0.49(0.58)	0.43(0.57)
PRT(low):Occ(post)	β_{12}	−2.13(3.75)	−2.28(3.69)
Sex(M):Age(old)	β_{34}	−12.68(7.55)	−11.28(7.44)
Variance components(`id`):			
sd(occ(pre))	$\sqrt{d_{11}}$	15.45(12.24,19.49)	15.60(12.40,19.64)
sd(occ(pos))	$\sqrt{d_{22}}$	14.19(11.14,18.08)	13.68(10.68,17.52)
cor(occ(pre),occ(pos))	ϱ_{12}	0.76(0.51, 0.89)	0.76(0.52, 0.89)
Variance function:			
power (μ^{δ})	δ		1.57(1.19,1.95)
Scale	σ	22.49(21.59,23.42)	0.01(0.00,0.06)

[a]Approximate SE for fixed effects and 95% CI for covariance parameters are included in parentheses

the two models (Sect. 13.7.2). The p value is statistically significant at the 5% significance level, indicating that we can reject the null hypothesis that $\delta = 0$.

Figure 17.2 compares the scatterplots of the conditional Pearson residuals *versus* the fitted values for models **M17.1** and **M17.2**. The syntax necessary to obtain the plot for model **M17.2** is similar to the one used in Panel R17.6 for model **M17.1** and, therefore, we do not show it. Compared to the corresponding plot for model **M17.1**, the variability of residuals for model **M17.2** appears to be more constant across the increasing values of the fitted values.

Figure 17.3 presents the normal Q-Q plot for the residuals for models **M17.1** and **M17.2**. Plots were constructed using the `qqnorm()`-function call similar to the syntax given in Panel R17.6. Again, it seems that the use of the power variance function (17.5) reduced the deviation from normality in model **M17.2** as compared to model **M17.1**, but it has not completely removed it.

Table 17.1 presents a summary of the parameter estimates for models **M17.1** and **M17.2**. In general, the estimated values of the fixed-effects coefficients are similar for the two models, with standard errors somewhat smaller for model **M17.2**. The latter point illustrates the gain in efficiency of the fixed-effects estimation when the residual variance structure is properly accounted for (Sect. 7.8.2).

The estimates of the matrix $\mathcal{D}$ for models **M17.1** and **M17.2** are also similar. A marked difference can be observed for the value of the scale parameter σ, but it is understandable, given different specifications of the residual error variance-covariance structures for the two models.

Finally, it is worth noting that, while getting an insight into the marginal variance-covariance and correlation structures, implied by model **M17.1**, posed no particular problem (see, e.g., Panel R17.4), it is more challenging for model **M17.2**. This is because the use of the variance function (17.5) implies that the marginal structures are different for different individuals.

17.4 A Model with Heteroscedastic Fiber-Type×Occasion-Specific Random Intercepts

The analyses of the type-1 fiber data, presented in Sects. 17.2 and 17.3, suggest that pre- and post-training observations for the same individual exhibit a positive correlation, which is slightly higher than the correlation between a pair of pre-training or post-training observations. This might be expected for type-2 fibers as well.

In this section, we extend the models, used in Sects. 17.2 and 17.3, so that we can analyze the data for both fiber types simultaneously. Toward this end, we consider the use of four random intercepts per individual, which allow us to account for the correlation between measurements obtained for different fiber types at different occasions.

17.4.1 Model Specification

As was mentioned in Sect. 17.2.1, Figs. 3.5 and 3.6 indicate a considerable between-subject variability with respect to the means and variances of the specific-force measurements for each of the four combinations of the occasions and fiber types. This suggests that an LMM with fiber-type×occasion-specific random intercepts for a subject might be reasonable. The model should also take into account the factors used in the experimental design.

Taking into account these considerations, we specify model **M17.3** as follows:

$$
\begin{aligned}
\text{SPEC.FO}_{ijtr} = {} & \beta_0 + \beta_1 \times \text{PRT}_i + \beta_2 \times \text{OCC}_{it} + \beta_3 \times \text{SEX}_i \\
& + \beta_4 \times \text{AGE}_i + \beta_5 \times \text{BMI}_i + \beta_6 \times \text{FIBER}_{ij} \\
& + \beta_{12} \times \text{PRT}_i \times \text{OCC}_{it} + \beta_{16} \times \text{PRT}_i \times \text{FIBER}_{ij} \\
& + \beta_{26} \times \text{OCC}_{it} \times \text{FIBER}_{ij} + \beta_{34} \times \text{SEX}_i \times \text{AGE}_i \\
& + b_{ijt} + \varepsilon_{ijtr} \\
\equiv {} & \mu_{ijt} + b_{ijt} + \varepsilon_{ijtr}.
\end{aligned} \tag{17.6}
$$

Compared to (17.1), (17.6) uses an extra index, j, which indicates fiber types ($j = 1,2$). Moreover, it includes the variable FIBER_{ij}, which is an indicator variable for the type-2 fibers for the i-th subject. Note that two-way interactions of variables PRT_i and OCC_{it} with FIBER_{ij} are included, to allow for different effects of the training intensity and measurement occasion for different fiber types. The effects of SEX_i and AGE_i are assumed to be the same for both fiber types, however.

Apart from the residual random errors ε_{ijtr}, which are assumed to be independent and normally distributed with mean zero and variance σ^2, the model equation specifies, for each subject, four fiber-type×occasion-specific random intercepts: b_{i11}, b_{i12}, b_{i21}, and b_{i22}. Let us define the vector $\mathbf{b}_i = (b_{i11}, b_{i12}, b_{i21}, b_{i22})'$, with the elements of the vector given in lexicographic order, i.e., first ordered by fiber type (first index) then by occasion (last index), so that the index, which corresponds to occasion, varies more quickly. This particular ordering will prove important for models defined later in this chapter. We assume that $\mathbf{b}_i$ is normally distributed with mean zero and variance-covariance matrix

$$\mathcal{D} \equiv \begin{pmatrix} d_{11,11} & d_{11,12} & d_{11,21} & d_{11,22} \\ d_{12,11} & d_{12,12} & d_{12,21} & d_{12,22} \\ d_{21,11} & d_{21,12} & d_{21,21} & d_{21,22} \\ d_{22,11} & d_{22,12} & d_{22,21} & d_{22,22} \end{pmatrix}, \tag{17.7}$$

where $d_{jt,j't'} = d_{j't',jt}$. The reason for using the somewhat nonstandard notation for the elements of the matrix $\mathcal{D}$ in (17.7) is that the four-index subscripts reflect the 2×2 factorial design of fiber types and occasions, for which the random intercepts are defined. Moreover, they clearly show that the ordering of rows and columns of the matrix $\mathcal{D}$ corresponds to the ordering of the elements of the vector $\mathbf{b}_i$. The notation will prove useful when, e.g., we will be comparing model **M17.3** with other models later in this chapter.

17.4.1.1 Marginal Interpretation

The model equation (17.6) implies that the marginal expected value of the SPEC.FO_{ijtr} measurement is equal to μ_{ijt}, defined in (17.6). The marginal variances and covariances can be expressed as follows:

$$\begin{aligned} \text{Var}(\text{SPEC.FO}_{ijtr}) &= d_{jt,jt} + \sigma^2, \\ \text{Cov}(\text{SPEC.FO}_{ijtr}, \text{SPEC.FO}_{ij't'r'}) &= d_{jt,j't'}, \end{aligned}$$

where $j \neq j'$ or $t \neq t'$ or $r \neq r'$.

To get more insight in the structure of the marginal variance-covariance and correlation matrices, let us define vectors:

$$\mathbf{y}_{ijt} \equiv (\text{SPEC.FO}_{ijt1}, \text{SPEC.FO}_{ijt2}, \ldots, \text{SPEC.FO}_{ijtn_{ijt}})',$$

and

$$\mathbf{y}_i \equiv (\mathbf{y}'_{i11}, \mathbf{y}'_{i12}, \mathbf{y}'_{i21}, \mathbf{y}'_{i22})'.$$

The marginal variance-covariance matrix $\mathcal{V}_i$ of $\mathbf{y}_i$, implied by model **M17.3**, has the following block structure:

$$\mathcal{V}_i \equiv \mathbf{Z}_i \mathcal{D} \mathbf{Z}'_i + \mathcal{R}_i = \begin{pmatrix} \mathcal{V}_{i,11} & \mathcal{V}_{i,11,12} & \mathcal{V}_{i,11,21} & \mathcal{V}_{i,11,22} \\ \mathcal{V}'_{i,11,12} & \mathcal{V}_{i,12} & \mathcal{V}_{i,12,21} & \mathcal{V}_{i,12,22} \\ \mathcal{V}'_{i,11,21} & \mathcal{V}'_{i,12,21} & \mathcal{V}_{i,21} & \mathcal{V}_{i,21,22} \\ \mathcal{V}'_{i,11,22} & \mathcal{V}'_{i,12,22} & \mathcal{V}'_{i,21,22} & \mathcal{V}_{i,22} \end{pmatrix}. \quad (17.8)$$

The diagonal blocks $\mathcal{V}_{i,jt}$ are matrices of dimension $n_{ijt} \times n_{ijt}$ and have the same structure as the matrix shown in (17.4), with all diagonal elements equal to $d_{jt,jt} + \sigma^2$ and all off-diagonal elements equal to $d_{jt,jt}$. The off-diagonal blocks $\mathcal{V}_{i,jt,j't'}$ are matrices of dimension $n_{ijt} \times n_{ij't'}$, with all elements equal to $d_{jt,j't'}$.

The corresponding marginal correlation matrix $\mathcal{C}_i$ has a blocked structure similar to that of matrix $\mathcal{V}_i$, shown in (17.8), with blocks of the same dimensions as the blocks of $\mathcal{V}_i$. The four diagonal blocks of $\mathcal{C}_i$ are correlation matrices with a compound-symmetry structure. They contain correlation coefficients that are all equal to $d^2_{jt,jt}/(d^2_{jt,jt} + \sigma^2)$ and that describe the correlation between any two different measurements taken for fiber type j at the same occasion t for a particular individual. The six off-diagonal blocks of the matrix $\mathcal{C}_i$ have all their elements equal to $d_{jt,j't'}/\sqrt{(d^2_{jt,jt} + \sigma^2)(d^2_{j't',j't'} + \sigma^2)}$. These elements correspond to the correlation coefficients between any two measurements taken for different fiber types (if $j \neq j'$) at the same occasion t, for the same fiber type j at different occasions (if $t \neq t'$), or for different fiber types at different occasions (if $j \neq j'$ and $t \neq t'$).

17.4.2 R *Syntax and Results*

Panel R17.8 displays the R syntax used to fit model **M17.3** with the help of the function `lme()`. First, in Panel R17.8a, we update the fixed-effects formula, used for model **M17.1**, by adding the factor `fiber.f` and its two-way interactions with factors `prt.f` and `occ.f`, as specified in (17.6). Then we use the updated formula in the call to the function `lme()`. Note that, in the `random` argument, we use the formula `fiber.f:occ.f-1|id`, i.e., we include neither the intercept nor the main effects of `fiber.f` and `occ.f`. As a result, for each individual, we define four random effects, which correspond to the levels of the `fiber.f:occ.f` interaction, i.e., to the four fiber-type $\times$ occasion combinations.

The fitted model is stored in the object `fm17.3`. The printout of `fm17.3` is extensive and we do not present it in Panel R17.8. We investigate, however, several of its components.

First, in Panel R17.8b, we show the estimates of the fixed-effects coefficients with their estimated standard errors, degrees of freedom, values of t-test statistic,

R17.8 *PRT Trial*: Model **M17.3** with four random intercepts fitted to the data for both fiber types using the function `lme()`. The formula-object `lme.spec.form1` was created in Panel R17.1

(a) *Model fitting*

```
> lme.spec.form3 <-
+    update(lme.spec.form1,                         # M17.3 ← M17.1
+           . ~ . + fiber.f + prt.f:fiber.f + occ.f:fiber.f)
> fm17.3 <-
+    lme(lme.spec.form3,                            # (17.6)
+        random = ~occ.f:fiber.f - 1|id,            # D:(17.7)
+        data = prt)
```

(b) *Fixed-effects estimates*

```
> fixed.D4 <- summary(fm17.3)$tTable           # β̂, se(β̂), t-test
> rnms <- rownames(fixed.D4)                   # β names (not shown)
> rnms[8:11] <-                                # Selected names shortened
+    c("Low:Pos", "Low:Type2", "Pos:Type2", "Male:Old")
> rownames(fixed.D4) <- rnms                   # Short names assigned
> printCoefmat(fixed.D4, digits = 3, zap.ind = 5)
                  Value Std.Error       DF t-value p-value
(Intercept)     129.611    14.288 2403.000   9.071    0.00
prt.fLow          1.951     4.313   57.000   0.452    0.65
occ.fPos          4.299     2.503 2403.000   1.717    0.09
sex.fMale        -2.037     5.021   57.000  -0.406    0.69
age.fOld          8.694     4.759   57.000   1.827    0.07
bmi               0.399     0.532   57.000   0.749    0.46
fiber.fType 2    25.302     2.404 2403.000  10.524    0.00
Low:Pos          -1.134     3.408 2403.000  -0.333    0.74
Low:Type2        -6.263     6.966   57.000  -0.899    0.37
Pos:Type2        -4.078     2.913 2403.000  -1.400    0.16
Male:Old          4.094     2.372 2403.000   1.726    0.08
```

and the corresponding p value. Toward this end, we refer to the component `tTable` of the object resulting from applying the function `summary()` to the model-fit object `fm17.3`. Before printing out the contents of the `tTable` array, we shorten the names of selected rows. To display the results, we use the function `printCoefmat()`, which allows more control over the format of the printout (see the syntax in Panel **M16.1** and its explanation in Sect. 16.2.2).

The fixed effect of most interest is the `Low:Pos` interaction term. The estimated coefficient is not significantly different from 0, with the p-value of the t-test equal to 0.74. It is worth noting that, for the fixed effects that are associated with covariates measured at the subject level, such as `bmi`, the number of degrees of freedom is equal to 57, as was the case for model **M17.1** (see Panel R17.2). This is because the number of subjects and the number of fixed effects, which are estimated at the subject level, is the same for both models. On the other hand, the number of degrees of freedom for the fixed-effects estimated at the observation, i.e., muscle-fiber level is larger for model **M17.3** compared to model **M17.1**. This is because the model is fitted to the data for both fiber types, i.e., to 2,471 observations in total, not just to type-1 fibers. Hence, the number of degrees of freedom for the observation-level coefficients is equal to $2,471 - (63+5) = 2,403$ (Sects. 14.7 and 17.2.2).

In Panel R17.9, we present the R syntax used to extract the results related to the variance-covariance matrix of the random effects and residual errors. In particular, in Panel R17.9a, the estimate of the matrix $\mathcal{D}$ is extracted from the object `fm17.3` by assigning the value `"random.effect"` to the argument `type` of the extractor-function `getVarCov()`. The vector `nms.` is used to replace the default (long) names for the rows and columns of the matrix displayed using the `rownames` function. The diagonal elements of the estimated matrix $\mathcal{D}$ suggest that we might consider simplifying the structure of the matrix $\mathcal{D}$ by assuming a constant variance for the random intercepts. However, such a structure is not supported by the standard *pdMat* classes available in R. We will attempt to address this issue in Sect. 17.7.

The estimated correlation matrix, corresponding to $\mathcal{D}$, is displayed with the help of the function `cov2cor()`. It is worth noting the similarity of, e.g., the elements $[1,2]$ and $[3,4]$, corresponding to the correlation coefficient between the random effects for different fiber types at the pre- and post-training measurements, respectively. This suggests a possibility of a more parsimonious representation of the matrix. We will investigate this issue in Sects. 17.6 and 17.7.

In Panel R17.9b, we extract information about matrices $\mathcal{R}_i$. Their dimensions depend on the number of repeated observations for a particular subject. For the individual with `id` equal to `"5"`, the matrix $\mathcal{R}_i$ is of dimension 41×41. Hence, it is not advisable to display the entire matrix. However, in accordance with the definition of model **M17.3**, the matrices $\mathcal{R}_i$ are diagonal with a constant element σ^2 on the diagonal. Thus, it is sufficient to display a few elements from the diagonal. As seen in Panel R17.9b, the first six diagonal elements of the matrix $\mathcal{R}_i$ for the subject "5" are all equal to 599.13.

Alternatively, given the constant, diagonal structure of the matrices $\mathcal{R}_i$, we can simply report the estimated value of the scale parameter σ^2. Note that, in model **M17.3**, the scale parameter σ can be interpreted as residual variance. To obtain its estimate, we refer to the component `sigma` of the model-fit object `fm17.3`. The estimate of σ^2 is equal to 599.13, which, obviously, corresponds to the values of the diagonal elements of the estimated matrix $\mathcal{R}_i$ for the subject "5".

Panel R17.10 shows the syntax to display the 95% CIs (Sect. 13.7.3) for the standard deviations and correlation coefficients, corresponding to the estimate of the matrix $\mathcal{D}$, which was presented in Panel R17.9. The intervals are obtained with the help of the function `intervals()` (see Table 14.5). Note that the default names, used by the function to identify the parameters, are too long. Thus, before displaying the confidence intervals, we modify the names. Toward this end, we store the result of application of the function `intervals()` to the model-fit object `fm17.3` in the object `CI`. The latter object is a list with two components: `reStruct`

R17.9 *PRT Trial*: The estimates of the matrix $\mathcal{D}$ and σ^2 for model **M17.3** with four random intercepts. The model-fit object `fm17.3` was created in Panel R17.8

(a) The $\mathcal{D}$-matrix estimate

```
> fm17.3cov <-                                      # D̂: (17.7) extracted
+    getVarCov(fm17.3, type = "random.effect")
> rownames(fm17.3cov)                               # Long names ...
  [1] "occ.fPre:fiber.fType 1" "occ.fPos:fiber.fType 1"
  [3] "occ.fPre:fiber.fType 2" "occ.fPos:fiber.fType 2"
> nms. <- c("T1.Pre", "T1.Pos", "T2.Pre", "T2.Pos")# ... abbreviated
> dimnames(fm17.3cov) <- list(nms., nms.)           # ... and reassigned.
> fm17.3cov                                         # D̂: (17.7) printed
  Random effects variance covariance matrix
         T1.Pre T1.Pos T2.Pre T2.Pos
  T1.Pre 248.78 175.32 212.49 155.38
  T1.Pos 175.32 184.61 109.99 172.97
  T2.Pre 212.49 109.99 241.63 133.74
  T2.Pos 155.38 172.97 133.74 247.03
    Standard Deviations: 15.773 13.587 15.544 15.717
> fm17.3cor <- cov2cor(fm17.3cov)                   # Corr(D̂) ...
> print(fm17.3cor, digits = 2,                      # ... printed.
+        corr = TRUE, stdevs = FALSE)
  Random effects correlation matrix
         T1.Pre T1.Pos T2.Pre T2.Pos
  T1.Pre   1.00   0.82   0.87   0.63
  T1.Pos   0.82   1.00   0.52   0.81
  T2.Pre   0.87   0.52   1.00   0.55
  T2.Pos   0.63   0.81   0.55   1.00
```

(b) *The $\mathcal{R}_i$-matrix estimate for the subject "5"*

```
> dim(R.5 <-                                        # Dims of R̂_i ...
+    getVarCov(fm17.3,
+               type = "conditional")[["5"]])       # ... for subject "5".
  [1] 41 41
> diag(R.5)[1:6]                           # First 6 diagonal elements
       1      2      3      4      5      6
  599.13 599.13 599.13 599.13 599.13 599.13
> (sgma <- fm17.3$sigma)                            # σ̂
  [1] 24.477
> print(sgma^2)                                     # σ̂^2
  [1] 599.13
```

R17.10 *PRT Trial*: Confidence intervals for the standard deviations and correlations corresponding to the matrix $\mathcal{D}$ for model **M17.3**. The model-fit object `fm17.3` was created in Panel R17.8

```
> CI <- intervals(fm17.3, which = "var-cov") # 95% CIs for θD
> interv <- CI$reStruct$id
> # rownames(interv)                       # Long names (not shown)
> thDnms <-
+     c("sd(T1Pre)", "sd(T1Pos)", "sd(T2Pre)", "sd(T2Pos)",
+       "cor(T1Pre,T1Pos)", "cor(T1Pre,T2Pre)", "cor(T1Pre,T2Pos)",
+                          "cor(T1Pos,T2Pre)", "cor(T1Pos,T2Pos)",
+                                              "cor(T2Pre,T2Pos)")
> rownames(interv) <- thDnms               # Short names assigned
> interv                                   # CIs printed
                    lower     est.    upper
 sd(T1Pre)       12.29101 15.77280 20.24091
 sd(T1Pos)       10.60119 13.58706 17.41391
 sd(T2Pre)       12.27121 15.54447 19.69085
 sd(T2Pos)       12.25335 15.71709 20.15996
 cor(T1Pre,T1Pos)  0.55111  0.81810  0.93314
 cor(T1Pre,T2Pre)  0.61009  0.86667  0.95874
 cor(T1Pre,T2Pos)  0.34280  0.62679  0.80581
 cor(T1Pos,T2Pre)  0.19870  0.52079  0.74134
 cor(T1Pos,T2Pos)  0.53362  0.80998  0.93005
 cor(T2Pre,T2Pos)  0.23955  0.54742  0.75526
```

and `sigma`. (The information of the structure of the object can be displayed using the command `str(CI)`.) The former component contains just a data frame named `id`, which includes three variables: `est`, `lower`, and `upper`. The variables provide, respectively, the point estimate and the lower and upper limits of the CI, for each parameter. Thus, we extract the data frame by referring to the component `reStruct$id` of the object `CI` and we store it in the data frame object `interv`. Then we modify the row names of `interv` by creating the vector `thDnms` with the new names and by assigning the new names to the rows of `interv` with the help of the `rownames()` function. Finally, we display the CIs. The lower limits of the CIs for the standard deviations are considerably larger than 0, which indicates that using all four random intercepts is justified. The intervals confirm that, as suggested earlier, we might consider simplifying the structure of the matrix $\mathcal{D}$ by assuming a constant variance for the random intercepts. Similarly, the lower limits of the CIs for the correlation coefficients are considerably larger than 0, indicating a positive correlation between the random intercepts.

Panel R17.11 shows the syntax to create plots of the conditional Pearson residuals for model **M17.3** and normal Q-Q plots. The residuals are obtained by applying the `residuals()` function (see Table 14.5) to the `fm17.3` model-fit object. Note that, in Sect. 13.6.2, it was argued in favor of using the normalized

R17.11 *PRT Trial*: Plots of the conditional Pearson residuals for model **M17.3**. The model-fit object `fm17.3` was created in Panel R17.8

```
> residP3 <- residuals(fm17.3, type =  "p") # Pearson residuals
> xyplot(residP3 ~ fitted(fm17.3)|     # Scatterplots ...
+          fiber.f:occ.f,              # ... per type*occasion (Fig. 17.4)
+          data = prt,
+          type = c("p", "smooth"))
> qqnorm(residP3); qqline(residP3)     # Q-Q plot (Fig. 17.5)
```

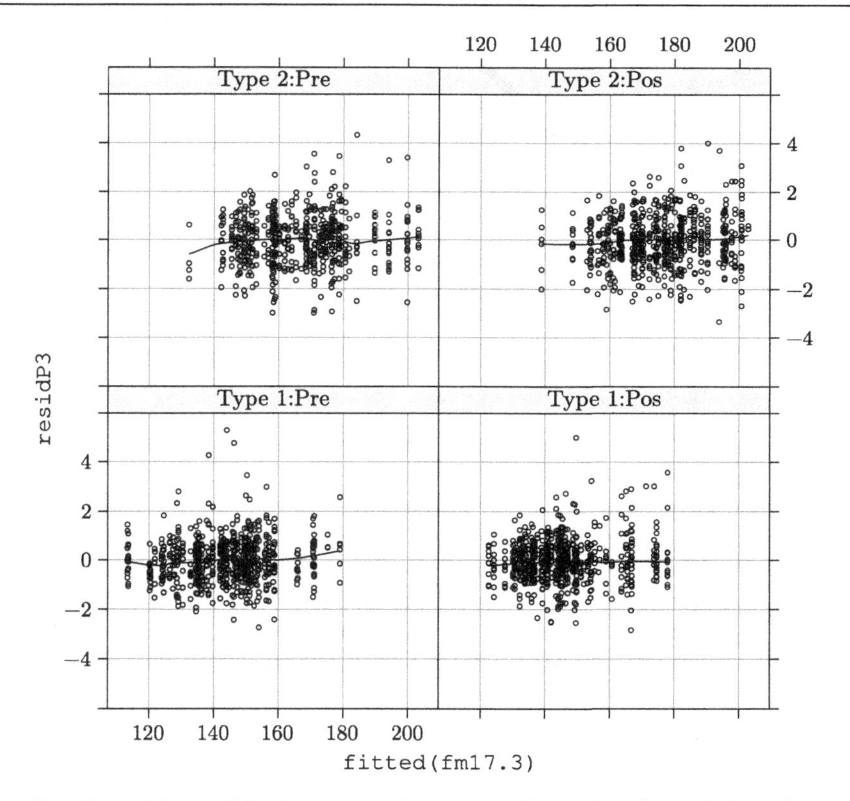

Fig. 17.4 Plots of the conditional Pearson residuals *versus* fitted values for model **M17.3**

residuals, which should be approximately independent and normally distributed. However, according to model **M17.3**, (conditional) residual errors are independent, so the normalized residuals are equivalent to Pearson residuals.

In Fig. 17.4, we use the `xyplot()` function to draw a scatterplot of the residuals against the fitted values for each combination of the fiber type and occasion. The syntax is similar to that shown in Panel R17.6. Similarly to the case of model **M17.1**, one could argue that the plots indicate a mean-dependent variance patterns. Thus, the

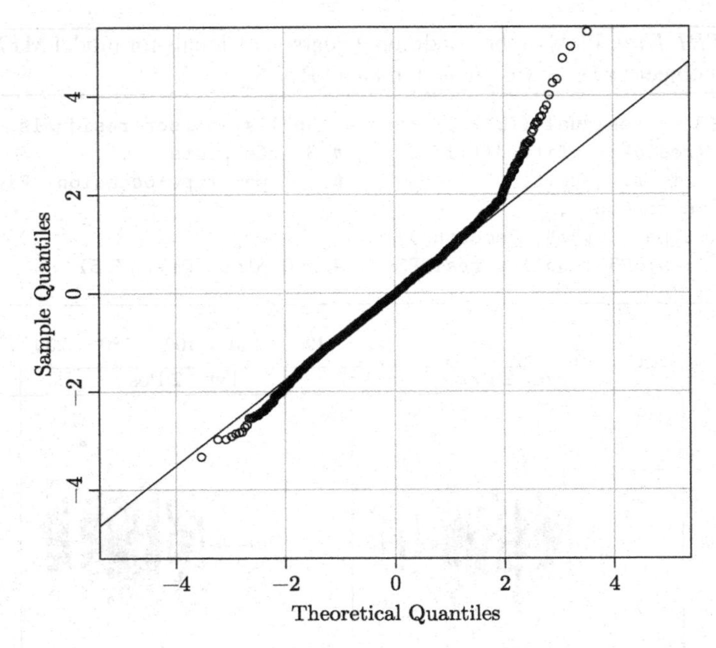

Fig. 17.5 Normal Q-Q plot of the conditional Pearson residuals for model **M17.3**

use of the power variance function (17.5) to address this issue might be considered. We leave it as an exercise to the reader.

At the bottom of Panel R17.11, we also show the use of the function `qqnorm()` to draw the normal Q-Q plot of the conditional residuals. The plot is shown in Fig. 17.5. A large part of the plot is reasonably linear. A deviation in the right tail is clear, though. It suggests a somewhat "thinner" right tail of the distribution of the residuals than that expected from a normal distribution. The pattern corresponds to the one seen in Fig. 17.3 for models **M17.1** and **M17.2**, which were only fitted to the type-1 fiber data.

17.4.2.1 Marginal Interpretation

The estimates of the fixed-effects coefficients, presented in Panel R17.8b, can be interpreted as the coefficients of the covariates involved in the marginal expected value μ_{ijt} (17.6) of the SPEC.FO$_{ijtr}$ measurements.

Panel R17.12 demonstrates the syntax for extracting information about the estimate of the marginal variance-covariance matrix $\mathcal{V}_i$, defined in (17.8).

In Panel R17.12a, we use the extractor-function `getVarCov()` (see Table 14.5) with arguments `type="marginal"` and `individual="5"` to extract the estimate of the marginal variance-covariance for the individual with identified by the grouping factor `id` equal to `"5"`. The result of the application of the function to the model-fit object `fm17.3` is stored in the object `Vx`. The object is a list with one component,

R17.12 *PRT Trial*: Estimates of the matrices $\mathcal{V}_i$ and $\mathcal{C}_i$ for model **M17.3**. The model-fit object `fm17.3` was created in Panel R17.8. The auxiliary function `auxF1()` was defined in Panel R17.4

(a) *The estimated matrix $\mathcal{V}_i$ for the subject "5"*

```
> Vx <-                                    # Vx is a list ...
+    getVarCov(fm17.3, type = "marginal",
+              individual = "5")           # ...  with one component.
> Vmtx.5 <- Vx$"5"                         # Vmtx.5 is V_i matrix: (17.8)...
> dim(Vmtx.5)                                # ...  with large dimensions.
  [1] 41 41
> dt5 <-                                   # Data with 41 rows ...
+    subset(prt,
+           select = c(id, fiber.f, occ.f), # ...  and 3 variables ...
+           id == "5")                       # ...  for subject "5".
```

(b) *Selected indices for rows and columns of $\widehat{\mathcal{V}}_i$*

```
> (i.u5  <- unlist(                        # Selected indices printed.
+    tapply(rownames(dt5),                 # Indices for subject "5" ...
+           list(dt5$fiber.f, dt5$occ.f), # ... by fiber.f and occ.f.
+           FUN = auxF1)))
  [1] "1"  "2"  "3"  "5"  "20" "22" "21" "25"
> dt.u5  <- dt5[i.u5, ]                    # Raw data for selected indices
> nms.u5 <-
+    paste(format(i.u5, 2, justify = "right"),
+          abbreviate(dt.u5$fiber.f, 2),      # Row names abbreviated
+          dt.u5$occ.f, sep = ".")
```

(c) *A submatrix of $\widehat{\mathcal{V}}_i$ and the corresponding correlation matrix*

```
> Vmtx.u5 <- Vmtx.5[i.u5, i.u5]       # Submatrix of V_i for subject "5"
> dimnames(Vmtx.u5) <- list(nms.u5, i.u5) # dimnames assigned
> Cmtx.u5 <- cov2cor(Vmtx.u5)         # Submatrix of Corr(V_i)
> uptri <- upper.tri(Cmtx.u5)         # Logical matrix
> Vmtx.u5[uptri] <- Cmtx.u5[uptri]
> print(Vmtx.u5, digits = 2)          # Submatrix printed
               1      2      3      5     20     22     21     25
   1.T1.Pre 848   0.29   0.25   0.25   0.22   0.22   0.18   0.18
   2.T1.Pre 249 847.91   0.25   0.25   0.22   0.22   0.18   0.18
   3.T2.Pre 212 212.49 840.76   0.29   0.14   0.14   0.16   0.16
   5.T2.Pre 212 212.49 241.63 840.76   0.14   0.14   0.16   0.16
  20.T1.Pos 175 175.32 109.99 109.99 783.74   0.24   0.21   0.21
  22.T1.Pos 175 175.32 109.99 109.99 184.61 783.74   0.21   0.21
  21.T2.Pos 155 155.38 133.74 133.74 172.97 172.97 846.16   0.29
  25.T2.Pos 155 155.38 133.74 133.74 172.97 172.97 247.03 846.16
```

which is a variance-covariance matrix. (The information of the structure of the object can be displayed using the command `str(CI)`.) We extract the component and save the matrix in the object `Vmtx.5`. The dimension of the matrix is large: it is 41×41, as can be seen in Panel R17.12a from the output of the `dim(Vmtx.5)` command. Printing of the whole matrix is thus not advisable. As seen from (17.8), however, the matrix should have a particular structure. Thus, printing a sub-matrix, corresponding to a set of a few measurements for different fiber types and occasions, should yield enough information.

Toward this end, we create the data frame `dt5`, which contains the subset of 41 rows for the variables `id`, `fiber.f`, and `occ.f` for the subject "5". In Panel R17.12b, with the help of the function `tapply()`, we apply the auxiliary function `auxF1()`, which was defined in Panel R17.4, to select two indices for each combination of levels of the factors `fiber.f` and `occ.f` from the vector of the row names of the data frame `dt5`. The resulting list contains indices for two different measurements for each combination of the levels of the `fiber.f` and `occ.f` factors for the subject "5". By applying the generic function `unlist()`, we simplify the list to a vector and store it in the object `i.u5`. We use the vector to select the corresponding rows from the data frame `dt5` and store them in the data frame `dt.u5`. Finally, we apply the generic function `paste()` to, first, format the width of the values of the vector `i.u5` to two characters; second, to abbreviate the values of the factor variable `fiber.f` from the data frame `dt.u5` to two characters; and third, to concatenate the so-formatted variables with the values of the variable `occ.f` from the data frame `dt.u5` using "." as separator. The concatenated strings, stored in the vector `nms.u5`, will be used to label the rows/columns in abbreviated printouts.

Finally, in Panel R17.12c, we select a submatrix of `Vmtx.5`, with rows and columns selected using the vector of indices `i.u5`, and store it in the object `Vmtx.u5`. Next, with the help of the function `dimnames`, we label the rows and columns of `Vmtx.u5` using the vectors `nms.u5` and `i.u5`, respectively. With the help of the function `cov2cor`, we also compute the correlation matrix corresponding to `Vmtx.5` and store the result in the object `Cmtx.u5`. By applying the `upper.tri()` function, we create a logical matrix `uptri`, which indicates the upper triangle of the matrix `Cmtx.u5`. With the help of the logical matrix, we replace the upper-triangle elements of `Vmtx.5` with the upper-triangle elements of `Cmtx.5`. In this way, we can compactly print out the information about the structure of matrices $\mathcal{V}_i$ and $\mathcal{C}_i$ for the subject "5". The printout clearly indicates that the matrices have got the structure described in Sect. 17.4.1.1. The estimates of the correlation-matrix elements suggest that, for instance, the correlation coefficient between any two pre-training measurements for type-1 and type-2 fibers, as well as for any two post-training measurements for type-2 fibers, is equal to 0.29. On the other hand, the correlation coefficient between any two post-training measurements for type-1 fibers is estimated to be equal to 0.24.

Model **M17.3** could be used as a ground for inference regarding the factors influencing the value of the SPEC.FO measurements. However, in the next sections, we consider several modifications of the model for purposes of illustration of various aspects of modeling of the variance-covariance structure of an LMM and of the related R tools and syntax.

17.5 A Model with Heteroscedastic Fiber-Type×Occasion-Specific Random Intercepts (Alternative Specification)

In this section, we consider an alternative parameterization of model **M17.3**, which enables further insight into the random structure of the data.

17.5.1 Model Specification

An equivalent form of model **M17.3**, which we will label as model **M17.3a**, can be obtained by using a different parameterization of random effects:

$$\begin{aligned}
\text{SPEC.FO}_{ijtr} = {} & \beta_0 + \beta_1 \times \text{PRT}_i + \beta_2 \times \text{OCC}_{it} + \beta_3 \times \text{SEX}_i \\
& + \beta_4 \times \text{AGE}_i + \beta_5 \times \text{BMI}_i + \beta_6 \times \text{FIBER}_{ij} \\
& + \beta_{12} \times \text{PRT}_i \times \text{OCC}_{it} + \beta_{16} \times \text{PRT}_i \times \text{FIBER}_{ij} \\
& + \beta_{26} \times \text{OCC}_{it} \times \text{FIBER}_{ij} + \beta_{34} \times \text{SEX}_i \times \text{AGE}_i \\
& + b_{i0} + b_{i1} \times \text{FIBER}_j + b_{i2} \times \text{OCC}_t \\
& + b_{i3} \times \text{OCC}_t \times \text{FIBER}_j + \varepsilon_{ijtr},
\end{aligned} \tag{17.9}$$

where the indices and symbols have a similar meaning as in (17.6). In (17.9), b_{i0} is an overall random intercept, shared by all measurements made for subject i, while b_{i1} is a random effect shared by all measurements for type-2 fibers (note that $\text{FIBER}_j = 1$ for type-2 fibers and 0 otherwise). Likewise, b_{i2} is a random effect shared by all post-training measurements ($\text{OCC}_t = 1$ for post-training measurements and 0 otherwise). Finally, b_{i3} is a random effect shared by all post-training measurements for type-2 fibers. We assume that the vector $\mathbf{b}_i \equiv (b_{i0}, b_{i1}, b_{i2}, b_{i3})'$ is normally distributed with mean zero and variance-covariance matrix

$$\mathcal{D} \equiv \begin{pmatrix} d_0 & d_{01} & d_{02} & d_{03} \\ d_{01} & d_1 & d_{12} & d_{13} \\ d_{02} & d_{12} & d_2 & d_{23} \\ d_{03} & d_{13} & d_{23} & d_3 \end{pmatrix}. \tag{17.10}$$

To establish the equivalence of models **M17.3** and **M17.3a**, we first note that the mean structure is the same. Thus, we only need to check the equivalence of the random structure. Toward this end, note that, according to model **M17.3a**, the random error in pre-training measurements for type-1 fibers includes, in addition to the residual error, the random intercept $b_{i,0}$; the error in pre-training measurements for type-2 fibers includes $b_{i,0} + b_{i,1}$; the error in post-training measurements for type-1 fibers includes $b_{i,0} + b_{i,2}$; and the error in post-training measurements for type-2 fibers includes $b_{i,0} + b_{i,1} + b_{i,2} + b_{i,3}$. Thus, upon defining

$$\begin{aligned} b_{i11} &\equiv b_{i0}, \\ b_{i12} &\equiv b_{i0} + b_{i1}, \\ b_{i21} &\equiv b_{i0} + b_{i2}, \\ b_{i22} &\equiv b_{i0} + b_{i1} + b_{i2} + b_{i3}, \end{aligned} \tag{17.11}$$

we can write the defining equation (17.9) of model **M17.3a** in the form of the equation (17.6) of model **M17.3**. Note that the transformation (17.11) can be compactly written as

$$\begin{pmatrix} b_{i11} \\ b_{i12} \\ b_{i21} \\ b_{i22} \end{pmatrix} = \begin{pmatrix} 1\,0\,0\,0 \\ 1\,0\,1\,0 \\ 1\,1\,0\,0 \\ 1\,1\,1\,1 \end{pmatrix} \begin{pmatrix} b_{i0} \\ b_{i1} \\ b_{i2} \\ b_{i3} \end{pmatrix} \equiv \boldsymbol{T_D} \begin{pmatrix} b_{i0} \\ b_{i1} \\ b_{i2} \\ b_{i3} \end{pmatrix}. \tag{17.12}$$

It follows that the matrix $\mathcal{D}$, specified in (17.7), can be obtained from the one given in (17.10) using the following transformation:

$$\begin{pmatrix} d_{11,11} & d_{11,12} & d_{11,21} & d_{11,22} \\ d_{12,11} & d_{12,12} & d_{12,21} & d_{12,22} \\ d_{21,11} & d_{21,12} & d_{21,21} & d_{21,22} \\ d_{22,11} & d_{22,12} & d_{22,21} & d_{22,22} \end{pmatrix} = \boldsymbol{T_D} \begin{pmatrix} d_0 & d_{01} & d_{02} & d_{03} \\ d_{01} & d_1 & d_{12} & d_{13} \\ d_{02} & d_{12} & d_2 & d_{23} \\ d_{03} & d_{13} & d_{23} & d_3 \end{pmatrix} \boldsymbol{T'_D}. \tag{17.13}$$

The transformation inverse to (17.12) allows to write the defining equation (17.6) in the form of (17.9).

17.5.2 R Syntax and Results

In Panel R17.13, we present the R syntax for fitting model **M17.3a** with the help of the function `lme()`. The main part of computations is given in Panel R17.13a. As compared to the syntax shown in Panel R17.8a, the main difference lies in the formula specified in the `random` argument. More specifically, the formula used in Panel R17.13a corresponds to the parameterization of the random-effects structure, applied in (17.9).

R17.13 *PRT Trial*: Model **M17.3a** fitted using the function `lme()`. The formula-object `lme.spec.form 3` was defined in Panel R17.8

(a) *Model fitting and extracting $\mathcal{D}$ estimate*

```
> fm17.3a <-
+    lme(lme.spec.form3,                        # M17.3a
+        random = ~1 + fiber.f + occ.f + fiber.f:occ.f|id,
+        data = prt)
> print(fm17.3a$sigma, digits = 4)              # σ̂
  [1] 24.48
> fm17.3acov <-                                 # D̂
+    getVarCov(fm17.3a,
+              type = "random.effect", individual = "5")
> dimnames(fm17.3acov)[[1]]                     # Row/col D̂ names ...
  [1] "(Intercept)"              "fiber.fType 2"
  [3] "occ.fPos"                 "fiber.fType 2:occ.fPos"
> nms <- c("(Int)", "T2", "Pos", "T2:Pos")  # ...  shortened
> dimnames(fm17.3acov) <- list(nms,nms)     # ...  and assigned.
> print(fm17.3acov, digits = 4)             # D̂ printed
  Random effects variance covariance matrix
          (Int)      T2     Pos T2:Pos
  (Int)  248.90 -36.34 -73.48  16.41
  T2     -36.34  65.45 -29.04 -21.75
  Pos    -73.48 -29.04  82.71  37.38
  T2:Pos  16.41 -21.75  37.38  63.70
    Standard Deviations: 15.78 8.09 9.094 7.981
```

(b) *Verification of (17.13)*

```
> td <-                                         # T_D: (17.12) created...
+    matrix(c(1, 0, 0, 0,
+             1, 0, 1, 0,
+             1, 1, 0, 0,
+             1, 1, 1, 1),
+           nrow = 4, ncol = 4, byrow = TRUE)
> mat.D4 <- td %*% fm17.3acov %*% t(td)        # ...  and applied.
> dimnames(mat.D4) <- list(nms., nms.)         # Row/col names shortened.
> print(mat.D4, digits = 5)                     # D̂: (17.7); see R17.9.
          T1.Pre T1.Pos T2.Pre T2.Pos
  T1.Pre 248.86 175.38 212.52 155.45
  T1.Pos 175.38 184.60 110.00 173.01
  T2.Pre 212.52 110.00 241.64 133.77
  T2.Pos 155.45 173.01 133.77 247.08
```

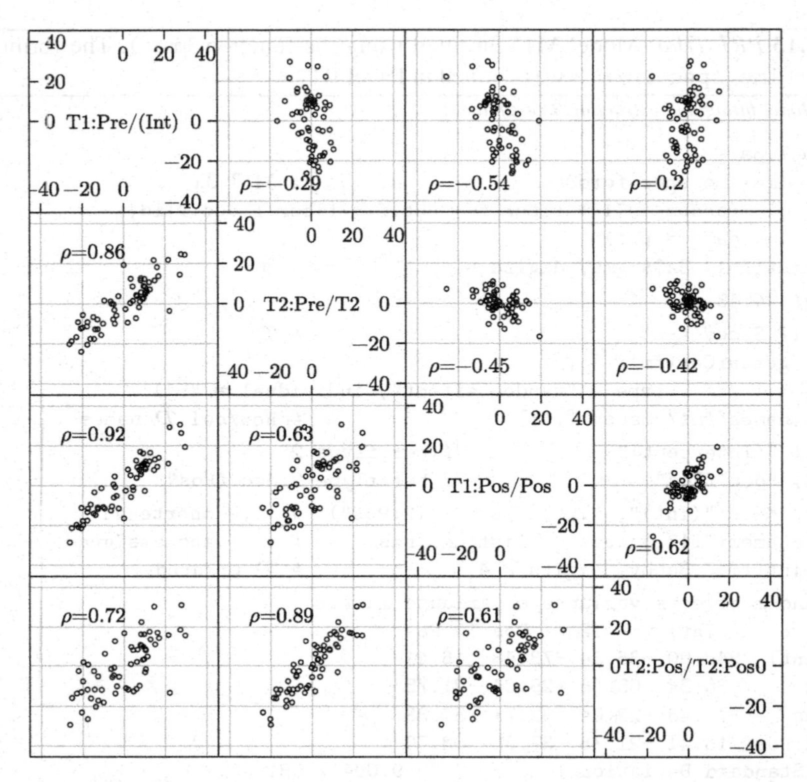

Fig. 17.6 *PRT Trial*: Scatterplot matrix of the predicted random effects (EBLUPs) for model **M17.3** (below diagonal) and **M17.3a** (above diagonal)

The fitted model is stored in the object `fm17.3a`. The estimates of the fixed-effects coefficients are the same as those for model **M17.3** and can be found in Panel R17.8b. In Panel R17.13a, we focus on selected results related to the random structure of model **M17.3a**. The estimate of residual SD, extracted from the model-fit object using its `$sigma` component, corresponds to the estimate obtained for model **M17.3** (see Panel R17.9). The estimate of the matrix $\mathcal{D}$, specified in (17.10), is extracted from the object `fm17.3a` using the `getVarCov()` function, stored as `fm17.3acov`, and printed (after shortening the row and column names). It shows comparable variances for b_{i1}, b_{i2}, and b_{i3}. This suggests that both fiber type and occasion contribute a similar amount of random variability to the data.

Additionally, in Panel R17.13b, the transformation (17.13) is applied to the estimated $\mathcal{D}$ matrix to show its equivalence (up to a rounding error) to the estimated $\mathcal{D}$ matrix for model **M17.3** (Panel R17.9).

Figure 17.6 presents scatterplots of the predicted random effects (EBLUPs), as defined in (13.50), for models **M17.3** and **M17.3a**. The EBLUPs were extracted from the relevant model-fit objects using the function `ranef()` (see Table 14.5).

Note that, to save the space, we do not present the R syntax, necessary to create Fig. 17.6. The scatterplots, presented in the figure, illustrate the correlation between the predictors of the various random effects for each model separately. The values of the correlation coefficients are indicated on the plots. It can be observed that, while there is a substantial correlation between the EBLUPs for model **M17.3**, there is much less association between the EBLUPs for model **M17.3a**. This is due to the different parameterizations of the random-effects structure, used in the two models. In particular, (17.11) shows that, for example, the random effects in model **M17.3** share the random component b_{i0}. Thus, the effects are bound to be correlated.

17.6 A Model with Heteroscedastic Fiber-Type×Occasion-Specific Random Intercepts and a Structured Matrix $\mathcal{D}$

As mentioned in Sect. 17.4, the estimated form of the random-effects variance-covariance matrix $\mathcal{D}$ of model **M17.3**, presented in Panel R17.9, suggests that we might consider simplifying the structure of the matrix. Toward this end, in Sect. 17.7, we will use a new *pdMat* class, *pdKronecker*, which we have developed and described in more detail in Sect. 20.2. In the current section, however, we will illustrate the use of the *pdKronecker* class.

17.6.1 Model Specification

First, we fit model **M17.4**, given by (17.6), but with the matrix $\mathcal{D}$ equal to the Kronecker product of two unstructured, 2×2 matrices, $\boldsymbol{A}$ and $\boldsymbol{B}$, say, corresponding to the fiber type and occasion, respectively. For reasons of implementation in R, the resulting matrix is represented as follows:

$$\mathcal{D} = \sigma^2 \otimes (c\boldsymbol{I}_1) \otimes \boldsymbol{A} \otimes \boldsymbol{B} \equiv \sigma^2 \otimes c \otimes \begin{pmatrix} 1 & a_{12} \\ a_{21} & a_{22} \end{pmatrix} \otimes \begin{pmatrix} 1 & b_{12} \\ b_{21} & b_{22} \end{pmatrix}$$

$$= \sigma^2 c \begin{pmatrix} 1 & b_{12} & a_{12} & a_{12}b_{12} \\ b_{21} & b_{22} & a_{12}b_{21} & a_{12}b_{22} \\ a_{21} & a_{21}b_{21} & a_{22} & a_{22}b_{12} \\ a_{21}b_{12} & a_{21}b_{22} & a_{22}b_{21} & a_{22}b_{22} \end{pmatrix}, \quad (17.14)$$

where $\otimes$ denotes the (right) Kronecker product. Note the use of the identity matrix $\boldsymbol{I}_1$ of dimension 1×1 in (17.14). The order of taking the Kronecker product of matrices $\boldsymbol{A}$ and $\boldsymbol{B}$ implies the ordering of the elements of the vector $\mathbf{b}_i$, explained in Sect. 17.4.1.

R17.14 *PRT Trial*: Model **M17.4** fitted using the functions `lme()` and `pdKronecker()`. The formula-object `lme.spec.form` was defined in Panel R17.8

```
> pdId  <- pdIdent(~1)                      # cI1 in (17.14)
> pd1UN <- pdLogChol(~fiber.f - 1)          # A for FiberType
> pd2UN <- pdLogChol(~occ.f - 1)            # B for PrePos
> pdL1  <-                                  # List of pdMat objects
+    list(X = pdId,
+         FiberType = pd1UN,
+         PrePos = pd2UN)
> (pdKnms <- names(pdL1))                   # Names saved for later use
  [1] "X"          "FiberType" "PrePos"
> fm17.4 <-                                 # M17.4
+    lme(lme.spec.form3,
+        random = list(id = pdKronecker(pdL1)),
+        data = prt)
```

The correlation matrix, corresponding to (17.14), is given by

$$
\boldsymbol{C} = \begin{pmatrix} 1 & \frac{a_1}{\sigma^2 c\sqrt{a_2}} & \frac{b_1}{\sigma^2 c\sqrt{b_2}} & \frac{b_1 a_1}{\sigma^2 c\sqrt{b_2 a_2}} \\ \frac{a_1}{\sigma^2 c\sqrt{a_2}} & 1 & \frac{b_1 a_1}{\sigma^2 c\sqrt{b_2 a_2}} & \frac{b_1}{\sigma^2 c\sqrt{b_2}} \\ \frac{b_1}{\sigma^2 c\sqrt{b_2}} & \frac{b_1 a_1}{\sigma^2 c\sqrt{b_2 a_2}} & 1 & \frac{a_1}{\sigma^2 c\sqrt{a_2}} \\ \frac{b_1 a_1}{\sigma^2 c\sqrt{b_2 a_2}} & \frac{b_1}{\sigma^2 c\sqrt{b_2}} & \frac{a_1}{\sigma^2 c\sqrt{a_2}} & 1 \end{pmatrix} \equiv \begin{pmatrix} 1 & x & y & z \\ x & 1 & z & y \\ y & z & 1 & x \\ z & y & x & 1 \end{pmatrix}. \quad (17.15)
$$

It is worth noting that the correlation matrix has a particular structure, presented at the right-hand side of (17.15). Thus, the Kronecker-product representation, specified in (17.14), allows for different variances of the four random intercepts included in model **M17.4**, but leads to a more parsimonious correlation structure of the $\mathcal{D}$ matrix, as compared to (17.7). In particular, it implies, e.g., equality of the elements $[1,2]$ and $[3,4]$ or of the elements $[1,4]$ and $[2,3]$. Also, elements on the anti-diagonal are equal. Note that an approximately similar structure can be observed for the estimated $\mathcal{D}$ matrix of model **M17.3** (see Panel R17.9).

17.6.2 R *Syntax and Results*

The main argument of the *pdKronecker*-class constructor function is a list of *pdMat* objects (see Sect. 14.2), representing the matrices used in the Kronecker product. The first object on this list needs to be a *pdIdent*-class object, representing 1×1 matrix corresponding to $c\boldsymbol{I}_1$ in (17.14).

Table 17.2 *PRT Trial*: REML-based parameter estimates[a] for models **M17.4** and **M17.5**

	Parameter	fm17.4	fm17.5
Model label		**M17.4**	**M17.5**
Log-REML value		−11523.18	−11523.84
Fixed effects:			
Intercept	β_0	126.22(14.28)	126.52(14.43)
PRT (low *vs.* high)	β_1	2.18(4.09)	2.09(4.18)
Occasion (post *vs.* pre)	β_2	3.90(2.74)	3.79(2.79)
Sex (M *vs.* F)	β_3	−1.62(5.01)	−1.70(5.09)
Age (old *vs.* yng)	β_4	8.32(4.77)	8.24(4.82)
BMI	β_5	0.56(0.53)	0.54(0.54)
Fiber (type 2 *vs.* 1)	β_6	25.29(2.49)	25.35(2.45)
PRT(Low):occ(post)	β_{12}	−0.05(3.60)	0.29(3.61)
Sex(M):age(Old)	β_{34}	−9.14(7.00)	−7.91(7.07)
PRT(low):fiber(T2)	β_{16}	−4.00(2.99)	−4.09(2.96)
Occ(post):fiber(T2)	β_{26}	3.97(2.34)	3.91(2.33)
Variance components(`id`):			
sd(fiber(T1), occ(pre))	$\sqrt{d_{11}}$	14.64(11.64,18.40)	15.15(12.89,17.81)
Fiber type (***A***)			
var(T2)/var(T1)	a_{22}:(17.14)	1.25(0.66, 2.20)	
cov(T2, T1)/var(T1)	a_{12}:(17.14)	0.93(0.70, 1.16)	
ϱ(T2, T1)	a_{12}:(17.16)		0.83(0.66, 0.92)
Occasion (***B***)			
var(pos)/var(pre)	b_{22}	0.92(0.44, 1.72)	
cov(pos, pre)/var(pre)	b_{12}	0.64(0.42, 0.87)	
ϱ(pos, pre)	b_{12}		0.67(0.45, 0.81)
Scale	σ	24.47(23.76,25.20)	24.47(23.77,25.20)

[a]Approximate SE for fixed effects and 95% CI for covariance parameters are included in parentheses

Panel R17.14 displays the syntax used to fit model **M17.3** with the $\mathcal{D}$ matrix given by (17.14). First, we define *pdMat* objects representing the matrices $c\boldsymbol{I}_1$, $\boldsymbol{A}$, and $\boldsymbol{B}$, as specified in (17.14), using the `pdIdent()` and `pdLogChol()` constructor functions (Sect. 14.2.1). The matrices are stored in objects `pdId`, `pd1UN`, and `pd2UN`, respectively. Then, the named list `pdL1`, containing the *pdMat*-class objects, is created. Also, the names of the components of `pdL1` are stored in an auxiliary vector `pdKnms` to facilitate labeling of the displayed output later. Finally, the model is fitted using the appropriate `lme()`-function call, with the value of the argument `random` provided in the form of a named list of *pdMat*-objects (Sects. 14.3 and 14.5). The results of fitting of the model are stored in the object `fm17.4`. The estimates of the fixed-effects coefficients are presented in Table 17.2.

Panel R17.15 displays the estimated $\mathcal{D}$ matrix stored in `DmtxKron1` object. To increase legibility of the printout, we modify the default row and column names.

The estimated $\mathcal{D}$ matrix has the structure defined in (17.14). It is close to its counterpart obtained for model **M17.3** (see Panel R17.9). Again, the estimated variances of the random intercepts suggest a possibility of a constant variance.

R17.15 *PRT Trial*: Evaluating the heteroscedastic Kronecker-product structure of the matrix $\boldsymbol{D}$ for model **M17.4**. The objects `fm17.4` and `pdKnms` were created in Panel R17.14

```
> DmtxKron1 <- getVarCov(fm17.4)                       # D̂: (17.14)
> rownames(DmtxKron1)                                  # Long row/col names
  [1] "fiber.fType 1:occ.fPre" "fiber.fType 1:occ.fPos"
  [3] "fiber.fType 2:occ.fPre" "fiber.fType 2:occ.fPos"
> nms <- c("T1:Pre", "T1:Pos", "T2:Pre", "T2:Pos")# Short names...
> dimnames(DmtxKron1) <- list(nms, nms)               # ...  assigned.
> DmtxKron1                                            # D̂ displayed.
  Random effects variance covariance matrix
         T1:Pre T1:Pos T2:Pre T2:Pos
  T1:Pre 214.24 137.81 199.57 128.37
  T1:Pos 137.81 197.43 128.37 183.92
  T2:Pre 199.57 128.37 267.58 172.12
  T2:Pos 128.37 183.92 172.12 246.60
    Standard Deviations: 14.637 14.051 16.358 15.703
> sgma    <- fm17.4$sigma                              # σ̂...
> (sgma2  <- sgma^2)                                   # ...  and σ̂^2
  [1] 598.68
> reSt <- fm17.4$modelStruct$reStruct          # Random-effects structure
> pdKron1 <- reSt[[1]]
> names(pdKron1) <- pdKnms                     # Component names reinstated
> (c1 <- as.numeric(as.matrix(pdKron1$X))) # Mandatory multiplier
  [1] 0.35785
> c1*sgma2                                             # Compare to d̂_11
  [1] 214.24
> (A1 <- as.matrix(pdKron1$FiberType))                 # Â for fiber type
                fiber.fType 1 fiber.fType 2
  fiber.fType 1       1.00000       0.93155
  fiber.fType 2       0.93155       1.24901
> (B1 <- as.matrix(pdKron1$PrePos))                    # B̂ for occasion
           occ.fPre occ.fPos
  occ.fPre  1.00000  0.64324
  occ.fPos  0.64324  0.92157
> # sgma2 * c1 %x% A1 %x% B1                   # D̂ ≡σ^2 cI_1⊗Â⊗B̂:(17.14)
```

Panel R17.15 also presents a method to extract the estimated components of the $\mathcal{D}$ matrix from the model-fit object `fm17.4`. The estimate of residual SD is obtained from the `sigma` component of the object. The corresponding estimate of the residual variance is very close to the estimate obtained for model **M17.3** and shown in Panel R17.8.

The other matrices, related to the Kronecker-product structure, defined in (17.14), are extracted from the `modelStruct$reStruct` component of the model-fit object. The component is of *reStruct* class, and we store it in the object `reSt`. The object contains only a single component of *pdMat* class, which, in turn, is extracted and stored in the object `pdKron1`. The latter object represents a Kronecker product and contains three components, which correspond to the matrices involved in the reperesentation, defined in (17.14). To facilitate referring to the components, we name them using the function "`names()<-`" and the vector `pdKnms`. We then extract the components and store them as matrix-objects `c1`, `A1`, and `B1`. It can be verified that, by taking the Kronecker product of the estimated residual variance and the three matrices, an estimate of the matrix $\mathcal{D}$ is obtained, which corresponds to the one stored in the `DmtxKron1` object.

17.7 A Model with Homoscedastic Fiber-Type×Occasion-Specific Random Intercepts and a Structured Matrix $\mathcal{D}$

In this section, we address the issue of simplifying the structure of the matrix $\mathcal{D}$, given by (17.7), by assuming a constant variance for the random intercepts.

17.7.1 Model Specification

Toward this end, we fit model **M17.5**, given by (17.6), but with the matrix $\mathcal{D}$ defined as a product of two 2×2 compound-symmetry matrices $\boldsymbol{A}$ and $\boldsymbol{B}$, corresponding to fiber type and occasion, respectively. The resulting matrix is represented as follows:

$$\mathcal{D} = \sigma^2 \otimes (c\boldsymbol{I}_1) \otimes \boldsymbol{A} \otimes \boldsymbol{B} \equiv \sigma^2 \otimes c \otimes \begin{pmatrix} 1 & a_{12} \\ a_{21} & 1 \end{pmatrix} \otimes \begin{pmatrix} 1 & b_{12} \\ b_{21} & 1 \end{pmatrix}$$

$$= \sigma^2 c \begin{pmatrix} 1 & b_{12} & a_{12} & a_{12}b_{12} \\ b_{21} & b_{22} & a_{12}b_{21} & a_{12}b_{22} \\ a_{21} & a_{21}b_{21} & a_{22} & a_{22}b_{12} \\ a_{21}b_{12} & a_{21}b_{22} & a_{22}b_{21} & a_{22}b_{22} \end{pmatrix}. \tag{17.16}$$

The corresponding correlation matrix is

$$\mathcal{C} = \begin{pmatrix} 1 & \frac{a}{\sigma^2 c} & \frac{b}{\sigma^2 c} & \frac{ba}{\sigma^2 c} \\ \frac{a}{\sigma^2 c} & 1 & \frac{ba}{\sigma^2 c} & \frac{b}{\sigma^2 c} \\ \frac{b}{\sigma^2 c} & \frac{ba}{\sigma^2 c} & 1 & \frac{a}{\sigma^2 c} \\ \frac{ba}{\sigma^2 c} & \frac{b}{\sigma^2 c} & \frac{a}{\sigma^2 c} & 1 \end{pmatrix} \equiv \begin{pmatrix} 1 & x & y & z \\ x & 1 & z & y \\ y & z & 1 & x \\ z & y & x & 1 \end{pmatrix}. \tag{17.17}$$

R17.16 *PRT Trial*: Model **M17.5** fitted using the functions `lme()` and `pdKronecker()`. The formula-object `lme.spec.form` was defined in Panel R17.8

```
> pdId  <- pdIdent(~1)                      # cI1 in (17.16)
> pd1CS <- pdCompSymm(~fiber.f -1)          # A for  FiberType
> pd2CS <- pdCompSymm(~occ.f-1)             # B for  PrePos
> pdL2  <-
+     list(X = pdId,
+          FiberType = pd1CS,               # A
+          PrePos = pd2CS)                  # B
> fm17.5 <-                                 # M17.5
+   lme(lme.spec.form3,
+        random = list(id=pdKronecker(pdL2)),
+        data = prt)
```

The structure of the correlation matrix is the same as the one implied by the Kronecker products of general matrices, shown in (17.15). However, in contrast to (17.15), variance components on the diagonal are constrained to be equal.

17.7.2 R *Syntax and Results*

Panel R17.16 displays the R syntax used to fit model **M17.5** with the $\mathcal{D}$ matrix specified by (17.16). Similarly to the syntax shown in Panel R17.14, the objects of *pdMat* class representing matrices $c\boldsymbol{I}_1$, $\boldsymbol{A}$, and $\boldsymbol{B}$, as specified in (17.16), are first defined using the `pdIdent()` and `pdCompSymm()` constructor functions (Sect. 14.2.1). The matrices are then stored in the objects `pdId`, `pd1CS`, and `pd2CS`, respectively. Next, the named list `pdL2` is created, with the three matrices as the components. Finally, the model is fitted using an appropriate `lme()`-function call, with the value of the argument `random` provided in the form of a named list of *pdMat*-class objects (Sects. 14.3 and 14.5). The results of the fitting of the model are stored in the object `fm17.5`.

Panel R17.17 displays the estimated matrix $\mathcal{D}$ for model **M17.5**. As implied by (17.16), the elements on the diagonal are equal. Note that the corresponding SD equals 15.15, which is perfectly compatible with the results shown in Panel R17.10 for model **M17.3** with a general matrix $\mathcal{D}$.

At this point, it is of interest to evaluate whether the simplifications of the $\mathcal{D}$ matrix structure, obtained using the Kronecker product, have impact on the model fit. Toward this end, we use the REML-based LR test (Panel R17.18), because the matrix $\mathcal{D}$ structures, specified by (17.7), (17.14), and (17.16), imply a sequence of nested models and all three models having the same mean structure. Panel R17.18 presents the output of the suitable `anova()`-function calls. Clearly, none of the

R17.17 *PRT Trial*: Evaluating the homoscedastic Kronecker-product structure of the matrix $\boldsymbol{D}$ for model **M17.5**. The objects `nms` and `fm17.5` were created in Panels R17.15 and R17.16, respectively

```
> DmtxKron2 <- getVarCov(fm17.5)              # D̂
> dimnames(DmtxKron2) <- list(nms,nms)       # Row/col names shortened
> DmtxKron2                                   # D̂ printed
  Random effects variance covariance matrix
          T1:Pre T1:Pos T2:Pre T2:Pos
  T1:Pre 229.57 153.35 191.48 127.91
  T1:Pos 153.35 229.57 127.91 191.48
  T2:Pre 191.48 127.91 229.57 153.35
  T2:Pos 127.91 191.48 153.35 229.57
    Standard Deviations: 15.152 15.152 15.152 15.152
```

R17.18 *PRT Trial*: Likelihood-ratio tests for models **M17.3–M17.5**, implied by the matrix $\boldsymbol{D}$ structures specified in (17.7), (17.14), and (17.16). The model-fit objects `fm17.3`, `fm17.4`, and `fm17.5` were created in Panels R17.8, R17.14, and R17.16, respectively

```
> anova(fm17.3, fm17.4, fm17.5)          # M17.3 ⊃ M17.4 ⊃ M17.5
         Model df   AIC   BIC logLik   Test L.Ratio p-value
  fm17.3     1 22 23084 23212 -11520
  fm17.4     2 17 23080 23179 -11523 1 vs 2  6.0852  0.2980
  fm17.5     3 15 23078 23165 -11524 2 vs 3  1.3176  0.5175
> anova(fm17.3, fm17.5)                  # M17.3 ⊃ M17.5
         Model df   AIC   BIC logLik   Test L.Ratio p-value
  fm17.3     1 22 23084 23212 -11520
  fm17.5     2 15 23078 23165 -11524 1 vs 2  7.4028  0.3882
```

tests is statistically significant at the 5% significance level. The result suggests that model **M17.4**, with the simplified structure of the matrix $\mathcal{D}$ resulting from the Kronecker product of two general variance-covariance matrices (17.14), can be used instead of the more complex model **M17.3**. Moreover, model **M17.4** can be further simplified to model **M17.5**, with the matrix $\mathcal{D}$ defined by the Kronecker product of two compound-symmetry matrices, as specified in (17.16). The result of the LR test comparing directly models **M17.3** and **M17.5**, shown also in Panel R17.18, is statistically not significant, which confirms that the latter model can be used instead of the former.

The estimates of the fixed-effects coefficients of model **M17.5** are presented in Table 17.2. They differ very little as compared to the estimates for model **M17.5**. This reflects the fact that the simplification of the structure of the $\mathcal{D}$ of the latter model did not worsen the fit, as implied by the results of the LR test. There is also not much difference between the estimates of the fixed-effects coefficients and their standard errors for model **M17.5** and the corresponding values shown in Panel R17.8b for model **M17.3**. Note that, as was the case for models **M17.3** and **M17.3a**, the estimate of the coefficient of interest, `PRT(Low):Occ(Post)`, is statistically not significant for model **M17.5**.

17.8 A Joint Model for Two Dependent Variables

In this section, we briefly illustrate how the new *pdKronecker* class can be used to fit an LMM jointly to two dependent variables. Toward this end, we consider two outcome variables from the PRT data: the specific force (SPEC.FO) and the isometric force (ISO.FO).

17.8.1 Model Specification

We fit model **M17.6**, defined as follows:

$$\begin{aligned} y_{ivjtr} = {} & \beta_{v,0} + \beta_{v,1} \times \text{PRT}_i + \beta_{v,2} \times \text{OCC}_{it} + \beta_{v,3} \times \text{SEX}_i \\ & + \beta_{v,4} \times \text{AGE}_i + \beta_{v,5} \times \text{BMI}_i + \beta_{v,6} \times \text{FIBER}_{ij} \\ & + \beta_{v,12} \times \text{PRT}_i \times \text{OCC}_{it} + \beta_{v,16} \times \text{PRT}_i \times \text{FIBER}_{ij} \\ & + \beta_{v,26} \times \text{OCC}_{it} \times \text{FIBER}_{ij} + \beta_{v,34} \times \text{SEX}_i \times \text{AGE}_i \\ & + b_{ivjt} + \varepsilon_{ivjtr}, \end{aligned} \tag{17.18}$$

where the notation corresponds to that used in (17.6) and y_{ivjtr} indicates the r-th measurement of the v-th variable ($v = 1$ for ISO.FO and 2 for SPEC.FO) for the subject i at the occasion t and the fiber type j. The residual random errors ε_{ijtvr} are assumed to be independent and normally distributed with mean zero and variance σ_v^2, where

$$\sigma_1 \equiv \sigma \quad \text{and} \quad \sigma_2 = \delta\sigma. \tag{17.19}$$

For each subject, the model equation specifies also eight variable×fiber-type×occasion-specific random intercepts b_{ijtv}. We define the vector

$$\mathbf{b}_i \equiv (b_{i111}, b_{i112}, b_{i121}, b_{i122}, b_{i211}, b_{i212}, b_{i221}, b_{i222})',$$

with the elements in a lexicographical order defined by indices for the dependent variable (index v, varying most slowly), fiber type (index j), and occasion (index t, varying most quickly). We assume that the vector $\mathbf{b}_i$ is normally distributed with mean zero and the variance-covariance matrix $\mathcal{D}$ that can be represented as follows:

$$\mathcal{D} = \sigma^2 \otimes (c\boldsymbol{I}_1) \otimes \boldsymbol{E} \otimes \boldsymbol{A} \otimes \boldsymbol{B} \tag{17.20}$$
$$= \sigma^2 c \begin{pmatrix} 1 & e_{12} \\ e_{21} & e_{22} \end{pmatrix} \otimes \begin{pmatrix} 1 & a_{12} \\ a_{21} & a_{22} \end{pmatrix} \otimes \begin{pmatrix} 1 & b_{12} \\ b_{21} & b_{22} \end{pmatrix}.$$

The matrices $\boldsymbol{E}$, $\boldsymbol{A}$, and $\boldsymbol{B}$ used in (17.21) are associated with factors having two levels each, representing two dependent variables, fiber type, and the occasion at which the measurement was taken, respectively. Note that the order of the matrices in the Kronecker product is compatible with the ordering of the elements of the vector $\mathbf{b}_i$. The blocked structure of the matrix $\mathcal{D}$, given in (17.21), implies a similar correlation structure of the four type×occasion-specific random intercepts for each of the two dependent variables. The correlation structure is identical to the one presented in (17.15).

17.8.2 R *Syntax and Results*

Panel R17.19 displays the R syntax used to fit model **M17.6** with the $\boldsymbol{D}$ matrix specified by (17.21). First, with the help of the function `melt()` from the package **reshape**, a new data frame, `prt.Dep2`, is created from the data frame `prt`. For each subject, the function "stacks" measurements from the variables `ISO.FO` and `SPEC.FO` in one variable, named by default `value`, and adds an extra variable, named `DV`, that contains labels ("iso.fo" or "spec.fo") that identify the contents of `value`. All other variables from the data frame `prt` are appropriately copied to the data frame `prt.Dep2`. Then, the model-formula object `lme.DV.form`, corresponding to the fixed-effects part of (17.18), is created (see Sect. 5.2). Basically, the formula uses a syntax similar to that used to create the formula-object `lme.spec.form` (Panel R17.8), but adds interactions of the factors `fiber.f` and `occ.f` with `DV` to include the dependent-variable-specific terms.

Next, the *pdMat*-class object `pd3UN`, corresponding to the matrix $\boldsymbol{E}$ in (17.21), is initialized using the `pdLogChol()` constructor function (Sect. 14.2.1). The object is stored together with the objects `pd1Id`, `pd1UN`, and `pd2UN` as named components of the list `pdL`. The names of the components are stored in the vector `pdKNms` for purposes of labeling printouts. Finally, model **M17.6** is fitted using an appropriate `lme()`-function call. Note that the value of the argument `random` is provided in the form of a named list of *pdMat*-class objects (Sects. 14.3 and 14.5). Also, the argument `weights` is specified using the *varIdent*-class constructor function (Sect. 8.2). This allows specifying different residual variances for the two dependent variables. Finally, the argument `control` is provided with the list `lmeC`, created by applying the function `lmeControl()` with the argument `msMaxIter=100`. This allows to increase the maximum number of iterations in the internal loop of the estimation algorithm (Sect. 13.8.1) from 50 (default) to 100. It appears that in the

R17.19 *PRT Trial*: Model **M17.6** for two dependent variables fitted using the functions `lme()` and `pdKronecker()`. Objects `pd1Id`, `pd1UN`, and `pd2UN` were created in Panel R17.14

```
> prt.Dep2 <-                                      # Raw data transposed
+     melt(prt, measure.var = c("iso.fo", "spec.fo"),
+          variable_name = "DV")                   # DV indicates dep. variable
> lme.DV.form <-                                   # Formula with interactions
+     formula(value ~ -1 + DV + DV:(-1 + (prt.f + occ.f
+             + fiber.f)^2
+             + sex.f + age.f + sex.f:age.f + bmi))
> pdUN <- pdLogChol(~DV-1)                         # E for DV
> pdL <-
+   list(X = pdId,
+        DV = pdUN,
+        FiberType = pd1UN,
+        PrePos = pd2UN)
> pdKnms   <- names(pdL)                           # Names saved for later use
> lmeC     <- lmeControl(msMaxIter = 100)  # Maximum iterations
> fm17.6 <-
+    lme(lme.DV.form,                              # Fixed part
+        random = list(id = pdKronecker(pdL)), # 8 random effects
+        data = prt.Dep2,
+        weights = varIdent(form = ~1|DV),         # DV-specific variance
+        control = lmeC)
```

case of model **M17.6** the increase is necessary to obtain convergence. The results of fitting of the model are stored in the object `fm17.6`.

Panel R17.20 displays the blocks of the estimated $\mathcal{D}$ matrix for model **M17.6**. First, the function `getVarCov()` (see Table 14.5) is used to extract the estimate of the matrix from the model-fit object. To simplify printouts, the default row and column names are replaced by abbreviated ones. Then, the two diagonal 4×4 blocks of the estimated matrix $\mathcal{D}$ are printed out, together with the off-diagonal block. The diagonal blocks show a clear difference in the scale of the dependent variables. Additionally, with the help of the `cov2cor()` function, the diagonal block of the correlation matrix, corresponding to $\mathcal{D}$, is displayed. As mentioned before, it gives the correlation between the four fiber-type$\times$occasion random intercepts for each of the two dependent variables. The correlation structure does resemble the one obtained for model **M17.3** (Panel R17.9), although the magnitude of the estimated correlation coefficients is smaller. This is most likely due to the fact that they are an "average" of associations for the two dependent variables.

In Panel R17.21, the estimated components of the Kronecker-product structure, defined in (17.21), are displayed. In particular, the estimated SD for the ISO.FO variable is extracted from the `sigma` component of the model-fit object `fm17.6`. The estimated values of the parameters of the *varIdent*-class residual variance function

R17.20 *PRT Trial*: The estimate of the matrix $\mathcal{D}$ for model **M17.6**. The model-fit object fm17.6 was created in Panel R17.19

```
> DmtxDV2Kron <- getVarCov(fm17.6)
> rownames(DmtxDV2Kron)                          # Row/col D̂ names
  [1] "DViso.fo:fiber.fType 1:occ.fPre"
  [2] "DViso.fo:fiber.fType 1:occ.fPos"
  [3] "DViso.fo:fiber.fType 2:occ.fPre"
  [4] "DViso.fo:fiber.fType 2:occ.fPos"
  [5] "DVspec.fo:fiber.fType 1:occ.fPre"
  [6] "DVspec.fo:fiber.fType 1:occ.fPos"
  [7] "DVspec.fo:fiber.fType 2:occ.fPre"
  [8] "DVspec.fo:fiber.fType 2:occ.fPos"
> nms <-                                          # ... shortened
+   c("is:T1:Pre", "is:T1:Pos", "is:T2:Pre", "is:T2:Pos",
+     "sp:T1:Pre", "sp:T1:Pos", "sp:T2:Pre", "sp:T2:Pos")
> dimnames(DmtxDV2Kron) <- list(nms, nms)      # ... and assigned.
> print(DmtxDV2Kron[1:4, 1:4], digits = 2)     # Block for iso.fo
          is:T1:Pre is:T1:Pos is:T2:Pre is:T2:Pos
is:T1:Pre     0.048     0.035     0.033     0.024
is:T1:Pos     0.035     0.043     0.024     0.030
is:T2:Pre     0.033     0.024     0.068     0.049
is:T2:Pos     0.024     0.030     0.049     0.061
> print(DmtxDV2Kron[5:8, 5:8], digits = 5)     # Block for spec.fo
          sp:T1:Pre sp:T1:Pos sp:T2:Pre sp:T2:Pos
sp:T1:Pre    206.37    148.90    141.26    101.93
sp:T1:Pos    148.90    186.45    101.93    127.63
sp:T2:Pre    141.26    101.93    290.84    209.86
sp:T2:Pos    101.93    127.63    209.86    262.77
> print(DmtxDV2Kron[1:4,5:8], digits = 3)      # Off-diagonal block
          sp:T1:Pre sp:T1:Pos sp:T2:Pre sp:T2:Pos
is:T1:Pre     1.342     0.968     0.919     0.663
is:T1:Pos     0.968     1.213     0.663     0.830
is:T2:Pre     0.919     0.663     1.892     1.365
is:T2:Pos     0.663     0.830     1.365     1.709
> print(cov2cor(DmtxDV2Kron)[1:4, 1:4],         # Corr. block for iso.fo
+       digits = 3)
          is:T1:Pre is:T1:Pos is:T2:Pre is:T2:Pos
is:T1:Pre     1.000     0.759     0.577     0.438
is:T1:Pos     0.759     1.000     0.438     0.577
is:T2:Pre     0.577     0.438     1.000     0.759
is:T2:Pos     0.438     0.577     0.759     1.000
```

R17.21 *PRT Trial*: Evaluating the Kronecker-product structure (17.21) of the matrix $\boldsymbol{D}$ for model **M17.6**. Objects `fm17.6` and `pdKnms` were created in Panel R17.19

```
> (sgma   <- fm17.6$sigma)                       # σ̂1 (iso.fo):17.19
  [1] 0.21513
> (vStDV2 <- fm17.6$modelStruct$varStruct) # δ̂
  Variance function structure of class varIdent representing
   iso.fo spec.fo
     1.00  113.49
> sgma*coef(vStDV2, unconstrained = FALSE) # σ̂2 (spec.fo)
  spec.fo
   24.414
> reStDV2 <- fm17.6$modelStruct$reStruct       # Random-effects structure
> DV2pdxKron   <- reStDV2[[1]]
> names(DV2pdxKron) <- pdKnms
> (c3 <- as.matrix(DV2pdxKron$X))              # Mandatory multiplier
              (Intercept)
  (Intercept)      1.0365
> (E3 <- as.matrix(DV2pdxKron$DV))             # Ê for dep. variables
            DViso.fo DVspec.fo
  DViso.fo     1.000    27.977
  DVspec.fo   27.977  4301.753
> (A3 <- as.matrix(DV2pdxKron$FiberType))      # Â for fiber type:
                                                  (17.21)
                   fiber.fType 1 fiber.fType 2
  fiber.fType 1          1.00000       0.68454
  fiber.fType 2          0.68454       1.40935
> (B3 <- as.matrix(DV2pdxKron$PrePos))         # B̂ for occasion
           occ.fPre occ.fPos
  occ.fPre  1.00000  0.72155
  occ.fPos  0.72155  0.90350
```

are extracted from the `modelStruct` component of the model-fit object and stored in the object `vStDV2`. In fact, the variance function involves only one parameter, δ, which is the ratio of the SDs for the SPEC.FO and ISO.FO variables, as defined in (17.19). The estimated value of the ratio is very large, confirming a difference in the measurement scale of the two dependent variables.

To extract the estimate of δ itself on the unconstrained scale, we used the `coef()` generic function with the `unconstrained` argument set to `FALSE`. By multiplying the result by the estimated value of the residual SD for the variable ISO.FO, stored in the object `sgma`, we obtain the estimated residual SD for the variable SPEC.FO. It corresponds to the value estimated for model **M17.3** (see Panel R17.9).

Other matrices, used in the Kronecker-product structure, defined in (17.21), are extracted from the `modelStruct$reStruct` component of the model-fit object.

R17.22 *PRT Trial*: Verification of the Kronecker-product structure (17.21) for model **M17.6**. The objects c3, A3, B3, and E3 were created in Panel R17.21 and the object nms was created in Panel R17.20

```
> cKron3 <- sgma^2 %x% c3 %x% E3 %x% A3 %x% B3 # Kronecker product ...
> rownames(cKron3) <- nms
> print(cKron3, digits = 2)                    # ... printed
              [,1]  [,2]  [,3]  [,4]   [,5]   [,6]   [,7]   [,8]
  is:T1:Pre 0.048 0.035 0.033 0.024   1.34   0.97   0.92   0.66
  is:T1:Pos 0.035 0.043 0.024 0.030   0.97   1.21   0.66   0.83
  is:T2:Pre 0.033 0.024 0.068 0.049   0.92   0.66   1.89   1.36
  is:T2:Pos 0.024 0.030 0.049 0.061   0.66   0.83   1.36   1.71
  sp:T1:Pre 1.342 0.968 0.919 0.663 206.37 148.90 141.26 101.93
  sp:T1:Pos 0.968 1.213 0.663 0.830 148.90 186.45 101.93 127.63
  sp:T2:Pre 0.919 0.663 1.892 1.365 141.26 101.93 290.84 209.86
  sp:T2:Pos 0.663 0.830 1.365 1.709 101.93 127.63 209.86 262.77
```

The component is of *reStruct* class, which we store in the object reStDV2. The object contains only a single component that inherits from the *pdMat* class, which we extract and store in the object DV2pdxKron. The latter object is an object of *pdKronecker* class with four components, which correspond to the matrices involved in the reperesentation, defined in (17.21). To facilitate referring to the components, we name them using the function "names()<-" and the vector pdKnms. We then extract the components and store them as the matrix-objects c3, A3, B3, and E3. As shown in Panel R17.22, it can be verified that, by taking the Kronecker product of the estimated value of the residual variance σ^2 and the three matrices, we obtain the result corresponding to the matrix stored in the DmtxKron1 object in Panel R17.20.

In Panel R17.23, a printout of the estimates of the fixed-effects coefficients for model **M17.6** is presented. Toward this end, we refer to the component tTable of the object resulting from applying the function summary() to the model-fit object fm17.6. Before printing out the contents of the tTable array, we shorten the row names. To display the results, we use the function printCoefmat(), which allows more control over the format of the printout (see the syntax in Panel **M16.1** and its explanation in Sect. 16.2.2).

The estimates of the fixed-effects coefficients for the variable SPEC.FO are comparable to the values obtained for model **M17.3** (see Panel R17.8b). The effects of interest, interactions iso:Low:Pos and spec:Low:Pos, are individually statistically not significant, indicating that the data do not support rejecting the null hypothesis of the lack of effect of the training intensity.

Note that conducting individual tests for the effect of training intensity on the two dependent variables might raise concerns about the need for a multiple comparison correction. Clearly, given the nonsignificance of the individual tests, in our case, such a correction would not change the conclusions. Formally, however, one could address this issue by conducting a joint test of significance

R17.23 *PRT Trial*: Estimates of the fixed-effects coefficients for model **M17.6**. The model-fit object `fm17.6` was created in Panel R17.19

```
> fixed.DV2.Kron1 <- summary(fm17.6)$tTable
> fxdNms3 <-
+  c("iso:1", "spec:1", "iso:Low", "spec:Low", "iso:Pos",
+    "spec:Pos", "iso:T2", "spec:T2", "iso:Male", "spec:Male",
+    "iso:Old", "spec:Old", "iso:BMI", "spec:BMI",
+    "iso:Low:Pos", "spec:Low:Pos", "iso:Low:T2", "spec:Low:T2",
+    "iso:Pos:T2", "spec:Pos:T2", "iso:Male:Old", "spec:Male:Old")
> rownames(fixed.DV2.Kron1) <- fxdNms3
> printCoefmat(fixed.DV2.Kron1, digits = 3, cs.ind = c(1, 2),
+              dig.tst = 2, zap.ind = 5)
                  Value Std.Error   DF t-value p-value
 iso:1          0.22321   0.20182 4858  1.1060    0.27
 spec:1       125.08991  13.73703 4858  9.1060    0.00
 iso:Low       -0.01715   0.05720 4858 -0.2998    0.76
 spec:Low       2.12290   4.00821 4858  0.5296    0.60
 iso:Pos        0.02793   0.03077 4858  0.9080    0.36
 spec:Pos       3.99249   2.46368 4858  1.6205    0.10
 iso:T2         0.03032   0.04252 4858  0.7130    0.48
 spec:T2       25.37114   3.18342 4858  7.9698    0.00
 iso:Male       0.05954   0.07093 4858  0.8395    0.40
 spec:Male     -2.11642   4.81216 4858 -0.4398    0.66
 iso:Old        0.01082   0.06746 4858  0.1604    0.87
 spec:Old       8.08777   4.58276 4858  1.7648    0.08
 iso:BMI        0.02184   0.00755 4858  2.8925    0.00
 spec:BMI       0.60113   0.51340 4858  1.1709    0.24
 iso:Low:Pos    0.02186   0.04061 4858  0.5383    0.59
 spec:Low:Pos  -0.16110   3.13993 4858 -0.0513    0.96
 iso:Low:T2     0.01656   0.05511 4858  0.3006    0.76
 spec:Low:T2   -4.12297   4.01129 4858 -1.0278    0.30
 iso:Pos:T2     0.06492   0.02744 4858  2.3663    0.02
 spec:Pos:T2    4.10831   2.47640 4858  1.6590    0.10
 iso:Male:Old   0.05800   0.09894 4858  0.5862    0.56
 spec:Male:Old -8.71976   6.72175 4858 -1.2972    0.20
> anova(fm17.6, type = "marginal")
                 numDF denDF F-value p-value
 DV                  2  4858  45.140  <.0001
 DV:prt.f            2  4858   0.283  0.7535
 DV:occ.f            2  4858   1.437  0.2376
 DV:fiber.f          2  4858  33.586  <.0001
 DV:sex.f            2  4858   0.708  0.4926
 DV:age.f            2  4858   1.730  0.1773
 DV:bmi              2  4858   4.184  0.0153
 DV:prt.f:occ.f      2  4858   0.168  0.8449
 DV:prt.f:fiber.f    2  4858   0.767  0.4645
 DV:occ.f:fiber.f    2  4858   3.610  0.0271
 DV:sex.f:age.f      2  4858   1.561  0.2101
```

of the two interaction terms, `iso:Low:Pos` and `spec:Low:Pos`. In the context of model **M17.6**, we can construct such a test by testing the significance of the interaction `DV:prt.f:occ.f`. Toward this end, in Panel R17.23, we use the `anova(fm17.6, type="marginal")` command. The p-value for the test of significance of the interaction is equal to 0.84, indicating that neither interaction `iso:Low:Pos` nor `spec:Low:Pos` is statistically significant.

The printout shown in Panel R17.23 suggests that the mean structure of the model could be simplified by removing, e.g., all interactions except of the occasion×type term for both dependent variables. Given the illustrative nature of the example, we do not pursue this direction, though.

17.9 Chapter Summary

In this chapter, we analyzed the PRT data by applying LMMs. The data had a more complex structure than the ARMD data, analyzed in the previous chapter. In particular, in the PRT dataset, multiple measurements for several variables for two fiber types at two different occasions (pre- and post-training) were available for each individual. This leads to a complex association structure between the measurements. By using LMMs, the hierarchical structure of the data and association structure were taken into account.

Table 17.3 provides information about the models defined in this chapter.

The main tool which was used to fit the models was the function `lme()` from the package **nlme**. The function capabilities were extended by introducing a new class of positive-definite matrices, defined by the Kronecker product of standard *pdMat*-class matrices, which can be used for defining the random effects structure of an LMM. At current, the solution is not available for the `lmer()` function from the package **lme4.0**.

Table 17.3 *PRT Trial*: Summary of linear mixed-effects models defined and fitted using REML in Chap. 17

Model label	Section	Syntax	R-object	Model eq.	Matrix $\mathcal{D}$ eq.	Residual variance
(**a**) Models for the specific force for type-1 fibers only						
M17.1	17.2	R17.1	`fm17.1`	(17.1)	(17.3)	constant
M17.2	17.3	R17.7	`fm17.2`	(17.1)	(17.3)	*varPower*
(**b**) Models for the specific force for type-1 & 2 fibers						
M17.3	17.4	R17.8	`fm17.3`	(17.6)	(17.7)	constant
M17.3a[a]	17.5	R17.13	`fm17.3a`	(17.9)	(17.10)	constant
M17.4	17.6	R17.14	`fm17.4`	(17.6)	(17.14)	constant
M17.5	17.7	R17.16	`fm17.5`	(17.18)	(17.16)	constant
(**c**) Joint model for the specific and isometric forces for type-1 & 2 fibers						
M17.6	17.8	R17.19	`fm17.6`	(17.18)	(17.21)	constant

[a] Model **M17.3a** is equivalent to model **M17.3**

We started by considering the data for the specific force as the dependent variable for type-1 fibers only. In Sect. 17.2, we fitted a conditional-independence LMM with two correlated, heteroscedastic, occasion-specific random effects for each individual (model **M17.1**). Based on the evaluation of the goodness-of-fit, we modified the model using the power-of-the-mean variance function (model **M17.2** in Sect. 17.3). In neither of the models, the effect of intensity of the training was statistically significant.

In the next step, we considered models for the specific force for both fiber types. In Sect. 17.4, we constructed model **M17.3**: a conditional-independence LMM with four correlated, heteroscedastic, fiber-type×occasion-specific random effects for each individual. We also considered an equivalent, re-parameterized form of the model, in which the random-effects structure was expressed by a linear model using fiber-type and occasion effects (model **M17.3a** in Sect. 17.5). The analysis of the random-effects structure suggested that it could be simplified by considering homoscedastic random effects. Toward this end, the newly developed *pdKronecker* class of positive-definite matrices was instrumental. First, for demonstration purposes, we used it in Sect. 17.6 to fit model **M17.4** with heteroscedastic random effects, but with a more structured variance-covariance matrix of the random effects than the one used for model **M17.3**. Then we used the *pdKronecker* class to fit model **M17.5** with four correlated, homoscedastic, fiber-type×occasion-specific random effects for each individual (Sect. 17.7). By using the LR test, we verified that the simplification of the random-effects structure did not statistically significantly influence the fit of the model, as compared to model **M17.3**.

In principle, parsimonious modeling of the variance-covariance structure of the data can increase the efficiency of estimation of the fixed effects. However, in our case, not much influence of the modeling on the point estimates and SEs of the fixed-effects coefficients could be observed. In particular, in none of the models the effect of training intensity was statistically significant.

Finally, in Sect. 17.8, we considered the most complex LMM for two dependent variables, the specific force and the isometric force, and the two fiber types. Model **M17.6** included eight correlated, heteroscedastic random effects, one for each combination of the dependent variable, fiber type, and occasion. It assumed different residual variances for the two dependent variables. The model allowed adjusting not only for the correlation between repeated measures for different fiber types at different occasions, but also for the association between measurements for the two different dependent variables. Also, it provided a natural way for a joint test of significance of the effect of training intensity on the two dependent variables. The result of the test was statistically not significant.

The presented sequence of models was aimed at extending the illustration of the capabilities of the function `lme()` for fitting LMMs. The models could be further modified by, for example, simplifying the fixed-effects structure. We did not pursue this direction, however, given the nonsignificance of the effect of training intensity.

Chapter 18
SII Project: Modeling Gains in Mathematics Achievement-Scores

18.1 Introduction

The SII Project was described in Sect. 2.4. In Sect. 3.4, an exploratory analysis of the data was presented. The data have a hierarchical structure, with pupils grouped in classes which, in turn, are grouped in schools. Thus, we deal with two levels of grouping in the data or, equivalently, with a three-level data hierarchy. In this chapter, we use LMMs to analyze the change in mathematics achievement-scores for pupils, MATHGAIN. In particular, we use models, which include random intercepts for schools and classes to account for the data hierarchy.

We begin in Sect. 18.2 with the model considered by West et al. (2007). In Sects. 18.3–18.7, we propose a series of models with different mean structures that allow investigating the influence of various school- and pupil-level covariates on the dependent variable. Section 18.8 illustrates the use of the function `lmer()` from the package **lme4.0** for fitting the final model from Sect. 18.7. Section 18.9 briefly summarizes the most important findings and topics presented in this chapter.

18.2 A Model with Fixed Effects for School- and Pupil-Specific Covariates and Random Intercepts for Schools and Classes

We begin the analysis of the data from the SII project by considering the model, proposed by West et al. (2007) (Model 4.4, Chap. 4, p. 159).

A. Gałecki and T. Burzykowski, *Linear Mixed-Effects Models Using R: A Step-by-Step Approach*, Springer Texts in Statistics, DOI 10.1007/978-1-4614-3900-4_18,

18.2.1 Model Specification

Model **M18.1** is defined as follows:

$$\begin{aligned}\text{MATHGAIN}_{sci} &= \beta_0 + \beta_1 \times \text{SES}_{sci} + \beta_2 \times \text{MINORITY}_{sci} \\ &\quad + \beta_3 \times \text{MATHKIND}_{sci} + \beta_4 \times \text{SEX}_{sci} + \beta_5 \times \text{HOUSEPOV}_s \\ &\quad + b_{0s} + b_{0sc} + \varepsilon_{sci} \\ &\equiv \mu_{sci} + b_{0s} + b_{0sc} + \varepsilon_{sci}, \qquad (18.1)\end{aligned}$$

where s indexes schools ($s = 1,2,\ldots,N$), c indexes classes ($c = 1,2,\ldots,n_s$), and i indexes individual pupils ($i = 1,\ldots,n_{sc}$). In (18.1), SEX and MINORITY are values of pupil-level indicator variables for, respectively, girls and minority pupils. MATHKIND and SES are continuous variables providing pupil's math score in kindergarten and pupil's socioeconomic status, respectively. Finally, HOUSPOV is a continuous, school-level variable containing the information about the percentage of neighboring households below the poverty level. Note that the mean structure of model **M18.1** does not include any class-specific covariates.

The residual random errors ε_{sci} in (18.1) are assumed to be independent and normally distributed with mean zero and a common variance σ^2. In addition to the residual errors, the model equation includes, for each pupil, two random intercepts: b_{0s} and b_{0sc}. The first one is the random effect of school, while the second one is the random effect of the class. We assume that the random intercepts b_{0c} and b_{0sc} are independent and normally distributed with means zero and variances $d_1 > 0$ and $d_2 > 0$, respectively. Moreover, they are independent of the residual random errors ε_{sci}.

18.2.1.1 Marginal Interpretation

Model **M18.1** implies that the marginal expected value of MATHGAIN_{sci} is equal to μ_{sci}, defined in (18.1).

Inclusion of the random intercepts b_{0s} and b_{0sc} in the model allows modeling of the correlation between the MATHGAIN measurements obtained for pupils from the same school, as well as for pupils from the same class. This can be seen from the resulting marginal variances and covariances:

$$\text{Var}(\text{MATHGAIN}_{sci}) = d_1 + d_2 + \sigma^2,$$

$$\text{Cov}(\text{MATHGAIN}_{sci}, \text{MATHGAIN}_{sc'i'}) = d_1, \qquad (18.2)$$

$$\text{Cov}(\text{MATHGAIN}_{sci}, \text{MATHGAIN}_{sci'}) = d_1 + d_2, \qquad (18.3)$$

where $c \neq c'$ and $i \neq i'$. It follows that gains in math scores of any two pupils from the same school are correlated with the within-school correlation coefficient equal to $d_1/(d_1 + d_2 + \sigma^2)$. Similarly, the within-class correlation coefficient for measurements of any two pupils from the same class (within the same school) is equal to $(d_1 + d_2)/(d_1 + d_2 + \sigma^2)$. Given that $d_1 > 0$ and $d_2 > 0$, it follows that the within-school correlation coefficient is assumed to be smaller than the within-class one.

18.2.2 R *Syntax and Results*

In Panel R18.1, we show the R syntax to fit model **M18.1**. First, we load the `SIIdata` data frame from the package **nlmeU**. Then we define the model-formula object `form1`, which corresponds to the fixed-effects part of (18.1). We use the formula object in the call to the function `lme()`. Note that we use the `random` argument in the form of a list with named components containing one-sided formulae (see the syntax (a) in Table 14.1). The formulae specify random intercepts for the levels of the nested grouping factors `schoolid` and `classid`. The model is fitted by using the ML estimation method, and the results are stored in the object `fm18.1`.

At the end of Panel R18.1, we present an alternate, simpler syntax for the `random` argument (see the syntax (d) in Table 14.1). As it was mentioned in Table 14.2, the simplified syntax requires the same form of the variance-covariance matrices of the random effects for the different levels of grouping. This is the case for model **M18.1**, as it includes only random intercepts at both the school and class levels of grouping.

In Panel R18.2, we extract information about the data hierarchy, implied by the syntax used in Panel R18.1. By using the `getGroupsFormula()` function (see Sect. 14.4), we verify that the hierarchy is defined by the grouping factors `schoolid` and `classid`, with the latter nested within the former. By applying the function `getGroups()` with the argument `level=1`, we extract the grouping factor, which defines the first (highest) level of the data hierarchy, from the model-fit object and store it in the object `grpF1`. We display the structure of `grpF1` by applying the generic function `str()`. The printout indicates that the grouping factor `schoolid` has 107 different levels.

By applying the function `getGroups()` without the use of the argument `level`, we extract the grouping factor, which defines the second (lowest) level of the data hierarchy, and store it in the object `grpF2`. The display of its structure indicates that this is the grouping factor `classid`, with 312 different levels. This information is consistent with the results presented in Sect. 3.4.

In Panel R18.3, we show the estimates of the fixed-effects coefficients along with their estimated variance-covariance matrix (see Sect. 13.5.5) for model **M18.1**. Toward this end, in Panel R18.3a, we use the function `fixef()` (see Table 14.5). The variance-covariance matrix is extracted from the model-fit object `fm18.1` in Panel R18.3b with the help of the function `vcov()` (see Table 14.5). Before

R18.1 *SII Project*: Model **M18.1** fitted to the data using the function `lme()`

```
> data(SIIdata, package="nlmeU")
> form1 <-
+    formula(mathgain ~ ses + minority  + # (18.1)
+              mathkind + sex + housepov)
> (fm18.1 <-
+    lme(form1,
+        random = list(schoolid = ~1,      # See Table 14.1, syntax (a)
+                      classid = ~1),
+        data = SIIdata,  method = "ML"))
 Linear mixed-effects model fit by maximum likelihood
   Data: SIIdata
   Log-likelihood: -5694.8
   Fixed: mathgain ~ ses + minority + mathkind + sex + housepov
 ...    [snip]
 Number of Observations: 1190
 Number of Groups:
              schoolid classid %in% schoolid
                   107                   312
> update(fm18.1,                           # An alternative syntax
+        random = ~1 | schoolid/classid)   # See Table 14.1, syntax (d)
 Linear mixed-effects model fit by maximum likelihood
   Data: SIIdata
   Log-likelihood: -5694.8
   Fixed: mathgain ~ ses + minority + mathkind + sex + housepov
      (Intercept)              ses minorityMnrt=Yes
        284.91086          5.23255         -7.74566
         mathkind             sexF         housepov
         -0.47061         -1.23071        -11.30141

 Random effects:
  Formula: ~1 | schoolid
         (Intercept)
 StdDev:      8.5881

  Formula: ~1 | classid %in% schoolid
         (Intercept) Residual
 StdDev:       9.018   27.056

 Number of Observations: 1190
 Number of Groups:
              schoolid classid %in% schoolid
                   107                   312
 ...    [snip]
```

R18.2 *SII Project*: Data grouping/hierarchy implied by model **M18.1**. The model-fit object `fm18.1` was created in Panel R18.1

```
> getGroupsFormula(fm18.1)                   # Grouping formula
  ~schoolid/classid
  <environment: 0x0000000007010718>
> str(grpF1 <- getGroups(fm18.1, level=1)) # Grouping factor at level 1
   Factor w/ 107 levels "1","2","3","4",..: 1 1 1 1 1 1 1 1 1 1...
   - attr(*, "label")= chr "schoolid"
> str(grpF2 <- getGroups(fm18.1))            # Grouping factor at level 2
   Factor w/ 312 levels "1/160","1/217",..: 1 1 1 2 2 2 2 2 2 2...
   - attr(*, "label")= chr "classid"
> grpF2
     [1] 1/160    1/160    1/160    1/217    1/217    1/217    1/217
     [8] 1/217    1/217    1/217    1/217    2/197    2/197    2/211
  ...       [snip]
  [1184] 107/96   107/96   107/96   107/96   107/96   107/239 107/239
  attr(,"label")
  [1] classid
  312 Levels: 1/160 1/217 10/178 10/208 10/278 10/303 ...  99/266
```

displaying the matrix, we abbreviate the names of the fixed-effects coefficients with the help of the function `abbreviate()` and use them instead of the full names assigned by default.

In Sect. 14.6, it was mentioned that the information about the estimated components of the variance-covariance structure of an LMM can be extracted from a model-fit object with the help of the function `getVarCov()` (see also Table 14.5). Unfortunately, as it is illustrated in Panel R18.4, the function does not work for models with multiple levels of grouping. Thus, we need to resort to other methods/functions to extract the information.

In particular, we can use the function `VarCorr()`. From the printout shown in Panel R18.4, we can observe that the estimated residual variance is an order of magnitude larger as compared to the estimated variances of the school- and class-specific random effects. By using the formulae (18.2) and (18.3), we can conclude that the estimated form of model **M18.1** implies that the correlation coefficient between the improvement in math scores for pupils from the same school is estimated to be equal to $73.755/(73.755+81.325+732.015) = 0.08$, while for the pupils from the same class it is estimated to be equal to $(73.755+81.325)/(73.755+81.325+732.015) = 0.17$. Thus, in accordance with the remark about (18.2)–(18.3) made earlier, the within-class correlation coefficient is larger than the within-school one.

Panel R18.5 presents the results of the tests for the fixed effects (Sect. 13.7.1) of model **M18.1**. The tests are obtained by applying the function `anova()` to the model-fit object `fm18.1`. In Panel R18.5a, F-tests for the individual effects are presented (Sect. 13.7.1). Note that, by using the argument `type="marginal"`, we

R18.3 *SII Project*: Estimates of the fixed-effects coefficients and their estimated variance-covariance matrix for model **M18.1**. The model-fit object `fm18.1` was created in Panel R18.1

(a) *Estimates of the fixed-effects coefficients.*

```
> (fxd <- fixef(fm18.1))                      # β̂
     (Intercept)              ses minorityMnrt=Yes
       284.91086          5.23255         -7.74566
        mathkind             sexF         housepov
        -0.47061         -1.23071        -11.30141
```

(b) *Estimated variance-covariance matrix of* $\widehat{\boldsymbol{\beta}}$.

```
> vcov1 <- vcov(fm18.1)                       # Var(β̂)
> nms <- abbreviate(names(fxd))               # Abbreviated β names ...
> dimnames(vcov1) <- list(nms, nms)           # ...   assigned.
> print(vcov1, digits = 2)
        (In)     ses    mM=Y     mthk    sexF     hspv
 (In) 120.71  1.6855 -6.9166 -0.23666 -0.7671 -18.5034
 ses    1.69  1.5421  0.4282 -0.00456  0.0386   0.9669
 mM=Y  -6.92  0.4282  5.6276  0.00814 -0.0619  -4.3187
 mthk  -0.24 -0.0046  0.0081  0.00049 -0.0012   0.0077
 sexF  -0.77  0.0386 -0.0619 -0.00117  2.7358  -0.1442
 hspv -18.50  0.9669 -4.3187  0.00767 -0.1442  96.4760
```

R18.4 *SII Project*: Extracting the estimates of the variances for the random intercepts of model **M18.1** using the `getVarCov()` and `varCorr()` functions. The model-fit object `fm18.1` was created in Panel R18.1

```
> getVarCov(fm18.1)
 Error in getVarCov.lme(fm18.1) :
   Not implemented for multiple levels of nesting
> VarCorr(fm18.1)
             Variance      StdDev
schoolid =   pdLogChol(1)
(Intercept)  73.755         8.5881
classid =    pdLogChol(1)
(Intercept)  81.325         9.0180
Residual    732.015        27.0558
```

R18.5 *SII Project*: Marginal-approach F-tests of the fixed effects for model **M18.1**. The model-fit object `fm18.1` was created in Panel R18.1

(a) *Tests for all fixed effects*

```
> anova(fm18.1, type = "marginal")
              numDF denDF F-value p-value
 (Intercept)      1   874  669.09  <.0001
 ses              1   874   17.67  <.0001
 minority         1   874   10.61  0.0012
 mathkind         1   874  446.27  <.0001
 sex              1   874    0.55  0.4582
 housepov         1   105    1.32  0.2537
```

(b) *Tests for selected effects*

```
> anova(fm18.1, Terms = c("housepov"))
 F-test for: housepov
   numDF denDF F-value p-value
 1     1   105  1.3172  0.2537
> anova(fm18.1, Terms = c("sex"))
 F-test for: sex
   numDF denDF F-value p-value
 1     1   874 0.55084  0.4582
```

(c) *The effects of* `sex` *and* `housepov` *tested jointly*

```
> anova(fm18.1, Terms = c("housepov", "sex"))
 Error in anova.lme(fm18.1, Terms = c("housepov", "sex")) :
   Terms must all have the same denominator DF
```

request the marginal-approach tests (Sect. 14.7). Results of the tests for the effects of the variables SEX and HOUSEPOV are statistically not significant at the 5% significance level. It is worth noting that the F-test for the variable HOUSEPOV is based on a different number of denominator degrees of freedom than the other tests. This is due to the fact that HOUSEPOV is a school-level variable, while the other variables are defined at the pupil level (Sect. 14.7).

Panel R18.5b presents the use of the argument `Terms` of the function `anova()` to obtain separate tests for the effects of variables SEX and HOUSEPOV (Sect. 14.7). Note that, by default, results of the marginal-approach tests are reported. Of course, the results are identical to those displayed in Panel R18.5a.

Finally, in Panel R18.5c, we attempt to perform a joint F-test for the effects of the variables SEX and HOUSEPOV. Because the variables are defined at different levels of the data hierarchy, the test fails. For this reason, in Sect. 18.4, we will employ the LR test to examine the joint effect of the two variables.

First, however, we check the fit of model **M18.1** to the data, because any tests based on a model are only meaningful as long as the model offers a reasonable representation of data. Thus, in Panel R18.6, we present the R syntax for residual diagnostics. We focus on marginal residuals, which allow us to examine whether the relationship between MATHGAIN and HOUSEPOV is linear.

In Panel R18.6a, we extract, with the help of the function `resid()` (see Table 14.5), residuals from the model-fit object `fm18.1`. Note that we use the argument `level=0` (see Table 14.5). Thus, the function returns the marginal residuals (Sect. 13.6.2), as required.

We store the residuals in the object `rsd1`. By applying the function `range()` to the object, we display the minimum and maximum values of the residuals. To identify observations with residuals larger, in absolute value, than 120, say, we create the logical vector `outi`, which identifies the rows in the data frame `SIIdata` that correspond to those residuals. We then use the vector to display the row numbers. We also use the vector to display the values of the residuals from the object `rsd1`. Note that the displayed values are labeled by levels of the `schoolid` grouping factor.

Panel R18.6b presents the code to construct plots of the marginal residuals against the values of the HOUSEPOV covariate for each sex. In the plot, the residuals larger, in absolute value, than 120 are to be identified. Toward this end, first, the `myPanel()` function is constructed, which selects the residuals and labels the corresponding symbols in the plot. By using the argument `pos=3` in the `ltext()` function, we indicate that the labels should be positioned above the data symbols. Then the `xyplot()` function is used to display the plot.

The resulting plot is shown in Fig. 18.1. The regression lines suggest that association between the mean of MATHGAIN and the HOUSEPOV variable may depend on sex. Thus, we might want to include an interaction between HOUSEPOV and SEX in model **M18.1**. This is what we consider next.

18.3 A Model with an Interaction Between School- and Pupil-Level Covariates

In this section, we consider model **M18.2**. Compared to **M18.1**, model **M18.2** includes an interaction between HOUSEPOV and SEX in the mean structure.

18.3.1 Model Specification

Model **M18.2** is defined as follows:

$$\begin{aligned}\text{MATHGAIN}_{sci} &= \beta_0 + \beta_1 \times \text{SES}_{sci} + \beta_2 \times \text{MINORITY}_{sci} \\ &\quad + \beta_3 \times \text{MATHKIND}_{sci} + \beta_4 \times \text{SEX}_{sci} + \beta_5 \times \text{HOUSEPOV}_s \\ &\quad + \beta_{4,5} \times \text{HOUSEPOV}_s \times \text{SEX}_{sci} \\ &\quad + b_{0s} + b_{0sc} + \varepsilon_{sci} \\ &\equiv \mu_{sci} + b_{0s} + b_{0sc} + \varepsilon_{sci}. \end{aligned} \tag{18.4}$$

R18.6 *SII Project*: Plots of the marginal residuals for model **M18.1**. The model-fit object `fm18.1` was created in Panel R18.1

(a) *Marginal residuals*

```
> rsd1 <-                              # Marginal residuals
+     resid(fm18.1, level = 0)
> range(rsd1)                          # Range
  [1] -176.86  124.84
> outi <- abs(rsd1) > 120              # Selection of outliers
> as.numeric(SIIdata$childid[outi]) # Outliers' ids
  [1]  41 665 754
> rsd1[outi]                           # Outliers' values and labels
        4       62       70
  -176.86  122.55   124.84
```

(b) *Plot of the marginal residuals vs.* `housepov` *by* `sex`.

```
> myPanel <- function(x,y, subscripts, ... ){
+   panel.xyplot(x,y,... )
+   outi <- abs(y) > 120
+   y1   <- y[outi]
+   x1   <- x[outi]
+   ltext(x1, y1, names(y1), pos=3)
+ }
> xyplot(rsd1 ~ housepov|sex, SIIdata,   # Fig. 18.1
+        type = c("p","r"),
+        panel = myPanel)
```

Note that model **M18.1** is nested within model **M18.2**. The random-effects structure of the two models is the same. Consequently, the marginal variances and covariances for both models are the same and given in (18.2) and (18.3).

18.3.2 R *Syntax and Results*

In Panel R18.7, we fit model **M18.2** and test the hypothesis about the added term. Toward this end, we first create, in Panel R18.7a, the model-formula object `form2`. Specifically, we add, with the help of the function `update()`, the interaction term `sex:housepov` to the model-formula object `form1`. Then we fit the model by updating the model-fit object `fm18.1` with the newly-created model formula. Note that the new model is fitted using the same estimation method that was applied when fitting the model represented by the object `fm18.1`, i.e., the ML estimation. The results are stored in the model-fit object `fm18.2`. Details of the results are shown in Table 18.1.

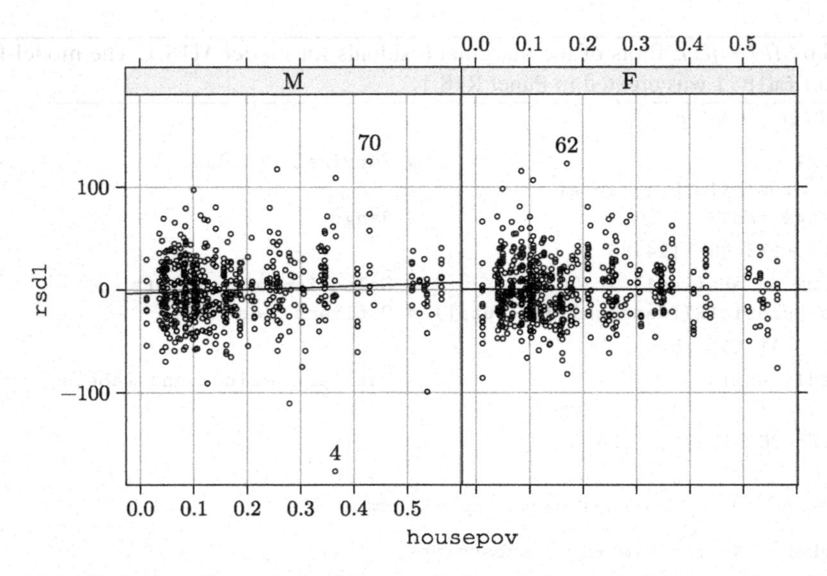

Fig. 18.1 *SII Project*: Scatterplots of the marginal residuals *versus* housepov by sex for model **M18.1**

Table 18.1 *SII Project*: ML estimates[a] of the parameters for models **M18.1–M18.3**

	Par.	fm18.1	fm18.2	fm18.3
Model label		**M18.1**	**M18.2**	**M18.3**
Log-ML value		−5694.82	−5693.35	−5695.77
Fixed effects:				
Intercept	β_0	284.91(11.01)	282.66(11.09)	282.34(10.84)
Ses	β_1	5.23(1.24)	5.16(1.24)	5.36(1.24)
Minority(Y *vs.* N)	β_2	−7.75(2.38)	−7.66(2.38)	−8.28(2.33)
Mathkind	β_3	−0.47(0.02)	−0.47(0.02)	−0.47(0.02)
Sex (F *vs.* M)	β_4	−1.23(1.66)	2.52(2.74)	
House pov.	β_5	−11.30(9.85)	−0.81(11.65)	
Sex(F) × housepov	$\beta_{4,5}$		−21.18(12.34)	
reStruct(schoolid):				
SD(b_{0s})	$\sqrt{d_1}$	8.59(6.15,11.99)	8.71(6.28,12.07)	8.52(6.04,12.00)
reStruct(classid):				
SD(b_{0sc})	$\sqrt{d_2}$	9.02(6.36,12.78)	8.86(6.19,12.69)	9.10(6.43,12.86)
Scale	σ	27.06(25.83,28.33)	27.03(25.81,28.31)	27.07(25.85,28.35)

[a]Approximate SE for fixed effects and 95% CI for covariance parameters are included in parentheses

Next, we apply the summary() function to extract information about the fitted model. Note that the resulting output was abbreviated. Selected lines indicate that the added interaction term is statistically not significant at the 5% significance level.

R18.7 *SII Project*: Model **M18.2** fitted using the function `lme()`. The objects `form1` and `fm18.1` were created in Panel R18.1

(a) *Fitting the model*

```
> form2 <- update(form1, . ~ . + sex:housepov)     # (18.4)
> fm18.2 <- update(fm18.1, form2)                  # M18.2 ← M18.1
> summary(fm18.2)                                  # Summary
  Linear mixed-effects model fit by maximum likelihood
   Data: SIIdata
      AIC   BIC  logLik
    11407 11458 -5693.3
  ...      [snip]
  Fixed effects: mathgain ~ ses + minority + mathkind + sex + ...
                          Value Std.Error  DF  t-value p-value
  (Intercept)           282.656   11.0936 873  25.4792  0.0000
  ...      [snip]
  sexF:housepov         -21.175   12.3367 873  -1.7165  0.0864
  ...      [snip]
```

(b) *Testing the hypothesis about the* `sex:housepov` *interaction term*

```
> anova(fm18.2, Terms = "sex:housepov")            # Approximate F-test
  F-test for: sex:housepov
    numDF denDF F-value p-value
  1     1   873  2.9462  0.0864
> anova(fm18.1, fm18.2)                            # M18.1 ⊂ M18.2
         Model df   AIC   BIC  logLik   Test L.Ratio p-value
  fm18.1     1  9 11408 11453 -5694.8
  fm18.2     2 10 11407 11458 -5693.3 1 vs 2  2.9477   0.086
```

In Panel R18.7b, we demonstrate two alternate ways to test the interaction. First, we apply the function `anova()` to the model-fit object with the argument `Terms="sex:housepov"` (Sect. 14.7). As a result, we obtain the F-test for the interaction term. Note that, formally speaking, it is a sequential-approach test, but because the interaction is specified as the last term in the model formula, the test is equivalent to the marginal-approach test.

Second, we apply the function `anova()` to the model-fit objects `fm18.1` and `fm18.2`. As a result, we obtain the LR test for the interaction between HOUSEPOV and SEX (see Sect. 14.7).

The results of both the F-test and the LR test are statistically not significant at the 5% significance level, but they are significant at the 10% significance level. Given the fact that tests for interaction terms have less power than tests for main effects, we might consider retaining the `sex:housepov` interaction term in the model. We will come back to this issue in the next section.

R18.8 *SII Project*: Model **M18.3** fitted using the function `lme()`. The model-fit objects `form1` and `fm18.1` were created in Panel R18.1; the object `fm18.2`, in Panel R18.7

```
> form3 <- update(form1, . ~ . - sex - housepov) # (18.5)
> fm18.3 <- update(fm18.1, form3)       # M18.3 ← M18.1
> anova(fm18.1, fm18.3, fm18.2)         # M18.3 ⊂ M18.1 ⊂ M18.2
       Model df   AIC   BIC  logLik   Test L.Ratio p-value
fm18.1     1  9 11408 11453 -5694.8
fm18.3     2  7 11406 11441 -5695.8 1 vs 2  1.8877  0.3891
fm18.2     3 10 11407 11458 -5693.3 2 vs 3  4.8355  0.1842
```

18.4 A Model with Fixed Effects of Pupil-Level Covariates Only

In this section, we consider model **M18.3**, which, as compared to model **M18.1**, excludes terms associated with HOUSEPOV and SEX variables from the mean structure.

18.4.1 Model Specification

Model **M18.3** is defined as follows:

$$\begin{aligned}\text{MATHGAIN}_{sci} &= \beta_0 + \beta_1 \times \text{SES}_{sci} + \beta_2 \times \text{MINORITY}_{sci}\\ &\quad + \beta_3 \times \text{MATHKIND}_{sci}\\ &\quad + b_{0s} + b_{0sc} + \varepsilon_{sci}\\ &\equiv \mu_{sci} + b_{0s} + b_{0sc} + \varepsilon_{sci}. \end{aligned} \tag{18.5}$$

Note that the model is nested within models **M18.1** and **M18.2**. The marginal variances and covariances for model **M18.3** are given in (18.2) and (18.3).

18.4.2 R *Syntax and Results*

To fit model **M18.3** we create, in Panel R18.8, the model-formula object `form3` by removing, with the help of the function `update()`, the terms `sex` and `housepov` from the model-formula object `form1`. Then we fit the model by updating the model-fit object `fm18.1` with the newly-created model formula. Note that the new model is fitted using the same estimation method that was applied to fit the model represented by the object `fm18.1`, i.e., the ML estimation. The results are stored in the model-fit

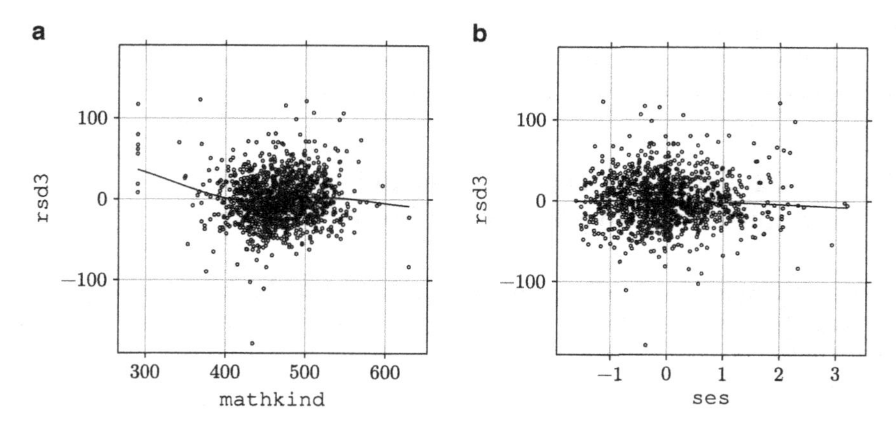

Fig. 18.2 *SII Project*: Scatterplots of the marginal residuals *versus* (**a**) `mathkind` and (**b**) `ses` for model **M18.3**

object `fm18.3`. Details of the results are presented in Table 18.1. We then apply the function `anova()` to the model-fit objects `fm18.1`, `fm18.2`, and `fm18.3`. As a result of using the most parsimonious model as the middle argument, we obtain two LR tests: one based on models **M18.1** and **M18.3**, and one based on models **M18.2** and **M18.3**.

The first test verifies the joint null hypothesis that the fixed-effects coefficients of the variables SEX and HOUSEPOV are equal to zero. Recall that the F-test for this hypothesis could not be obtained (see Panel R18.5). The result of the LR test is statistically not significant and implies that the variables could be removed from model **M18.1**, leading to model **M18.3**.

The second LR test verifies the joint null hypothesis that the effects of the variables SEX and HOUSEPOV, including their interaction, are equal to zero. Note that the F-test for the interaction alone was significant at the 10% significance level (Sect. 18.3.1). However, the result of the LR test is statistically not significant, suggesting that the variables SEX and HOUSEPOV, including their interaction, can be removed from model **M18.2**. Again, this leads to the choice of model **M18.3**.

Figure 18.2 presents the plots of the marginal residuals *versus* MATHKIND and SES for model **M18.3**. The vector of the residuals, `rsd3`, was created in a way similar to the vector `rsd1` in Panel R18.6. The two plots in Fig. 18.2 were created by using the following syntax:

```
> xyplot(rsd3 ~ mathkind, SIIdata,             # Fig. 18.2a
+        type = c("p", "smooth"))
> xyplot(rsd3 ~ ses, SIIdata,                  # Fig. 18.2b
+        type = c("p", "smooth"))
```

The smoothed line added to the residuals shown in Fig. 18.2a clearly indicates a nonlinear effect of the variable MATHKIND on the mean value of the dependent variable. This suggests that we might consider modifying the mean structure of model **M18.3**, given by (18.5), by using a different functional form of MATHKIND. Such a modification will be considered in the next section.

18.5 A Model with a Third-Degree Polynomial of a Pupil-Level Covariate in the Mean Structure

In this section, we consider a model similar to **M18.3**, but with the linear effect of variable MATHKIND replaced by a third-degree polynomial function. We will refer to this model as model **M18.4**.

18.5.1 Model Specification

Model **M18.4** is defined as follows:

$$\begin{aligned}\text{MATHGAIN}_{sci} &= \beta_0 + \beta_1 \times \text{SES}_{sci} + \beta_2 \times \text{MINORITY}_{sci}\\ &\quad + \beta_{3,p_1} \times p_1(\text{MATHKIND}_{sci}) + \beta_{3,p_2} \times p_2(\text{MATHKIND}_{sci})\\ &\quad + \beta_{3,p_3} \times p_3(\text{MATHKIND}_{sci})\\ &\quad + b_{0s} + b_{0sc} + \varepsilon_{sci}\\ &\equiv \mu_{sci} + b_{0s} + b_{0sc} + \varepsilon_{sci}, \end{aligned} \tag{18.6}$$

where $p_1(\text{MATHKIND}_{sci})$, $p_2(\text{MATHKIND}_{sci})$, and $p_3(\text{MATHKIND}_{sci})$ are orthogonal polynomials of degree 1, 2, and 3, respectively. Note that models **M18.1** and **M18.3** are nested within model **M18.4**. The random-effects structure of the models is the same.

As compared to the use of an "ordinary" linear combination of powers of variable MATHKIND, the use of the orthogonal polynomials gives the advantage of removing the multicollinearity between the covariates corresponding to the coefficients β_{3,p_1}, β_{3,p_2}, and β_{3,p_3}. The disadvantage is that, to obtain the final form of the third-degree polynomial for MATHKIND, the coefficients of the orthogonal polynomials need to be calculated and combined with β_{3,p_1}, β_{3,p_2}, and β_{3,p_3}.

18.5.2 R *Syntax and Results*

Panel R18.9 presents the R code for fitting model **M18.4**. The formula, assigned to the formula-object `form4`, includes the result of applying the function `poly()` to the variable `mathkind`. The function returns orthogonal polynomials of a particular

R18.9 *SII Project*: Model **M18.4** fitted using the function `lme()`. The model-fit object `fm18.3` was created in Panel R18.8

```
> form4 <-                                              # (18.6)
+     formula(mathgain ~ ses + minority + poly(mathkind, 3))
> fm18.4 <- update(fm18.3, form4)                       # M18.4 ← M18.3
> anova(fm18.3, fm18.4)                                 # M18.3 ⊂ M18.4
       Model df   AIC   BIC  logLik   Test L.Ratio p-value
fm18.3     1  7 11406 11441 -5695.8
fm18.4     2  9 11352 11397 -5666.7 1 vs 2  58.067  <.0001
```

Table 18.2 *SII Project*: ML-based parameter estimates[a] for models **M18.4–M18.6**

	Par.	fm18.4	fm18.5	fm18.6
Model label		**M18.4**	**M18.5**	**M18.6**
Log-ML value		−5666.73	−5665.71	−5663.91
Fixed effects:				
Intercept	β_0	62.12(2.07)	207.48(10.51)	61.35(2.08)
Ses	β_1	5.22(1.21)	5.22(1.21)	8.84(1.95)
Minority(Y *vs.* N)	β_2	−7.19(2.29)	−7.09(2.29)	−6.86(2.29)
mathkind poly(3)				
Linear	β_{3,p_1}	−658.17(31.14)		−660.46(31.08)
Quadratic	β_{3,p_2}	128.94(28.32)		124.46(28.32)
Cubic	β_{3,p_3}	−175.73(28.09)		−178.34(28.06)
mathkind bs(df=4)				
bs1	β_{3,f_1}		−92.38(15.05)	
bs2	β_{3,f_2}		−159.72(12.35)	
bs3	β_{3,f_3}		−169.96(13.85)	
bs4	β_{3,f_4}		−271.89(19.46)	
Ses × Mnrty(Y/N)	$\beta_{1,2}$			−5.82(2.45)
reStruct(`schoolid`):				
SD(b_{0s})	$\sqrt{d_1}$	8.46(6.01,11.91)	8.45(6.00,11.91)	8.17(5.70,11.71)
reStruct(`classid`):				
SD(b_{0sc})	$\sqrt{d_2}$	9.07(6.49,12.69)	9.05(6.46,12.66)	9.27(6.70,12.82)
Scale	σ	26.36(25.17,27.61)	26.34(25.15,27.58)	26.28(25.10,27.53)

[a]Approximate SE for fixed effects and 95% CI for covariance parameters are included in parentheses

degree computed over the specified set of points (Sect. 5.3.1). In our case, we use a polynomial of the third degree. By updating the model-fit object `mth3` with the newly-defined formula object, we fit model **M18.4** to the data using the ML estimation. Note that we do not display the results; they are shown in Table 18.2. Then, with the help of the `anova()` function, we calculate the log-likelihoods for models **M18.3** and **M18.4**. As a result, we obtain the LR test for the null hypothesis that $\beta_{3,p_2} = \beta_{3,p_3} = 0$ in (18.6). The result of the test is statistically significant at the 5% significance level and indicates that including the third-degree polynomial of the variable MATHKIND improves the fit of model **M18.3**.

R18.10 *SII Project*: Predicted values of mathgain for model **M18.4**. The model-fit object fm18.4 was created in Panel R18.9

```
> auxL <-                                                # Auxiliary list
+    list(ses = 0,
+         minority = factor(c("Mnrt=No", "Mnrt=Yes")),
+         mathkind = seq(290, 625, by = 5))
> dim (auxDt <-  expand.grid(auxL))                      # Data frame created
  [1] 136   3
> names(auxDt)
  [1] "ses"        "minority" "mathkind"
> prd   <- predict(fm18.4, auxDt, level = 0)             # Predicted values
> prd4Dt <- data.frame(auxDt, pred4 = prd)
> head(prd4Dt)
    ses minority mathkind  pred4
  1   0  Mnrt=No      290 212.38
  2   0 Mnrt=Yes      290 205.19
  3   0  Mnrt=No      295 203.41
  4   0 Mnrt=Yes      295 196.21
  5   0  Mnrt=No      300 194.83
  6   0 Mnrt=Yes      300 187.64
> xyplot (pred4 ~ mathkind, groups = minority,   # Fig. 18.3a
+         data = prd4Dt, type = "l", grid = TRUE)
```

Panel R18.10 presents the syntax for creating a plot of predicted values for model **M18.4**. Toward this end, we first create an auxiliary list auxL. The names of the components of the list correspond to the names of the covariates included in the model. The component ses is set to the numeric value of 0, minority is a factor with two levels, and mathkind is a numeric vector containing values from 290 to 625 in steps of 5. The list is used as an argument in the function expand.grid() to create a data frame from all combinations of the values of the vectors and factors contained in the components of the list. The resulting data frame, auxDt, contains 136 rows and three variables. Note that the number of rows is equal to $1 \times 68 \times 2$, i.e., it corresponds to the number of combinations of the values of the vectors ses and mathkind and factor minority from the list auxL.

The data frame auxDt is then used in the argument newdata of the function predict() (see Table 14.5). The function is applied to the model-fit object fm18.4 to compute the predicted values for model **M18.4**. By specifying the argument level=0, we obtain the population-level predicted values, i.e., estimates of the mean-values μ_{sci}, as defined in (18.7).

We store the resulting numeric vector in the object prd. We add the vector as the variable pred4 to the data frame auxDt and store the result in the data-frame object prd4Dt. Finally, we use the latter data frame in the argument data of the function xyplot() to construct the plot of the predicted values against the values of the

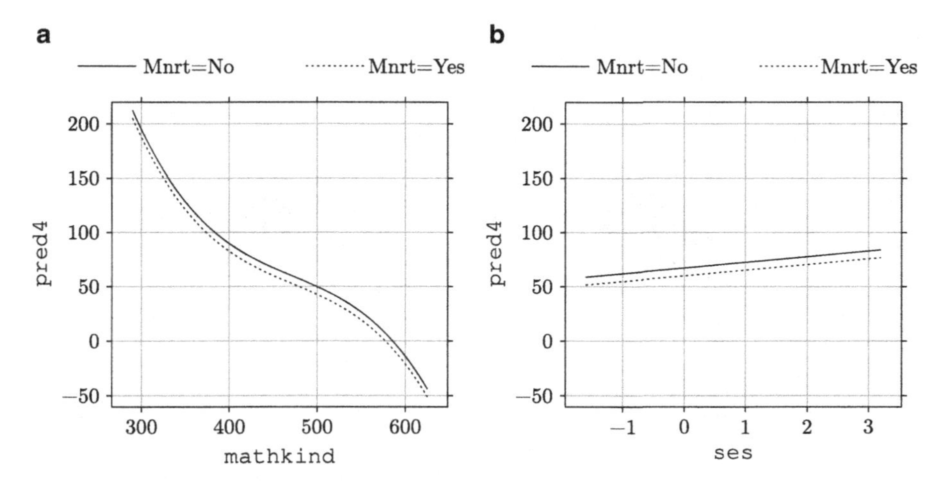

Fig. 18.3 *SII Project*: Plots of the predicted values of `mathgain` for model **M18.4** *versus* (**a**) `mathkind` and (**b**) `ses` for the two `minority` levels

variable `mathkind` within the groups defined by the levels of the factor `minority`. Note that, per definition of the data frame `prd4Dt`, the plot corresponds to the value of `ses` equal to 0.

The resulting plot is shown in Fig. 18.3a. It indicates that the mean value of the gain in the math score decreases with increasing pupil's math score in the spring of the kindergarten year.

Figure 18.3b presents the plot of the predicted values for model **M18.4** *versus* the values of the variable `ses` within the groups defined by the levels of the factor `minority`. The value of the `mathkind` covariate is assumed to be equal to 450. The `xyplot()`-function call, necessary to create the plot, is very similar to the one used to construct the plot in Fig. 18.3a (see Panel R18.10).

The plot in Fig. 18.3b indicates a linear increase of the mean value of the gain in the math score with increasing pupil's socioeconomic status. Of course, the linearity of the increase stems from the assumption made about the functional form of the effect of the variable SES on the expected value of MATHGAIN in (18.6).

Figure 18.4 presents the plot of the marginal residuals *versus* the covariates MATHKIND and SES for model **M18.4**. We do not present the necessary code, as it is similar to the one used to create, e.g., Fig. 18.2. The smoothed line shown in the plot suggests that the inclusion of the third-degree polynomial of the variable does not fully remove the nonlinearity with respect to MATHKIND, observed in Fig. 18.2a. To address this issue, we could consider using a smooth function of MATHKIND. A model including such a function is presented in the next section.

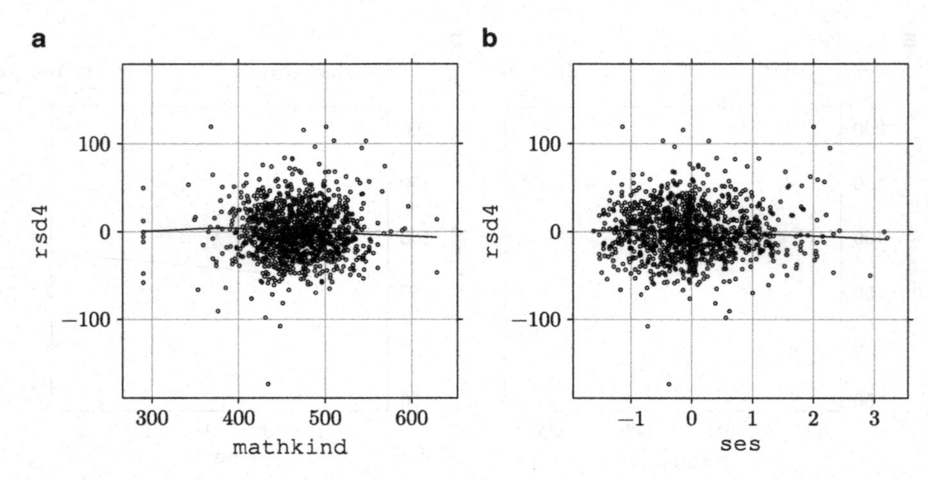

Fig. 18.4 *SII Project*: Scatterplots of the marginal residuals *versus* (**a**) `mathkind` and (**b**) `ses` for model **M18.4**

18.6 A Model with a Spline of a Pupil-Level Covariate in the Mean Structure

In this section, we consider model **M18.5**, which results from replacing the third-degree polynomial of the variable MATHKIND in model **M18.4** by a spline.

18.6.1 Model Specification

Model **M18.5** is defined as follows:

$$\begin{aligned} \text{MATHGAIN}_{sci} &= \beta_0 + \beta_1 \times \text{SES}_{sci} + \beta_2 \times \text{MINORITY}_{sci} \\ &\quad + f(\text{MATHKIND}_{sci}) \\ &\quad + b_{0s} + b_{0sc} + \varepsilon_{sci} \\ &\equiv \mu_{sci} + b_{0s} + b_{0sc} + \varepsilon_{sci}, \end{aligned} \tag{18.7}$$

where $f(\text{MATHKIND}_{sci})$ denotes a smooth function (spline) of MATHKIND_{sci}. Note that the random-effects structure of the model is the same as for models **M18.1–M18.4**. However, the models are not nested within model **M18.5**.

R18.11 *SII Project*: Model **M18.5** fitted using the function `lme()`. The model-fit objects `fm18.3` and `fm18.4` were created in Panels R18.8 and R18.9, respectively

```
> require(splines)
> form5 <-                                              # (18.7)
+    formula(mathgain ~ ses + minority + bs(mathkind, df = 4))
> fm18.5 <- update(fm18.4, form5)                       # M18.5 ← M18.4
> AIC(fm18.3, fm18.4, fm18.5)
       df      AIC
fm18.3  7 11405.53
fm18.4  9 11351.47
fm18.5 10 11351.42
> detach(package:splines)
```

18.6.2 R *Syntax and Results*

Panel R18.11 presents the R syntax for fitting model **M18.5** to the data. Note that we need to attach the package **splines**. The formula-object `form5` represents a formula corresponding to the fixed-effects structure of (18.7). In particular, it includes the function `bs()` applied to the variable `mathkind`. The function returns a B-spline (Sect. 5.3.1). In our case, we use a cubic spline (the default) with four degrees of freedom. As a result, we obtain two splines joined at one knot located at the median of the values of the `mathkind` variable. More information about the use of the function `bs()` can be obtained from the R help-system by issuing the command `?bs` after attaching the package **splines**.

By updating the model-fit object `fm18.3` with the newly-defined formula object, we fit model **M18.5** to the data by using the ML estimation.

The results of fitting of model **M18.5** are presented in Table 18.2. They suggest that, e.g., the improvement in math scores for minority pupils was, on average, lower by about -7.1, as compared to nonminority pupils. On the other hand, the improvement in math scores increased, on average, by 5.2 with a unit increase of the SES variable.

It might be of interest to compare the fits of models **M18.3** and **M18.4** with model **M18.5** to check if the use of the smooth function of the variable MATHKIND improves the fit of the models. As models **M18.3** and **M18.4** are not nested within model **M18.5**, we cannot use the LR test to compare their fits. Instead, we may use an information criterion (Sect. 13.7.1). In Panel R18.11, we apply the function `AIC()` (see Table 14.5) to compute AIC (see Sect. 4.7.2) for the three models. The lowest value of the criterion is obtained for model **M18.5**. Formally speaking, this suggests that the model fits the data better than the other two models. Note, however, that the difference in AIC between models **M18.4** and **M18.5** is minimal and we could consider the description of the data offered by the former model as well.

18.7 The Final Model with Only Pupil-Level Variables in the Mean Structure

In this section, we construct model **M18.6**, which includes an interaction between variables SES and MINORITY. Given the minimal difference in AIC between models **M18.4** and **M18.5** (see Sect. 18.6), we use the former as a basis for model **M18.6**.

18.7.1 Model Specification

Model **M18.6** is specified as follows:

$$\begin{aligned}
\text{MATHGAIN}_{sci} &= \beta_0 + \beta_1 \times \text{SES}_{sci} + \beta_2 \times \text{MINORITY}_{sci} \\
&\quad + \beta_{1,2} \times \text{SES}_{sci} \times \text{MINORITY}_{sci} + \beta_{3,p_1} \times p_1(\text{MATHKIND}_{sci}) \\
&\quad + \beta_{3,p_2} \times p_2(\text{MATHKIND}_{sci}) + \beta_{3,p_3} \times p_3(\text{MATHKIND}_{sci}) \\
&\quad + b_{0s} + b_{0sc} + \varepsilon_{sci} \\
&\equiv \mu_{sci} + b_{0s} + b_{0sc} + \varepsilon_{sci},
\end{aligned} \tag{18.8}$$

where, as in (18.6), $p_1(\text{MATHKIND}_{sci})$, $p_2(\text{MATHKIND}_{sci})$, and $p_3(\text{MATHKIND}_{sci})$ are orthogonal polynomials of degree 1, 2, and 3, respectively. Note that model **M18.4** is nested within model **M18.6**. The random-effects structure of the models is the same, with marginal variances and covariances given in (18.2) and (18.3).

18.7.2 R Syntax and Results

Panel R18.12 presents the R syntax for fitting model **M18.6** to the data. It also contains the result of the LR test for the comparison of models **M18.5** and **M18.6**. The result of the test is statistically significant at the 5% significance level. It allows to reject the null hypothesis that the fixed-effect coefficient of the interaction between SES and MINORITY is equal to zero.

The estimated value of the coefficient for the interaction can be found in Table 18.2. It indicates that, for minority pupils, the increase of the improvement in math scores, corresponding to a unit increase of the SES variable, was smaller by about −5.8 than the increase for nonminority pupils.

In Panel R18.13 we present the code used to create plots of various types of residuals shown in Figs. 18.5 and 18.6.

R18.12 *SII Project*: Model **M18.6** fitted using the function `lme()`. The model-fit object `fm18.4` was created in Panel R18.9

```
> form6 <-                                        # (18.8)
+    formula(mathgain ~ ses + minority + poly(mathkind, 3) +
+              ses:minority)
> fm18.6 <- update(fm18.4, form6)                 # M18.6 ← M18.4
> anova(fm18.4, fm18.6)                           # M18.4 ⊂ M18.6
       Model df   AIC   BIC  logLik   Test L.Ratio p-value
  fm18.4     1  9 11352 11397 -5666.7
  fm18.6     2 10 11348 11399 -5663.9 1 vs 2  5.6455  0.0175
```

R18.13 *SII Project*: Plots of residuals for model **M18.6**. The model-fit object `fm18.6` was created in Panel R18.12

(a) *Plots of the marginal residuals*

```
> rsd6 <- resid(fm18.6, level = 0)
> xyplot(rsd6 ~ ses | minority, SIIdata,
+        type = c("p", "smooth"))             # Fig.18.5
```

(b) *The normal Q-Q plots of the class-level conditional Pearson residuals*

```
> qqnorm(fm18.6)                              # Fig. 18.6a
> qqnorm(fm18.6,                              # Equivalent call
+        form = ~resid(., type = "p", level = 2))
> qqnorm(fm18.6,                              # Fig. 18.6b
+        form = ~resid(., type = "p")         # Residuals...
+                 | sex*minority,              # ... by sex and minority.
+        id = 0.0005)                         # Outliers identified.
```

(c) *The normal Q-Q plot of the school-level conditional Pearson residuals*

```
> qqnorm(fm18.6,                              # Plot not shown
+        form = ~resid(., type = "p",
+                      level = 1))            # School level
```

Figure 18.5 presents the scatterplots of the marginal residuals *versus* the values of the variable SES for the two levels of the variable MINORITY for model **M18.6**. The figure was created by using the syntax shown in Panel R18.13a. In particular, we extract the residuals from the model-fit object `fm18.6` with the help of the function `resid()` and store them in the object `rsd6`. Subsequently, we use the object in the formula provided to the function `xyplot()` to construct the plot of the residuals against the values of the variable `ses` within the groups defined by the levels of the factor `minority`.

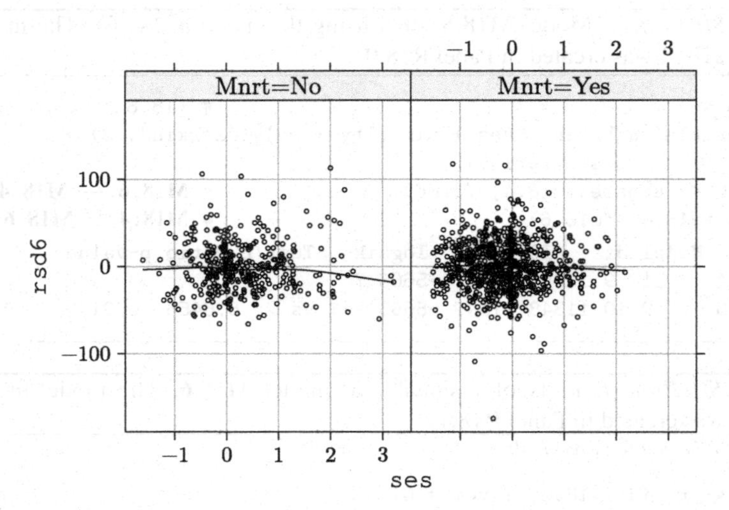

Fig. 18.5 *SII Project*: Scatterplots of the marginal residuals *versus* `ses` for the two `minority` levels for model **M18.6**

It may be noted that both panels of Fig. 18.5 suggest that the effect of SES is not linear. Thus, model **M18.6** might be further modified by using a smooth function to capture the effect of the variable. We leave investigation of this modification as an exercise to the reader.

At this point, we take a closer look at the conditional residuals (Sect. 13.6.2) of the underlying model **M18.6**. Because there are two levels of grouping in the data, there are also two types of conditional residuals: class-level ones, which are the differences between the observed values of the dependent variable and the estimated means $\widehat{\mu}_{sci} + \widehat{b}_{0s} + \widehat{b}_{0sc}$, and school-level ones, which are based on the deviations from the estimated means $\widehat{\mu}_{sci} + \widehat{b}_{0s}$. Note that the class-level residuals are predictors of the residual errors ε_{sci}.

In Panel R18.13b, we present the syntax for constructing the normal Q-Q plot of the class-level conditional Pearson residuals (see Table 7.5; Sect. 13.6.2; and Table 14.5). Toward this end, the function `qqnorm()` is used. The results are displayed in Fig. 18.6. Note that we present two equivalent forms of the `qqnorm()`-function call for the construction of Fig. 18.6a. The second form shows explicitly the arguments of the function. In particular, it is worth noting that the argument `level=2` is used in the function `resid()` (Sect. 14.6). This is the highest grouping level, i.e., the `classid` level (see Panel R18.2). As a result, the class-level residuals are created. On the other hand, in Panel R18.13c, we create school-level conditional residuals by using the argument `level=1` in the call to the function `resid()`.

The plot, shown in Fig. 18.6a, is reasonably linear, with a few deviations from linearity in the tails.

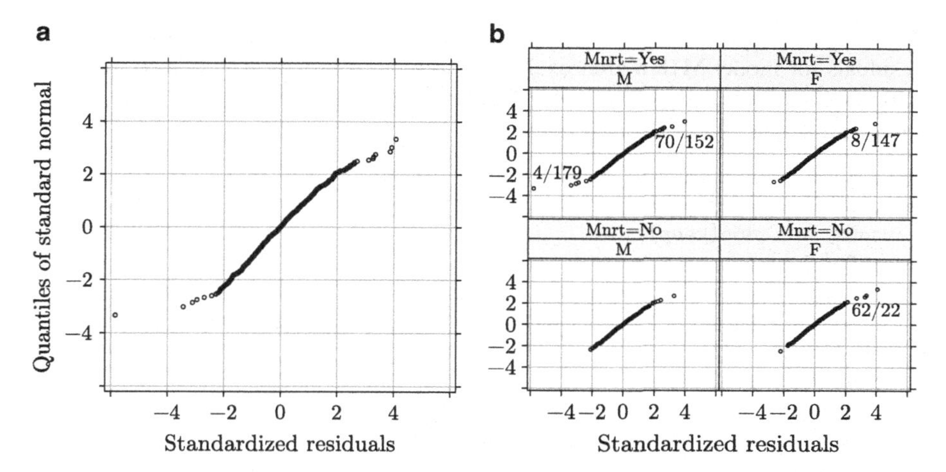

Fig. 18.6 *SII Project*: Normal Q-Q plots of the conditional Pearson residuals for model **M18.6**. (**a**) Overall (**b**) By sex and minority

Panel R18.13b also includes the `qqnorm()`-function call necessary to construct the normal Q-Q plot for the four combinations of the levels of the SEX and MINORITY variables. In the call, we use the argument `id=0.0005`. As a result, the observations with the absolute standardized residuals greater than the $1 - 0.0005/2$ quantile of the standard normal distribution are identified in the plot. By default, the outlying observations are labeled using the levels of the innermost grouping factor; in our case, this is `classid`. More information about the arguments of the `qqnorm()` function for the *lme*-class objects can be obtained from the R help-system by issuing the command `?qqnorm.lme`.

The plots, shown in Fig. 18.6b, are reasonably linear. Note that only four observations are labeled in the plots. This is because we used a relatively low probability value in the argument `id`; this choice was made for illustrative purposes in order to limit the number of labeled points in the plots. In practice, we might have used a larger value, like 0.05, that would correspond to selecting the observations with the absolute standardized residuals larger than 1.96.

As has just been mentioned, the plots, shown in Fig. 18.6, do not raise substantial doubts about the assumption of the normality of residual error. However, there may be situations when we might want to investigate the influence of outliers on the assumption in more detail. In Panel R18.14, we illustrate how such an investigation can be conducted.

In particular, in Panel R18.14a, we create a logical vector `keep`, which identifies the rows of the vector `rsd6` that contain the Pearson residuals with an absolute value smaller than say 3. We then store the selected residuals in the vector `rsd6x`. By displaying the rows of `rsd6`, which correspond to the logical negation of `keep`, we check that there are ten observations with residuals larger than or equal, in absolute value, to 3. These ten observations are not included in the vector `rsd6x`.

R18.14 *SII Project*: Normal Q-Q plots of the class-level conditional Pearson residuals for model **M18.6** after excluding outlying residuals. The objects `fm18.6` and `rsd6` were created in Panels R18.12 and R18.13, respectively

(a) *Identifying and excluding outlying residuals*

```
> rsd6p <- resid(fm18.6, type = "p")
> keep <- abs(rsd6p) < 3
> rsd6x <- rsd6p[keep]
> rsd6p[!keep]
   4/179    8/147   27/104     40/9    53/14    62/22   70/152    75/42
 -5.8391   3.8563   3.2514  -3.4276   3.0977   4.0555   3.9099   3.2727
  85/196   86/132
  3.3377  -3.1083
```

(b) *Normal Q-Q plot of the extracted residuals (Fig. 18.7a)*

```
> qqDtx <- qqnorm(rsd6x, plot.it = FALSE)
> xp1 <-  xyplot(x ~ y, data.frame(qqDtx))                # Draft plot
> update(xp1,                                              # Plot updated
+         ylab = "Quantiles of standard normal",
+         xlab = "Standardized residuals",
+         grid = TRUE)
```

(c) *Normal Q-Q plots of the residuals by* `sex` *and* `minority` *(Fig. 18.7b)*

```
> qqDtx2 <- cbind(SIIdata[keep, ], qqDtx)
> xp2 <-                                  # See R18.14b how to update xp2
+     xyplot(x ~ y | sex*minority, data = data.frame(qqDtx2))
```

In Panel R18.14b, we apply the function `qqnorm()` to the vector `rsd6x`, and store the result in the object `qqDtx`. The object is then used in a call to the function `xyplot()` to create the normal Q-Q plot, presented in Fig. 18.7a. As compared to Fig. 18.6a, the plot shows less deviations from linearity.

Finally, in Panel R18.14c, we merge the object `qqDtx` with the part of the data frame `SIIdata`, which corresponds to the observations with residuals smaller, in absolute value, than 3. We store the resulting data frame in the object `qqDtx2`. In this way, we can use the covariates, contained in the data frame `SIIdata`, in a plot of the residuals. The object `qqDtx2` is then used in a call to the function `xyplot()` to create separate normal Q-Q plots for the different combinations of the SEX and MINORITY levels. The plots are presented in Fig. 18.7b. As compared to Fig. 18.6b, they show less deviations from linearity.

In Panel R18.15, we present the code which can be used to create plots of the predicted random effects (Sect. 13.6.1). In particular, in Panel R18.15a, we use the function `ranef()` (Sect. 14.6) to extract the estimates of the random effects from

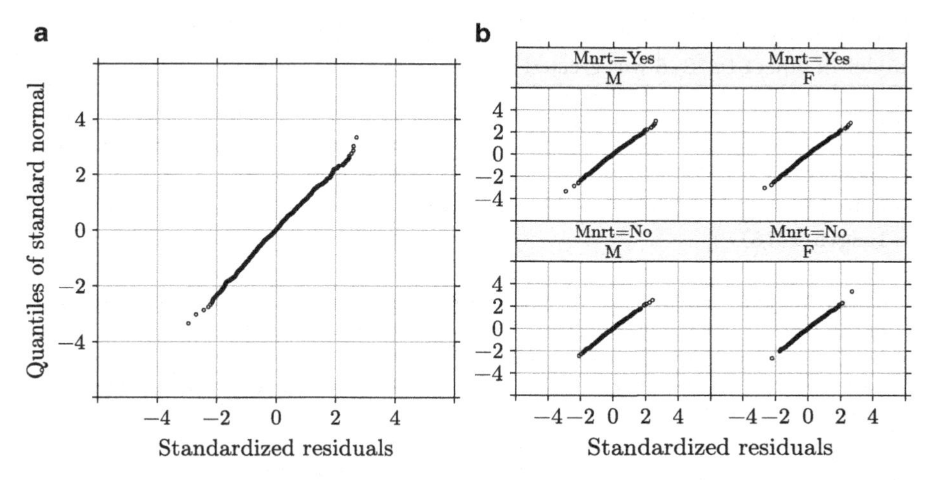

Fig. 18.7 *SII Project*: Normal Q-Q plots of the conditional Pearson residuals for model **M18.6** with ten outlying residuals omitted. (**a**) Overall (**b**) By sex and minority

the model-fit object `fm18.6`. We store the result in the object `ref6`. The object is a list with two components corresponding to the two grouping factors, `schoolid` and `classid`. The components are data frames which contain the EBLUPs of the random effects for the levels of the two factors.

Applying the function `plot()` to the object `ref6` would, by default, result in a plot of the predicted random effects for the grouping factor at the highest level of the data hierarchy, i.e., class. However, the labels of the levels of the factor at the vertical axis of the plot would appear nonlegible. Thus, before displaying the plot, we may want to simplify the labels. Toward this end, we first save the result of the application of the function `plot()` to `ref6` in the object `pref6`. We then extract the *y*-axis labels by referring to the component `Y.limits` of `pref6` and store it in the object `pref6lims`. With the help of the function `length()`, we get the number of the labels, store it as the scalar `len`, and create the index-vector `sel` which contains every 15th integer from 1 to `len`. We then set all elements of the character-vector `pref6lims`, except those corresponding to the values contained in `sel`, to an empty string. Finally, we use the function `update()` to update the plot-object `pref6` by assigning the modified vector of labels to the `ylim` component. We also modify the `ylab` component, which corresponds to the *y*-axis label. The resulting plot is shown in Fig. 18.8a. The plot could be used to, e.g., select predicted random effects with extremely large or small values. Arguably, there are no such values in the graph presented in Fig. 18.8a.

Panel R18.15b shows the syntax to create the plot, corresponding to the one shown in Fig. 18.8a, for the random effects for schools. To this end, we first use the extractor-function `ranef()` with the argument `level=1` (Sect. 14.6) to extract the

R18.15 *SII Project*: Predicted random effects (EBLUPs) for model **M18.6**. The model-fit object `fm18.6` was created in Panel R18.12

(a) *Predicted random effects for classes (Fig. 18.8a)*

```
> ref6 <- ranef(fm18.6)                        # Random effects for classes.
> mode(ref6)                                   # A list ...
  [1] "list"
> length(ref6)                                 # ...  with two components.
  [1] 2
> pref6 <- plot(ref6)              # Default plot for classes; not legible.
> pref6lims <- pref6$y.limits                  # Y-labels extracted
> len  <- length(pref6lims)                    # No. of labels
> sel  <- seq(1, len, by = 15)                 # Select every 15-th label.
> pref6lims[-sel] = ""                         # Other labels set to blank.
> update(pref6, ylim = pref6lims,              # Assign new Y-labels.
+        ylab = "classid %in% schoolid") # Y-axis label
```

(b) *Predicted random effects for schools (Figure 18.8b)*

```
> ref61 <- ranef(fm18.6, level = 1)           # Random effects for schools.
> plot(ref61)                                  # Plot the random effects.
```

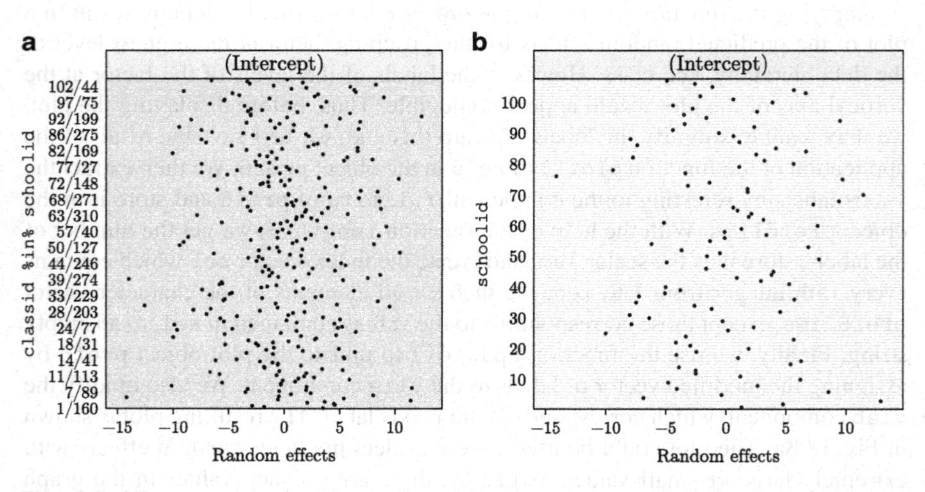

Fig. 18.8 *SII Project*: Dotplots of the predicted random effects (EBLUPs) for the two grouping factors (**a**) class and (**b**) school for model **M18.6**

predicted random effects, and then we apply the function `plot()`. The resulting plot is shown in Fig. 18.8b. Based on the plot, one could conclude that, e.g., the predicted random effect for school `76` has a relatively large negative value.

As was mentioned in Sect. 13.6.1, Q-Q plots and histograms of predicted random effects are of limited value when checking the normality of the effects is of interest. However, the plots of can be used to detect, e.g., outlying values. In Panel R18.16, we present the R syntax to construct normal Q-Q plots for the predicted school and class random effects for model **M18.6**. Toward this end, the function `qqnorm()` is used (see also Panels R18.13 and R18.14). We first apply it to construct the plot for the EBLUPs of the random effects for classes. By using the argument `id=0.2`, we identify in the plot the observations with the predicted random effects greater, in absolute value, than the $1-0.2/2 = 0.90$ quantile of the standard normal distribution. Finally, we modify the label of the x-axis using the argument `xlab`. The resulting normal Q-Q plot is shown in Fig. 18.9a. We use a similar syntax to construct a normal Q-Q plot for the predicted random effects for schools. The plot is shown in Fig. 18.9b.

Figure 18.10 presents the plots of the predicted values for model **M18.6** *versus* the values of the variables `mathkind` (with the value of `ses` assumed to be equal to 0) and `ses` (with the value of `mathkind` assumed to be equal to 450) within the groups defined by the levels of the factor `minority`. The plots are very similar to the those presented in Fig. 18.3. The main modification is the addition of the 95% CIs for the predicted values. The syntax, necessary to create the plots in Fig. 18.10, is more complex than the one used to construct the plots shown in Fig. 18.3 (see Panel R18.10); thus, we do not present it here.

18.8 Analysis Using the Function `lmer()`

In this section, we briefly illustrate how to fit model **M18.6** using the function `lmer()` from the package **lme4.0**.

In Panel R18.17, we present three different calls to the function `lmer()`, which fit model **M18.6** to the `SIIdata`. The first call uses a general (recommended) syntax for specifying the random-effects structure for a two-level LMM with nested effects (see syntax (2*b*) in Table 15.1). In particular, the nesting of grouping factors, i.e., `classid` within `schoolid`, is explicitly expressed using the crossing operator `:` (see Sect. 5.2.1) in the Z-term `(1|schoolid:classid)`, included in the model formula along with the `(1|schoolid)` term. To shorten the printout of the results of fitting of the model, we first save the application of the function `summary()` to the model-fit object `fm18.6mer` in the object `summ`. We then print `summ` using the function `print()` with the argument `corr=FALSE`. In this way, we omit the correlation matrix of the estimated fixed-effects coefficients from the printout. The displayed results are essentially the same as those shown in Table 18.2.

R18.16 *SII Project*: Normal Q-Q plots of the predicted random effects (EBLUPs) for model **M18.6**. The model-fit object `fm18.6` was created in Panel R18.12

```
> qqnorm(fm18.6, ~ranef(., level = 2), # Random effects for classes
+        id = 0.2,                     # Fig. 18.9a
+        xlab = "Random effects for classes")
> qqnorm(fm18.6, ~ranef(., level=1),   # Random effects for schools
+        id = 0.2,                     # Fig. 18.9b
+        xlab = "Random effects for schools")
```

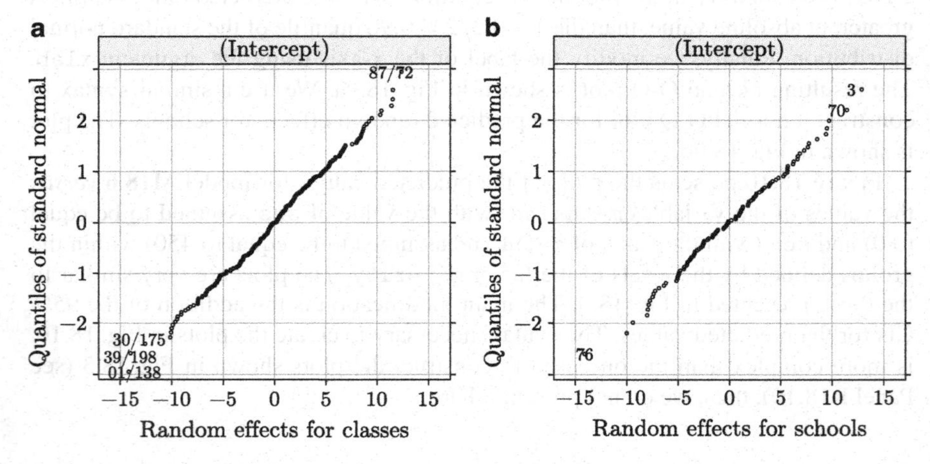

Fig. 18.9 *SII Project*: Normal Q-Q plots of the predicted random effects (EBLUPs) for the two grouping factors (**a**) class and (**b**) school for model **M18.6**

In the data frame `SIIdata`, the levels of the factor `classid` have been coded as explicitly nested within the levels of `schoolid` (Sect. 2.4.3). This approach is actually recommended in the case of representing factors with nested levels. Hence, it is possible to fit model **M18.6** using a simpler syntax, namely, `(1|schoolid) + (1|classid))` for the random-effects structure (see syntax (2*a*) in Table 15.2). In particular, in the second syntax shown in Panel R18.17, the Z-term for the factor `classid`, `(1|classid)`, does not use the crossing operator `:` and, therefore, does not explicitly indicate the nesting. However, given that the nesting is explicitly reflected in the data, the syntax also fits model **M18.6**.

Finally, the third form of the `lmer()`-function call, shown in Panel R18.17, uses the nonessential operator `/` (see Table 5.3) in the Z-term `(1|schoolid/classid)` to abbreviate the specification of the random-effects part of the `lmer()` model formula (see syntax (2*c*) in Table 15.2).

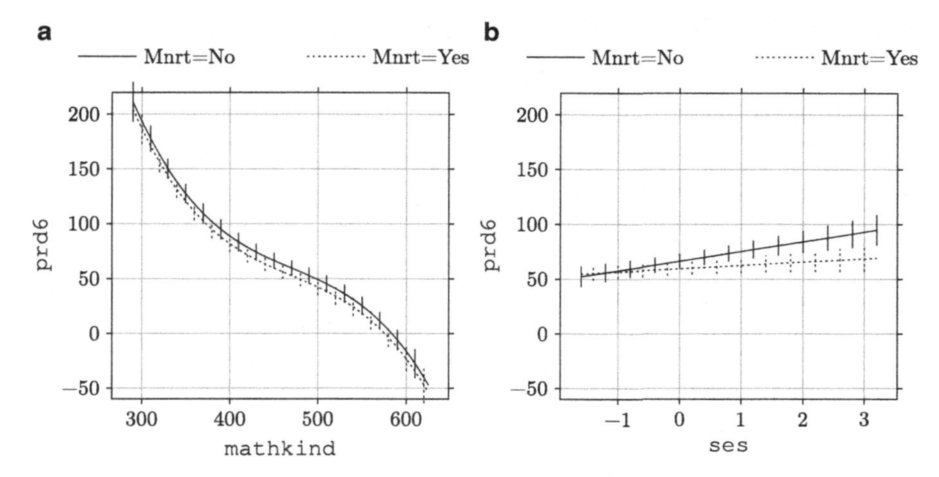

Fig. 18.10 *SII Project*: Plots of the predicted values with 95% confidence intervals *versus* `mathkind` and `ses` for the two `minority` levels for model **M18.6**. Predicted values (**a**) *versus* `mathkind` (`ses` set to 0) (**b**) *versus* `ses` (`mathkind` set to 450)

In Panel R18.18, we extract selected additional information from the model-fit object `fm18.6mer`. In particular, we use the `anova()` function to extract the results of F-tests for the fixed effects. As it was mentioned in Sect. 15.5, for *mer*-class model-fit objects, the function does not provide p-values for the tests. To obtain the p-values, extra calculations are needed. These were illustrated in, e.g., Panel R16.21.

By applying the function `logLik()` (see Table 15.3), we extract the value of the log-likelihood (Sect. 13.5.3) for model **M18.6**. The obtained value is equivalent to the one reported for the `lme()` function in, e.g., Panel R18.12.

The estimated values of the variances of the school- and class-level random effects are obtained with the help of the function `VarCorr()` (see Table 15.3). Note that, to make the display legible, we apply the function `unlist()` to the result of the application of the function `VarCorr()` to the object `fm18.6mer`. The obtained values are corresponding to the SDs reported for the `lme()` function in Table 18.2. The estimate of the scale parameter σ is obtained by applying the `sigma()` extractor-function to the model-fit object (see Table 15.3). It correspond to the value shown, e.g., in Table 18.2. Note that, given that model **M18.6** is a conditional-independence LMM with homoscedastic residual errors, σ can be interpreted as the residual SD.

In Panel R18.19, we demonstrate how to create normal Q-Q plots for conditional residuals and predicted random effects. Specifically, in Panel R18.19a, we show the code used to extract the raw class-level residuals (Sect. 13.6.2) and to construct the normal Q-Q plot for them (Sect. 15.4). Note that we do not show the plot itself, as it corresponds to the one shown in Fig. 18.6a. The only difference is that the latter shows the Pearson residuals (Sect. 7.5.1).

R18.17 *SII Project*: Model **M18.6** fitted using the function `lmer()`. Three forms of syntax for nested random effects presented in Table 15.2 are illustrated

```
> library(lme4.0)
> fm18.6mer <-
+    lmer(mathgain ~ ses + minority + poly(mathkind, 3) + ses:minority +
+           (1|schoolid) + (1|schoolid:classid), # Syntax #1 (general)
+         data = SIIdata, REML = FALSE)
> summ  <- summary(fm18.6mer)
> print(summ, corr = FALSE)
  Linear mixed model fit by maximum likelihood
  Formula: mathgain ~ ses + minority + poly(mathkind, 3) + ...
     Data: SIIdata
     AIC   BIC logLik deviance REMLdev
   11348 11399  -5664    11328   11291
  Random effects:
   Groups           Name        Variance Std.Dev.
   schoolid:classid (Intercept)  86.0     9.27
   schoolid         (Intercept)  66.8     8.17
   Residual                     690.9    26.28
  Number of obs: 1190, groups: schoolid:classid, 312; schoolid, 107

  Fixed effects:
                       Estimate Std. Error t value
   (Intercept)            61.35       2.07   29.65
   ses                     8.84       1.94    4.55
   minorityMnrt=Yes       -6.86       2.28   -3.01
   poly(mathkind, 3)1   -660.46      30.99  -21.31
   poly(mathkind, 3)2    124.46      28.24    4.41
   poly(mathkind, 3)3   -178.34      27.97   -6.37
   ses:minorityMnrt=Yes   -5.82       2.44   -2.39
> update(fm18.6mer,
+        mathgain ~ ses + minority + poly(mathkind, 3) + ses:minority +
+          (1|schoolid) + (1|classid))                 # Syntax #2
  ...    [snip]
> update(fm18.6mer,
+        mathgain ~ ses + minority + poly(mathkind, 3) + ses:minority +
+          (1|schoolid/classid))                       # Syntax #3
  ...    [snip]
```

Panel R18.19b presents the code used to extract the predicted random effects (Sect. 13.6.1) and to construct their normal Q-Q plots for schools and classes. First, we use the function `ranef()` (see Table 15.3) to extract the predictors (EBLUPs) and save the result in the object `rnf6qn`. Application of the function `plot()` to the object results in a display of two normal Q-Q plots, one for the school-specific predictors and one for class-specific predictors. To allow processing the plots, we store the result in the object `rnf6qn`. The object is a list with two components named `schoolid` and `classid` which contain the Q-Q plots for the respective factors. We then select each of the components in turn and, with the help of the function

R18.18 *SII Project*: Extracting information about the estimated fixed- and random-effects structure of model **M18.6** from the *mer*-class model-fit object. Object `fm18.6mer` was created in Panel R18.17

```
> anova(fm18.6mer)                # Approximate F-test statistics
  Analysis of Variance Table
                     Df Sum Sq Mean Sq F value
  ses                 1    481     481    0.70
  minority            1      8       8    0.01
  poly(mathkind, 3)   3 368141  122714  177.62
  ses:minority        1   3936    3936    5.70
> logLik(fm18.6mer)                           # ML value
  'log Lik.' -5663.9 (df=10)
> unlist(VarCorr(fm18.6mer))                  # d̂1 and d̂2
  schoolid:classid          schoolid
            85.981            66.768
> sigma(fm18.6mer)                            # σ̂
  [1] 26.285
```

R18.19 *SII Project*: Plots of the raw class-level conditional residuals and predicted random effects for model **M18.6**. The model-fit object `fm18.6mer` was created in Panel R18.17

(a) *Normal Q-Q plot of the raw class-level residuals*

```
> rsd6 <- resid(fm18.6mer)
> qqnorm(rsd6)
```

(b) *Normal Q-Q plot of predicted random effects*

```
> rnf6mer <- ranef(fm18.6mer)          # Random effects
> rnf6qn  <- plot(rnf6mer, grid = TRUE)# Q-Q plot for random effects
> update(rnf6qn[["schoolid:classid"]], # For classid (see Fig. 18.9a)
+        ylab = c("Random effects for classes"))
> update(rnf6qn[["schoolid"]],         # For schoolid (see Fig. 18.9b)
+        ylab = c("Random effects for schools"))
```

`update()`, we modify the label of the vertical axis. The resulting plots correspond to those shown in Fig. 18.9, but with switched axes and without an indication of the outlying values of the predicted random effects.

Table 18.3 *SII Project*: Summary of linear mixed-effects models defined and fitted in Chap. 18

Model[a]	Section	Syntax	R-object	Mean	Comment
M18.1[b]	18.2	R18.1	`fm18.1`	(18.1)	West et al. (2007)
M18.2	18.3	R18.7	`fm18.2`	(18.4)	`sex:housepov` added to **M18.1**
M18.3	18.4	R18.8	`fm18.3`	(18.5)	`sex, housepov` removed
M18.4	18.5	R18.9	`fm18.4`	(18.6)	`poly(mathkind, 3)`
M18.5	18.6	R18.11	`fm18.5`	(18.7)	`bs(mathkind, df = 4)`
M18.6	18.7	R18.12	`fm18.6`	(18.8)	`minority:ses` added to **M18.4**
	18.8	R18.17	`fm18.6mer`		

[a]All models were fitted using the ML estimation

[b]The mean structure for model **M18.1** is defined by the following formula:
`mathgain ~ ses + minority + mathkind + sex + housepov`

18.9 Chapter Summary

In this chapter, we analyzed the SII data by applying LMMs. The dataset is an example of data with a three-level hierarchy, with pupils grouped in classes, and classes grouped in schools. To deal with the hierarchy, we used models which included random intercepts for schools and classes. Marginally, the models allowed for correlation between the improvements in math scores for pupils from the same school and/or from the same class. In particular, the models implied that the within-class correlation coefficient was larger than the within-school one (Sect. 18.2).

Table 18.3 provides a summary information about all models considered in this chapter. We focused on the modeling of the mean structure. As a starting point, we assumed, in Sect. 18.2, the model proposed by West et al. (2007) (Model 4.4, Chap. 4, p. 159). In Sects. 18.3 and 18.4, we investigated the need for the inclusion of the school-level variables HOUSEPOV and SEX and their interaction in the mean structure. It turned out that the corresponding terms could be left out from the model. In Sects. 18.5 and 18.6, we modified the functional form of the effect of the variable MATHKIND. In particular, in Sect. 18.5, we considered the use of a third-degree polynomial, while in Sect. 18.6, we used a spline function. We found that the latter provided only a minimally better fit to the data. Finally, in Sect. 18.7, we considered an inclusion of an interaction between MINORITY and SES in the mean structure of the model. It turned out that the effect of interaction was statistically significant.

To fit the models to the data, we primarily used the function `lme()` from the package **nlme**. This function is a natural choice for fitting LMMs with nested random effects. The presented sequence of models allowed us to illustrate several additional features of the function `lme()`, like the specification of multilevel LMMs, the use of spline functions for fixed effects (Sects. 18.6 and 18.7), and the use of marginal residuals to assess linear relationship with respect to covariates.

All the models could also be fitted by using the function `lmer()` from the **lme4.0** package. For comparison purposes, in Sect. 18.8, we briefly showed how to fit the final model by applying the function. An important point in this respect was the proper specification of the nesting of the random effects in the syntax.

The results of fitting of all models were summarized in Tables 18.1 and 18.2. The final model, model **M18.6**, suggests that, on average, the gain in math score is smaller for minority pupils. The mean gain increases linearly with pupil's socioeconomic status, but the increase is much slower for minority pupils. On the other hand, the mean gain decreases as a third-degree polynomial of pupil's math score in the spring of the kindergarten year. According to the estimated random structure of model **M18.6**, improvements in math scores for pupils from the same school are correlated with the correlation coefficient of 0.08. For pupils from the same class, the correlation coefficient is equal to 0.18.

Chapter 19
FCAT Study: Modeling Attainment-Target Scores

19.1 Introduction

The FCAT study was described in Sect. 2.5. An exploratory analysis of the data from the study was presented in Sect. 3.5. In this chapter, we use LMMs with crossed random effects to analyze the data. In particular, we consider the models proposed by Tibaldi et al. (2007).

First, we analyze the total target scores. In Sect. 19.2, we consider, for illustration purposes, a simple two-way ANOVA model with crossed fixed effects. The fixed-effects estimates for this model are used as a reference for the random-effects estimates obtained for LMMs. Next, we consider an LMM with crossed random effects for targets and pupils and with independent, homoscedastic residual errors. In Sect. 19.3, we fit the model using the function `lmer()` from the package **lme4.0**, while in Sect. 19.4 we fit it using the function `lme()` from the package **nlme**. In this way, we can illustrate the merits of both functions when applying them for fitting LMMs with crossed random effects.

We then conduct an alternative analysis of the FCAT data by considering the average target score instead of a total score as the dependent variable. In Sect. 19.5, we analyze the score using an LMM with crossed random effects and independent, heteroscedastic residual errors. To fit the model, we use the function `lme()`.

We conclude this chapter with Sect. 19.6, where we briefly summarize our findings.

19.2 A Fixed-Effects Linear Model Fitted Using the Function `lm()`

We begin the analysis of the data with a simple two-way ANOVA model applied to the total target scores. In particular, we consider effects of pupils and targets as *fixed*.

A. Gałecki and T. Burzykowski, *Linear Mixed-Effects Models Using R: A Step-by-Step Approach*, Springer Texts in Statistics, DOI 10.1007/978-1-4614-3900-4_19,

19.2.1 Model Specification

We consider model **M19.1**, defined as follows:

$$\text{SCORE}_{st} = \mu + \beta_{1,s} + \beta_{2,t} + \varepsilon_{st}, \tag{19.1}$$

where t indexes targets ($t = 1,2,\ldots,n$) and s indexes pupils ($s = 1,2,\ldots,m$), with $n = 9$ and $m = 539$. In (19.1), μ is the overall mean of the total attainment-target score, $\beta_{1,s}$ is the coefficient corresponding to the fixed effect for the s-th pupil, and $\beta_{1,t}$ is the coefficient corresponding to the fixed effect for the t-th target. The residual random errors ε_{st} are assumed to be independent and normally distributed with mean zero and variance σ^2. Thus, (19.1) can be seen as an example of an equation for an LM with homogeneous variance, specified at the level of the observation unit, as given by (4.1).

We should stress that model **M19.1** is considered only for reference purposes and should not be treated as a reasonable model for the FCAT data. In particular, it involves a large number of coefficients for the fixed effects, because it includes a separate effect for each of the pupils and targets. Moreover, it ignores the difference in the scales for the different scores and the correlation between responses for different attainment targets obtained for the same pupil.

19.2.2 R *Syntax and Results*

In Panel R19.1, we show the R syntax to fit model **M19.1**. First, we load the `fcat` data frame from the package **nlmeU**. We then change the default values of the `contrasts` option to `contr.sum` and `contr.poly` for unordered and ordered factors, respectively (Sect. 5.3.2). The use of the `contr.sum` contrasts implies that the estimated fixed effects can be interpreted as *fixed* deviations from the overall mean.

Subsequently, we fit model **M19.1** to the data using the function `lm()`, which is included in the set of basic functions in R (Sect. 5.4). The model formula `scorec ~id + target` implies that the levels of factors `id` and `target` define the fixed effects in the mean structure. Note that, by default, an intercept is included in the formula (Sect. 5.2.1). Moreover, by default, the OLS estimation is used (Sect. 4.4.1).

The results of fitting of the model are stored in the object `fm19.1`. An abbreviated display of the results, presented in Panel R19.1, shows the estimated coefficients for the fixed effects defined by the `id` and `target` factors. The estimated mean value of responses over all pupils and attainment targets is equal to 3.9033. For the pupil with, e.g., `id=1`, the mean response over all targets is larger by 0.9856 from the mean value. On the other hand, for the fifth attainment target (structuring of a comic-strip text, see Table 2.1), the mean response for all pupils was larger by 2.0707 than the overall mean value.

R19.1 *FCAT Study*: Model **M19.1** with crossed fixed-effects of targets and pupils fitted using the function lm()

```
> data(fcat, package = "nlmeU")
> opts <- options()                        # Global options saved
> options(contrasts =                      # Default contrasts changed
+           c("contr.sum", "contr.poly"))
> options("contrasts")                     # Changes verified
  $contrasts
  [1] "contr.sum"  "contr.poly"
> (fm19.1 <-                               # M19.1: (19.1)
+    lm(scorec ~ id + target, data = fcat))
  Call:
  lm(formula = scorec ~ id + target, data = fcat)

  Coefficients:
  (Intercept)          id1          id2          id3          id4
       3.9033       0.9856       0.3189       0.9856       0.9856
  ...    [snip]
        id535        id536        id537        id538      target1
      -2.6811       0.0967      -1.5700      -1.1255      -1.3300
      target2      target3      target4      target5      target6
       0.1932       1.3082      -1.1649       2.0707      -0.2966
      target7      target8
       1.1282      -1.5026

> options(opts)                            # Global options restored
```

We will use the estimated fixed-effects coefficients for reference purposes. Thus, in Panel R19.2, we extract the estimates from the model-fit object fm19.1 with the help of the function coef() (see Table 5.5) and store them in the vector fxd. Note that the order of elements of the vector corresponds to the printout of the estimates presented in Panel R19.1. To extract the estimated coefficients corresponding to the factor id, we create a logical vector idx. The vector identifies those elements of fxd which names begin with the string "id". To identify the elements, we apply the function substr() to the character vector which contains the names of the rows of fxd. In particular, the function extracts a substring of length 2, starting at the first character of each name. We then compare the extracted substrings with the character string "id".

By using the vector idx, we select the elements which correspond to the estimated coefficients for the levels of the factor id from the vector fxd. We store the elements in the vector fxi and print out their names to check the correctness of the selection. Finally, we add to the vector the negative of the sum of all the elements. Note that, according to the definition of the contr.sum contrast, this is the estimated value of the coefficient for the last level of the factor id. Finally, we store the updated vector in the object fxd.id.

R19.2 *FCAT Study*: Extracting the estimated fixed-effects coefficients for model **M19.1**. The model-fit object `fm19.1` was created in Panel R19.1

(a) *Extracting the estimates for the factor* `id`

```
> fxd <- coef(fm19.1)
> idx <- substr(names(fxd), 1, 2) == "id"          # Logical vector
> names(fxi <- fxd[idx])
> (fxd.id <- c(fxi, "id539" = -sum(fxi)))                # β̂_{2,s}
       id1        id2        id3        id4        id5        id6
  0.985570   0.318903   0.985570   0.985570   0.874459   0.541126
...       [snip]
     id535      id536      id537      id538      id539
 -2.681097   0.096681  -1.569986  -1.125541  -0.681097
```

(b) *Extracting the estimates for the factor* `target`

```
> idx <- substr(names(fxd), 1, 6) == "target"
> names(fxi <- fxd[idx])
 [1] "target1" "target2" "target3" "target4" "target5" "target6"
 [7] "target7" "target8"
> (fxd.trgt <- c(fxi, "target9" = -sum(fxi)))       # β̂_{1,t}
  target1  target2  target3  target4  target5  target6  target7
 -1.33004  0.19316  1.30818 -1.16491  2.07071 -0.29664  1.12822
  target8  target9
 -1.50258 -0.40610
```

In Panel R19.2b, we extract the estimates of the coefficients corresponding to the levels of the factor `target`. The R code is very similar to the one used in Panel R19.2a. The estimates are stored in the vector `fxd.trgt`, which will be used, together with the vector `fxd.id`, for comparison purposes with the results of the analysis conducted in the next section.

19.3 A Linear Mixed-Effects Model with Crossed Random Effects Fitted Using the Function `lmer()`

Pupils participating in the FCAT study can be considered a random representation of all pupils in Flanders. Similarly, the evaluated attainment targets might be considered a random sample of many possible targets that could be measured. Hence, we could consider the effects of pupils and targets as *random* (see, e.g., Van den Noortgate et al. (2003) or Tibaldi et al. (2007)). Consequently, we could analyze the data using an LMM. Note that each pupil provided a response for all targets.

Thus, we can consider an LMM in which the random effects of pupils and attainment targets are crossed (Sect. 15.2). In this section, we fit such a model to the FCAT data using the function `lmer()` from the package **lme4.0**.

19.3.1 Model Specification

Following Tibaldi et al. (2007), we consider model **M19.2**, defined as follows:

$$\text{SCORE}_{st} = \mu + b_{1,s} + b_{2,t} + \varepsilon_{st}, \tag{19.2}$$

where $b_{1,s} \sim \mathcal{N}(0, d_S)$ is the random effect corresponding to the pupil s ($s = 1,2,\ldots,n$), $b_{2,t} \sim \mathcal{N}(0, d_T)$ is the random effect corresponding to the target t ($t = 1,2,\ldots,m$) and independent of $b_{1,s}$, and $\varepsilon_{st} \sim \mathcal{N}(0, \sigma^2)$ is the residual (measurement) error independent of both $b_{1,s}$ and $b_{2,t}$. Note that (19.2) has the form corresponding to (15.4).

19.3.1.1 Marginal Interpretation

Model **M19.2** implies that the marginal expected value of SCORE_{ts} is equal to μ, defined in (19.2). The marginal variances and covariances are as follows:

$$\text{Var}(\text{SCORE}_{st}) = d_T + d_S + \sigma^2,$$

$$\text{Cov}(\text{SCORE}_{st}, \text{SCORE}_{s't}) = d_T, \tag{19.3}$$

$$\text{Cov}(\text{SCORE}_{st}, \text{SCORE}_{st'}) = d_S, \tag{19.4}$$

$$\text{Cov}(\text{SCORE}_{st}, \text{SCORE}_{s't'}) = 0,$$

where $t \neq t'$ and $s \neq s'$. Equation (19.3) implies that responses for different targets for the same pupil are correlated with the correlation coefficient equal to $d_T/(d_T + d_S + \sigma^2)$. On the other hand, (19.4) implies that responses of different pupils for the same target are also correlated with the correlation coefficient equal to $d_S/(d_T + d_S + \sigma^2)$.

19.3.2 R Syntax and Results

In Panel R19.3, we fit model **M19.2** to the FCAT data using the function `lmer()` from the package **lmer4** (Chap. 15). The model formula used in the function call specifies two crossed random effects corresponding to the levels of factors `target` and `id` (Sect. 15.3.1). As was mentioned in Sect. 15.1, the function `lmer()` uses the sparse-matrix representations, which facilitates the large-matrix manipulations

R19.3 *FCAT Study*: Model **M19.2** with crossed random effects of targets and pupils fitted using the function `lmer()`

```
> library(lme4.0)
> system.time(
+    fm19.2mer <-
+       lmer(scorec ~ (1|target) + (1|id),
+            data = fcat))
   user  system elapsed
   0.13    0.00    0.13
> fm19.2mer                                        # M19.2: (19.2)
  Linear mixed model fit by REML
  Formula: scorec ~ (1 | target) + (1 | id)
     Data: fcat
     AIC   BIC logLik deviance REMLdev
   16204 16230  -8098    16196   16196
  Random effects:
   Groups   Name        Variance Std.Dev.
   id       (Intercept) 0.686    0.828
   target   (Intercept) 1.616    1.271
   Residual             1.347    1.161
  Number of obs: 4851, groups: id, 539; target, 9

  Fixed effects:
              Estimate Std. Error t value
  (Intercept)    3.903      0.425    9.18
```

necessary when fitting LMMs. For purposes of comparison with the function `lme()`, in Panel R19.3, we execute the `lmer()`-function call while applying the function `system.time()`. The latter computes the execution time of the former. From the output, it can be seen that the execution took a fraction of a second.

The estimated values of d_S, d_T, and σ^2, shown in Panel R19.3, are equal to, respectively, 0.686, 1.616, and 1.347. The estimate of the overall mean value is equal to 3.903, as for the fixed-effects model **M19.1** (see Panel R19.1). This is due to the balanced nature of the dataset.

Panel R19.4 presents the R code to extract basic information from the model-fit object `fm19.2mer`. First, we create the object `summ.merFit`, which is of class *summary.mer* and represents a summary of the *mer*-class model-fit object. Information about the slots, available in a *summary.mer*-class object, can be obtained, after loading the package **lme4.0**, by issuing the command `help("mer-class")` (see also Sect. 15.4).

By applying the `isREML()` extractor function (see Table 15.3) to the model-fit object, we obtain the information whether REML estimation (Sect. 13.5.2) was used. Note that the call to the `lmer()` function, applied in Panel R19.3, did not

R19.4 *FCAT Study*: Extracting information about the fitted form of model **M19.2**. The model-fit object `fm19.2mer` was created in Panel R19.3

```
> summ.merFit <- summary(fm19.2mer)          # Summary of the model-fit
> isREML(fm19.2mer)                          # REML used?
  [1] TRUE
> (cl <- getCall(summ.merFit))               # Function call
  lmer(formula = scorec ~ (1 | target) + (1 | id), data = fcat)
> cl$data                                    # The name of data frame
  fcat
> formula(fm19.2mer)                         # Formula
  scorec ~ (1 | target) + (1 | id)
> fixef(fm19.2mer)                           # β̂
  (Intercept)
       3.9033
> coef(summ.merFit)                          # β̂, se(β̂), t-test
              Estimate Std. Error t value
  (Intercept)   3.9033    0.42536  9.1764
> (VCorr <- unlist(VarCorr(fm19.2mer)))      # d̂_S, d̂_T
       id  target
  0.68637 1.61578
> sigma(fm19.2mer)                           # σ̂
  [1] 1.1608
```

change the default value of the REML argument, which is REML=TRUE (Sect. 15.3). Hence, the REML estimation was used, as confirmed in Panel R19.4.

The call to the function which led to the creation of the *mer*-class model-fit object is obtained by applying the `getCall()` extractor function (see Table 15.3) and stored in the object `cl`. Obviously, the result shown in Panel R19.4 is the same as the call used in Panel R19.3 to create the model-fit object `fm19.2mer`. By referring to the `data` component of the `cl` object (Table 15.3), we obtain the value of the `data` argument used in the call, i.e., the name of the data frame used to fit the model. Similarly, by referring to the `formula` component, we could extract the value of the formula argument. Alternatively, as shown in Panel R19.4, we can extract the model formula by applying the generic function `formula()` (Table 15.3) directly to the model-fit object.

With the help of the function `fixef()` (Table 15.3), we extract the estimates of the fixed-effects coefficients from the model-fit object `fm19.2mer`. For model **M19.2**, it is only the value of the overall intercept. To obtain the matrix of the estimates together with their SEs and values of the t-test statistics, we use the `coef(summ.merFit)` command.

R19.5 *FCAT Study*: Normal Q-Q plots of the predicted random effects (EBLUPs) and the corresponding random coefficients for model **M19.2**. The model-fit object `fm19.2mer` was created in Panel R19.3

```
> rnf <- ranef(fm19.2mer)                 # ranef.mer-class object
> names(rnf)
  [1] "id"      "target"
> length(plx <- plot(rnf))                # Two Q-Q plots saved.
  [1] 2
> plx[1]                                  # Fig. 19.1a
  $id
> plx[2]                                  # Fig. 19.1b
  $target
> plot(coef(fm19.2mer))                   # Fig. 19.2
```

Information about the estimated variance-covariance matrices of the random effects is extracted from the model-fit object with the help of the function `VarCorr()` (Table 15.3). Note that the object, resulting from the application of the function, is a list with components named after the factors used in the model formula to define the random-effects structure. Thus, in the case of the model-fit object `fm19.2mer`, it is a list with two components named `id` and `target`. To compactly display the information about the variance-covariance matrices of the random effects corresponding to the levels of the two factors, we `unlist()` the object resulting from the application of the `VarCorr()` function. We then obtain the estimates of the variances of the random intercepts for `id` and `target`. Note that the two random intercepts are uncorrelated, as implied by the model formula used in the call to the function `lmer()` in Panel R19.3 (Sect. 15.3.1).

Finally, the estimate of the scale parameter, σ, is obtained by using the `sigma()` extractor function (Table 15.3). Note that model **M19.2** is a conditional-independence LMM with homoscedastic residual errors; hence, σ can be also interpreted as the residual SD.

According to the results presented in Panel R19.4, the estimated values of the random-effects variances d_S and d_T are equal to 0.6864 and 1.6158, respectively. The estimate of the residual variance σ^2 is equal to $1.1608^2 = 1.3475$. Thus, the total variability of the total target scores equals $0.6864 + 1.6158 + 1.3475 = 3.6497$. The between-pupil and between-target variability constitutes, respectively, 18.8% and 44.3% of the total variability. Note that these percentages give also, in accordance with (19.3) and (19.4), the estimated values of the correlation coefficients between responses for different targets for the same pupil and between responses of different pupils for the same target, respectively.

In Panel R19.5, we present the R code to extract the predicted random effects (EBLUPs; see Sect. 13.6.1) for model **M19.2**. To extract the estimates, we apply the function `ranef()` (Table 15.3) to the model-fit object `fm19.2mer` and store

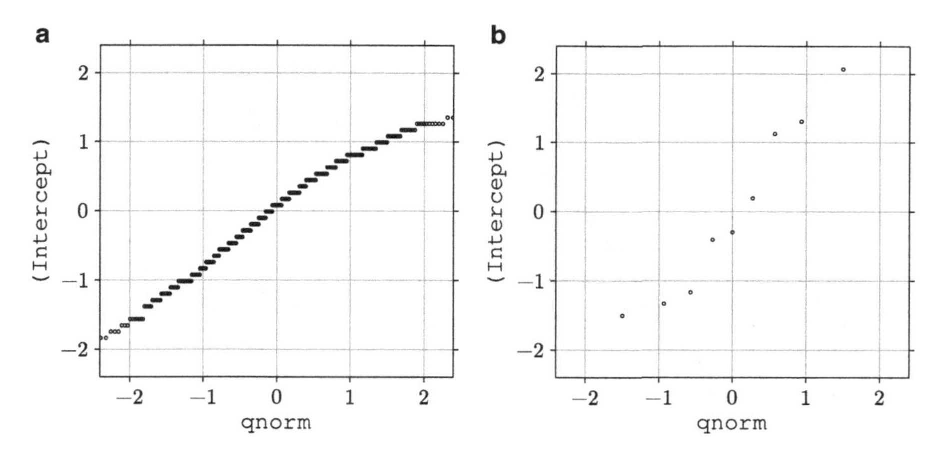

Fig. 19.1 *FCAT Study*: Normal Q-Q plots of the predicted random effects (intercepts) for (**a**) `id` (**b**) `target` in model **M19.2**

the result in the object `rnf`. The object is a list with two named components, `id` and `target`, which are data frames that contain the predicted random effects. The default plot for the object produces two graphs with normal Q-Q plots of the effects, one for each factor. This is because model **M19.2** includes a single random effect for each of the factors (Sect. 15.4). In Panel R19.5, we store the result of applying of the `plot()` function to the object `fm19.2mer` in the object `plx`. The latter is a list with two components named `id` and `target`. By displaying the components, we obtain separate normal Q-Q plots for the estimated random intercepts associated with two factors. The plots are presented in Fig. 19.1.

Figure 19.2 presents two related Q-Q plots. They are constructed by plotting the predicted random coefficients for model **M19.2**, obtained by applying the `coef()` function to the object `fm19.2mer`. The coefficients result from summing the fixed effects and the "coupled" random effects (see Sects. 14.6 and 15.4). Note that, because model **M19.2** includes intercept as the only fixed effect, the Q-Q plots shown in Fig. 19.2 have essentially the same shape as the ones presented in Fig. 19.1.

As was mentioned in Sect. 13.6.1, Q-Q plots and histograms of predicted random effects are of limited value when checking the normality of the effects is of interest. Thus, the Q-Q plots shown in Figs. 19.1 and 19.2 should be treated with caution. Their shapes do indicate a possible deviation from normality, however. This might be due to, e.g., the fact that the responses for the attainment targets were, strictly speaking, not continuous.

In Panel R19.6, we present the R code to create dotplots of the predicted random effects for model **M19.2**. Toward this end, we use the function `dotplot()` (Sect. 15.4). To obtain separate plots for the factors `id` and `target`, we first store the result of applying the function to the object `rnf`, which contains the EBLUPs, in

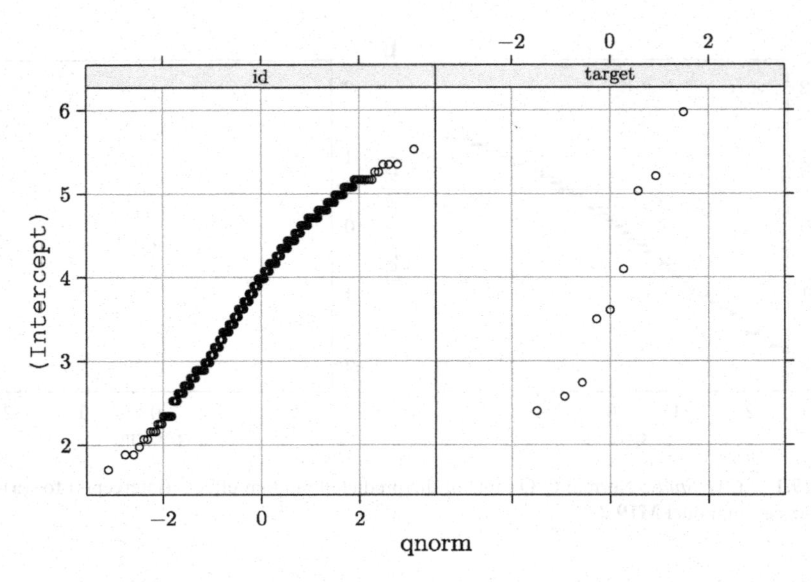

Fig. 19.2 *FCAT Study*: Normal Q-Q plots of the predicted random coefficients for model **M19.2**

R19.6 *FCAT Study*: Dotplots of the predicted random effects (EBLUPs) for model **M19.2**. Objects `fm19.2mer` and `rnf` were created in Panels R19.3 and R19.5, respectively

(a) *Without confidence intervals*

```
> dpx <- dotplot(rnf)
> # dpx[1]                                  # Dotplot for id (not shown)
> dpx[2]                                    # Fig. 19.3a
  $target
```

(b) *With confidence intervals*

```
> rnf.pVar <- ranef(fm19.2mer, postVar = TRUE) # ranef.mer-class object
> dpx.pVar <- dotplot(rnf.pVar)
> # dpx.pVar[1]                             # Dotplot for id (not shown)
> dpx.pVar[2]                               # Fig. 19.3b
  $target
```

the object dpx. The latter is a list with two components named `id` and `target`. By displaying the components, we can obtain separate dotplots for the random effects corresponding to the two factors. In Fig. 19.3a, we show the plot for `target`. Note that the levels of the factor, shown on the y-axis, are ordered according to increasing values of the predicted random intercepts. The default plot for `id` is illegible and

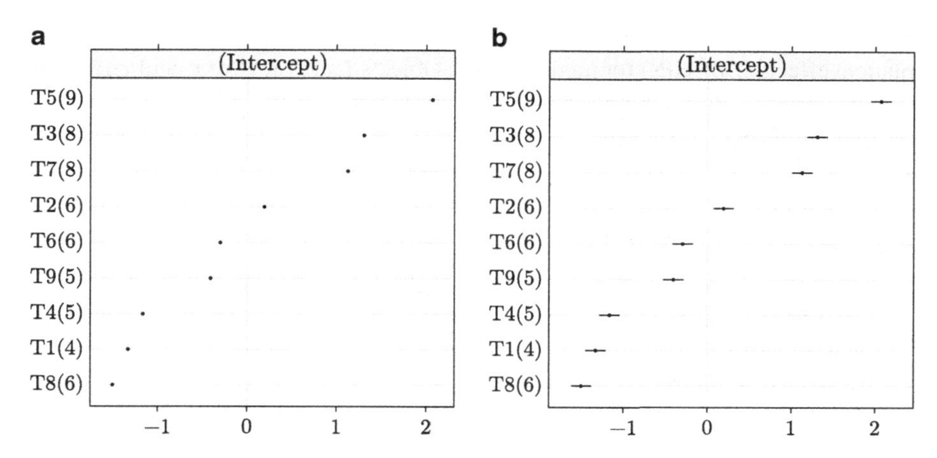

Fig. 19.3 *FCAT Study*: Dotplots of the predicted random effects for the factor `target` for model **M19.2**. Effects (**a**) without confidence intervals (**b**) with confidence intervals

would need to be preprocessed before displaying. To save space, we do not present the necessary code, nor the plot itself.

The dotplot, shown in Fig. 19.3a, can be enhanced by adding CIs to the point predictors of the random effects. In Panel R19.6b, we present the necessary R code. In particular, we extract the random effects from the `fm19.2mer` model-fit object using the function `ranef()` with the argument `postVar=TRUE`. In this way, the resulting object `rnf.pVar` is extended by incorporating the conditional variance-covariance matrices, also called the "posterior variances", of the random effects. Note that the default value of the argument is `FALSE`. More information about the arguments of the function `ranef()` can be obtained by issuing the command `?ranef`.

We then apply the function `dotplot()` to the object `rnf.pVar`. To obtain separate plots for the factors `id` and `target`, we first store the result of the application of the function in the object `dpx.pVar`, which is a list with two components bearing the names of the factors. By displaying the components, we obtain dotplots with 95% CIs for the random effects corresponding to the two factors. In Fig. 19.3b, we show the plot for `target`. The plot can be seen as an example of a "caterpillar plot." It can be used to, e.g., detect observations with extreme values of the predicted random effects. In our case, a clustering of targets, reflecting dependence on the number of items, can be clearly observed. For instance, target 5, which includes nine items, has got the largest random effect, and it is followed by targets 3 and 7, each based on eight items.

In Panel R19.7, we present the R code to reproduce the results, which illustrate the "shrinkage" phenomenon (see, e.g., (13.51) and Sect. 7.5 in Verbeke and Molenberghs (2000)) for the estimated random effects for model **M19.2**.

R19.7 *FCAT Study*: Various ways of illustrating "shrinkage" of the predicted random effects (EBLUPs) for model **M19.2**. Objects `fxd.id`, `VCorr`, and `rnf` were created in Panels R19.2, R19.4, and R19.5, respectively

(a) $var(\widehat{\mathbf{b}}_i) \leq \widehat{Var}(\widehat{\mathbf{b}}_i)$ *(see* (13.51))

```
> (eVCorr <-  sapply(rnf, var))       # var(b̂_{1,s}), var(b̂_{2,t}),
       id  target
  0.56346 1.61458
> VCorr                               # Var^(b_{1,s}) ≡ d̂_S, Var^(b_{2,t}) ≡ d̂_T
       id  target
  0.68637 1.61578
> all(eVCorr < VCorr)
  [1] TRUE
```

(b) $abs(\widehat{\beta}_{1,s}) - abs(b_{1,s}) \geq 0$ *for intercepts associated with the factor* `id`

```
> rnf.id     <- rnf$id                # Data frame with b̂_{1,s}
> arnf.id    <- abs(rnf.id)           # abs(b̂_{1,s})
> afxd.id    <- abs(fxd.id)           # abs(β̂_{1,s})
> range(afxd.id - arnf.id)
  [1] 0.002584 0.480100
```

(c) *Same as in the part (b) above, for* `target`

```
> rnf.trgt  <- rnf$target
> arnf.trgt <- abs(rnf.trgt)
> afxd.trgt <- abs(fxd.trgt)
> range(afxd.trgt - arnf.trgt)
  [1] 0.00021946 0.00235273
```

(d) *Graphical illustration of the relation in part (b) above for the factor* `id`

```
> names(dt  <- data.frame(afxd.id, arnf.id))
  [1] "afxd.id"         "X.Intercept."
> names(dt)[2] <- "arnf.id"
> myPanel <- function(x, y, ... ){
+   panel.grid(h = -1, v = -1)
+   panel.lines(c(0, 3), c(0, 3), lty = 2)
+   panel.xyplot(x, y, ... )
+ }
> xyplot(arnf.id ~ afxd.id,           # Fig. 19.4a
+         data = dt, panel = myPanel)
```

More specifically, in Panel R19.7a, we compute the sample variances of the predicted random effects for model **M19.2**. Toward this end, with the help of the function `sapply()`, we apply the function `var()`, which calculates sample variance, to each element of the list `rnf`. Note that, as was mentioned in the description of the code in Panel R19.5, `rnf` is a list with two named components which are data frames that contain the EBLUPs for factors `id` and `target`. We store the sample variances for the two factors in the vector `eVCorr`. We then compare the variance to the estimated variances d_S and d_T of the (assumed) normal distribution of the random effects, which were computed earlier and stored in the vector `VCorr`. The comparison shows that the sample variances of the random-effects predictors are smaller than the estimated variances of the random effects. This illustrates the shrinkage, as defined by, e.g., (13.51) and (7.7) in Verbeke and Molenberghs (2000).

In Panel R19.7b, we store the data frame, which is contained in the component `id` of the object `rnf` and which contains EBLUPs of the random effects for the factor `id`, in the object `rnf.id`. We then compute the absolute values of the EBLUPs and store them in the object `arnf.id`. Similarly, we compute the absolute values of the estimated fixed-effects coefficients for the factor `id` for model **M19.1**. We then calculate the range of the differences between the absolute values of the two sets of estimates. It can be seen that the range includes only positive real values. Thus, the absolute values of the predicted random effects for the factor `id` in LMM **M19.2** are smaller than the absolute values of the estimates of the corresponding fixed-effects coefficients in LM **M19.1**. Hence, the former can be seen as being "shrunk" relative to the latter. Panel R19.7c presents similar code and results for the factor `target`.

Finally, the R code, shown in Panel R19.7d, is used to construct a graphical illustration of the results presented in Panel R19.7b. First, we create the data frame `dt` by merging the data frames `afxd.id` and `arnf.id`. The resulting data frame contains two variables, `afxd.id` and `X.intercept`, corresponding to the fixed-effects estimates and random-effects predictors for the factor `id` for models **M19.1** and **M19.2**, respectively. For the sake of consistency, we change the name of the second variable to `arnf.id` with the help of the `names()` function. Then, we define the function `myPanel()`, which adds a grid aligned with the axis labels (`panel.grid(h=-1,v=-1)`) and a dashed line (`lty=2`) connecting the points with the coordinates $(0,0)$ and $(3,3)$. We use the function in the `panel` argument of the `xyplot()` function, which produces a scatterplot of variables `afxd.id` and `arnf.id` from the data frame `dt`. The plot is shown in Fig. 19.4a. It suggests that the values of the `arnf.id` variable, i.e., the absolute values of the random-effects EBLUPs for the factor `id` for model **M19.2**, are smaller than the absolute values for the corresponding fixed-effects estimates for model **M19.1**. Again, this can be seen as the result of "shrinkage."

Figure 19.4b shows a scatterplot similar to the one shown in Fig. 19.4a, but for the factor `target`. The plot was constructed using an R code corresponding to the one presented in Panel R19.7d. Thus, we do not show the code. For `target`, there is only a minimal amount of "shrinkage."

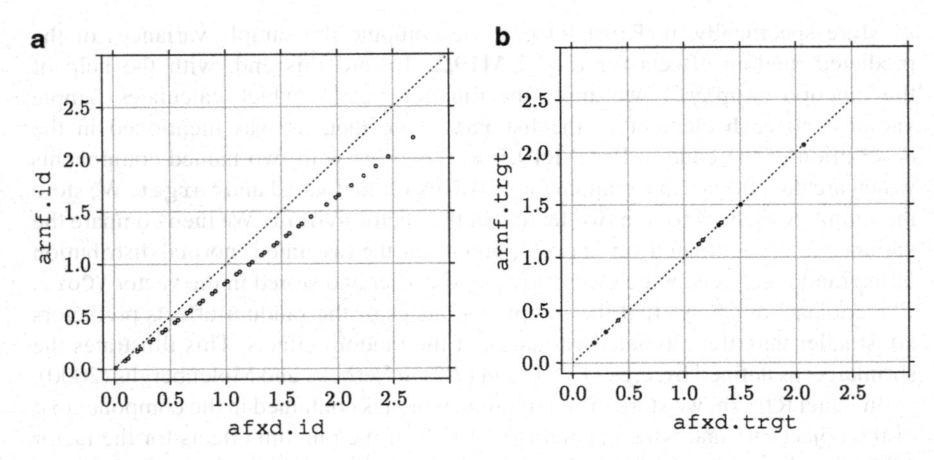

Fig. 19.4 *FCAT Study*: Illustration of "shrinkage". Plots on the absolute scale of the predicted random effects for model **M19.2** *versus* the corresponding fixed-effects estimates for model **M19.1** for factors (**a**) `id` and (**b**) `target`

In the next section, we will fit model **M19.2** to the FCAT data using the function `lme()` from the package **nlme** (see Chap. 14). As was mentioned in Sect. 15.1, definitions of several functions from the packages **nlme** and **lme4.0** differ. Hence, before attempting to use the former package, we should detach the latter to avoid masking functions' and names' conflicts. We may also want to detach the package **Matrix**, which is required by **lme4.0**, but not by **nlme**. In addition, we remove the objects `rnf` and `plx`, as we would like to reuse them in R sessions to follow. To these aims, the following R commands should be used:

```
> detach(package:lme4.0)
> detach(package:Matrix)
> rm(rnf, plx)
```

19.4 A Linear Mixed-Effects Model with Crossed Random Effects Fitted Using the Function `lme()`

In this section, we fit model **M19.2** to the FCAT data using the function `lme()` from the package **nlme** (see Chap. 14).

In Panel R19.8, we present the suitable syntax. Obviously, we first have to attach the package **nlme**.

To fit model **M19.2**, we first need to add to the data two auxiliary variables, `one1` and `one2`, with all values equal to 1. We use the function `within()` to add the variables to the data frame `fcat`, and we store the result in a new data frame,

R19.8 *FCAT Study*: Model **M19.2** fitted using the function lme()

```
> library(nlme)
> fcat1 <- within(fcat, one1 <- one2 <- 1L)
> system.time(
+    fm19.2 <-
+       lme(scorec ~ 1,
+           random = list(one1 = pdIdent(~target - 1),
+                         one2 = pdIdent(~id - 1)),
+           data = fcat1))
    user  system elapsed
   45.47    0.53   46.04
> fm19.2                                        # M19.2: (19.2)
  Linear mixed-effects model fit by REML
    Data: fcat1
    Log-restricted-likelihood: -8097.8
    Fixed: scorec ~ 1
  (Intercept)
       3.9033

  Random effects:
   Formula: ~target - 1 | one1
   Structure: Multiple of an Identity
          targetT1(4) targetT2(6) targetT3(8) targetT4(5)
  StdDev:      1.2711      1.2711      1.2711      1.2711
          targetT5(9) targetT6(6) targetT7(8) targetT8(6)
  StdDev:      1.2711      1.2711      1.2711      1.2711
          targetT9(5)
  StdDev:      1.2711

   Formula: ~id - 1 | one2 %in% one1
   Structure: Multiple of an Identity
              id1     id2     id3     id4     id5     id6     id7
  StdDev: 0.82847 0.82847 0.82847 0.82847 0.82847 0.82847 0.82847
  ...      [snip]
            id533   id534   id535   id536   id537   id538   id539
  StdDev: 0.82847 0.82847 0.82847 0.82847 0.82847 0.82847 0.82847
          Residual
  StdDev:   1.1608

  Number of Observations: 4851
  Number of Groups:
            one1 one2 %in% one1
               1              1
```

`fcat1`. We then apply the function `lme()` with the `fixed` argument (Sect. 14.5) specified using a formula with `scorec` as the dependent variable and an intercept as the only fixed effect. The random-effects structure is defined using the `random` argument in the form of a named list (Table 14.1). In the list, the variables `one1` and `one2` are considered to be the grouping factors. Note that the syntax implies that the levels of the factor `one2` are assumed to be nested within the levels of `one1`. However, as both factors have got only *one* level, the nesting is irrelevant, because all observations are treated, in fact, as coming from a single group. As a result, the random-effects structures, defined for both factors by the `pdIdent()` constructor functions (Sect. 14.2.1), are crossed. In particular, the formulae which are used in the constructors specify that random effects are defined by the levels of factors `id` and `target`. In both cases, the variance-covariance matrix of the random effects is defined by a positive-definite matrix of class *pdIdent* (Sect. 14.2.1), i.e., by a multiple of an identity matrix. Thus, the random effects are independent and normally distributed with a constant variance.

Note that treating all the observations as coming from a single group has an important implication for the size of the design matrices which are used in the numerical calculations. The `lme()`-function syntax, used in Panel R19.8, defines a model which formally corresponds to an LMM with nested random effects, specified by (15.7), with

$$\mathbf{y} \equiv (\text{SCORE}_{11}, \ldots, \text{SCORE}_{19}, \text{SCORE}_{21}, \ldots, \text{SCORE}_{539,9})',$$

the design matrix $\boldsymbol{X} \equiv \mathbf{1}_{4851}$, the random-effects matrices $\boldsymbol{Z}_1 \equiv \boldsymbol{I}_{539} \otimes \mathbf{1}_9$ and $\boldsymbol{Z}_{12} = \mathbf{1}_{539} \otimes \boldsymbol{I}_9$, and the random-effects vectors $\mathbf{b}_1 \equiv (b_{1,1}, b_{1,2}, \ldots, b_{1,539})'$ (for pupils) and $\mathbf{b}_{12} \equiv (b_{12,1}, b_{12,2}, \ldots, b_{12,9})'$ (for targets). It follows that the matrices $\boldsymbol{Z}_1$ and $\boldsymbol{Z}_{12} \equiv \boldsymbol{Z}_2$ are of dimensions 4851×539 and 4851×9, respectively. From the structure of the Z-matrices, it can be seen that, in fact, the random effects are crossed and that the specified model is equivalent to the LMM defined by (15.8).

As was mentioned in Sect. 15.1, the function `lme()` does not use the sparse-matrix representations, which are employed in the function `lmer()`. Hence, `lme()` requires longer computation time than `lmer()`. To show this, in Panel R19.8, we execute the `lme()`-function call while applying the function `system.time()`. From the output, it can be seen that the execution took more than 45 s. This is substantially longer than the execution time obtained for the function `lmer()`, which was equal to a fraction of a second (see Panel R19.3).

Note that, in the `lme()`-function call used in Panel R19.8, the argument `method` was left unspecified. Thus, the REML estimation was used (Sects. 13.5.2 and 14.5).

The results, shown in Panel R19.8, correspond to those displayed in Panel R19.3.

In Panel R19.9, we show the R syntax which allows us to extract some basic information from the model-fit object `fm19.2`

In particular, in Panel R19.9a, we extract the name of the data frame used for fitting of the model. Toward this end, we refer to the `data` component of the `call` component of the model-fit object (Table 14.6). We confirm that the model was fitted to the extended data frame `fcat1`, created in Panel R19.8.

R19.9 *FCAT Study*: Extracting information about the fitted form of model **M19.2**. The model-fit object fm19.2 was created in Panel R19.8

(a) *Basic information*

```
> fm19.2$call$data              # Data name
  fcat1
> logLik(fm19.2)                # REML value
  'log Lik.' -8097.8 (df=4)
> fixef(fm19.2)                 # β̂
  (Intercept)
       3.9033
> fm19.2$dims$N                 # Number of observations
  [1] 4851
```

(b) *Estimated variances of the random effects (intercepts)*

```
> getVarCov(fm19.2)
  Error in getVarCov.lme(fm19.2) :
    Not implemented for multiple levels of nesting
> VarCorr(fm19.2)
                Variance              StdDev
  one1 =        pdIdent(target - 1)
  targetT1(4) 1.61575               1.27112
  ...        [snip]
  targetT9(5) 1.61575               1.27112
  one2 =        pdIdent(id - 1)
  id1           0.68637               0.82847
  ...        [snip]
  id538         0.68637               0.82847
  Residual      1.34745               1.16080
```

By applying the function logLik() (Table 14.5), we print out the value of the log-restricted-likelihood for the fitted model. Note that in the lme()-function call, shown in Panel R19.8, the default value of the method argument (Table 14.4) was used. Thus, model **M19.2** was fitted to the fcat1 data frame using the REML estimation.

With the help of the function fixef() (Table 14.5), we extract the estimated value of the intercept from the model-fit object fm19.2. By referring to the N component of the dims component of the model-fit object, we confirm that the model was fitted to the 4,851 observations from the data frame fcat1.

In Panel R19.9b, we extract information about the estimated variance-covariance structure of the random effects. As can be seen from the printouts presented in the panel, the default tool, which can be used for this purpose, i.e., the function

R19.10 *FCAT Study*: Confidence intervals for the fixed-effects coefficients and the variance-covariance parameters of model **M19.2**. The model-fit object `fm19.2` was created in Panel R19.8

```
> intervals(fm19.2)
  Approximate 95% confidence intervals

   Fixed effects:
                lower   est.  upper
  (Intercept) 3.0691 3.9033 4.7376
  attr(,"label")
  [1] "Fixed effects:"

   Random Effects:
    Level: one1
                      lower   est.  upper
  sd(target - 1) 0.77274 1.2711 2.0909
    Level: one2
                 lower    est.   upper
  sd(id - 1) 0.77044 0.82847 0.89088

   Within-group standard error:
   lower   est.  upper
  1.1366 1.1608 1.1855
```

`getVarCov()` (Table 14.5), fails to produce results. Thus, we use the alternative solution, i.e., the function `VarCorr()`. The obtained results are exactly the same as those shown in Panel R19.3.

Panel R19.10 presents 95% CIs (Sect. 13.7.3) for the fixed-effects coefficients and the variance-covariance parameters. The intervals are obtained by applying the function `intervals()` (Table 14.5) to the model-fit object `fm19.2`. Note that, for a proper performance of the function, it is necessary to fit model **M19.2** using the two auxiliary variables `one1` and `one2`, as was done in Panel R19.8.

Panel R19.11 presents the R code for extracting and plotting the predicted random effects (Sect. 13.6.1) and residuals (Sect. 13.6.2). In particular, in Panel R19.11a, we use the function `ranef()` (Table 14.5) to obtain the random-effects predictors (EBLUPs) and store them in the object `rnf`. Note that the object is a list with two components, named `one1` and `one2`. The components are data frames with one row which provide the predictors of the random effects for the factors `target` and `id`, respectively (see the `lme()`-function call in Panel R19.8). Next, we attempt to use the `plot()` function to obtain normal Q-Q plots of the estimates. Unfortunately, the use of this method results in error. To overcome this issue, in Panel R19.11b, we present an alternative method to construct the plots. Toward this end, we first use the function `lapply()` to transpose the two components of the object `rnf` and store them in the object `rnft`. Next, we use `lapply()` to apply the function

R19.11 *FCAT Study*: Extracting and plotting predicted random effects and residuals for model **M19.2**. The model-fit object fm19.2 was created in Panel R19.8

(a) *Default plot() method does not work*

```
> rnf <- ranef(fm19.2)
> plot(rnf)
  Error in eval(expr, envir, enclos) : object '.pars' not found
```

(b) *Alternative method for the normal Q-Q plots of the predicted random effects*

```
> rnft <- lapply(rnf, t)          # Transpose components
> names(plxLis  <-                # Auxiliary list ...
+    lapply(rnft, qqnorm,         # ...  with two components
+           plot.it = FALSE))
  [1] "one1" "one2"
> plx <-
+    lapply(plxLis,
+       FUN = function(el) xyplot(y ~ x, data = el, grid = TRUE))
> plx[["one1"]]               # Q-Q plot for id (see Fig. 19.1a)
> plx[["one2"]]               # Q-Q plot for target (see Fig. 19.1b)
```

(c) *Extracting and plotting conditional Pearson residuals*

```
> rsd2 <-                              # Equivalent to raw residuals
+    resid(fm19.2, type = "pearson")
> xyplot(rsd2 ~ target, data = fcat1)  # Fig. not shown
> bwplot(rsd2 ~ target, data = fcat1,  # Fig. 19.5
+        panel = panel.bwxplot)        # User defined panel (not shown)
```

qqnorm() to the transposed components and store the prepared Q-Q plots in the list-object plxLis. Note that, while creating the object plxLis, we use the argument plot.it=FALSE to suppress displaying the plots. The object has two components, named one1 and one2. We apply the function lapply() to create normal Q-Q plots for each of the components with the help of the function xyplot() and store the plots in the object plx. Finally, by "displaying" each of the two components of the object plx, we can display the separate Q-Q plots of predicted random effects for target and id. Note that we do not show the resulting plots, as they are identical to the graphs presented in Fig. 19.1.

In Panel R19.11c, with the help of the function resid() (see Table 14.5), we extract conditional Pearson residuals (Sect. 13.6.2) from the model-fit object fm19.2 and store them in the object rsd. Note that, given the structure of model **M19.2**, the residuals are equivalent to the raw residuals. By using the xyplot()-function call, presented in Panel R19.11c, we would obtain a stripplot of the residuals for each target. We do not show the resulting graph. Instead, to

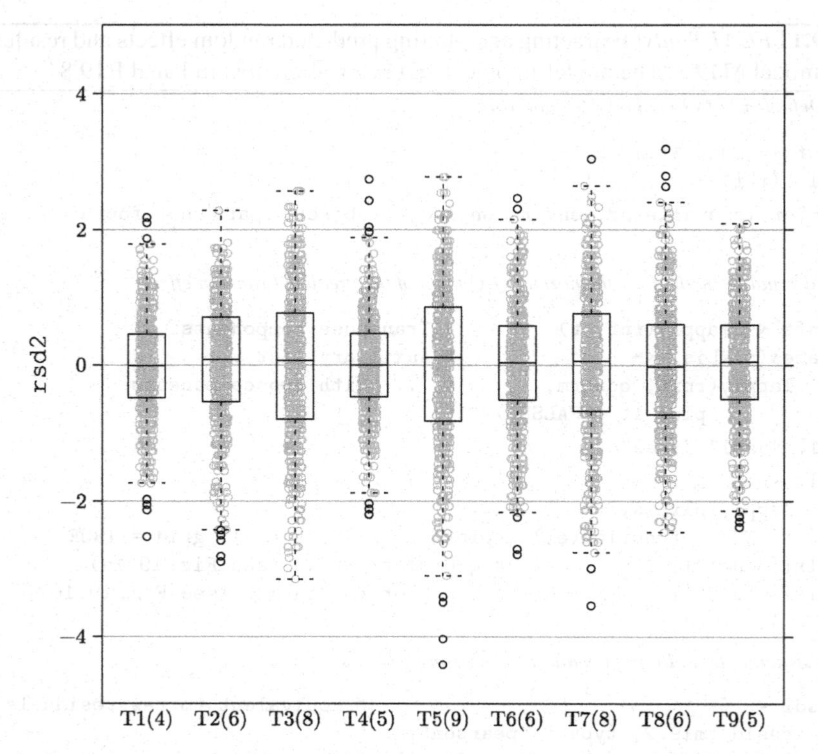

Fig. 19.5 *FCAT Study*: Stripplots of the conditional Pearson residuals for different targets for model **M19.2**

enhance its interpretation, we use the use function `bwplot()` from the package **lattice** to superimpose a box-and-whiskers plot over each stripplot (see also, e.g., Panel R12.9). Note that in the `panel` argument we use an auxiliary panel function which combines a stripplot with a box-and-whiskers plot, adds a grid of horizontal lines aligned with the axis labels, and adds a bit of jitter to the x-axis coordinates of the points between the whiskers. For the sake of brevity, we do not show the code for the function (but we include it in the package **nlmeU**).

The resulting plot is shown in Fig. 19.5. It suggests a slightly nonconstant variability of residuals across different targets. It may be due to the difference in the measurement scale for different targets, resulting from the different number of items (Sect. 2.5). We will address this issue in the next section.

19.5 A Linear Mixed-Effects Model with Crossed Random Effects and Heteroscedastic Residual Errors Fitted Using lme()

As mentioned at the end of the previous section, the scores for different targets were measured on different scales due to the different number of items per target. In this section, we try to address this issue by fitting an LMM similar to model **M19.2**, but using the average target response as the dependent variable.

19.5.1 Model Specification

Let us denote by ITEMSCORE_{sti} the response of the s-th pupil to the i-th item for the target t. Note that $i = 1, \ldots, n_t$, with $n_1 = 4$, $n_2 = 6$, $n_3 = 8$, $n_4 = 5$, $n_5 = 9$, $n_6 = 6$, $n_7 = 8$, $n_8 = 6$, and $n_9 = 5$. Consider the following model, proposed by Tibaldi et al. (2007):

$$\text{ITEMSCORE}_{sti} = \widetilde{\mu} + \widetilde{b}_{1,s} + \widetilde{b}_{2,t} + \widetilde{\varepsilon}_{sti}, \tag{19.5}$$

where $\widetilde{b}_{1,s} \sim \mathcal{N}(0, \widetilde{d}_S)$ is the random effect corresponding to the pupil s, $\widetilde{b}_{2,t} \sim \mathcal{N}(0, \widetilde{d}_T)$ is the random effect corresponding to the target t and independent of $\widetilde{b}_{1,s}$, and $\widetilde{\varepsilon}_{sti} \sim \mathcal{N}(0, \widetilde{\sigma}^2)$ is the residual (measurement) error independent of both $\widetilde{b}_{1,s}$ and $\widetilde{b}_{2,t}$. Equation (19.5) implies that, for the total target score,

$$\text{SCORE}_{st} \equiv \sum_{i=1}^{n_t} \text{ITEMSCORE}_{sti},$$

the following holds:

$$\text{SCORE}_{st} = n_t \cdot \widetilde{\mu} + n_t \cdot \widetilde{b}_{1,s} + n_t \cdot \widetilde{b}_{2,t} + \sum_{i=1}^{n_t} \widetilde{\varepsilon}_{sti}. \tag{19.6}$$

Formally speaking, (19.6) implies that, with a varying number of items per target, the mean and variance of the total score SCORE_{st} are different for different targets. However, if the number of items per target is similar, the mean and variance are approximately constant. This is the case of the FCAT data. Thus, the use of model **M19.2**, defined in (19.2), could be justified.

However, if we were concerned about the variability of the mean and variance due to the varying number of items, we could perform an alternative analysis of the FCAT data. Toward this end, we might consider the average target score,

$$\text{SCORE}_{st}/n_t = \sum_{i=1}^{n_t} \text{ITEMSCORE}_{sti}/n_t.$$

Equation (19.5) implies that

$$\begin{aligned} \text{SCORE}_{st}/n_t &= \widetilde{\mu} + \widetilde{b}_{1,s} + \widetilde{b}_{2,t} + \sum_{i=1}^{n_t} \widetilde{\varepsilon}_{sti}/n_t \\ &= \widetilde{\mu} + \widetilde{b}_{1,s} + \widetilde{b}_{2,t} + \widetilde{\varepsilon}^{*}_{st}. \end{aligned} \tag{19.7}$$

Consequently, the average target score is normally distributed with the following (marginal) mean and variance:

$$\text{E}(\text{SCORE}_{st}/n_t) = \widetilde{\mu}, \tag{19.8}$$

$$\text{Var}(\text{SCORE}_{st}/n_t) = \widetilde{d}_S + \widetilde{d}_T + \widetilde{\sigma}^2/n_t. \tag{19.9}$$

Thus, the mean of the average target score is constant and equal to the mean of the item response, defined in (19.5). On the other hand, the variance of the average target score depends on the number of items through the rescaled residual-variance component, $\widetilde{\sigma}^2/n_t$.

We label the model, defined by (19.7)–(19.9), as model **M19.3**. Note that its form is very similar to model **M19.2**, defined in (19.2). In particular, it includes the intercept as the only fixed effect and contains two crossed random effects for pupils and targets. An important formal difference is the presence of the target-specific residual variance, as shown in (19.9). A fundamental difference, however, is that all the fixed and random effects of model **M19.3** are specified at the item level, as defined by (19.5). On the other hand, the effects of model **M19.3** were defined on the target (total score) level, as seen from (19.6).

In the next section, we fit model **M19.3** to the FCAT data.

19.5.2 R Syntax and Results

Panel R19.12 presents the R code to fit model **M19.3** to the FCAT data using the function `lme()` from the package **nlme** (Chap. 14).

Before fitting the model, we need a few steps to prepare data. First, we create the vector `nItms` that contains the number of items per target. We use it to construct the variable `nItems` which, in turn, is used to create the mean target score. Both variables are created within the data frame `fcat1` with the help of the function `within()`. The resulting data frame is stored in the object `fcatm`.

Additionally, we use the vector `nItms` to create the vector `varWghts` which contains the inverse of the square root of the number of items for each target. Next, we transform the elements of `varWghts` into the ratios relative to the first element which is equal to 0.5. The first element of the transformed vector is necessarily equal

R19.12 *FCAT Study*: Model **M19.3** fitted using the function lme()

```
> nItms <- c(4, 6, 8, 5, 9, 6, 8, 6, 5)  # Number of items per target
> (nms <- levels(fcat1$target))           # Names extracted ...
 [1] "T1(4)" "T2(6)" "T3(8)" "T4(5)" "T5(9)" "T6(6)" "T7(8)"
 [8] "T8(6)" "T9(5)"
> names(nItms) <- nms                     # ...  and assigned
> fcatm <-                               # Add to the data frame...
+    within(fcat1,
+          {
+           nItems <- nItms[as.numeric(target)] #...  no. of items...
+           scorem <- scorec/nItems          # ...  mean target-score.
+          })
> (varWghts <- 1/sqrt(nItms))           # Variance function weights
  T1(4)   T2(6)   T3(8)   T4(5)   T5(9)   T6(6)   T7(8)   T8(6)
0.50000 0.40825 0.35355 0.44721 0.33333 0.40825 0.35355 0.40825
  T9(5)
0.44721
> (fxdW <- varWghts[-1]/0.5)            # Ratios wrt the 1st element
  T2(6)   T3(8)   T4(5)   T5(9)   T6(6)   T7(8)   T8(6)   T9(5)
0.81650 0.70711 0.89443 0.66667 0.81650 0.70711 0.81650 0.89443
> fm19.3 <-                              # M19.3
+    lme(scorem ~ 1,
+        random = list(one1 = pdIdent(~target - 1),
+                      one2 = pdIdent(~id - 1)),
+        weight = varIdent(form = ~1|target, fixed = fxdW),
+        data = fcatm)
```

to 1, so we remove it. The result is stored in the vector fxdW. The reason for such a construction of the vector will become clear in the sequel.

Finally, we fit model **M19.3** by applying the function lme() to the data frame fcatm. The values of the fixed and random arguments are similar to the values used in the lme()-function call to fit model **M19.2** (see Panel R19.8). Of course, the formula used in the fixed argument specifies the mean target score, scorem, as the dependent variable. Note that we do not specify the argument method, so that the REML estimation is used (Sect. 14.5).

To define the target-specific residual variance, as implied by (19.9), we use the weight argument (Sect. 14.5). In particular, we use the *varIdent*-class constructor function (Sect. 8.2) to specify different variances per strata defined by levels of the target factor. Moreover, we use the fixed argument of the constructor function to fix the values of the variance parameters. In this case, the parameters are the ratios of residual SDs for various targets relative to the first target (Sect. 7.3.2). In particular, according to model **M19.3**, SD for the t-th target is assumed to be equal to $\widetilde{\sigma}/\sqrt{n_t}$. Hence, the corresponding variance-function parameter δ_t is equal to

R19.13 *FCAT Study*: Results of fitting model **M19.3**. The model-fit object `fm19.3` was created in Panel R19.12

```
> summary(fm19.3)$tTable
              Value Std.Error   DF t-value   p-value
(Intercept) 0.61319  0.031088 4850  19.724 2.285e-83
> VarCorr(fm19.3)
             Variance              StdDev
one1 =       pdIdent(target - 1)
targetT1(4)  0.0083463             0.091358
...     [snip]
targetT9(5)  0.0083463             0.091358
one2 =       pdIdent(id - 1)
id1          0.0172872             0.131481
...     [snip]
id538        0.0172872             0.131481
Residual     0.0506879             0.225140
```

$$(\widetilde{\sigma}/\sqrt{n_t})/(\widetilde{\sigma}/\sqrt{n_1}) = (1/\sqrt{n_t})/(1/\sqrt{n_1}) = (1/\sqrt{n_t})/0.5,$$

with $\delta_1 \equiv 1$ (Sect. 7.3.2). This is exactly how the elements of the vector `fxdW` have been defined. Hence the use of the vector in the `fixed` argument of the `lme()` function in Panel R19.12.

As a side remark, we note that, instead of the `varIdent()` constructor function, we could also use the `varFixed()` one (Sect. 8.2). Toward this end, we would need to add to the `fcatm` data frame a variable containing the inverse of the number of items for each target response. The variable would then be used as the variance covariate in the `varFixed()` constructor function. We leave this as an exercise to the reader.

Results of fitting of the model are shown in Panel R19.13. The mean value of the item response is estimated to be equal to 0.6132. The estimates of the variances of the target- and pupil-specific random effects are equal to 0.0083 and 0.0173, respectively. Note that they are much smaller than the estimates obtained for model **M19.2** (see Panel R19.9). As mentioned before, the difference is due to the fact that the dependent variables, used in the two models, are measured on different scales.

The scale parameter, σ, is estimated to be equal to 0.2251. The parameter is equal to the SD for the first target, which is given by $\widetilde{\sigma}/\sqrt{n_1} = \widetilde{\sigma}/2$. Hence, the estimate of $\widetilde{\sigma}^2$ is equal to $2 \cdot (0.2251^2) = 0.2028$.

The estimated values of $\widetilde{d}_S$, $\widetilde{d}_T$, and $\widetilde{\sigma}^2$ imply that the total variance of item responses is equal to $0.0173 + 0.0083 + 0.2028 = 0.2284$. The between-pupil and between-target variability constitutes, respectively, 7.6% and 3.6% of the total variance.

It is worth noting that model **M19.3** cannot be fitted using the function `lmer()` from the package **lme4.0**. This is because the function allows to fit only LMMs with independent, homoscedastic residual errors (Sect. 15.3.1).

19.6 Chapter Summary

In this chapter, we analyzed the FCAT data by applying LMMs. In particular, we used LMMs with crossed random effects.

First, we analyzed the total target score as the dependent variable. For illustration purposes, in Sect. 19.2, we fitted a two-way, fixed-effects ANOVA model. Toward this end, we used the function `lm()`. In Sect. 19.3, we fitted an LMM with crossed random effects and independent, homoscedastic residual errors by using the function `lmer()` from the package **lme4.0**. By comparing the predicted random effects with the estimates of the fixed-effects coefficients from the ANOVA model, we could illustrate the "shrinkage" of EBLUPs. In Sect. 19.4, we fitted the same LMM by using the function `lme()` from the package **nlme**. Toward this end, we had to perform additional data manipulations. Moreover, the required computation time was substantially longer than for the function `lmer()`. This illustrated the advantage of using the sparse-matrix representations in the latter function.

In Sect. 19.5, we conducted an alternative analysis of the FCAT data. It was based on considering the average target score as the dependent variable. We analyzed the score by using an LMM with crossed random effects and independent, heteroscedastic residual errors. To fit the model, we used the function `lme()`. Note that the model could not be fitted using the function `lmer()`, because the function does not allow for heteroscedastic residual errors.

Although the FCAT data set has got a relatively simple structure, its analysis is not as straightforward. For instance, as illustrated by models **M19.2** and **M19.3**, the data can be modeled on the target or item scale. Clearly, the choice of the scale is important and influences the results. It also entails different assumptions regarding, e.g., homoscedasticity of residual errors. Also, treating the set of targets and pupils, included in the study, as random samples from a population, may be debatable. Such an approach was considered by, e.g., Van den Noortgate et al. (2003) or Tibaldi et al. (2007). In fact, models **M19.2** and **M19.3** were proposed by Tibaldi et al. (2007). However, an alternative would be to, e.g., consider the effects of targets as fixed and the effects of pupils as random. We leave the pursuit of this line of analysis to the reader.

Chapter 20
Extensions of the R Tools for Linear Mixed-Effects Models

20.1 Introduction

In this chapter, we present selected tools and functions introduced in the **nlmeU** package.

In particular, in Sect. 20.2, we describe in more detail the experimental class *pdKronecker*, which represents Kronecker products of positive-definite matrices. The class is useful to define variance-covariance matrices of random effects in LMMs in cases when the effects exhibit a factorial structure. Note that the class was applied in the analysis of the PRT trial data in Chap. 17. Section 20.3 presents the tools for conducting influence diagnostics for a fitted LMM. In Sect. 20.4, we introduce the function `simulateY()`, which can be used to simulate the values of the dependent variable based on a fitted LMM. The function is useful in, e.g., computing the empirical p-values of the tests based on the LMM. Finally, in Sect. 20.5, we consider the issue of computing the power of a test of a fixed effect in an LMM using `Pwr()` function and a simulation technique.

20.2 The New *pdMat*Class: *pdKronecker*

In Sect. 14.2, we described several *pdMat* classes available in the package **nlme**, which can be applied to represent the positive-definite matrix $\boldsymbol{D}$ involved in the definition of the distribution of random effects in LMMs. These classes were designed primarily for models in which the random effects could be presented as a vector $\mathbf{b}_i$ without any factorial structure imposed on its elements. Sometimes, however, such a structure can be present. This was the case for the PRT trial data, described in Sects. 2.3 and 3.3 and analyzed in Chap. 17.

When specifying an LMM, the presence of a factorial structure in the random effects can be ignored. For the PRT data, such a strategy was applied, e.g., in model **M17.5** in Sect. 17.4. The drawback is that, unless constraints are imposed

A. Gałecki and T. Burzykowski, *Linear Mixed-Effects Models Using R: A Step-by-Step Approach*, Springer Texts in Statistics, DOI 10.1007/978-1-4614-3900-4_20,

R20.1 A hypothetical example illustrating the concept of the new *pdKronecker* class

```
> D1 <- c(3, 9,
+         9, 30)
> dim(D1) <- c(2, 2)          # D1 for factor  f1
> D2 <- c(2, 4,
+         4, 10)
> dim(D2) <- c(2, 2)          # D2 for factor f2
> D1 %x% D2                   # D1 ⊗ D2
      [,1] [,2] [,3] [,4]
 [1,]    6   12   18   36
 [2,]   12   30   36   90
 [3,]   18   36   60  120
 [4,]   36   90  120  300
```

on the variance-covariance matrix of the random effects, it may require estimation of a relatively large number of variance-covariance parameters. In the case of model **M17.5**, the matrix was of dimension 4×4 and involved ten parameters. Although we were successful in fitting the model to the PRT data, it may happen that the need to estimate a large number of variance-covariance parameters may cause problems in fitting an LMM.

The issue can be addressed by a parsimonious specification of the matrix $\boldsymbol{D}$, resulting from the recognition of the factorial structure in the random effects. Examples of such strategy were provided in the analysis of the PRT data in Sects. 17.6–17.8. Toward this end, we defined a new *pdMat* class, i.e., *pdKronecker*, which represents the matrix $\boldsymbol{D}$ in terms of a Kronecker product of matrices of lower dimensions. As a result, the matrix is represented with a smaller number of parameters. The *pdKronecker* class implements the methodology developed by Galecki (1994).

In this section, we present further details on the use of the *pdKronecker* class. In the presentation, we use a hypothetical, simple example of the structure implied by the class. The example is shown in Panel R20.1.

For simplicity, we assume that the factorial structure in the random effects is imposed by two crossed factors, `f1` and `f2`, say, each with two levels. They will be defined in more detail later in this section. Similar to the PRT example, we assume that each factor contributes its own factor-specific matrix, $\boldsymbol{D}_1$ and $\boldsymbol{D}_2$, say, shown in Panel R20.1. The two matrices define the overall matrix $\boldsymbol{D} = \boldsymbol{D}_1 \otimes \boldsymbol{D}_2$, where $\otimes$ stands for the (right) Kronecker product. This conceptual framework has an attractive interpretation, in that each of the underlying factors contributes independently to the overall $\boldsymbol{D}$ matrix. It is worth noting that other pairs of matrices, e.g., $\boldsymbol{D}_1/c$ and $c\boldsymbol{D}_2$, where c is a constant greater than zero, result in the same matrix $\boldsymbol{D}$. Thus, to assure the identifiability of the model for the matrix $\boldsymbol{D} = \boldsymbol{D}_1 \otimes \boldsymbol{D}_2$, and to uniquely define the decomposition of the $\boldsymbol{D}$ matrix, constraints on the elements of $\boldsymbol{D}_1$ and $\boldsymbol{D}_2$ need to be imposed. We will deal with this issue later in this section.

R20.2 Construction of the main argument of the pdKronecker() constructor function. Matrices D1 and D2 were created in Panel R20.1

```
> library(nlme)
> (pdId <- pdIdent(as.matrix(1), form = ~1)) # Mandatory
  Positive definite matrix structure of class pdIdent representing
       [,1]
  [1,]    1
> (pd1 <- pdLogChol(D1, form = ~f1-1))        # D1
  Positive definite matrix structure of class pdLogChol representing
       [,1] [,2]
  [1,]    3    9
  [2,]    9   30
> (pd2 <- pdLogChol(D2, form = ~f2-1))        # D2
  Positive definite matrix structure of class pdLogChol representing
       [,1] [,2]
  [1,]    2    4
  [2,]    4   10
> pdL1 <-                                     # The main argument
+     list(X = pdId,
+          pD1 = pd1,
+          pD2 = pd2)
```

20.2.1 Creating Objects of Class pdKronecker

In Panels R20.2 and R20.3, we illustrate how to use the pdKronecker() constructor function to create an example of an object representing the corresponding class.

First, in Panel R20.2, we illustrate the most convenient way to define the main argument of the constructor function by creating a list of objects of the *pdMat* class. Toward this end, after attaching the **nlme** package, we define the objects. In general, they do not have to be initialized. However, in Panel R20.2, we use the matrices D1 and D2 (Sect. 14.2.1) to initialize objects of *pdMat* class. In this way, we can simultaneously illustrate how to construct an initialized object of class *pdKronecker* representing the matrix $\boldsymbol{D}$.

Note that the first object created in Panel R20.2, pdId, is a mandatory object of class *pdIdent* (Sect. 14.2.1). It represents a 1×1 matrix needed to address the identifiability issue mentioned earlier. The two other objects, pd1 and pd2, are of class *pdLogChol* and represent the matrices $\boldsymbol{D}_1$ and $\boldsymbol{D}_2$, respectively. By using pdId, pd1, and pd2, we create the list pdL1 which will be used as the main argument of the pdKronecker() construction-function call.

In Panel R20.3, we illustrate the construction of an object of class *pdKronecker*. Toward this end, in Panel R20.3a, we formally define the factors f1 and f2 by using the gl() function. We store the factors in an auxiliary data frame named dt.

R20.3 Construction of an object of class *pdKronecker*. The object pdL1 was created in Panel R20.2

(a) *Auxiliary data containing the definitions of factors* f1 *and* f2

```
> f1 <- gl(2, 1, labels = c("A","B"))
> f2 <- gl(2, 1, labels = c("a","b"))
> (dt <- data.frame(f1, f2))
  f1 f2
1  A  a
2  B  b
```

(b) *Constructing an object of class pdKronecker*

```
> library(nlmeU)
> (pdK <- pdKronecker(pdL1, data = dt))          # D1 ⊗ D2
 Positive definite matrix structure of class pdKronecker ...
         f1A:f2a f1A:f2b f1B:f2a f1B:f2b
 f1A:f2a       6      12      18      36
 f1A:f2b      12      30      36      90
 f1B:f2a      18      36      60     120
 f1B:f2b      36      90     120     300
> (nms <- Names(pdK))
 [1] "f1A:f2a" "f1A:f2b" "f1B:f2a" "f1B:f2b"
```

Subsequently, in Panel R20.3b, we use the pdKronecker() constructor function from the package **nlmeU** to create the object pdK representing the matrix $\boldsymbol{D}_1 \otimes \boldsymbol{D}_2$. We use the list pdL1 and the data frame dt as the first and second argument of the pdKronecker() function, respectively. The auxiliary data are used to provide the information needed to assign the names to the rows and columns of the matrix represented by the object pdK. At the bottom of Panel R20.3b, we extract the names with the help of the function Names(), print them, and store in the vector nms for later use.

The construction of an initialized object of class *pdKronecker* can be useful if, e.g., there is a need to provide initial values for the $\boldsymbol{\theta}_D$ parameters for the function lme(). The use of uninitialized *pdMat*-class objects to construct the main argument of the pdKronecker() constructor function was illustrated in Panels R17.14, R17.16, and R17.19.

20.2.2 Extracting Information from Objects of Class pdKronecker

In Panel R20.4, we show how to extract the components used to represent a Kronecker-product matrix represented by a *pdKronecker*-class object. For

R20.4 Extracting component matrices from an object of class *pdKronecker*. The object pdK was created in Panel R20.3

```
> (c0x <- as.matrix(pdK[[1]]))
            (Intercept)
(Intercept)           6
> (D1x <- as.matrix(pdK[[2]]))                # Proportional to D_1
    f1A f1B
f1A   1   3
f1B   3  10
> (D2x <- as.matrix(pdK[[3]]))                # Proportional to D_2
    f2a f2b
f2a   1   2
f2b   2   5
> Dx <-c0x %x% D1x %x% D2x                    # D = D_1 ⊗ D_2
> dimnames(Dx) <- list(nms, nms)
> Dx
        f1A:f2a f1A:f2b f1B:f2a f1B:f2b
f1A:f2a       6      12      18      36
f1A:f2b      12      30      36      90
f1B:f2a      18      36      60     120
f1B:f2b      36      90     120     300
```

illustration, we use the object pdK. It is a list with three components, which are *pdMat*-class objects. The classes of those objects correspond to the classes of the objects pdId, pd1, and pd2, used to define pdK. In particular, the first component is an object of class *pdIdent*, while the second and third components are objects of class *pdLogChol*. In Panel R20.4, we extract the components from the list pdK, and we apply the function `as.matrix()` to store them as the matrices c0x, D1x, and D2x, respectively. From the printout of the matrices, we can see that they are *not* equal to the matrices $c\boldsymbol{I}_1$, $\boldsymbol{D}_1$, and $\boldsymbol{D}_2$, represented by pdId, pd1, and pd2. This is due to the fact, which was mentioned earlier, that the matrices used to create a Kronecker product cannot be uniquely identified from the product itself. To resolve the nonidentifiability, the matrices D1x and D2x are constrained to contain 1 as the upper-left element. As a result, the matrices are *proportional* to $\boldsymbol{D}_1$ and $\boldsymbol{D}_2$, respectively. In Panel R20.4, we show that the Kronecker product of the matrices represented by c0x, D1x, and D2x is equal to the Kronecker product of $\boldsymbol{D}_1$ and $\boldsymbol{D}_2$.

In Panel R20.5, we explain how the formula associated with an object of the *pdKronecker* class can be extracted. We also explain how it is built. This formula plays a critical role in defining the random-effects design matrix $\boldsymbol{Z}$, when an object of class *pdKronecker* is used in the `random` argument of the `lme()` function. Not surprisingly, the formula is created from the formulae, which we call *component formulae*, applied in the definition of the *pdMat*-class objects used to define the main argument of the `pdKronecker()` constructor function.

R20.5 Extracting the formula from an object of class *pdKronecker*. The object pdK was created in Panel R20.3

(a) *Formula for the pdKronecker-class object*

```
> formula(pdK, asList = TRUE)              # List of component formulae
  [[1]]
  ~1

  [[2]]
  ~f1 - 1

  [[3]]
  ~f2 - 1
> formula(pdK)                             # One-sided formula default)
  ~f2:f1 - 1
  <environment: 0x000000000be54b40>
```

(b) *Explaining how the formula for an object of the pdKronecker class is created*

```
> (pdKform <- formula(~(f2-1):(f1-1)-1))
  ~(f2 - 1):(f1 - 1) - 1
> pdKterms <- terms(pdKform)               # Terms object
> labels(pdKterms)                         # Expanded formula
  [1] "f2:f1"
> attr(pdKterms, "intercept")              # Intercept omitted
  [1] 0
```

In Panel R20.5a, we use the `formula()` function to extract two representations of the formula associated with the pdK object. The first representation is in the form of a list. It allows identinfying the component formulae which correspond to the formulae used to define objects pdId, pd1, and pd2 in Panel R20.2. The second (default) representation is in the form of a one-sided formula used to create the matrix $\boldsymbol{Z}$. The formula contains the interaction of the factors `f1` and `f2` as the only term (without an intercept). The order of `f1` and `f2` in the interaction term is reversed to accommodate the order of rows in the $\boldsymbol{D}$ matrix resulting from the use of (right) Kronecker product of the component matrices.

In Panel R20.5b, we explain how the matrix-$\boldsymbol{Z}$ formula is constructed from the component formulae. More specifically, the formulae corresponding to the second and third component of the object pdK are "multiplied" using the `:` operator, and the intercept is removed from the obtained result. Note that the first component formula `~1`, associated with the mandatory *pdIdent()*-class object, is not used in this operation. The resulting formula is stored in the object `pdKform`.

By applying the `terms()` function (Sect. 5.2.2), we construct the object `pdK-terms`, which contains the information about all the terms in the formula `pdKform`. In particular, by applying the function `labels()`, we check that `pdKform`, in its expanded form, contains only one term, i.e., `"f2:f1"`. Moreover, by extracting the `intercept` attribute of the `pdKterms` object, we verify that the formulae do not contain the intercept. Thus, `pdKform` is indeed equivalent to the matrix-$\mathbf{Z}$ formula of the object `pdK`.

20.3 Influence Diagnostics

In Sect. 4.5.3, we briefly discussed the issue of investigating the influence of individual observations on the estimates of the parameters of an LM. Two measures that can be used toward this end are Cook's distance, defined in (4.26), and the likelihood displacement, defined in (4.27). In Sects. 7.5.2, 10.5.2, and 13.6.3, we indicated how these measures can be adapted to the case of LMs with heterogeneous variance, LMs with fixed effects for correlated data, and LMMs, respectively. In this section, we present an implementation of the likelihood displacement and Cook's distance for LMMs in R.

As an illustration, we consider model **M16.5**, which was fitted to the `armd` data in Sect. 16.4.3. The results of fitting of the model were stored in the model-fit object `fm16.5`.

20.3.1 Preparatory Steps

In this section, we present preparatory steps for influence diagnostics. We start with the extraction of selected results for the fitted model **M16.5**. We then introduce an auxiliary function `logLik1()` designed to calculate a contribution of a given subject to the overall likelihood for a given model.

20.3.1.1 Selected Results for Model M16.5

In Panel R20.6, we fit the model **M16.5** and extract basic results from the model fit. First, in Panel R20.6a, we update the model-fit object `fm16.5` to obtain the ML estimates. The updated model fit is stored in the object `mf16.5ml`. With the help of the function `formula()`, we recall the formula defining the mean structure of the model. We also extract the name of the data frame used to fit the model. Finally, we apply the `logLik()` function to obtain the value of the log-likelihood function for model **M16.5**. Note that the number of degrees of freedom reported by `logLik()` is equal to 8. It corresponds to the total number of the parameters in the model,

R20.6 *ARMD Trial*: Extracting selected results for model **M16.5**. The model-fit object fm16.5 was created in Panel R16.13

(a) *Basic information*

```
> fm16.5ml <- update(fm16.5, method = "ML")# ML estimation
> formula(fm16.5ml)                        # Recall model formula.
  visual ~ visual0 + time + treat.f
> fm16.5ml$call$data                       # Recall data name.
  armd
> logLik(fm16.5ml)                         # Log-ML value
  'log Lik.' -3210.7 (df=8)
```

(b) *Fixed-effects estimates and their variance-covariance matrix*

```
> beta0  <- fixef(fm16.5ml)                # β̂
> names(beta0)                             # Long names
  [1] "(Intercept)" "visual0"         "time"
  [4] "treat.fActive"
> names(beta0) <- abbreviate(names(beta0), minlength = 7) # Short names
> beta0                                    # β̂ printed.
   (Intrc)  visual0     time  trt.fAc
   5.44721  0.89973 -0.24155 -2.65638
> vcovb  <- vcov(fm16.5ml)                  # Var(β̂)
> colnames(vcovb) <- names(beta0)           # Short names
> vcovb                                     # Var(β̂) printed.
                  (Intrc)     visual0        time     trt.fAc
  (Intercept)   5.0475640 -7.9651e-02 -3.8602e-03 -6.8078e-01
  visual0      -0.0796512  1.4407e-03  1.5213e-06  1.1239e-03
  time         -0.0038602  1.5213e-06  5.6988e-04 -6.1566e-05
  treat.fActive -0.6807838  1.1239e-03 -6.1566e-05  1.2568e+00
```

i.e., four fixed-effects coefficients ($\boldsymbol{\beta}$), four variance-covariance parameters ($\boldsymbol{\theta}_D$) describing the diagonal matrix $\boldsymbol{D}$, one parameter (δ) related to the power variance function describing the diagonal matrix $\boldsymbol{R}_i$, and the scale parameter σ.

In Panel R20.6b, we extract the $\boldsymbol{\beta}$ estimates and their estimated variance-covariance matrix. Toward this end, we use the functions `fixef()`) and `vcov()`, respectively. We save the estimates and the matrix in the objects `beta0` and `vcovb`, respectively. They will be needed for influence diagnostics performed in Sect. 20.3.2. Note that, with the help of the `abbreviate()` function, the names of the $\boldsymbol{\beta}$ coefficients in the vector `beta0` are shortened to (at least) seven characters, to simplify the display of the contents of the vector. The abbreviated names are also used to label the columns of the `vcovb` matrix.

20.3.1.2 An Auxiliary Function `logLik1()`

At the bottom of Panel R20.6a, we used the `logLik()` function to obtain the value of the log-likelihood for the fitted model **M16.5**. It should be noted that the function returns the log-likelihood evaluated at the set of the estimated fixed effects and variance-covariance parameters and for the dataset, to which the model was fitted. In the context of influence diagnostics, we need a more general function which allows to evaluate the log-likelihood function for an arbitrary set of values of the model parameters and with respect to data different than the ones used to fit the model.

Toward this end, we can use the auxiliary function `logLik1()` which has been included in the package **nlmeU**. The primary use of the function is to calculate the contribution of *one* subject in the data to the overall log-likelihood, defined in (13.27) for a given model. The use of the function in the context of influence diagnostics will be presented in Sect. 20.3.2.

The function `logLik1()`, illustrated in Panel R20.7, has three arguments:

`modfit` An object of class *lme* representing an LMM fitted to a given dataset using the ML estimation

`dt1` A data frame with data for *one* subject, for whom the log-likelihood function is to be evaluated

`dtInit` An optional auxiliary data frame

The data frame provided in the argument `dt1` is typically created by choosing a subset with one subject from the data used to obtain the model-fit object specified in the `modfit` argument. However, in general, any plausible data for one subject, not necessarily from the dataset used to fit the model, can be used.

The auxiliary data provided in the argument `dtInit` is temporarily appended to the `dt1` data during the `logLik1()`-function execution. It may be necessary in a situation when the information, contained in the data defined by the argument `dt1`, is not sufficient to properly construct the objects needed to calculate the log-likelihood. This may occur if, e.g., `dt1` does not contain information about all levels of a factor needed to construct the design matrix, variance function, or correlation matrix. In most cases, the data frame used in the `dtInit` argument is obtained by selecting a small subset of the data used to fit the model.

The `logLik1()` function returns the numeric contribution of the single subject, with the data specified in the `dt1` argument, to the log-likelihood for the model specified in the `modfit` argument.

20.3.1.3 Contributions of Individual Subjects to the Log-Likelihood for Model M16.5

Panel R20.7a illustrates how to calculate contributions of individual subjects to the log-likelihood for a given model. In particular, we first create the data frame `df1` with the data for the subject "1" from the data frame `armd`. Then, we apply the function `logLik1()` to calculate the contribution of the subject to the log-likelihood

R20.7 *ARMD Trial*: Contributions of individual subjects to the log-likelihood for model **M16.5**. The model-fit object `fm16.5ml` was created in Panel R20.6

(a) *Examples of using the function* `logLik1()`

```
> require(nlmeU)
> df1 <- subset(armd, subject %in% "1")        # Data for subject "1"
> logLik1(fm16.5ml, df1)                        # logLik_i for subject "1"
  [1] -6.6576
> lLik.i <- by(armd, armd$subject,
+    FUN = function(dfi) logLik1(fm16.5ml, dfi))
> lLik.i <- as.vector(lLik.i)   # Coerce array to vector
> lLik.i[1:5]                   # logLik_i for the first five subjects
  [1]  -6.6576 -13.4708 -11.1361 -13.3109 -12.9930
> sum(lLik.i)                   # sum_i logLik_i; compare to Panel R20.6a
  [1] -3210.7
```

(b) *Plot of individual contributions to the log-likelihood (traditional graphics)*

```
> nx <- by(armd, armd$subject, nrow)              # n_i
> lLik.n <- lLik.i/as.vector(nx)                  # logLik_i/n_i
> outL <- lLik.n < -6                             # TRUE for values < -6
> lLik.n[outL]                                    # logLik_i/n_i   < -6
  [1] -6.0775 -7.2559 -6.2956 -6.9220 -6.3644 -6.7953 -6.5079
> subject.c <- levels(armd$subject)
> subject.x <- as.numeric(subject.c)
> plot(lLik.n ~ subject.x, type = "h")            # Fig. 20.1
> points(subject.x[outL], lLik.n[outL], type = "p", pch = 16)
> text(subject.x[outL], lLik.n[outL], subject.c[outL])
```

for model **M16.5**. Note that we use the data frame `df1` as the second argument, i.e., `dt1`, of the function. The contribution to the log-likelihood for the first subject is equal to -6.6576.

Next, we use the function `logLik1()` to compute the log-likelihood contributions for all subjects from the data frame `armd`. Toward this end, we use the function `by()`, which splits `armd` into data frames containing data for each subject separately and applies the function `logLik1()` to each of the data frames. As a result, we obtain the one-dimensional array `lLik.i` of class *by* with the log-likelihood contributions for all 234 subjects. Although not critical, we convert the array to a vector with the same name using the `as.vector()` function. In Panel R20.7a, we display the first five elements of the vector. We also compute the sum of all log-likelihood contributions by applying the function `sum()` to the vector. The result, -3210.7, corresponds to the value of the log-likelihood function for model **M16.5**, obtained with the help of the function `logLik()` in Panel R20.6a.

In Panel R20.7b, we present the syntax to plot the per-observation individual log-likelihood contributions. First, with the help of the `by()` function, we create the array `nx`, which contains the number of observations for each subject from the data frame `armd`. Next, we compute the per-observation individual-subject log-likelihood contributions by dividing the contributions by the number of observations for each subject. In this way, we adjust for the difference in the number of observations for different subjects. The per-observation contributions are stored in the vector `lLik.n`. We then select and print out the elements of the vector corresponding to the per-observation contributions less than, say, −6. Subsequently, we create the character vector `subject.c` with subjects' identifiers. We also create the corresponding numeric vector `subject.x`. The latter is used in the formula used in the call to the traditional graphics function `plot()` to obtain a plot of the per-observation log-likelihood contributions for all subjects. By using the function `points()`, we add points identifying the values of the contributions smaller than −6. Moreover, with the help of the function `text()`, we label the points with the corresponding subjects' identifiers from the character vector `subject.c`. The resulting plot is shown in Fig. 20.1 and should be primarily treated as an illustration of how the `logLik1()` function works. Influence-diagnostics calculations will be presented in the next section.

20.3.2 Influence Diagnostics

In this section, we use the results of the preparatory steps, conducted in Sect. 20.3.1, to perform influence-diagnostics calculations for model **M16.5**. More specifically, we evaluate the influence of every subject included in the dataset `armd`.

20.3.2.1 Fitting the Model to the "Leave-One-Subject-Out" Data

In Panel R20.8, we create a list containing the results of fitting model **M16.5** to the "leave-one-subject-out" (LOO) datasets and explore its contents. In particular, in Panel R20.8a, we define the function `lmeU()`, which fits the model to the data from the `armd` data frame with a particular subject removed. The identifier of the subject to be removed is passed as the only argument, `cx`, of the function. When the function `lmeU()` is executed, an LOO data frame, named `dfU`, is created with the subject, indicated by the `cx` argument, omitted. Subsequently, model **M16.5** is fitted to `dfU` by applying the `update()` function to the model-fit object `fm16.5ml`. As a result, the function `lmeU()` returns an object of class *lme*.

Next, with the help of the function `lapply()`, we apply the function `lmeU()` to the consecutive elements of the character vector `subject.c`. As a result, we obtain the list `lmeUall`, with *lme*-class model-fit objects as elements. The model-fit objects

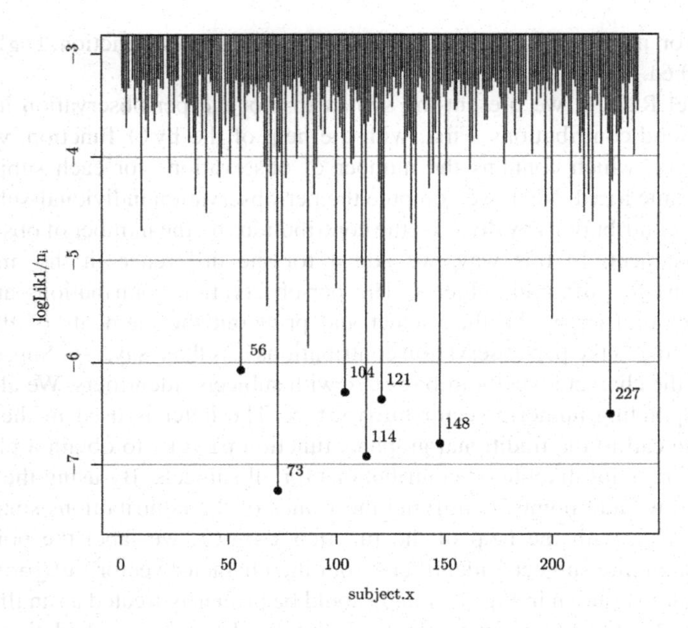

Fig. 20.1 *ARMD Trial*: Plot of the per-observation individual-subject log-likelihood contributions for model **M16.5** using traditional graphics

contain the results of fitting model **M16.5** to the `armd` data frame, while removing the data for each subject in turn. Finally, we name the components of the `lmeUall` list using the subject identifiers stored in the vector `subject.c`.

Note that the creation of the `lmeUall` list involves refitting model **M16.5** many times, and it therefore takes a long time. The execution time can be shortened if we decide to perform, e.g., a reduced number of likelihood iterations, instead of performing iterations until convergence as we did here. This may be a reasonable choice given that the starting values for the numerical optimization used by the `lmeU()` function are taken from the model-fit object `fm16.5ml`. The values, based on the first few iterations, are expected to give a fairly good approximation of the LOO estimates.

In Panel R20.8b, we explore the contents of the `lmeUall` list. The names of the first six components are printed out using the function `names()`. To extract the LOO data frame for, e.g., the subject "6", we refer to the `"6"` component of the `lmeUall` list. The extracted data frame is stored in the object `dataU6`. By using the function `dim()`, we check that the data frame has 863 observations and 8 variables. Note that the data frame `armd` has 867 observations (see Panel R2.5), while the subject "6" had four visual-acuity measurements (see, e.g., Panel R16.2). Thus, by leaving the subject out, we reduce the number of observations from 867 to 863. Moreover, by displaying the first six unique values of the variable `subject` from the `dataU6` data frame, we check that the subject "6" was, indeed, omitted from the data.

R20.8 *ARMD Trial*: Model **M16.5** fitted to a sequence of "leave-one-subject-out" (LOO) datasets. The objects `fm16.5ml` and `subject.c` were created in Panels R20.6 and R20.7, respectively

(a) *Creating the object* `lmeUall` *containing fitted models*

```
> lmeU <- function(cx) {
+     dfU <- subset(armd, subject != cx)            # LOO data
+     update(fm16.5ml, data = dfU)                  # LOO fit
+ }
> lmeUall        <- lapply(subject.c, lmeU)         # List with LOO fits
> names(lmeUall) <- subject.c                       # Names assigned
```

(b) *Exploring the contents of the* `lmeUall` *object*

```
> names(lmeUall)[1:6]
  [1] "1" "2" "3" "4" "6" "7"
> dataU6 <- lmeUall[["6"]]$data      # LOO data for subject "6"
> dim(dataU6)                        # Number of rows is 863
  [1] 863   8
> unique(dataU6$subject)[1:6]        # Subject no. 6 omitted
  [1] 1 2 3 4 7 8
  234 Levels: 1 2 3 4 6 7 8 9 10 11 12 13 14 15 16 17 18 19 ...  240
```

20.3.2.2 Likelihood Displacement for Model M16.5

The likelihood displacement was defined in (4.27). For an LMM, it requires the computation of the full log-likelihood (13.27) for $\widehat{\boldsymbol{\Theta}}$, the ML estimate of $\boldsymbol{\Theta} \equiv (\boldsymbol{\beta}', \boldsymbol{\theta}', \sigma^2)'$ obtained by fitting the model to all data, and for $\widehat{\boldsymbol{\Theta}}_{(-i)}$, the ML estimate obtained by fitting the model to the data with the *i*-th subject excluded (Sect. 13.6.3). Note that both values of the log-likelihood, used in the definition of likelihood displacement, should be calculated taking into account *all* observations.

In Panel R20.9, we present the code used to calculate and to plot individual-subject likelihood displacements for model **M16.5**. Toward this end, in Panel R20.9a, we define an auxiliary function `lLik()` which, for a given subject indicated by the main argument, `cx`, extracts the *lme*-class model-fit object for the corresponding LOO data. The model-fit object is stored in the object `lmeU`. The corresponding log-likelihood value is extracted from `lmeU` with the help of the function `logLik()` and stored in the object `lLikU`. Note that the stored log-likelihood value does not include the contribution from the excluded subject. Hence, the function `logLik1()` is invoked to calculate the contribution of subject `cx` and to store it in the object `lLik.s`. The returned value `llikU + lLik.s` is the log-likelihood evaluated for

R20.9 *ARMD Trial*: Likelihood displacements for model **M16.5**. The objects `subject.c` and `lmeUall` were created in Panels R20.7 and R20.8, respectively

(a) *Calculation of the likelihood displacements*

```
> lLik <- function(cx){
+     lmeU    <- lmeUall[[cx]]              # LOO fit extracted
+     lLikU  <- logLik(lmeU, REML = FALSE)  # LOO log-likelihood
+     df.s    <-                            # Data for subject cx...
+        subset(armd, subject == cx)
+     lLik.s <- logLik1(lmeU, df.s)        # ... and log-likelihood.
+     return(lLikU + lLik.s)                # "Displaced" log-likelihood...
+ }
> lLikUall <- sapply(subject.c, lLik)      # ... for all subjects.
> dif.2Lik <- 2*(logLik(fm16.5ml) - lLikUall) # Vector of LD_i
> summary(dif.2Lik)
    Min.  1st Qu.   Median     Mean  3rd Qu.     Max.
 0.00285 0.00948 0.01490 0.05280 0.03300 0.82200
```

(b) *Plot of the likelihood displacements with an indication of outlying values*

```
> names(dif.2Lik) <- subject.c                    # Subjects' ids assigned
> outL  <-   dif.2Lik > 0.5                       # Outlying LD_i's
> dif.2Lik[outL]
      73      75     104     114     121     227     231
 0.57543 0.56786 0.56269 0.66459 0.82188 0.55467 0.59549
> library(lattice)
> subject.f <- factor(subject.c, levels = subject.c)
> myPanel <- function(x, y, ... ){
+    x1 <- as.numeric(x)
+    panel.xyplot(x1, y, ... )
+    ltext(x1[outL], y[outL], subject.c[outL]) # Label outlying LD_i's
+ }
> dtp <-                                          # Fig. 20.2
+    dotplot(dif.2Lik ~ subject.f, panel = myPanel, type = "h")
> lxlims <- length(dtp$x.limits)
> update(dtp, xlim = rep("", lxlims), grid = "h")
```

all observations from the `armd` data frame using the "displaced" estimates of the model parameters, i.e., $\widehat{\boldsymbol{\Theta}}_{(-i)}$.

The function `lLik()` is then sequentially applied to all subjects in the `armd` data frame with the help of the `sapply()` function. The resulting vector `lLikUall` contains the values of the "displaced" log-likelihood for all subjects. Subsequently, it is used in the calculation of the values of likelihood displacement for all subjects, as discussed in Sect. 13.6.3. The displacements are stored in the vector `dif.2Lik`.

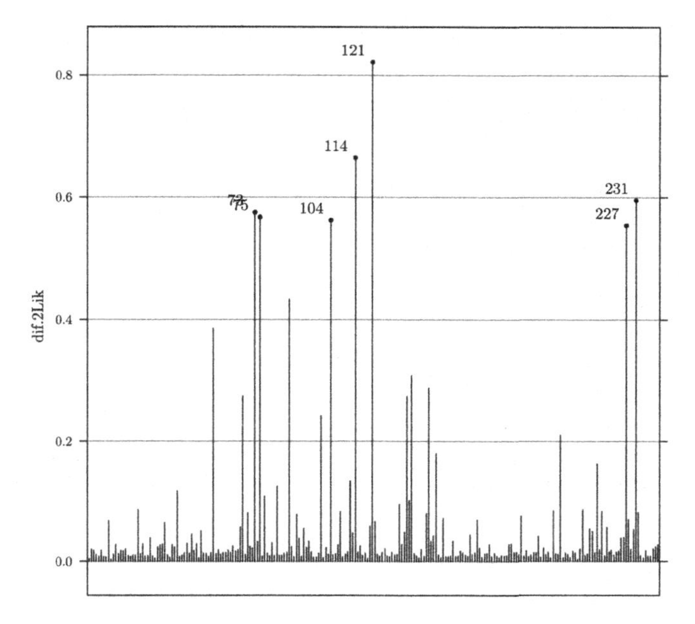

Fig. 20.2 *ARMD Trial*: Plot of the likelihood-displacement values *versus* subjects' rank for model **M16.5** for all subjects from the `armd` dataset

By applying the `summary()` function to the vector, we obtain summary statistics of the computed likelihood-displacement values.

In Panel R20.9b, we create the logical vector `outL` which indicates the subjects with the values of the likelihood displacement exceeding, say, 0.5. From the printout of the selected elements of the vector `dif.2Lik` it follows that there are seven such subjects.

We then use the function `dotplot()` from the package **lattice** to plot the likelihood-displacement values for all subjects. The x-axis of the plot is constructed using numeric representation of the `subject.f` factor, containing values ranging from 1 to 234, rather than the subject labels stored in the `subject.c` vector. In the `panel` argument we apply a user-defined `myPanel()` function, which labels the outlying likelihood displacements using the corresponding subject's identifier. Note that we save the plot in the object `dtp`, which we then update by adding a grid of horizontal lines and by removing illegible labels on x-axis.

The resulting plot is shown in Fig. 20.2. The seven subjects with the likelihood-displacement values larger than 0.5 are clearly identified.

20.3.2.3 Cook's Distance for the β Estimates

Cook's distance for the $\boldsymbol{\beta}$ estimates was defined in (4.26) for the classical LM. The definition can be extended to LMMs in a straightforward manner.

R20.10 *ARMD Trial*: Calculation of Cook's distances for model **M16.5**. The objects vcovb and beta0 were created in Panel R20.6, while the objects subject.c and lmeUall were created in Panels R20.7 and R20.8, respectively

(a) *Calculation of Cook's distances*

```
> betaUall <- sapply(lmeUall, fixef)            # Matrix with β̂(−i)
> vb.inv <- solve(vcovb)
> CookDfun <- function(betaU){
+    dbetaU <- betaU - beta0                    # β̂(−i)−β̂
+    CookD.value <- t(dbetaU) %*% vb.inv %*% dbetaU
+ }
> CookD.num <- apply(betaUall, 2, CookDfun)
> (n.fixeff <- length(beta0))                   # Number of fixed effects
  [1] 4
> rankX <- n.fixeff                             # Rank of matrix X
> CookD <- CookD.num/rankX                      # Cook's distance Di
```

(b) *Plot of Cook's distances using traditional graphics. Outlying values annotated*

```
> outD <- CookD > 0.03                          # Outlying Di's
> subject.c[outD]                               # Subjects' ids
  [1] "75"  "114" "145" "227" "231"
> plot(CookD ~ subject.c,
+      ylab = "Cook's D", type = "h")          # Fig. 20.3
> text(as.numeric(subject.c[outD]),
+      CookD[outD], subject.c[outD])            # Annotation
> points(subject.c[outD], CookD[outD])
```

In Panel R20.10, we present calculations of Cook's distance for the $\boldsymbol{\beta}$ estimates for model **M16.5**. The calculations are somewhat simpler compared to that for the likelihood displacement presented in the previous section. In Panel R20.10a, we present the syntax that can be used to perform the calculations.

We begin by creating a matrix containing the LOO estimates of $\boldsymbol{\beta}$. Toward this end, with the help of function `sapply()`, we apply the function `fixef()` to each of the *lme*-class objects contained in `lmeUall`.

Next, we compute the inverse of the variance-covariance matrix of $\widehat{\boldsymbol{\beta}}$ with the help of the function `solve()`. We store the resulting matrix in the object `vb.inv`.

Subsequently, we define the function `CookDfun()` which, for a vector given in the `betaU` argument, computes the value of the numerator of Cook's distance, as in (4.26). The function is then applied sequentially to all columns of the matrix `betaUall` with the help of the `sapply()` function. The resulting vector is divided by the number of the fixed-effects coefficients, which, under the assumption that the design matrix is of full rank, is equivalent to the rank of the design matrix. The

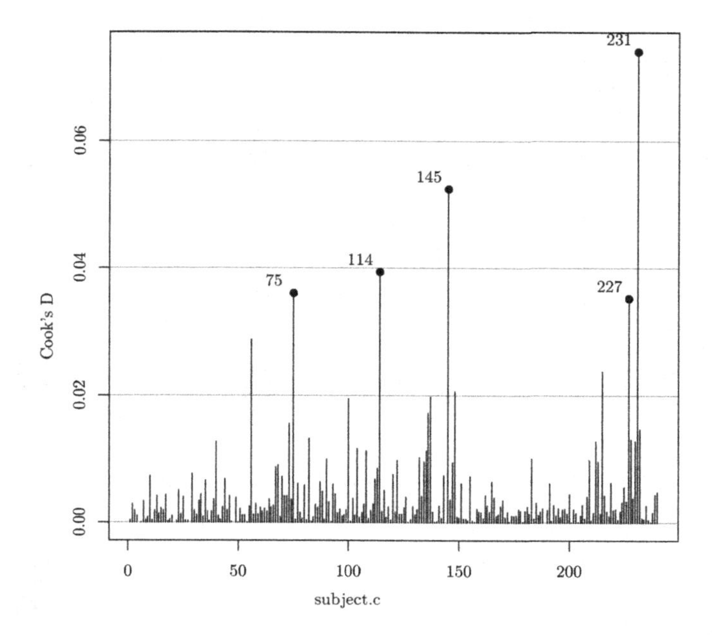

Fig. 20.3 *ARMD Trial*: Plot of Cook's distances *versus* subjects' identifiers for model **M16.5** for all subjects from the `armd` dataset

outcome is stored in the vector `CookD` and contains the values of Cook's distance for all subjects.

In Panel R20.10b, we create the logical vector `outD` which indicates the subjects with the values of Cook's distance exceeding, say, 0.03. From the printout of the selected elements of the vector `subject.c`, it follows that there are five such subjects.

We then use the function `plot()` to plot the Cook's distance values for all subjects. The plot is enhanced by adding the labels and symbols (closed circles) for the subjects with outlying values of the distance. The result is shown in Fig. 20.3. Note that subjects `"75"`, `"114"`, `"227"`, and `"231"` are present both in Figs. 20.2 and 20.3.

Figure 20.4 presents the scatterplot matrix of the two-dimensional projections of the $(\widehat{\boldsymbol{\beta}}_{(-i)} - \widehat{\boldsymbol{\beta}})/\widehat{\text{se}}(\widehat{\boldsymbol{\beta}})$ differences for all pairs of the fixed-effects coefficients.

The plot was generated using the `splom()` function. The main argument of the function was obtained by subtracting the `beta0` vector from the rows of the transposed `betaUall` matrix, created in Panel R20.10. Note that, to conserve space, we do not present the details of the syntax used to create Fig. 20.4.

The labels used in the panels located on the diagonal of the figure provide the estimates of the fixed-effects coefficients of model **M16.5** and their estimated SEs. The panels above diagonal include points for all subjects. The points for nonoutlying values are plotted using the small-size open circles. The five outlying values are displayed using different plotting symbols defined in the legend of the figure at the

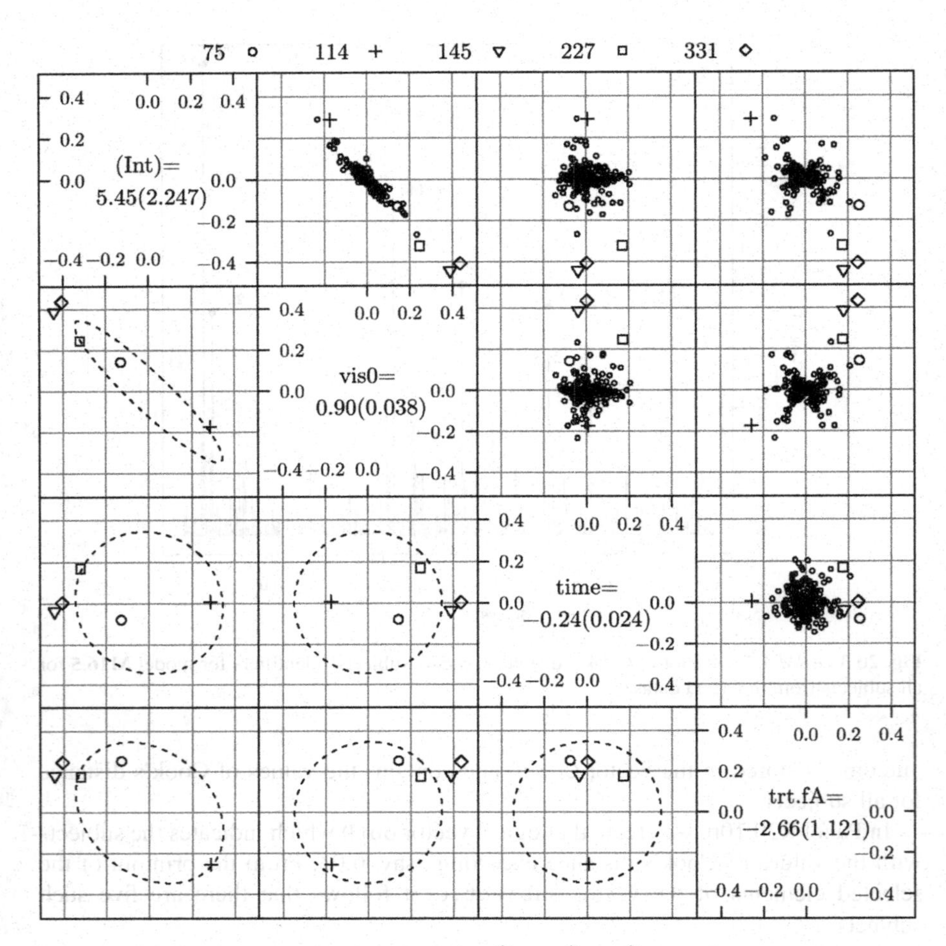

Fig. 20.4 *ARMD Trial*: Standardized differences $(\widehat{\boldsymbol{\beta}}_{(-i)} - \widehat{\boldsymbol{\beta}})/\widehat{\text{se}}(\widehat{\boldsymbol{\beta}})$ for model **M16.5**

top of the graph. The panels below the diagonal focus on the five outlying values. The ellipses represent the two-dimensional projections of an ellipsoid corresponding to Cook's distance of 0.03.

The plots shown in Fig. 20.4 suggest that, for instance, removing the subject "75" attenuates the estimate of the coefficient of the `treat.fActive` variable, i.e., the treatment effect. The effect of removal of this subject on the estimates of the remaining fixed-effect coefficients is relatively small. In contrast, removing the subject "227" affects the estimates of all fixed-effects coefficients to a different degree and in different directions. More specifically, the intercept is driven toward lower values; the positive slope associated with visual acuity at baseline, `visual0`, is further increased; the negative slope associated with `time` is brought closer to zero; and the treatment effect is attenuated.

Overall, we note that the effect of removing any of the subjects on the fixed-effects estimates is fairly small, as it does not exceed 0.4 of the SE of any of the estimates.

20.4 Simulation of the Dependent Variable

In this section, we consider simulation of the dependent variable, based on the marginal distribution implied by the fitted model. Toward this end, we have developed the function `simulateY()`, which can be used for objects of class *lme*. We note that the function is different from `simulate.lme()`, available in the **nlme** (Sect. 14.7), in that the latter returns simulation-based REML and/or ML values and not the values of the dependent variable.

In Panel R20.11, we demonstrate the use of the `simulateY()` function to create the empirical distribution of the $\boldsymbol{\beta}$ estimates. As an example, we consider model **M16.5**, which was fitted to the `armd` data in Sect. 16.4.3. Note, however, that the presented syntax is fairly general and can be used for other LMMs as well.

We apply the function `simulateY()` to the object `fm16.5ml`. Recall that this object was created in Panel R20.6a and stores the results of the ML estimation of model **M16.5**. We note that the object `fm16.5ml` is used as the main (first) argument

R20.11 *ARMD Trial*: The use of the `simulateY()` function to create the empirical distribution of $\widehat{\boldsymbol{\beta}}$ for model **M16.5**. The object `fm16.5ml` was created in Panel R20.6a

```
> library(nlmeU)
> simY <- simulateY(fm16.5ml, nsim=1000)   # Simulated y from M16.1
> auxDt <- subset(armd,                    # Auxiliary data
+    select = c(subject, visual, visual0, time, treat.f))
> simYsumm <-
+    apply(simY,
+          MARGIN = 2,                     # Over columns
+          FUN = function(y){
+             auxDt$visual <- y            # Dependent variable updated
+             auxFit <-                    # Update M16.1 with new y
+                update(fm16.5ml, data = auxDt)
+             summ <- summary(auxFit)      # Summary
+             beta <- fixef(summm)
+             list(beta = beta)
+ })
> simYsumm[[1]]                            # β̂ for the 1st simulation
  $beta
   (Intercept)       visual0        time treat.fActive
       4.59395       0.90295    -0.24768      -3.94995
```

of the `simulateY()` function. The value of the second argument, `nsim`, requests generation of 1,000 simulations. They are drawn from the marginal distribution of the dependent variable implied by the fitted model **M16.5**. The generated data are stored in the matrix `simY` with 867 rows and 1,000 columns.

In the next step, we prepare an auxiliary data frame `auxDt`. It contains a subset of variables from the `armd` data frame. The selected variables are those that are needed to fit model **M16.5**. Next, with the help of the `apply()` function, we sequentially use every column of the matrix `simY` as the argument `y` of an auxiliary function defined in the argument `FUN` of `apply()`. The auxiliary function performs the following steps:

- The dependent variable `visual` in the `auxDt` data frame is replaced with a new set of simulated values contained in the vector `y`.
- Model **M16.5** is fitted to the modified data frame.
- The vector `beta` with the estimates of fixed-effects coefficients is extracted from the summary of the model-fit object with the help of the `fixef()` function.
- The vector of estimates is returned as a list with one component named `beta`.

The `apply()`-function call returns a list named `simYsumm` with 1,000 components, each corresponding to one simulation. This list contains all the information about the empirical distribution of the $\widehat{\boldsymbol{\beta}}$ estimates. For the reader's reference, the first component with the estimates obtained for the first simulation is presented at the bottom of Panel R20.11.

It should be mentioned that the creation of the `simYsumm` list involves refitting model **M16.5** many times, and it therefore takes a long time. The execution time can be shortened, if we decide to perform, e.g., a reduced number of likelihood iterations, instead of performing iterations until convergence (see also a similar comment for Panel R20.8a in Sect. 20.3.2).

It is also worth noting that the creation of the auxiliary data frame `auxDt` with selected variables for use within the `apply()` function is important for at least three reasons: to avoid overwriting the original data stored in the data frame `armd`, to reduce the amount of internal memory used, and to shorten the execution time of the `update()` function used to refit model **M16.5**.

In Panel R20.12, we explore the basic characteristics of the empirical distribution of the $\boldsymbol{\beta}$ estimates obtained in Panel R20.11.

Toward this end, with the help of the `sapply()` function, we extract the vectors with the values of $\widehat{\boldsymbol{\beta}}$ for each simulation from the list-object `simYsumm` and bind them column-wise into the matrix `betaE`. Then, we use the function `rowMeans()` to compute the mean values of the columns, i.e., across the rows, of `betaE`. As a result, we obtain the sample means of the 1,000 (simulated) estimates of the fixed-effects coefficients for the intercept and the variables `time` and `treat.f`. The mean values are close to the ML estimates obtained in Panel R20.6b for model **M16.5** fitted to the data frame `armd`.

R20.12 *ARMD Trial*: Summary statistics of the empirical distribution of $\widehat{\boldsymbol{\beta}}$ for model **M16.5**. The object simYsumm was created in Panel R20.11

```
> betaE <- sapply(simYsumm,                    # Matrix with β̂
+    FUN = function(x) x$beta)
> rowMeans(betaE)                              # Empirical β̂ (see Panel R20.6b)
    (Intercept)         visual0              time treat.fActive
        5.37646         0.90113          -0.24115      -2.66659
> cov(t(betaE))                                # Empirical Var(β̂)
                 (Intercept)       visual0         time treat.fActive
(Intercept)        4.7447721 -7.5371e-02 -3.5957e-03   -0.64417519
visual0           -0.0753712  1.3785e-03 -1.6726e-05    0.00072322
time              -0.0035957 -1.6726e-05  5.5794e-04    0.00183248
treat.fActive     -0.6441752  7.2322e-04  1.8325e-03    1.22061704
```

By applying the function cov() to the transpose of the matrix betaE, we obtain the empirical estimate of the variance-covariance matrix of $\widehat{\boldsymbol{\beta}}$. Again, the estimate is close to the ML estimate obtained in Panel R20.6b for the model **M16.5** fitted to the data frame armd.

It is worth noting that the calculations performed in Panels R20.11 and R20.12 can be extended, in a straightforward manner, to create empirical distributions for the estimates of other parameters of the model **M16.5** like, e.g., the parameter δ defined in (16.8) or parameters d_{11} and d_{22} defined in (16.16). In Sect. 20.5.3, the syntax to simulate values of the F-statistics is given.

20.5 Power Analysis

Power analysis is an important step in the design of any experiment. In this section, we present R tools that can be used to compute the power of an F-test of a fixed effect in an LMM. In particular, we consider two approaches. One is based on the method proposed by Helms (1992) and its implementation described in Litell et al. (2006). A description of the approach is also presented in Verbeke and Molenberghs (2000). The second approach is based on simulations. It uses the function simulateY(), introduced in Sect. 20.4.

For illustration purposes, we consider the power analysis for the test of the treatment effect in model **M16.5**. The model was defined for the ARMD data in Sect. 16.4.3. First, in Sect. 20.5.1, we consider a *post hoc* power analysis based on the fitted model **M16.5**. Then, in Sects. 20.5.2 and 20.5.3, we consider an *a priori* power analysis.

R20.13 *ARMD Trial*: Preparatory steps for the *post hoc* power calculations for the treatment effect (**M16.5**). The model-fit object `fm16.5` was created in Panel R16.13

(a) *Extracting the basic information about the model*

```
> formula(fm16.5)                                 # Recall formula
 visual ~ visual0 + time + treat.f
> fixef(fm16.5)                                   # β̂
  (Intercept)       visual0          time treat.fActive
      5.44156       0.89983      -0.24156      -2.65528
```

(b) *F-test for the treatment effect using the* `anova()` *function*

```
> anova(fm16.5)                                   # Default call
            numDF denDF F-value p-value
 (Intercept)     1   632  8471.1  <.0001
 visual0         1   231   558.6  <.0001
 time            1   632   102.1  <.0001
 treat.f         1   231     5.5  0.0195
> anova(fm16.5, Terms = "treat.f")                # Terms argument
 F-test for: treat.f
  numDF denDF F-value p-value
 1     1   231  5.5345  0.0195
> anova(fm16.5, L = c("treat.fActive" = 1)) # L argument
 F-test for linear combination(s)
 [1] 1
  numDF denDF F-value p-value
 1     1   231  5.5345  0.0195
```

20.5.1 Post Hoc *Power Calculations*

In this section, we illustrate a *post hoc* power analysis for the treatment effect in model **M16.5**. The model was fitted to the ARMD data in Sect. 16.4.3 and is represented by the model-fit object `fm16.5`. The mean structure and the model equation are given in (12.9) and (16.17), respectively. The coefficient of interest is β_3, as it describes the "Active" treatment effect. We are interested in testing the null hypothesis $\mathrm{H}_0 : \beta_3 = 0$ against the alternative hypothesis $\mathrm{H}_A : \beta_3 \neq 0$. In particular, we want to compute the power for the alternative hypothesis $\mathrm{H}_{A,c} : \beta_3 = c$, where c is the value of β_3 estimated from the data.

In Panel R20.13, we present several preparatory steps for the *post hoc* power calculations. The actual calculations are deferred to Panel R20.14.

First, in Panel R20.13a, we recall the formula for the fixed effects, used in fitting of model **M16.5**. The term of interest is `treat.f`. Then, with the help of the

R20.14 *ARMD Trial*: The *post hoc* power calculations for the treatment effect in model **M16.5**. The model-fit object `fm16.5` was created in Panel R16.13

(a) *Detailed calculations "by hand"*

```
> alpha <- 0.05                                  # α
> df1 <- 1                                       # numDF
> df2 <- 231                                     # denDF
> Fvalue <- 5.5345                               # F-value (from R20.13b)
> (Fcrit <-              # Critical value for the F-test under H0
+    qf(1 - alpha,
+       df1 = df1, df2 = df2, ncp =0))
  [1] 3.882
> nc <- Fvalue * df1                             # Noncentrality parameter
> pf(Fcrit,                                      # Power
+    df1 = df1, df2 = df2,
+    ncp = nc, lower.tail = FALSE)
  [1] 0.64907
```

(b) *Post hoc power calculations using the* `Pwr()` *function*

```
> library(nlmeU)
> Pwr(fm16.5)                                          # Default call
  Power calculations:
              numDF denDF F-value     nc  Power
  (Intercept)     1   632  8471.1 8471.1 1.0000
  visual0         1   231   558.6  558.6 1.0000
  time            1   632   102.1  102.1 1.0000
  treat.f         1   231     5.5    5.5 0.6491
> Pwr(fm16.5,  L = c("treat.fActive" = 1))    # The L argument
  Power calculations for a linear combination:
       treat.fActive
  [1,]             1
    numDF denDF F-value     nc  Power
  1     1   231  5.5345 5.5345 0.6491
```

function `fixef()`, we extract the estimates of the fixed-effects coefficients from the model-fit object `fm16.5`. In this way, we learn about the values of the estimates and about the names used to identify the estimates. In particular, we note that the fixed-effect estimate named `treat.fActive` corresponds to the `treat.f` term. We also note that c, used in specification of $H_{A,c}$, is equal to `-2.65`.

In Panel R20.13b, we explore three different forms of the syntax for the `anova()`-function call to obtain the F-test for `treat.f`. All of these forms use the model-fit object `fm16.5` as the main argument. The most commonly used syntax, which was also applied in earlier chapters, does not involve any additional arguments. By default, it returns results of the sequential-approach F-tests for *all* terms in the mean structure of the model. The last two forms use the `Terms` and `L` arguments, respectively, and return the F-test just for the desired term, i.e., `treat.f`. All three calls to the `anova()` function return the value of 5.54 for the F-test statistic for the `treat.f` factor. The value will be used later in the power calculations. At this point, we want to bring reader's attention to the syntax employing the `L` argument. This syntax is the most general, because it allows specifying any linear hypothesis involving the $\boldsymbol{\beta}$ elements. Typically, the argument `L` is set to a matrix, $\boldsymbol{L}$, introduced in (4.30) in the context of the classical LM. In our example, the argument `L = c("treat.fActive" = 1))` defines the vector $\mathbf{L} \equiv (0,0,0,1)$.

The reason why we present, in Panel R20.13b, the three different forms of syntax for the `anova()`-function call is that they can be adopted by the `Pwr()` function, which will be introduced later in this section.

In Panel R20.14, we demonstrate how to perform the *post hoc* power calculations for the treatment effect in model **M16.5**.

First, in Panel R20.14a, we perform detailed calculations "by hand." We specify all the information needed to perform the calculations, including the level of significance α, the numerator and denominator degrees of freedom, and the value of the F-test statistic from the ANOVA table displayed in Panel R20.13b. We store these values in the objects `alpha`, `df1`, `df2`, and `Fvalue`, respectively.

Then, we use the first three objects as the arguments for the function `qf()`, which is the quantile function for the F-distribution, to calculate the critical value, `Fcrit`, of the F-test for the treatment effect under the null hypothesis H_0. Based on the value of the F-test statistic and the number of the numerator degrees of freedom, we calculate the noncentrality parameter `nc` and use it as an argument in `pf()`, the cumulative distribution function for a noncentral F-distribution, to calculate the *post hoc* power for the observed treatment effect. The obtained power is equal to 0.65.

In Panel R20.14b, we present the *post hoc* power calculations using the generic function `Pwr()` that has been included in the package **nlmeU**. The syntax of the `Pwr()`-function call is similar to that used for `anova()` in Panel R20.13b. The calculations performed by the function `Pwr()` proceed in a similar way to the one presented in Panel R20.14a. Thus, the result is identical. We conclude that, at the significance level $\alpha = 0.05$, the *post hoc* power to detect the observed difference of -2.65 in visual acuity is 0.65.

The *post hoc* power is a re-expression of the p-value (Lenth 2001). Hence, it does not add any new information about the current study. It could be of some value if we contemplated repeating the current study with exactly the same sample size. The *post hoc* power would then provide information about the probability of rejecting the null hypothesis if the treatment effect were of the same magnitude as the one we have observed. In practice, however, it is more important and meaningful to perform *a priori* power calculations, i.e., to compute the power of a newly-designed study before the study commences. In the next section, we present the use of the `Pwr()` function toward this end.

20.5.2 A Priori *Power Calculations for a Hypothetical Study*

In this section, we perform *a priori* power calculations using the analytical formulae proposed by Helms (1992).

For the sake of the presentation, let us imagine that we intend to design a two-group randomized clinical trial to investigate the effect of a new intervention on visual acuity in patients with age-related macular degeneration.

Based on the cost considerations, the tentative sample size is set to $n = 20$ subjects per group. Similarly to the ARMD trial, we plan to have four-time measurements at the same time points, i.e., at 4, 12, 24, and 52 weeks.

We assume that, in the analysis of the data, the following simplified model will be employed:

$$\begin{aligned} \text{VISUAL}_{it} = {} & \beta_0 + \beta_2 \times \text{TIME}_{it} + \beta_3 \times \text{TREAT}_i \\ & + b_{0i} + b_{2i} \times \text{TIME}_{it} + \varepsilon_{it}. \end{aligned} \tag{20.1}$$

The mean structure of the model is defined by the effects of continuous time and treatment factor, represented by the `time` and `treat.f` terms in the `lme()`-function model formula. The random-effects structure is defined in the same way as in model **M16.5**, with a diagonal matrix $\boldsymbol{D}$ (16.16) and the power-of-time variance function (16.8).

For the purpose of the new study, we assume that the clinically meaningful and attainable treatment effect, i.e., the mean difference in visual acuity between the two treatment groups, is equal to $\beta_3 = 10$ letters at each timepoint between 4 and 52 weeks. We also assume that, every 10 weeks, visual acuity declines in both groups, on average, by $\beta_2 = 1$ letter. Consequently, we specify the null hypothesis $\text{H}_0 : \beta_3 = 0$, with the alternative $\text{H}_{A,c} : \beta_3 = 10$.

For the power calculations, we set the values of the variance-covariance parameters as follows: $d_{11} = 100$, $d_{22} = 0.09$, $\delta = 0.15$, and $\sigma = 5$. These values are slightly higher as compared to those obtained for model **M16.5** and should result in a more conservative power estimate.

The *a priori* power calculations will be performed in the following steps:

- Construction of an exemplary dataset (Panel R20.15)
- Construction of an object of class *lme* containing all the information about the alternative model (Panel R20.16)
- Calculation of the power using the function `Pwr()` (Panel R20.17)

In Panel R20.15, we illustrate how to create an exemplary dataset for an *a priori* power analysis for a treatment effect in a newly-designed, hypothetical study.

First, we define the numeric vectors `npg` and `subject` containing, respectively, the number of subjects per treatment group and subjects' identifiers. Then, with the help of the function `gl()`, we create the `treat.f` factor for the subject-level data. The first argument of the function `gl()` specifies the desired number of levels (in our case, two); the second indicates the number of replications of each level (in our

R20.15 *ARMD Trial*: Constructing an exemplary dataset for an *a priori* power analysis of the treatment effect in a hypothetical study

(a) *Create an exemplary dataset*

```
> npg <- 20                                        # No of subjects per group
> subject <- 1:(2*npg)                             # Subjects' ids
> treat.f <- gl(2, npg, labels = c("Placebo", "Active"))
> dts <- data.frame(subject, treat.f)              # Subject-level data
> dtL <-
+    list(time = c(4, 12, 24, 52),
+         subject = subject)
> dtLong <- expand.grid(dtL)                       # "Long" format
> mrgDt <- merge(dtLong, dts, sort = FALSE) # Merged
> exmpDt <-
+    within(mrgDt,
+           {
+            m0 <- 65 - 0.1 * time       # Under H0
+            mA <- 85 - 0.1 * time       # Under HA
+            mA <- ifelse(treat.f %in% "Active", mA, m0)
+           })
```

(b) *The syntax to create Fig. 20.5 illustrating the exemplary data*

```
> selDt <-
+    with(exmpDt,
+         {
+          lvls <- levels(treat.f)               # "Placebo", "Active"
+          i <- match(lvls, treat.f)             # 1, 81
+          subj <- subject[i]                    # 1, 21
+          subset(exmpDt, subject %in% subj)
+         })
> library(lattice)
> xyplot(mA ~ time,                              # Fig. 20.5
+        groups = treat.f,
+        data = selDt,
+        type = "l",
+        auto.key = list(lines = TRUE, points = FALSE),
+        grid = TRUE)
```

case, 20); and the third provides the labels for the levels (in our case, "Placebo" and "Active").

Next, we construct the data frame dts with two variables, subject and treat.f, with the first 20 subjects coming from the "Placebo" group and the subsequent 20 subjects coming from the "Active" group. In addition, we create the list dtL with two components named "time" and "subject". Both are numeric vectors; the first one contains the values of 4, 12, 24, 52, while the second one contains the values from 1 to 40. Then, with the help of the function expand.grid() we construct a "long"-format data frame dtLong, which contains

R20.16 *ARMD Trial*: Constructing an object of class *lme* representing the alternative model. The object exmpDt was created in Panel R20.15a

(a) *Objects of class pdMat and varPower to define variance-covariance structures*

```
> D0 <- diag(c(100, 0.09))                    # 𝒟
> sgma  <- 5                                  # σ
> (D  <- D0/(sgma*sgma))                      # D
      [,1]   [,2]
 [1,]    4 0.0000
 [2,]    0 0.0036
> (pd1 <- pdDiag(D, form = ~time, data = armd))
 Positive definite matrix structure of class pdDiag representing
             (Intercept)   time
 (Intercept)           4 0.0000
 time                  0 0.0036
> (vF <- varPower(form = ~time, fixed = 0.15))
 Variance function structure of class varPower representing
 power
  0.15
```

(b) *Fitting the model to the exemplary data using the* `lme()` *function with 0 iterations*

```
> cntrl <-
+    lmeControl(maxIter = 0, msMaxIter = 0, niterEM = 0,
+               returnObject = TRUE, opt = "optim")
> fmA <-
+    lme(mA ~ time + treat.f,
+        random = list(subject = pd1),
+        weights = vF,
+        data = exmpDt,
+        control = cntrl)
> fixef(fmA)                                  # β verified
    (Intercept)          time treat.fActive
           65.0          -0.1          20.0
> sigma(fmA)                                  # σ ≈ 0
 [1] 2.2947e-14
```

the combinations of all values of the two vectors from the dtL list. Subsequently, we merge dtLong and dts by applying the merge() function. The result is stored as the data frame mrgDt. Finally, with the help of the function within(), we add to the data frame two variables, m0 and mA. They contain the expected values of the dependent variable under the null and the alternative hypothesis, respectively. The variable mA is the key variable in the context of the power

R20.17 *ARMD Trial*: The use of the Pwr() function to perform the *a priori* power calculations for the treatment effect in a hypothetical study. The objects sgma and fmA were created in Panels R20.16a and R20.16b, respectively

(a) *Power calculations using the Pwr() function*

```
> Pwr(fmA, sigma = sgma, L = c("treat.fActive" = 1))
  Power calculations for a linear combination:
       treat.fActive
  [1,]             1
    numDF denDF F-value    nc  Power
  1     1    38   8.043 8.043 0.7892
```

(b) *Use of the* altB *argument to create data for plotting the power curve*

```
> dif <- seq(1, 15, by = 0.1)                        # Δ
> dim(dif) <- c(length(dif), 1)
> colnames(dif) <- "treat.fActive"
> dtF <-                                             # Data for Fig. 20.6
+    Pwr(fmA, sigma = sgma,
+        L = c("treat.fActive" = 1), altB = dif)
> dtF[ ,1:4]                                         # Four variables
      (Intercept) time treat.fActive    Power
  1            65 -0.1           1.0 0.058804
  2            65 -0.1           1.1 0.060664
  ...     [snip]
  91           65 -0.1          10.0 0.789248
  ...     [snip]
  140          65 -0.1          14.9 0.984490
  141          65 -0.1          15.0 0.985534
```

(c) *Plotting the power curve*

```
> xyplot(Power ~ treat.fActive,                      # Fig. 20.6
+        data = dtF, type="l",
+        auto.key = list(lines = TRUE, points = FALSE),
+        grid = TRUE)
```

calculations. On the other hand, m0 plays an auxiliary role and is not essential for the calculations.

It is worth noting that the exemplary dataset exmpDt is created solely using the assumed values of the fixed effects. The random structure of the model is not reflected by any means in the exemplary data.

In Panel R20.15b, we demonstrate how to graphically represent the expected values of the dependent variable under the model implied by the exemplary data. Toward this end, we create an auxiliary data frame selDt by extracting two subjects,

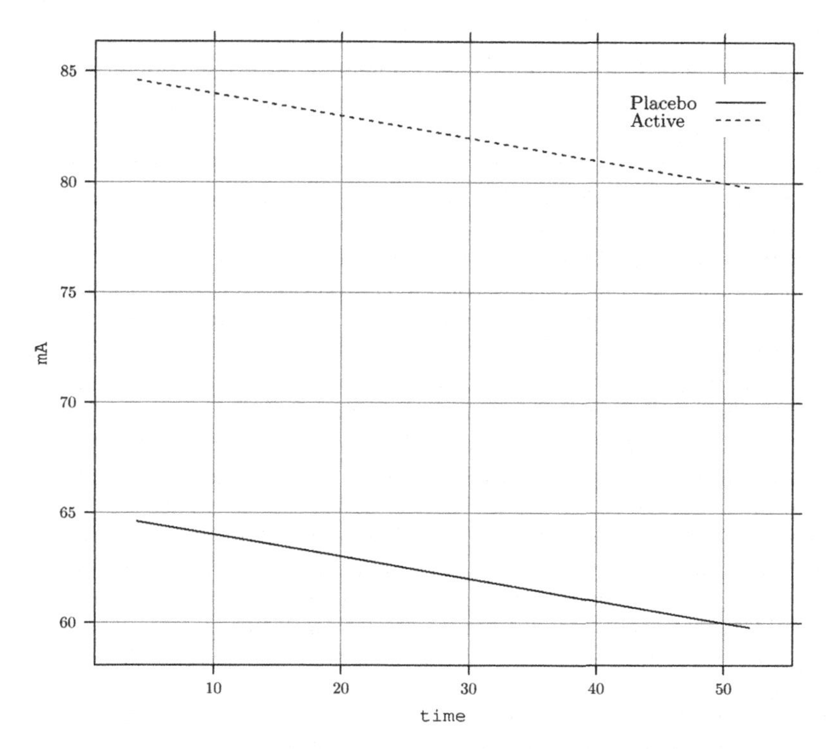

Fig. 20.5 *ARMD Trial*: Mean values for the alternative model for the *a priori* power analysis, as specified by the exemplary data

one from the `"Active"` group and another from the `"Placebo"` group. We then use these data to plot, with the help of the `xyplot()` function from the **lattice** package, the values of the variable `mA` against `time` separately for each of the treatment groups. The resulting plot is shown in Fig. 20.5.

In Panel R20.16, we create an object of class *lme* containing all the information about the alternative model. Toward this end, first, in Panel R20.16a, we initialize the object `pd1` representing the diagonal variance-covariance matrix $\mathcal{D}$ for the random effects. The object is initialized with the use of the *pdDiag*-class constructor function (Sect. 14.2.1). In addition, with the help of the *varPower*-class constructor function (Sect. 8.2.1), we create the object `vF`, which represents the assumed variance function. Note that we use the argument `fixed` of the constructor function, which implies that the variance-function parameter δ is fixed at the assumed value of 0.15 and is not to be changed during any numerical-optimization routine.

In Panel R20.16b, we use the `lme()` function to create an object of class *lme* representing the alternative model. The underlying idea is to incorporate all the variance-covariance parameters, defined in Panel R20.16a, into the model-fit object and hold the parameters unchanged. We also require that the fixed-effects coefficients, implied by the exemplary dataset `exmpDt`, remain unchanged. Toward this end, we set the components maxIter, msMaxIter, and nIterEM of the lmeControl list to zero so that the lme() function is executed without performing any iterations, i.e., without any change of the initial values of the model parameters. To avoid irrelevant error messages, we indicate, by setting the appropriate value of the opt component of the list, the use of the optim() optimizer instead of nlinmb() used by default. Finally, we set the logical value of the returnObject component to TRUE, indicating that lme() should return the model-fit object when the maximum number of iterations is reached without convergence of the algorithm.

We store the lmeControl list with the appropriately modified components in the object cntrl. We then use the object in the control argument of the lme() function to fit the assumed LMM to the exemplary dataset exmpDt. By extracting the values of the fixed-effects coefficients from the model-fit object fmA, we confirm that they correspond to the values assumed for the alternative model. Although we do not show it in Panel R20.16b, it is straightforward to demonstrate that also the variance-covariance parameters incorporated in fmA did not change as compared to their initial values, as intended. We note, however, that in contrast to the other variance-covariance parameters, the value of the scale parameter σ has become very close to zero, i.e., has changed from the starting value $\sigma = 5$. This is not surprising, given that in the exemplary dataset no residual random error was introduced. The fact that the value of σ is close to zero is a confirmation that the exemplary data and the fixed-effects formula, used in the lme()-function call, define the same mean structure, as intended.

In Panel R20.17, we present various arguments used in the Pwr()-function call. In Panel R20.17a, we employ a typical set of arguments to perform *a priori* power calculations. The syntax is similar to the one employing the L argument in Panel R20.14b for the *post hoc* power analysis. The only, but critical, difference is the use of the sigma argument. The use of the argument replaces the close-to-zero value of the scale parameter σ in the model-fit object fmA with the intended value of 5, stored in the object sgma.

The resulting output of the Power() indicates that the power of the F-test for the assumed alternative hypothesis, which specifies the mean difference in visual acuity between the two treatment groups equal to 10, is equal to 0.789.

In Panel R20.17b, we demonstrate the use of the additional argument altB of the Pwr() function to create data for the plotting of power curves. First, we create a one-column matrix dif with different values of the treatment effect ranging from 1 to 15. To assure the proper merging of the information contained in the matrix, we apply the name "treat.fActive" for the matrix column corresponding to fixed effect that is being tested. By applying the function Pwr() with the argument altB, we create the data frame named dtF, which can be used to plot a power curve showing the relationship between the treatment effect and the power of the F-test. For the readers's reference, a few selected rows of the data frame dtF are printed.

Finally, in Panel R20.17c, we use the xyplot() function from the **lattice** package to plot the power curve. The resulting plot is shown in Fig. 20.6.

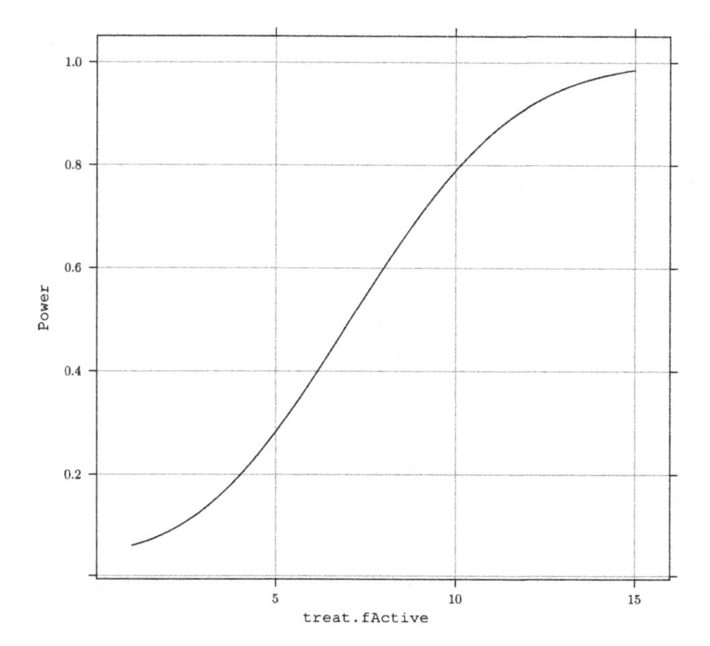

Fig. 20.6 *ARMD Trial*: The power curve resulting from the *a priori* power calculations

20.5.3 Power Evaluation Using Simulations

In Sect. 20.5.2, we performed *a priori* power calculations based on the analytical formulae proposed by Helms (1992). In this section, we demonstrate how to evaluate power using simulations.

Typically, to evaluate the power, we would simulate the empirical distribution of the test statistic under the null and alternative hypotheses. For the sake of simplicity of presentation, we will use a simplified approach. More specifically, we will use the well-established analytical approximation of the null distribution of the F-test statistic by an F-distribution with an appropriate number of degrees of freedom. In contrast, the empirical F-test-statistic distribution under the alternative hypothesis will be obtained using simulations. We note, however, that our presentation extends in a straightforward way to the simulation of the test statistics both under the null *and* alternative hypotheses.

To illustrate the simulation approach, we will use the same example of a hypothetical study as in Sect. 20.5.2. In particular, for calculations, we use the objects `exmpDt`, `sgma`, and `fmA`, created in Panels R20.15a, R20.16a, and R20.16b, respectively. Recall that these objects contain the exemplary data, the scale parameter, and the information about the LMM needed for the power calculations.

In Panel R20.18, we present the essential steps needed to simulate the distribution of the F-test statistics under the alternative hypothesis.

R20.18 *ARMD Trial*: Simulation of the F-test statistics based on the model-fit object `fmA`. The objects `exmpDt`, `sgma`, and `fmA` were created in Panels R20.15a, R20.16a, and R20.16b, respectively

```
> simA <- simulateY(fmA, sigma = sgma, nsim = 1000) # Simulation
> dt <- exmpDt                                      # Working copy
> simfmA <-
+    apply(simA,
+          2,                                       # Over columns
+          function(y){
+             dt$mA <- y                            # mA over-written
+             auxFit <- update(fmA, data = dt)
+             anova(auxFit)                         # ANOVA table
+             })
> simfmA[[1]]                                       # First ANOVA
            numDF denDF F-value p-value
(Intercept)     1   119  4186.2  <.0001
time            1   119     4.8  0.0307
treat.f         1    38     3.4  0.0738
```

First, we use the `simulateY()` function from the package **nlmeU** (Sect. 20.4) to simulate the values of the dependent variable. Toward this end, we employ the marginal distribution of the dependent variable implied by the model-fit object `fmA`. Recall that this object contains the information about all the parameters necessary for the power calculations, except of the scale parameter σ. To provide the value of the parameter, we use the argument `sigma`. The simulated values of the dependent variable are stored in the columns of a 160×1000 matrix `simA`. Next, with the help of the `apply()` function, we sequentially use every column of the matrix `simA` as the argument `y` of an auxiliary function. The auxiliary function performs the following steps:

- The dependent variable `mA` in a working copy `dt` of the exemplary dataset `exmpDt` is replaced with a new set of simulated values contained in the vector `y`.
- The model-fit object `mA` is updated by re-fitting the model to the modified data frame and stored as the object `auxFit`.
- The function `anova()` is applied to the model-fit object `auxFit`.

The `apply()`-function call returns a list named `simfmA` with 1,000 components, each corresponding to one simulation. This list contains the information about the F-tests for the fixed effects of the alternative LMM. For the reader's reference, the first component of the list is presented at the bottom of Panel R20.18.

It is straightforward to note that, to obtain empirical *null* distribution of the F-test statistics, we would proceed in a very similar way. The only difference would be that, instead of the model-fit object `fmA`, we would use a similar model-fit object, but fitted to the `m0` dependent variable (see Panel R20.15a), instead.

R20.19 *ARMD Trial*: Empirical power of the F-test for the treatment effect based on the simulated values of the F-test statistics. The object `simfmA` was created in Panel R20.18

```
> FstatE <-                       # Empirical F-test statistics under H_A
+    sapply(simfmA, function(x) x["treat.f", "F-value"])
> summary(FstatE)
    Min. 1st Qu.  Median    Mean 3rd Qu.    Max.
   0.056   4.650   8.240   9.270  12.600  38.200
> Fcrit <- qf(1- 0.05, 1, 38, ncp =0)
> (nsim <- length(FstatE))
  [1] 1000
> (powerE <- sum(FstatE > Fcrit)/nsim)          # Empirical power
  [1] 0.783
```

In Panel R20.19, we demonstrate the syntax allowing to finalize the simulation-based power calculations.

First, we use the `sapply()` function to extract the values of the F-test statistic for the `"treat.f"` fixed effect from the ANOVA tables contained in the components of the `simfmA` list. These values represent the empirical distribution of the test statistic under the alternative hypothesis H_A. They are stored in the numeric vector `FstatE`. By applying the `summary()` function to the vector, we obtain the summary statistics of the empirical distribution. As we mentioned earlier, the critical value `Fcrit` of the F-test statistic at the significance level $\alpha = 0.05$ under the null hypothesis is obtained based on the well-established approximation involving a central F-distribution. The empirical power is simply obtained by calculating the proportion of the simulated values of the F-test statistic larger than the critical value. The empirical power of 0.783, shown in Panel R20.19, is very close to the value obtained based on the analytical formulae in Panel R20.17a.

In summary, the simulation approach to power calculations is attractive, due to its flexibility. It can prove especially useful if the impact of various mechanisms of missing values needs to be investigated. In that case, the missing data patterns to be evaluated need to be reflected in the exemplary data used to create the model-fit objects underlying the simulations.

Acronyms

AIC	Akaike's information criterion
ANOVA	Analysis of variance
ANCOVA	Analysis of covariance
BIC	Bayesian information criterion
BLUP	Best linear unbiased predictor
BMI	Body mass index
CI	Confidence interval
EB	Empirical Bayes
EBLUP	Empirical BLUP
EM	Expectation-minimization
GLIM	Generalized linear model
GLMM	Generalized linear mixed-effects model
GLS	Generalized least squares
IRLS	Iteratively reweighted least squares
LM	Linear model
LMM	Linear mixed-effects model
LR	Likelihood ratio
MAR	Missing at random
MIVQUE	Minimum-variance quadratic unbiased estimation
ML	Maximum likelihood
MLE	Maximum likelihood estimate
NLMM	Nonlinear mixed-effects model
OLS	Ordinary least squares
O-O	Object-oriented
PL-GLS	Pseudo-likelihood generalized least squares
PnIRLS	Penalized iteratively reweighted least squares
PnLS	Penalized least squares

A. Gałecki and T. Burzykowski, *Linear Mixed-Effects Models Using R: A Step-by-Step Approach*, Springer Texts in Statistics, DOI 10.1007/978-1-4614-3900-4,

PWLS	Penalized weighted least squares
REML	Restricted maximum likelihood
SD	Standard deviation
SE	Standard error
SVD	Singular value decomposition

References

Baayen, R., Davidson, D., & Bates, D. (2008). Mixed-effects modeling with crossed random effects for subjects and items. *Journal of Memory and Language*, *59*(4), 390–412.

Bates, D. (2012). *Computational Methods for Mixed Models*. R Foundation for Statistical Computing.

Bates, D., & Maechler, M. (2012). *Matrix: Sparse and Dense Matrix Classes and Methods*. R package version 1.0–10. `http://CRAN.R-project.org/package=Matrix`.

Bates, D., Maechler, M., & Bolker, B. (2012). Fitting linear mixed-effects models using lme4. *Journal of Statistical Software (forthcoming)*.

Box, G. E. P., Jenkins, G. M., & Reinsel, G. C. (1994). *Time Series Analysis*. Prentice Hall Inc., third ed. Forecasting and control.

Cantrell, C. (2000). *Modern Mathematical Methods for Physicists and Engineers*. Cambridge University Press.

Carroll, R., & Ruppert, D. (1988). *Transformation and Weighting in Regression*. Chapman & Hall/CRC.

Chambers, J., & Hastie, T. (1992). *Statistical Models in S*. Wadsworth & Brooks/Cole Advanced Books & Software.

Chatterjee, S., Hadi, A., & Price, B. (2000). *The Use of Regression Analysis by Example*. John Wiley & Sons.

Claflin, D.R., Larkin, L.M., Cederna, P.S., Horowitz, J.F., Alexander, N.B., Cole, N.M., Galecki, A.T., Chen, S., Nyquist, L.V., Carlson, B.M., Faulkner, J.A., & Ashton-Miller, J.A. (2011) Effects of high- and low-velocity resistance training on the contractile properties of skeletal muscle fibers from young and older humans. *Journal of Applied Physiology*, *111*, 1021–1030.

Crainiceanu, C., & Ruppert, D. (2004). Likelihood ratio tests in linear mixed models with one variance component. *Journal of the Royal Statistical Society: Series B*, *66*, 165–185.

Cressie, N., & Hawkins, D. (1980). Robust estimation of the variogram: I. *Mathematical Geology*, *12*(2), 115–125.

Cressie, N. A. C. (1991). *Statistics for Spatial Data*. Wiley Series in Probability and Mathematical Statistics: Applied Probability and Statistics. John Wiley & Sons.

Dahl, D. B. (2009). *xtable: Export tables to LaTeX or HTML*. R package version 1.5-6. `http://CRAN.R-project.org/package=xtable`.

Dalgaard, P. (2008). *Introductory Statistics with R*. Springer.

Davidian, M., & Giltinan, D. (1995). *Nonlinear Models for Repeated Measurement Data*. Chapman & Hall.

Demidenko, E. (2004). *Mixed Models: Theory and Applications*. Wiley-Interscience, first ed.

A. Gałecki and T. Burzykowski, *Linear Mixed-Effects Models Using R: A Step-by-Step Approach*, Springer Texts in Statistics, DOI 10.1007/978-1-4614-3900-4,

Fai, A., & Cornelius, P. (1996). Approximate F-tests of multiple degree of freedom hypotheses in generalized least squares analyses of unbalanced split-plot experiments. *Journal of Statistical Computation and Simulation*, *54*(4), 363–378.

Fitzmaurice, G. M., Laird, N. M., & Ware, J. H. (2004). *Applied Longitudinal Analysis*. Wiley Series in Probability and Statistics. John Wiley & Sons.

Galecki, A. (1994). General class of covariance structures for two or more repeated factors in longitudinal data analysis. *Communications in Statistics-Theory and Methods*, *23*(11), 3105–3119.

Gelman, A., Carlin, J. B., Stern, H. S., & Rubin, D. B. (1995). *Bayesian Data Analysis*. Texts in Statistical Science Series. Chapman & Hall.

Golub, G. H., & Van Loan, C. F. (1989). *Matrix Computations*, vol. 3 of *Johns Hopkins Series in the Mathematical Sciences*. Johns Hopkins University Press, second ed.

Gurka, M. (2006). Selecting the best linear mixed model under REML. *The American Statistician*, *60*(1), 19–26.

Hastie, T., Tibshirani, R., & Friedman, J. (2009). *The Elements of Statistical Learning: Data Mining, Inference and Prediction*. Springer.

Helms, R. (1992). Intentionally incomplete longitudinal designs: I. Methodology and comparison of some full span designs. *Statistics in Medicine*, *11*(14-15), 1889–1913.

Henderson, C. (1984). *Applications of Linear Models in Animal Breeding*. University of Guelph.

Hill, H., Rowan, B., & Ball, D. (2005). Effect of teachers' mathematical knowledge for teaching on student achievement. *American Educational Research Journal*, *42*, 371– 406.

Janssen, R., Tuerlinckx, F., Meulders, M., & De Boeck, P. (2000). A hierarchical IRT model for criterion-referenced measurement. *Journal of Educational and Behavioral Statistics*, *25*(3), 285.

Jones, R. (1993). *Longitudinal Data with Serial Correlation: A State-space Approach*. Chapman & Hall/CRC.

Kenward, M., & Roger, J. (1997). Small sample inference for fixed effects from restricted maximum likelihood. *Biometrics*, *53*(3), 983–997.

Laird, N., & Ware, J. (1982). Random-effects models for longitudinal data. *Biometrics*, *38*(4), 963–974.

Leisch, F. (2002). Sweave: Dynamic generation of statistical reports using literate data analysis. In: W. Härdle, & B. Rönz (eds.) *Compstat 2002 — Proceedings in Computational Statistics*, (pp. 575–580). Physica Verlag, Heidelberg.

Lenth, R. (2001). Some practical guidelines for effective sample size determination. *The American Statistician*, *55*(3), 187–193.

Liang, K.-Y., & Self, S. G. (1996). On the asymptotic behaviour of the pseudolikelihood ratio test statistic. *Journal of the Royal Statistical Society: Series B*, *58*(4), 785–796.

Litell, R., Milliken, G., Stroup, W., Wolfinger, R., & Schabenberger, O. (2006). *SAS for Mixed Models*. SAS Publishing.

Molenberghs, G., & Verbeke, G. (2007). Likelihood ratio, score, and Wald tests in a constrained parameter space. *The American Statistician*, *61*(1), 22–27.

Murrell, P. (2005). *R Graphics*. Chapman & Hall/CRC.

Murrell, P., & Ripley, B. (2006). Non-standard fonts in postscript and pdf graphics. *The Newsletter of the R Project*, *6*(2), 41.

Neter, J., Wasserman, W., & Kutner, M. (1990). *Applied Linear Statistical Models*. Irwin.

Pharmacological Therapy for Macular Degeneration Study Group (1997). Interferon α-IIA is ineffective for patients with choroidal neovascularization secondary to age-related macular degeneration. Results of a prospective randomized placebo-controlled clinical trial. *Archives of Ophthalmology*, *115*, 865–872.

Pinheiro, J., & Bates, D. (1996). Unconstrained parametrizations for variance-covariance matrices. *Statistics and Computing*, *6*(3), 289–296.

Pinheiro, J., & Bates, D. (2000). *Mixed-effects Models in S and S-PLUS*. Springer.

R Development Core Team (2010). *R: A Language and Environment for Statistical Computing*. R Foundation for Statistical Computing.

Rothenberg, T. (1984). Approximate normality of generalized least squares estimates. *Econometrica, 52*(4), 811–825.

Santos Nobre, J., & da Motta Singer, J. (2007). Residual analysis for linear mixed models. *Biometrical Journal, 49*(6), 863–875.

Sarkar, D. (2008). *Lattice: Multivariate Data Visualization with R*. Springer.

Satterthwaite, F. E. (1941). Synthesis of variance. *Psychometrika, 6*, 309–316.

Schabenberger, O. (2004). Mixed model influence diagnostics in proceedings of the twenty-ninth annual sas users group international conference. *Proceedings of the Twenty-Ninth Annual SAS Users Group International Conference, 189*, 29.

Schabenberger, O., & Gotway, C. (2005). *Statistical Methods for Spatial Data Analysis*, vol. 65. Chapman & Hall.

Scheipl, F., Greven, S., & Kuechenhoff, H. (2008). Size and power of tests for a zero random effect variance or polynomial regression in additive and linear mixed models. *Computational Statistics & Data Analysis, 52*(7), 3283–3299.

Schluchter, M., & Elashoff, J. (1990). Small-sample adjustments to tests with unbalanced repeated measures assuming several covariance structures. *Journal of Statistical Computation and Simulation, 37*(1-2), 69–87.

Searle, S., Casella, G., & McCulloch, C. (1992). *Variance Components*. John Wiley & Sons.

Self, S. G., & Liang, K.-Y. (1987). Asymptotic properties of maximum likelihood estimators and likelihood ratio tests under nonstandard conditions. *Journal of the American Statistical Association, 82*(398), 605–610.

Shapiro, A. (1985). Asymptotic distribution of test statistics in the analysis of moment structures under inequality constraints. *Biometrika, 72*(1), 133–144.

Stram, D., & Lee, J. (1994). Variance components testing in the longitudinal mixed effects model. *Biometrics, 50*(4), 1171–1177.

Tibaldi, F., Verbeke, G., Molenberghs, G., Renard, D., Van den Noortgate, W., & De Boeck, P. (2007). Conditional mixed models with crossed random effects. *British Journal of Mathematical and Statistical Psychology, 60*(2), 351–365.

Van den Noortgate, W., De Boeck, P., & Meulders, M. (2003). Cross-classification multilevel logistic models in psychometrics. *Journal of Educational and Behavioral Statistics, 28*(4), 369–386.

Venables, W., & Ripley, B. (2010). *Modern Applied Statistics with S*. Springer.

Verbeke, G., & Molenberghs, G. (2000). *Linear Mixed Models for Longitudinal Data*. Springer.

Verbeke, G., & Molenberghs, G. (2003). The use of score tests for inference on variance components. *Biometrics, 59*(2), 254–262.

Vonesh, E., & Chinchilli, V. (1997). *Linear and Nonlinear Models for the Analysis of Repeated Measurements*. CRC.

West, B. T., Welch, K. B., & Gałecki, A. T. (2007). *Linear Mixed Models: A Practical Guide Using Statistical Software*. Chapman and Hall/CRC.

Wickham, H. (2007). Reshaping data with the reshape package. *Journal of Statistical Software, 21*(12).

Wilkinson, G., & Rogers, C. (1973). Symbolic description of factorial models for analysis of variance. *Applied Statistics, 22*, 392–399.

Function Index

A. Gałecki and T. Burzykowski, *Linear Mixed-Effects Models Using R: A Step-by-Step Approach*, Springer Texts in Statistics, DOI 10.1007/978-1-4614-3900-4,

Subject Index

A. Gałecki and T. Burzykowski, *Linear Mixed-Effects Models Using R: A Step-by-Step Approach*, Springer Texts in Statistics, DOI 10.1007/978-1-4614-3900-4,

S

T

V